Emulsions

This book presents a comprehensive investigation of the complex scientific principles and practical applications of emulsions, offering valuable references for researchers, industry professionals, and advanced students in the fields of physical chemistry, material sciences, and engineering. The book is divided into three sections, which allow the readers to explore the fundamental principles of interfacial phenomena, the methodologies for emulsion preparation, and the diverse applications of emulsions in various industrial branches in depth.

The initial section establishes the fundamental principles, with a particular focus on the physicochemical characterization of interfaces pertinent to emulsion stabilization. The book meticulously examines several topics, including adsorption dynamics, particle-laden interfaces, and thin liquid films. It supports these discussions with a combination of experimental techniques and theoretical insights. The second section shifts focus to the practical aspects of emulsion production, examining state-of-the-art methods, the role of stabilizers, and strategies for enhancing their stability, including innovations in nanoemulsions and aqueous two-phase systems. The final section presents several real-world applications, demonstrating the significant role that emulsions play in a number of fields, including the petroleum industry, pest control, and pharmaceutical formulations.

This text, featuring contributions from leading experts, synthesizes foundational knowledge and cutting-edge research, thereby bridging the gap between theory and practice. With an interdisciplinary scope, the book addresses both academic and industrial perspectives, and it is therefore an indispensable addition to the library of anyone working with emulsions. It provides the tools and understanding required to advance work in the development of new materials, optimization of formulations, and investigation of the mechanics of interfacial systems.

Progress In Colloid And Interface Science

Series Editors

Reinhard Miller and Eduardo Guzmán
(formerly Reinhard Miller and Libero Liggieri)

The ***Progress in Colloid and Interface Science*** series consists of edited volumes and monographs dedicated to the relevant topics in the field of colloid and interface science. The series aims to present the most recent research developments and progress made on particular topics. It covers fundamental and/or applied subjects including any type of interfaces, their theoretical description and experimental analysis; formation and characterization of disperse systems; bulk properties of solutions; experimental methodologies; self-assembling materials at interfaces; and applications in various areas including agriculture, biological systems, chemical engineering, coatings, cosmetics, flotation processes, food processing, materials processing, and household products.

Emulsions: From Single Interfaces to Applications
Reinhard Miller and Eduardo Guzmán-Solís

Interfacial Rheology
Reinhard Miller and Libero Liggieri

Bubble and Drop Interfaces
Reinhard Miller and Libero Liggieri

Drops and Bubbles in Contact with Solid Surfaces
Michele Ferrari, Libero Liggieri, and Reinhard Miller

Colloid and Interface Chemistry for Nanotechnology
Peter Kralchevsky, Reinhard Miller, and Francesca Ravera

Computational Methods for Complex Liquid–Fluid Interfaces
Mohammad Taeibi Rahni, Mohsen Karbaschi, and Reinhard Miller

Physical-Chemical Mechanics of Disperse Systems and Materials
Eugene D. Shchukin and Andrei S. Zelenev

Foam Films and Foams
Fundamentals and Applications
Dotchi Exerowa, Georgi Gochev, Dimo Platikanov, Libero Liggieri, and Reinhard Miller

For more information about this series, please visit: https://www.crcpress.com/Progress-in-Colloid-and-Interface-Science/book-series/CRCPROCOLINT

Emulsions

From Single Interfaces to Applications

Edited by

Reinhard Miller and Eduardo Guzmán-Solís

CRC Press
Taylor & Francis Group
Boca Raton London New York

CRC Press is an imprint of the
Taylor & Francis Group, an **informa** business

First edition published 2025
by CRC Press
2385 NW Executive Center Drive, Suite 320, Boca Raton FL 33431

and by CRC Press
4 Park Square, Milton Park, Abingdon, Oxon, OX14 4RN

CRC Press is an imprint of Taylor & Francis Group, LLC

ISBN: 978-1-032-63610-8 (hbk)
ISBN: 978-1-032-63612-2 (pbk)
ISBN: 978-1-032-63618-4 (ebk)

DOI 10.1201/9781032636184

Typeset in Times
by Deanta Global Publishing Services, Chennai, India

Contents

SECTION II Emulsion preparation and characterization

SECTION III Emulsion Applications

About the Editors

Reinhard Miller was born on 10 July 1950 in Biesenthal, Germany. He is Associate Senior Scientist at the Institute for Condensed Matter Physics, Technical University Darmstadt, Germany.

He studied Mathematics at the University of Rostock (1969–1973) and Colloid Chemistry at the Technical University Dresden (1977–1978), Germany. He got his PhD in 1978 and Habilitation in 1988, both at the Academy of Sciences in Berlin, Germany, on problems of adsorption kinetics and dilational viscoelasticity of surfactants. From 1992 to 2015 Dr Miller worked as Group-Leader at the Max Planck Institute for Colloids and Interfaces in Potsdam/Golm in the field of thermodynamics, kinetics, and mechanics of surfactant and protein adsorption layers at liquid–fluid interfaces. From 2015 to 2019 he was Senior Scientist at the same Max Planck Institute for Colloids and Interface. Since 2020 he has been Associate Senior Scientist at the Institute for Soft Matter Physics, TU Darmstadt, Germany.

He is presently the Editor of the book series with Taylor and Francis 'Progress in Colloid and Interface Science' (so far seven volumes).

Dr Reinhard Miller served as President of the 'European Colloid and Interface Society' ECIS (2012–2014) and as President of the 'International Association of Colloid and Interface Scientists' IACIS (2015–2018).

His scientific interests are dynamics and mechanics of liquid interfaces, thermodynamics of adsorption of surfactants, proteins, polymers, particles and their mixtures, interfacial interactions and 2D shear and dilational rheology, and the formation and stability of foams and emulsions. He published about 715 papers in referenced journals (Web of Science), about 50 chapters in books, and about 50 contributions to printed conference proceedings. He has an h-index of 71 (Web of Science), 73 (Scopus), and 86 (Google Scholar). ORCID: 0000-0001-8943-7521.

Eduardo Guzmán-Solís was born in 1981 in Madrid (Spain) and is Associate Professor of Physical Chemistry at the University Complutense of Madrid (UCM), Spain. He received his PhD in Chemistry at the University Complutense of Madrid in 2009 under the supervision of Prof. Ramón González Rubio and Francisco Ortega, where he carried out studies on the fabrication and physicochemical properties of polyelectrolyte multilayers obtained by the layer-by-layer method. From 2010 to 2014, he worked as a postdoctoral researcher with Dr Libero Liggieri and Dr Francesca Ravera (Istituto per l'Energetica e le Interfasi, Italy) on the effect of nanoparticles in the interfacial properties of models, mimicking the behavior of the lung surfactant layer. In 2014, he returned to the UCM as Juan de la Cierva Fellow, obtaining an Assistant Professor position in February 2016, and then in March 2020 he was appointed as Associate Professor. He has co-supervised three PhD theses. He has been involved in several EU- and Spanish-funded I+D grants and has been scientifically responsible for six cooperation projects between academia and industry. He is a member of the editorial board of different scientific journals, including *Coatings* (Editor-in-Chief of the Section 'Liquid–Fluid Interfaces'), *Polymers*, *Advances in Colloid and Interface Science*, and *Colloids and Interfaces*, and since 2023 he has been the co-editor of the book series 'Progress in Colloid and Interface Science'.

Foreword

The subject of this new volume in the series 'Progress in Colloid and Interface Science' is *Emulsions*. It is comprehensively entitled *Emulsions: From Single Interfaces to Applications* and represents a cornerstone of colloid and interface science, with implications ranging from everyday consumer products, such as milk, mayonnaise, or cosmetics, to advanced industries, including tertiary recovery, pesticides, or biomedicine. It can therefore be claimed that emulsions play a central role in modern society.

Emulsions are colloidal dispersions of one immiscible liquid in another (usually water and an organic liquid called oil) and have long been of interest to scientists. However, the production of emulsions is not trivial and, in most cases, requires an energy input, usually in the form of shaking, stirring, or ultrasonic homogenization, to break one of the liquids into small droplets and ensure its homogeneous dispersion in the other. This process creates many unstable interfaces between the two immiscible liquids, which must be stabilized by the addition of suitable surface-active molecules (surfactants, polymers, proteins, or colloidal particles). Their adsorption on the resulting droplet/continuous phase interfaces provides stability to the emulsions and hinders the dynamic mechanisms driving the phase separation. It should be noted that the stabilization of emulsions is the result of a complex interplay between bulk and interfacial properties. Understanding the main physicochemical aspects governing the stability of emulsions is very important for the development of their many applications.

This volume takes a deep immersion into the complex world of emulsion science from a multidisciplinary perspective. To this end, the book is divided into three distinct parts, including an up-to-date overview of various physicochemical and technological aspects of interest in the study of emulsions. Beginning with a fundamental examination of individual interfaces and progressing to practical applications, the reader is guided to a comprehensive exploration of emulsions and their multifaceted nature. The contributors – experts in their respective fields – offer unique perspectives and cutting-edge findings that illuminate the complexities of emulsion behavior and its impact on diverse industries.

At the heart of emulsion science is the study of interfaces, where molecules at the liquid–liquid interface orchestrate fascinating phenomena. The first part presents a broad perspective on the properties of liquid–liquid interfaces and the techniques commonly used to characterize them. This is important because it is recognized that interfacial properties are closely related to the formation and stabilization of emulsions.

In the second part, a series of chapters explore the molecular mechanisms underlying emulsification, examining how surfactants and emulsifiers influence stability and structure, and describing the current state of the art in emulsion preparation and characterization. By elucidating the physicochemical principles that govern emulsion formation and stabilization, the authors enable the reader to understand the underlying forces that drive the formation and stabilization of these complex systems.

Moreover, this volume does not limit itself to theoretical discussions but extends its impact to practical applications. This will be the focus of the third part of the book. Emulsions serve as versatile drug delivery platforms, enabling targeted therapies and controlled-release formulations. In cosmetics, emulsions form the basis of luxurious creams and lotions, enhancing skin feel and efficacy. In food technology, emulsions are used for flavor encapsulation, texture modification, and shelf-life extension. The contributions in this volume highlight the transformative potential of emulsions to address today's challenges in a variety of areas.

As we progress through the chapters, we encounter innovative approaches to emulsion design and characterization. Advanced techniques such as interfacial rheology, confocal microscopy, and microfluidics provide unprecedented insight into the emulsion's behavior at the microscale. The fusion of experimental and computational methods is opening new avenues for tailoring emulsions with desired properties, paving the way for next-generation materials and formulations.

In navigating the complex landscape of emulsion science, it is clear that collaboration and interdisciplinary synergy are essential. This book exemplifies the collective efforts of researchers from chemistry, physics, biology, engineering, and beyond, coming together to unravel the mysteries of emulsions and harness their potential for societal benefits.

We would like to thank the authors for their scientific contributions to this volume. Their dedication and expertise have enriched our understanding of emulsion science and inspired future avenues of exploration. We hope that this book will serve as a valuable resource for researchers, students, and practitioners alike, fostering a deeper appreciation of the fascinating world of emulsions.

We invite you to join us on this enlightening journey through *Emulsions: From Single Interfaces to Applications*. May this volume stimulate curiosity, spark innovation, and catalyze transformative discoveries in the dynamic field of colloid and interface science.

Reinhard Miller, Institute for Condensed Matter Physics, TU Darmstadt.

Eduardo Guzmán, Departamento de Química Física and Instituto Pluridisciplinar, Universidad Complutense de Madrid

Editors of the Progress in Colloid and Interface Science Series

Contributors

Eugene V. Aksenenko
Institute of Colloid Chemistry and Chemistry of Water
National Academy of Sciences of Ukraine
Kyiv, Ukraine

Pablo G. Argudo
Department of Molecular Spectroscopy
Max Planck Institute for Polymer Research
Mainz, Germany

Carlo Carbone
Departamento de Química Física, Facultad de Ciencias Químicas
Universidad Complutense de Madrid
Madrid, Spain

Luigi Cristofolini
Department of Mathematical, Physical and Computer Sciences
University of Parma
Parma, Italy

Jan Czarnecki (retired)
Department of Chemical and Materials Engineering
University of Alberta
Edmonton, AB, Canada

Jordi Esquena
Institute of Advanced Chemistry of Catalonia
Consejo Superior de Investigaciones Científicas (IQAC-CSIC)
Barcelona, Spain
and
Networking Research Center on Bioengineering
Biomaterials and Nanomedicine (CIBER-BBN)
Barcelona, Spain

Valentin B. Fainerman (passed away in December 2021 in Odessa, Ukraine)
SINTERFACE Technologies
Berlin, Germany

Grupo de Manejo
Integrado y Bioecología de Artrópodos Dulceacuícolas (MIyBAD) - Instituto de Ecología y Desarrollo Sustentable (INEDES, CONICET-UNLu)
Luján, Buenos Aires, Argentina
and
Departamento de Ciencias Básicas
Universidad Nacional de Luján
Luján, Buenos Aires, Argentina

Georgi Gochev
Jerzy Haber Institute of Catalysis and Surface Chemistry
Polish Academy of Sciences
Krakow, Poland
and
Institute of Physical Chemistry
Bulgarian Academy of Sciences
Sofia, Bulgaria

Eduardo Guzmán
Departamento de Química Física, Facultad de Ciencias Químicas
Universidad Complutense de Madrid
Madrid, Spain
and

Instituto Pluridisciplinar
Universidad Complutense de Madrid
Madrid, Spain

Aliyar Javadi
Institute of Process Engineering and Environmental Technology
Technical University of Dresden
Dresden, Germany
and
World-Academies Startup
Technical University of Dresden
Dresden, Germany

Volodymyr I. Kovalchuk
Institute of Biocolloid Chemistry
National Academy of Sciences of Ukraine
Kyiv, Ukraine

Libero Liggieri
CNR-ICMATE – Institute of Condensed Matter Chemistry and Technologies for Energy
UOS Genova
Genova, Italy

Giuseppe Loglio
CNR-ICMATE – Institute of Condensed Matter Chemistry and Technologies for Energy
UOS Genova
Genova, Italy

Grupo de Manejo
Integrado y Bioecología de Artrópodos Dulceacuícolas (MIyBAD) - Instituto de Ecología y Desarrollo Sustentable (INEDES, CONICET-UNLu)
Luján, Buenos Aires, Argentina
and
Laboratorio de Investigaciones en Madera (LIMAD), Facultad de Ciencias Agrarias y Forestales
UNLP
Buenos Aires, Argentina

Armando Maestro
Centro de Física de Materiales (CSIC, UPV/EHU)
San Sebastián, Spain
and
IKERBASQUE – Basque Foundation for Science
Bilbao, Spain

LEPAMAP-PRODIS Research Group, Department of Chemical Engineering, Agricultural and Agrifood Technology
University of Girona
Girona, Spain

Elena Mileva
Institute of Physical Chemistry
Bulgarian Academy of Sciences
Sofia, Bulgaria

Reinhard Miller
Institute for Condensed Matter Physics
Technical University Darmstadt
Darmstadt, Germany

Ljubomir Nikolov
Institute of Physical Chemistry
Bulgarian Academy of Sciences
Sofia, Bulgaria

Jesús F. Ontiveros
Univ. Lille, CNRS, Centrale Lille, Univ. Artois, UMR 8181, UCCS, Unité de Catalyse et Chimie du Solide
Lille, France

Davide Orsi
Department of Mathematical, Physical and Computer Sciences
University of Parma
Parma, Italy

Francisco Ortega
Departamento de Química Física,
Facultad de Ciencias Químicas
Universidad Complutense de Madrid
Madrid, Spain
and
Instituto Pluridisciplinar
Universidad Complutense de Madrid
Madrid, Spain

Francesca Ravera
CNR-ICMATE – Institute of
Condensed Matter Chemistry and
Technologies for Energy
UOS Genova
Genova, Italy

Ramón G. Rubio
Departamento de Química Física,
Facultad de Ciencias Químicas
Universidad Complutense de Madrid
Madrid, Spain
and
Instituto Pluridisciplinar
Universidad Complutense de Madrid
Madrid, Spain

Eva Santini
CNR-ICMATE – Institute of
Condensed Matter Chemistry and
Technologies for Energy
UOS Genova
Genova, Italy

Emanuel Schneck
Institute for Condensed Matter Physics
Technical University Darmstadt
Darmstadt, Germany

Plamen Tchoukov
Institute of Physical Chemistry
Bulgarian Academy of Sciences
Sofia, Bulgaria

Catherine P. Whitby
School of Food Technology and Natural
Sciences
Massey University
Palmerston North, New Zealand

Jan Zawala
Jerzy Haber Institute of Catalysis and
Surface Chemistry
Polish Academy of Sciences
Krakow, Poland

Section I

Physico-chemical characterization of interfaces with relevance in emulsion stabilization

1 Fundamentals of Liquid/ Fluid Interfaces

Volodymyr I. Kovalchuk, Eduardo Guzmán, Ramón G. Rubio, Aliyar Javadi, Emanuel Schneck, Reinhard Miller, and Valentin B. Fainerman

1.1 INTRODUCTION TO LIQUID INTERFACES

We speak of an interface whenever two bodies, media, or phases are in contact with each other. Interfaces exist between all types of media. They are commonly called surfaces when solids or liquids are in contact with a gas phase, while the more general term 'interface' is used for all other combinations, such as two liquids or a liquid and a solid. There are even interfaces between two aqueous phases, which do not mix with each other, such as the solutions of two different polymers, between a polymer solution and a salt solution, or even between solutions of very different concentrations.[1]

Liquid interfaces can play various roles, as it is discussed for example in reference[2]. Many technologies are based on the impact of respective interfaces with their particular properties. By the use of surface-active components, such as surfactants,[3] polymers,[4] or biopolymers like proteins,[5] the behavior of interfaces can be modified in a very defined way. Aerosols, foams, and emulsions as liquid disperse systems, for example, are thermodynamically unstable and can exist only due to the action of surface-active compounds. Dispersed systems are produced in industry but they also appear in nature. For example, foams appear in rivers or at ocean beaches when sufficient amounts of biosurfactants are produced by plants or microorganisms in the water. A prominent example of emulsions in nature is milk, the main food of infants but also the origin of many products in the human diet. Natural aerosols are for example mists, fogs, and fumes.

A key quantity for characterizing a liquid surface or interface is the surface or interfacial tension γ. This parameter corresponds to the work that is required to create a surface/interface of a unit area. The molecular origin of this work is the asymmetry of intermolecular interactions in an interfacial region. To generate for example an emulsion, i.e., to disperse one liquid in an immiscible second one, droplets have to be formed. The sum of the surface areas of all droplets multiplied by the corresponding interfacial tension is the minimum thermodynamic work required to form such a system. Hence, reducing the interfacial tension is a way to save enormous amounts of energy in technologies that produce emulsions.

DOI: 10.1201/9781032636184-2

Although these colloidal dispersions are thermodynamically unstable systems, their use is generally based on kinetic stability, the general basis of which is the so-called DLVO theory. This famous theory was named after the four scientists who contributed most significantly: Derjaguin, and Landau,[6] and Verwey and Overbeek,[7] The DLVO theory considers the combined effects of the van der Waals attraction and the electrostatic repulsion between two particles in a dispersed system and can be applied to suspensions as well as to foams, emulsions, and aerosols. More details of this theory are given in later parts of this book.

The present chapter serves as a general introduction to the subject 'emulsions' and provides the basic definitions of parameters and phenomena, including the definition of an interface and its properties, the impact of surface-active agents on the interfacial properties, and their role in the formation and stabilization of water-in-oil (W/O) and oil-in-water (O/W) emulsions, as well as of W/O/W or O/W/O multiple emulsions.

1.2 GENERAL DEFINITION OF A FLUID INTERFACE

The easiest way to understand the idea of an interface is presented schematically in Figure 1.1. In the left scheme (a), a very simplified view of the transition from phase α to a second phase β is shown. An arbitrarily chosen property c (which remains unspecified for a moment) changes abruptly at the interface from c^α to c^β .

In reality, there is a smooth transition of any property from one phase to the other within the so-called interfacial region (scheme b). The thickness of the interfacial layer depends on the type of the two phases, but it is in any case in the range of nm.[8] If property c is for example the concentration of a surfactant, which accumulates in the interfacial region, we observe a profile as shown in the right scheme (c), i.e., c is changing from c^α to c^β with a maximum at the interface, depending on the surface activity of the surfactant, i.e., its tendency to accumulate in this region.

The pink area refers to the adsorbed amount Γ (which has the units of a number per unit area) and is dedicated as property to the dividing surface, i.e., the three-dimensional interfacial region with its finite thickness is mapped onto two

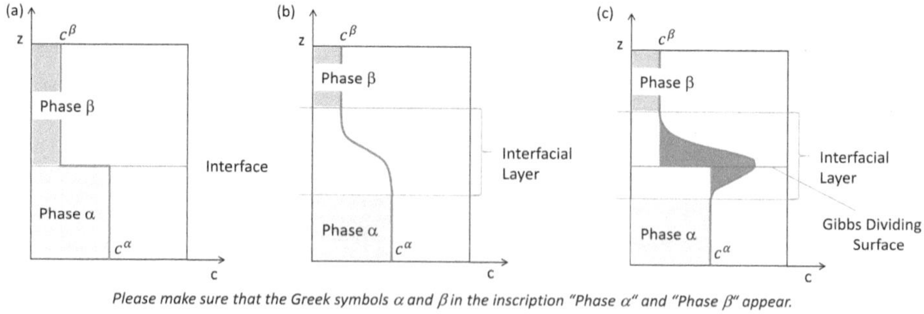

FIGURE 1.1 Interfacial region between the phases α and β: (a) with an abrupt transition of the property c, (b) with a smooth transition of the property c, and (c) with a transition of the property c including an excess amount (adsorption Γ) at the interface.

dimensions. This way of describing the properties of the interfacial region and the adsorbed amount was introduced by Gibbs and proved to be a useful approach still until today.[9]

Milner[10] was probably the first who considered the adsorption of surface-active compounds at interfaces to be a time-dependent process. A very extensive description of the fundamentals of adsorption dynamics was given in reference [11]. Although this was three decades ago, the principles are still valid; however, many quantitative details have been added to the knowledge available at that time. Thus, for many adsorption processes at fluid interfaces, meanwhile a semi-quantitative or even quantitative description has been established.

For adsorption processes at liquid/liquid interfaces, however, an additional complication appears when the solutes, for example surfactant molecules, are soluble in both adjacent liquid phases. Then, the surfactant can start to adsorb at the interface from one liquid phase and subsequently transfer partly into the second liquid phase, controlled by the respective distribution coefficient.[12] Moreover, the molecules of the oil phase are included in the structure of the interfacial adsorption layer, which will be discussed in more detail in Chapter 2. These mixed interfacial layers composed of surfactant and solvent molecules show cooperative and competitive effects, depending on the concentration of the surfactants.[13]

1.3 THERMODYNAMICS OF INTERFACIAL LAYERS

The Gibbs fundamental equation is the main relationship that combines the adsorbed amount Γ with the interfacial tension γ and the surfactant bulk concentration c,

$$\Gamma = -\frac{1}{\upsilon RT}\frac{d\gamma}{dlnc}. \tag{1.1}$$

Here, R is the ideal gas constant and T is the absolute temperature. The factor υ has a value of 1 for non-ionic and 2 for ionic surfactants, or a value in between 1 and 2 depending on the coupling of cations and anions within the surface layer.[14]

General relations for describing the properties of bulk phases and interfacial layers of surfactants are commonly derived starting with the chemical potential of the solute and the solvent. The chemical potentials of components i in phase α are functions of pressure P and temperature T and activity of the component in the bulk solution,

$$\mu_i^\alpha = \mu_i^{0\alpha}(T,P) + RT\ln f_i^\alpha x_i^\alpha, \tag{1.2}$$

where $x_i = m_i / \sum m_j$ are the corresponding molar fractions and m_i are the number of moles of the ith component, while $\mu_i^{0\alpha}$ and f_i^α are the chemical standard potentials and activity coefficients of the component i, respectively. The standard chemical potentials in the bulk, $\mu_i^{0\alpha}(T,P)$, are functions of pressure P and temperature T. For the same component i, the standard potential in the interfacial layer depends not only on P and T but also on the interfacial tension γ. Thus, for the chemical potential of the interfacial layer (marked by the superscript s), we obtain

$$\mu_i^s = \mu_i^{0s}(T,P,\gamma) + RT\ln f_i^s x_i^s. \tag{1.3}$$

With the help of an additional relationship containing the interfacial tension γ, the standard chemical potentials μ_i^{0s} can be expressed as a function of P and T only. This was successfully done in literature by using the variation of the free enthalpy of the Gibbs‘ dividing surface at constant pressure and temperature,[3]

$$dG = -Ad\gamma + \sum_{i=1}^{n} \mu_i^s dm_i^s, \tag{1.4}$$

with A being the interfacial area and n the number of components in the system. By combining Equations (1.3) and (1.4), and some thermodynamic standard assumptions as described in detail in reference [3], we finally end up with a relationship called Butler's equation,[15]

$$\mu_i^s = \mu_i^{0s} + RTlnf_i^s x_i^s - \gamma\omega_i. \tag{1.5}$$

The applicability of this equation was discussed many times and often in a controversial way. However, until now all adsorption models derived on its basis were very successful and neither in conflict with experimental findings nor with any basic thermodynamic principles. Various adsorption models of increasing complexity will be discussed further in Chapter 2, starting with the simplest model of Henry followed by the model of Langmuir[16] and further by the model of Frumkin.[17] We will also show that the reorientation model[18] as well as the most recently developed multistate model,[19] both developed by Fainerman, are systematic refinements of the classical models by Henry, Langmuir, and Frumkin.

1.4 DYNAMICS OF LIQUID ADSORPTION LAYERS

The adsorption of amphiphilic molecules at an interface is a self-assembling process as it leads to a gain in free energy. After an interface between two fluids is freshly formed, it is initially free of adsorbed molecules. The number of adsorbed molecules then gradually increases with time until an equilibrium state is reached between adsorbed molecules and those in the sublayer adjacent to the interface. In this process, the negative concentration gradient toward the interface causes a flux of surfactant molecules from the bulk to the interface. The change in adsorption with time, $\Gamma(t)$, and the concentration profile $c(x,t)$ are shown schematically in Figure 1.2. At $t = 0$, it is assumed that there is a homogeneous distribution of surfactants in the bulk and no molecules are adsorbed at the interface located at $x = 0$. With the creation of the new interface, few surfactant molecules being close enough to the interface can adsorb and create a steep concentration gradient adjacent to the interface (a), which generates a diffusive surfactant flux toward the interface. In steps (b)–(d), with

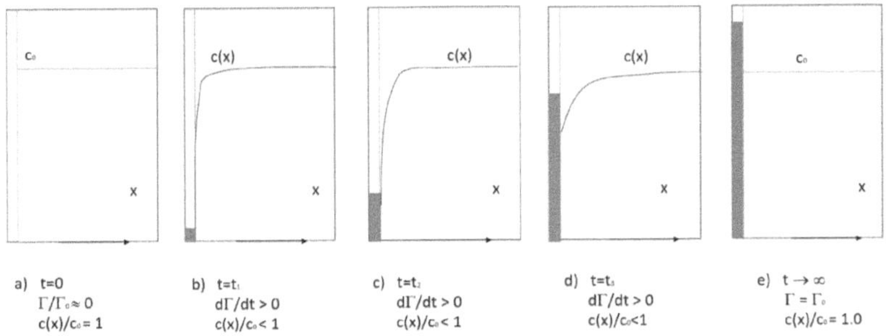

FIGURE 1.2 Change of adsorbed molecules at the interface and the surfactant concentration profile in the solution bulk (schematic).

increasing time $t_1 < t_2 < t_3$ the concentration gradient relaxes due to diffusion of the surfactant molecules, i.e., it gets less steep but reaches deeper into the solution bulk. Simultaneously, some new molecules adsorb at the interface and Γ increases with time. Finally, the concentration gradient disappears (e), and we get again a homogeneous concentration distribution, and the number of adsorbed molecules per unit area saturates at Γ_0 at equilibrium, i.e., at $t = \infty$.

The model derived in 1946 by Ward and Tordai[20] was the first quantitative description of the adsorption process. The authors considered a diffusional transport of surfactant molecules from the solution bulk to the subsurface, a place close to the interface. From here, the surfactant molecules can adsorb at the interface without any additional transport. In reference [11], the history of all types of adsorption models was presented, including many of the earlier very simple attempts to adequately describe the process of adsorption. The result obtained by Ward and Tordai is historically most significant, although they derived only an integral equation to describe the time dependence of the adsorbed amount $\Gamma(t)$ at a given surfactant bulk concentration c_0,

$$\Gamma(t) = 2\sqrt{\frac{D}{\pi}}\left[c_0\sqrt{t} - \int_0^{\sqrt{t}} c(0, t-\tau)\, d\sqrt{\tau}\right]. \tag{1.6}$$

This result for a so-called diffusion-controlled adsorption kinetics model is known since then; however, due to its complexity, it was not suitable for the analysis of experimental data. Therefore, it was simplified by deleting the second term on the right-hand side, which leads to a relation that can help to analyze experimental data most easily,

$$\Gamma(t) = 2c_0\sqrt{\frac{Dt}{\pi}}. \tag{1.7}$$

The slope of the plot Γ vs. $\sqrt{t}$ is proportional to $\sqrt{D}$ so that the value of the diffusion coefficient D of the studied surfactant can be estimated. Note, however, Equation (1.7) is a very rough approximation and valid only for small adsorption times t as shown in reference [21]. In Chapter 2, we will present a short history and then discuss the most recent state-of-the-art theoretical models to describe the adsorption kinetics of surfactants, polymers, proteins, and their mixtures at liquid/fluid interfaces. This will include also the case of surfactants soluble in both liquid phases, which requires the consideration of a distribution or partitioning coefficient as an impact parameter influencing the adsorption kinetics remarkably.[12] In addition to the adsorption kinetics, also the dynamics of interfacial layers after external perturbations is of importance. The interfacial response to generated harmonic perturbation will also be discussed in Chapter 2, as such experiments provide important information on the dilational interfacial viscoelasticity of the adsorbed layers,[22] discussed as important parameter for the stabilization of emulsions.

1.5 SUMMARY AND OUTLOOK

The adsorption equilibrium and its time evolution are essential for the formation and stabilization of emulsions. The present state of the art allows that most of the essential properties can be understood qualitatively, in many cases even quantitatively. While we have a good knowledge on transport processes in liquid systems, the processes happening between bulk phases and at interfaces or within interfacial layers are still only partially understood due to their considerable complexity. There are presently many attempts to improve the situation by including molecular dynamics simulations. Although these calculations are unable to simulate complete emulsions, they can look into molecular details at interfaces and help to analyze the structure of the formed adsorption layers.

ACKNOWLEDGMENTS

The work was supported by the ESA MAP project 'Emulsion Dynamics and Drop Interfaces (EDDI)' with the project number 4000128643/19.

REFERENCES

1. Nicolai, T.; Machado. J.P.E., Effect of the interfacial tension on droplet association in aqueous multiphase systems. *Langmuir* **2021**, *37,* 5909–5915.
2. Aveyard, B. *Surfactants—In Solution, at Interfaces and in Colloidal Dispersions*, Oxford University Press: Oxford, UK, 2019.
3. Fainerman, V.B.; Möbius, D.; Miller, R. (Eds.). Surfactants: Chemistry, interfacial properties and applications. In *Studies in Interface Science*, Elsevier: Amsterdam, The Netherlands, 2001; Volume 13.
4. Fleer, G.J.; Cohen Stuart, M.A.; Scheutjens, J.M.H.M.; Cosgrove, T.; Vincent, B. *Polymers at Interfaces*, Springer: Berlin, Germany, 1993.
5. Fainerman, V.B.; Miller, R. Adsorption isotherms of proteins at liquid interfaces. In *Proteins at Liquid Interfaces*, in "Studies in Interface Science", Möbius, D., Miller, R. Eds.; Elsevier: Amsterdam, The Netherlands, 1998; Volume 7, pp. 51–102.
6. Derjaguin, B.; Landau, L. Theory of the stability of strongly charged lyophobic sols and of the adhesion of strongly charged particles in solutions of electrolytes. *Zh. Exp. Theor. Phys.* **1945**, *15*, 663–682.

7. Verwey, E.J.W.; Overbeek, J.Th.G. Theory of the stability of lyophobic colloids. *J. Phys. Colloid Chem.* **1948**, *51*, 631–636.
8. Kanduč, M.; Miller, R.; Stubenrauch, C.; Schneck, E. Interface adsorption versus bulk micellization of surfactants: Insights from molecular simulations. *J. Chem. Theory Comp.* **2024**, *20*, 1568–1578.
9. Gibbs, J.W. *The Collected Works*, Vol. 1, Thermodynamics, Longmans Green & Co.: London, UK, 1928.
10. Milner, S.R.V. On surface concentration, and the formation of liquid films. *Phil. Mag.* **1907**, *13*, 96–110.
11. Dukhin, S.S.; Kretzschmar, G.; Miller, R. Dynamics of adsorption at liquid interfaces: Theory, experiment, application. In *Studies in Interface Science*, Möbius, D., Miller, R., Eds.; Elsevier: Amsterdam, The Netherlands, 1995; Volume 1.
12. Ravera F.; Ferrari M.; Liggieri L. Adsorption and partitioning of surfactants in liquid–liquid systems. *Adv. Colloid Interface Sci.* **2000**, *88*, 129–177.
13. Fainerman, V.B.; Aksenenko, E.V.; Kovalchuk, V.I.; Mucic, N.; Javadi, A.; Liggieri, L.; Ravera, F.; Loglio, G.; Makievski, A.V.; Schneck, E.; Miller, R. New view of the adsorption of surfactants at the water/alkane interface – competitive and cooperative effects of surfactant and alkane molecules. *Adv. Colloid Interface Sci.* **2020**, *279*, 102143.
14. Bae, S.; Haage, K.; Wantke, K.; Motschmann, H. On the factor in Gibbs equation for ionic surfactants. *J. Phys. Chem. B* **1999**, *103*, 1045–1050.
15. Butler, J.A.V. The thermodynamics of the surfaces of the solutions. *Proc. Roy. Soc. Ser. A* **1932**, *138*, 348–375.
16. Langmuir, I. The construction and fundamental properties of solids and liquids part II liquids. *J. Am. Chem. Soc.* **1917**, *39*, 1848–1906.
17. Frumkin, A. Electrocapillary curve of higher aliphatic acids and the state equation of the surface layer. *Z. Phys. Chem.* **1925**, *116*, 466–485.
18. Fainerman, V.B.; Miller, R.; Makievski, A.V. Reorientation of polyethylene glycol oxyethylene ether in non-equilibrium adsorption layers at the water/air interface. Role of molecular weight and temperature. *Langmuir* **1995**, *11*, 3054–3060.
19. Fainerman, V.B.; Kovalchuk, V.I.; Aksenenko, E.V.; Ravera, F.; Liggieri, L.; Loglio, G.; Makievski, A.V.; Mishchuk, N.O.; Schneck, E.; Miller R. A multistate adsorption model for the adsorption of $C_{14}EO_4$ and $C_{14}EO_8$ at the solution/air interface. *Colloids Interfaces* **2021**, *5*, 39.
20. Ward, A.F.H.; Tordai, L., Time-dependence of boundary tensions of solutions I. The role of diffusion in time-effects. *J. Chem. Phys.* **1946**, *14*, 453–461.
21. Makievski, A.V.; Fainerman, V.B.; Miller, R.; Bree, M.; Liggieri, L.; Ravera, F. Determination of equilibrium surface tension values by extrapolation via long time approximations. *Colloids Surf. A* **1997**, *122*, 269–273.
22. Miller, R.; Liggieri, L. (Eds.). Interfacial rheology. In *Progress in Colloid and Interface Science*, Brill Publication: Leiden: The Netherlands, 2009; Volume 1.

2 Adsorption at Liquid/Fluid Interfaces

Equilibrium and Dynamic Aspects

Eugene V. Aksenenko, Volodymyr I. Kovalchuk, Georgi Gochev, Libero Liggieri, Giuseppe Loglio, Aliyar Javadi, Emanuel Schneck, Reinhard Miller, and Valentin B. Fainerman

2.1 INTRODUCTION

To modify the properties of liquid interfaces, which is usually required to reduce the interfacial tension or to stabilize foams or emulsions, the addition of surface-active molecules is needed.[1] Due to their amphiphilicity, these molecules adsorb at interfaces, reduce the interfacial tension, and form layers of a certain thickness with respective interfacial viscoelastic properties. While classical low-molecular-weight surfactants adsorb typically as monolayers, polymers or proteins can form double layers or even multilayers.

The basis for the characterization of surfactants at liquid/fluid interfaces is their adsorption isotherm $\Gamma(c)$, i.e., a relation between the adsorbed amount Γ as a function of the bulk concentration c. As shown by the Gibbs fundamental Equation (1.1), for soluble surfactants, Γ and c are interrelated in a very general way, and Γ can therefore be determined via measurements of the interfacial tension $\gamma(c)$ as a function of the bulk concentration c. This shows the importance of the experimentally most easily accessible interfacial tension $\gamma(c)$ or interfacial pressure isotherm $\Pi(c) = \gamma_0 - \gamma(c)$ with γ_0 being the interfacial tension of the surfactant-free system. The relationship which is more of thermodynamic importance, however, is the so-called equation of state, which gives a relation between the two quantities of the interfacial layer $\Pi(\Gamma)$.

In most applications, the adsorption layers are not at equilibrium but under dynamic conditions.[2] This is the reason why just knowing the adsorption isotherms or equations of state is usually not sufficient to evaluate the applicability of certain surfactants or polymers in practice. Therefore, in addition to the determination of the adsorption isotherm for a given surfactant at a respective interface as fundamental information, also the corresponding dynamic properties are essential for many technical applications. This includes the adsorption kinetics, in most

 DOI: 10.1201/9781032636184-3

cases in terms of the dynamic interfacial tension $\gamma(t)$, and the dilational interfacial viscoelasticity.[3] These points are discussed further below in more detail.

Particularly important for surfactants in liquid/liquid systems is the fact that they can be soluble in both adjacent phases. This means, for example, that in emulsions the added surfactant can be partially soluble in both the aqueous phase and the oil phase. The partitioning of the surfactant among both liquid phases depends on the partition coefficient $k_p = c_{oil}/c_{water}$,[4] i.e., the ratio of the surfactant concentration in oil to that in water at equilibrium. In most cases, from the value of k_p we can conclude which type of emulsion is formed. Following the Bancroft rule,[1] for $k_p < 1$ we expect to obtain an O/W emulsion, while in the opposite case, $k_p > 1$, the surfactant is more soluble in the oil phase, which typically leads to W/O emulsions. As we will see in Section 2.4, the partition coefficient has a strong impact on the adsorption dynamics as well as the dilational interfacial viscoelasticity.

In the subsequent sections of this chapter, we will discuss the adsorption of polymers and proteins, and also of mixed protein/surfactant solutions. Finally, a short discussion of the dilational viscoelasticity of liquid adsorption layers is discussed.

2.2 SURFACE-ACTIVE COMPONENTS

The main group of substances for the modification of interfacial layers, required for example to generate emulsions and to stabilize them, are surfactants, an artificial word created from the expression **surf**(ace)-**act**(ive) **a**(ge)**nts**. Until surfactants were synthesized, they have been named either as 'surface-active materials' or by their name, mainly fatty acids and fatty alcohols. Since the mid of the 20th century, the word 'surfactant' is used in all respective fields of science and applications.

Depending on their electrical state in aqueous solutions, we distinguish between nonionic, anionic, cationic, and amphoteric surfactants (Figure 2.1). There are many homologous series of nonionic surfactants, i.e., surface-active molecules with an aliphatic chain and an uncharged headgroup. Among these, the alkyl dimethyl phosphine oxides are very often used as model surfactants due to their simplicity, chemical purity and good solubility in water, including even those with longer alkyl chains.

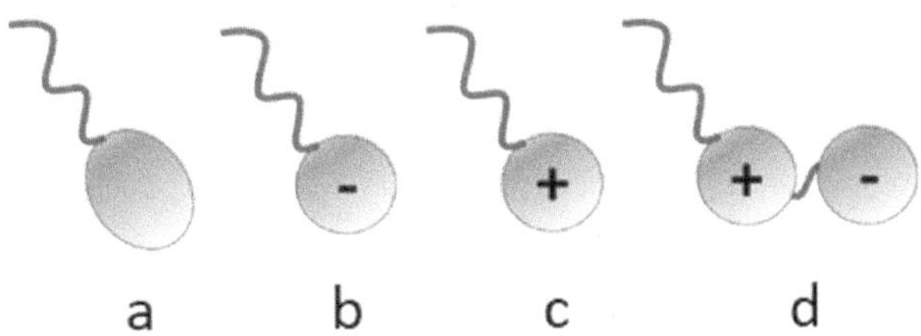

FIGURE 2.1 General structure of nonionic (a), anionic (b), cationic (c), and amphoteric surfactants (d).

The most commonly used nonionics, however, are the oxyethylated alcohols (C_nEO_m). The properties of these surfactants can be somehow tailored with respect to their interfacial behavior by tuning the alkyl chain length (parameter *n*) and the number of ethylene units as hydrophilic part (parameter *m*).[5] This affects also their solubility in water and oil.

Anionic surfactants have been the first synthetic compounds used as detergents. The typical structure is an aliphatic chain and a negatively charged headgroup with a positively charged counterion. The most prominent representative of anionic surfactants is sodium dodecyl sulfate shown in Figure 2.2.[6]

Note, although Na^+ is only the positively charged counterion, the surface properties of the aqueous solutions depend also on this ion, so that, for example, the critical micelle concentration (CMC) values decrease for the analogous compounds formulated with different cations in the order $Li^+ > Na^+ > K^+ > Cs^+$, according to the Hofmeister series.[7]

The most frequently studied cationic surfactants are the alkyl trimethyl ammonium halogenides (see Figure 2.3), which have been systematically studied not only at the water/air[8] but also at the water/alkane interface.[9]

Amphoteric surfactants contain a hydrophobic aliphatic chain and two hydrophilic headgroups, a positive and a negative one. The charged groups are connected via a spacer group. At the isoelectric point pK, they are zwitterionic and behave like nonionic surfactants, below the pK they are cationics, and above the pK they are anionic surfactants.[10] Their surface activity depends not only on the length of the aliphatic chain but also on the spacer length between the two charged head groups, and also on the solution's pH.

It is worth mentioning the Traube rule,[11] which, when applied to the surface activity of surfactants, provides an excellent tool to select the proper surfactant with required surface behavior for respective applications. With respect to the interfacial

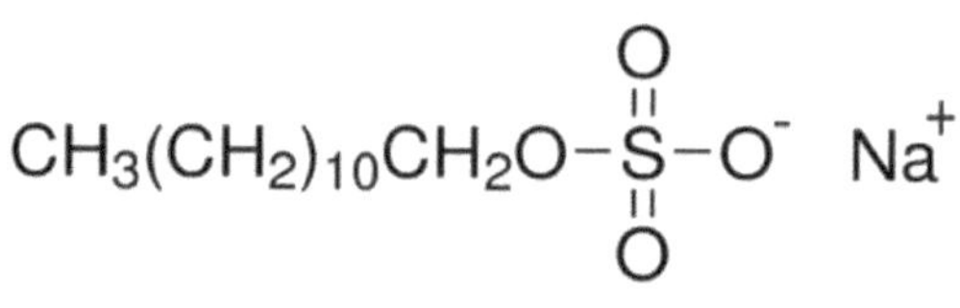

FIGURE 2.2 Chemical formula of the anionic surfactant sodium dodecyl sulfate (SDS).

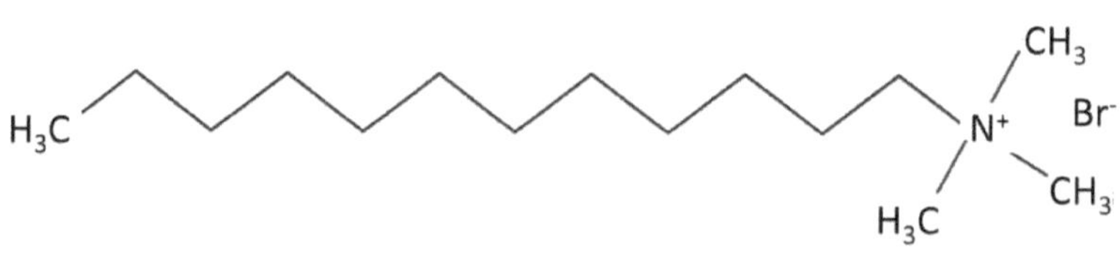

FIGURE 2.3 Structure formula of the cationic surfactant dodecyl trimethyl ammonium bromide (DoTAB).

tension isotherms, for example, with each additional CH_2 group in the alkyl chain, the isotherm is shifted toward lower concentrations, and two of these groups produce a shift of about one order of magnitude to lower concentrations (this will be explained in more detail in Section 2.3.1). The same is true with the other properties, for example for the corresponding CMC values.

Instead of classical surfactants, also amphiphilic polymers are used in practice for stabilizing foams and emulsions. These polymers are typically di-block or tri-block copolymers, consisting of hydrophilic and hydrophobic segments that make them amphiphilic. The adsorption happens in a way similar to that of ordinary surfactants; however, due to their size, the molar concentrations and the corresponding diffusion coefficients D are rather low. Thus, the time for reaching the adsorption equilibrium can be rather long. The establishment of an equilibrium adsorption layer includes not only the diffusional transport in the solution bulk, but also the arrangement of the chains at the interface, controlled by a respective dynamics of chain conformations.[12] The surface activity of block copolymers depends on the length of the hydrophilic and hydrophobic segments. There is a huge number of different amphiphilic block copolymers as discussed for example in the book by Alexandridis and Lindman,[13] and for many of them the interfacial behavior was studied extensively.[14]

For proteins (biopolymers), the sequence of their building blocks, the amino acids, is given by nature. Most proteins are highly surface-active, i.e., they significantly adsorb at fluid interfaces even at extremely low bulk concentrations. A comprehensive overview of the adsorption of proteins at liquid interfaces was given by Fainerman and Miller.[15] Since then, the adsorption model for proteins at liquid interfaces was further developed and even refined for the application to measurements of the dilational interfacial viscoelastic properties of protein layers.[16]

Since the publications of Ramsden[17] and Pickering,[18] we know that also microscopic solid particles are able to act similarly to surfactants, accumulating at the liquid interface and stabilizing foams and emulsions. Nowadays, particles are synthesized with tunable size, shape, surface chemistry, and hydrophobicity, for respective types of applications in various fields, including food, cosmetics, etc.[19] In many applications, mixtures of surface-active compounds are needed to reach the required effects on the interfacial properties, such as surfactant–surfactant,[20] polymer–surfactant,[21] protein–surfactant,[22] or particle–surfactant mixtures.[23] The generally accepted functions of the two (or even more) components are that the initially adsorbing species help to generate the emulsion while the slower adsorbing molecules typically serve as stabilizers. Particle–surfactant mixtures are also used to modify the nature of the particles, which, for example, turn from hydrophilic to amphiphilic after the adsorption of surfactant molecules at their surface.[24]

2.3 ADSORPTION MODELS FOR SURFACTANTS

The characterization of surfactants by their main properties at particular fluid interfaces includes first their adsorption isotherms $\Gamma(c)$. As direct measurements of the adsorbed amounts Γ are experimentally not easy, it can be replaced by measuring the interfacial tension isotherms $\gamma(c)$. Then, via the Gibbs fundamental Equation (1.1), we can obtain the function $\Gamma(c)$. In addition to these equilibrium interfacial properties, also the dynamics of adsorption is of paramount importance, including the dynamics of the adsorption process and the response of an interface to external compressions or expansions, measured by dynamic interfacial tensions and dilational interfacial viscoelasticity, respectively. The theories behind these measurements with some experimental data as examples are described here while the corresponding experimental details are the subject of Chapter 3.

2.3.1 Adsorption Isotherms and Equations of State for Surfactants

The main adsorption isotherms for surfactants were derived and discussed in much more detail elsewhere,[25] while here we give just a brief overview. The Henry adsorption model,[26] which is the simplest available one, reads

$$\Gamma = Kc = \frac{\gamma_0 - \gamma}{RT} = \frac{\Pi}{RT}. \tag{2.1}$$

This equation is equivalent to that for ideal gases, in which the pressure is proportional to the number of molecules.

The adsorbed amount Γ is proportional to the surfactant concentration through the constant K, which is a measure of the surface activity of the surfactant, and to the surface pressure $\Pi(c) = \gamma_0 - \gamma(c)$. In this oversimplified view, the area required for the adsorption of the surfactant molecules is not considered, so that adsorption is point-wise. Thus, the use of this linear adsorption model is allowed only at very low surface coverage, when all adsorbed molecules have sufficient space at the interface. Experiments show that $\Pi(c)$ in this range is actually well approximated by a linear dependence.

The main idea of the Langmuir adsorption model[27] is a localized adsorption, which means that each molecule, even when not interacting with its neighbors, occupies a finite area at the interface. That defines the maximum number of adsorbed molecules per unit area, Γ_∞, based on the geometrical area covered by a monolayer of surfactant molecules and leads to the following relationships,

$$\gamma = \gamma_0 + RT\Gamma_\infty \ln\left(1 - \frac{\Gamma}{\Gamma_\infty}\right) = \gamma_0 - RT\Gamma_\infty \ln\left(1 + bc\right), \tag{2.2}$$

where Γ_∞ is the maximum number of adsorbed molecules per unit area, and the surface activity coefficient b has the dimension of an inverse concentration and is an excellent measure to characterize the surface activity of a surfactant. From Equations (1.1) and (2.2), it follows that at $c = 1/b$ we obtain $\Gamma = \Gamma_\infty/2$, i.e., at this concentration half of the maximum surface coverage is reached. Note also, at $c \ll 1/b$, Equation (2.2) can be simplified to Equation (2.1) for the Henry model with $K = b\Gamma_\infty$. An equation equivalent to the Langmuir adsorption model was already proposed in 1908 by von Szyszkowski,[28] however, on a completely empirical way. Thus, we should better name this model after both pioneers, Szyszkowski and Langmuir.

A further refinement of the Langmuir model was proposed by Frumkin.[29] He realized that there is a remarkable deviation between experimental data for some surfactants and the $\gamma(c)$ dependence calculated from the Langmuir model. Therefore, Frumkin added an interaction term leading to the following equation of state:

$$\gamma = \gamma_0 + RT\Gamma_\infty \left[\ln\left(1 - \frac{\Gamma}{\Gamma_\infty}\right) + a\left(\frac{\Gamma}{\Gamma_\infty}\right)^2 \right] \text{ or } \gamma = \gamma_0 + \frac{RT}{\omega_0}\left[\ln(1-\theta) + a\theta^2\right], \quad (2.3)$$

where $\omega_0 = 1/\Gamma_\infty$ is the molar area of adsorbate at the state when the maximum possible adsorption is achieved, i.e., each of the adsorbed molecules requires a finite area at the interface, the surface coverage $\theta = \Gamma/\Gamma_\infty$ is introduced, and it is assumed that adsorbed molecules can interact with each other depending on the value of a. The adsorption isotherm expressed as the relation between the surfactant concentration in the solution and the surface coverage for this model reads

$$bc = \frac{\theta}{1-\theta}\exp(-2a\theta). \quad (2.4)$$

This equation is physically reasonable only for values of the interaction parameter $a < 2$, otherwise the equation is not unambiguous, predicting that in a certain concentration range the derivative $d\Gamma/dc$ is negative, which would correspond to an unstable monolayer. The presence of an interaction term makes this isotherm particularly suited to model adsorption of ionic surfactants, which, in fact, may interact electrostatically within the adsorption layer.

In Figure 2.4, the interfacial tension isotherms of four members of the homologous series of the cationic surfactants alkyl trimethyl ammonium bromides C_nTAB (for $n = 10$, 12, 14, and 16) are presented.[30] The best fit of the data by the Frumkin isotherm is given by the green dashed lines. It becomes clear that, in the framework of the classical thermodynamic description, this is an excellent agreement between experimental and modeled data.

A refinement of this model for surfactants adsorbed at water/oil interfaces is given further in Section 2.7 including the impact of the oil molecules (solid red lines). According to recent findings, the oil molecules are part of the interfacial surfactant layers, and at low surface coverages even cooperative effects between surfactant and oil molecules are observed.

A further refinement of the Frumkin model has been derived most recently by Fainerman et al.,[31] combining it with the reorientation or, more generally, by the multistate adsorption model, which assumes different adsorption states of the surfactant molecules in the interfacial layer. For surfactants with a simple structure, these states may correspond to different tilt angles with respect to the interface. For surfactants with a more complex structure, these states may describe different conformations, such as those of the ethylene oxide chain as hydrophilic headgroup for the class of nonionic surfactants poly(oxyethylene) monoalkyl ethers (or oxyethylated alcohols) C_nEO_m,

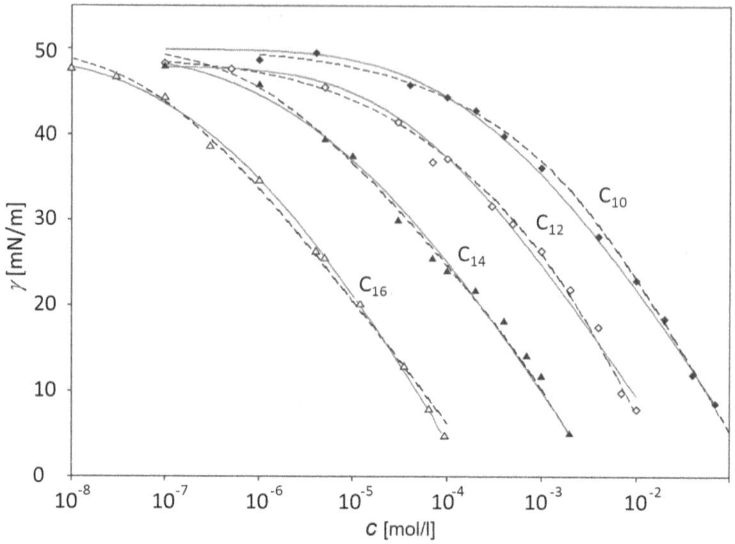

FIGURE 2.4 Interfacial tension isotherms of C_{10}TAB (◆), C_{12}TAB (◇), C_{14}TAB (▲), and C_{16}TAB (△) at the solution/hexane interface in the presence of phosphate buffer (10 mM, pH 7); solid lines – model for mixed interfacial layers including surfactants and hexane, dashed lines – Frumkin model. Adapted with permission from Fainerman et al.,[30] Copyright (2013), American Chemical Society.

More specifically, the model assumes s possible adsorbed states with minimum and maximum molar areas ω_1 and ω_s, respectively. Also, the effect of an intrinsic compressibility is taken into account by the model: molar areas in each state (except the sth) depend on the coverage θ of the adsorbed layer and the surface pressure Π. This could be thought of as the dependence of tilt angles of adsorbed molecules on the increasing crowding at the interface with the increase in θ and Π. For the state with minimum molar area, this dependence is $\omega_1 = (1 - \varepsilon \Pi\theta) \bullet \omega_{10}$, where ω_{10} is the minimum molar area at infinite dilution ($\Pi = 0$), and ε is the intrinsic compressibility coefficient. The molar area of the jth state ($1 \leq j \leq s$) is related to that of the state 1 and state s as $\omega_j = \omega_1 + (j - 1)\omega_0$, where the increment $\omega_0 = \omega_j - \omega_{j-1} = (\omega_s - \omega_1)/(s - 1)$ does not depend on j. The assumption of a constant increment simplifies the model strongly, reducing the number of the required model parameters while still preserving the most significant features of the system behavior. The average molar area ω is determined from the total coverage θ, partial adsorption in the jth state Γ_j and with the total adsorption Γ as $\theta = \omega\Gamma = \sum_{j=1}^{s} \Gamma_j \omega_j$, where $\Gamma = \sum_{j=1}^{s} \Gamma_j$. The equation of state for the multistate model reads,[30]

$$-\frac{\Pi\omega_1}{RT} = \ln(1-\theta) + \theta\left(1 - \frac{\omega_1}{\omega}\right) + a\theta^2, \tag{2.5}$$

and the corresponding adsorption isotherm for each jth adsorption state has the form:

$$b_j c = \frac{\omega \Gamma_j}{(1-\theta)^{\omega_j/\omega}} \exp\left[-2a\frac{\omega_j}{\omega}\theta\right] \tag{2.6}$$

Here a is Frumkin's intermolecular interaction parameter and b_j is the activity coefficient for the molecules adsorbed in state j. The second term on the right-hand side of Equation (2.5) represents the entropy contribution caused by the s different adsorption states. Note that this model is a systematic extension of earlier models and with the simplification of $s = 1$, i.e., with the assumption of only one adsorption state, the set of equations leads directly back to the Frumkin model as given by Equations (2.3) and (2.4).

In Figure 2.5 an example is given for the application of the multistate adsorption model to equilibrium surface tension data measured with the profile analysis tensiometer (PAT) in the buoyant bubble mode.

The use of air bubbles formed inside the surfactant solutions makes sure that no depletion effects affect the accuracy of the measured data (see Chapter 3, Section 3.3). The different types of curves in the figure are calculated with the multistate adsorption model described earlier, using different numbers s of adsorption states. The best fit of the model to the experimental data shows that for $C_{14}EO_4$ we should assume at least five conformations in the adsorption layer, while for $C_{14}EO_8$ we must even assume at least ten different adsorption states for establishing an optimum structure of the adsorption layer.[31] This conclusion was not drawn just on the basis of

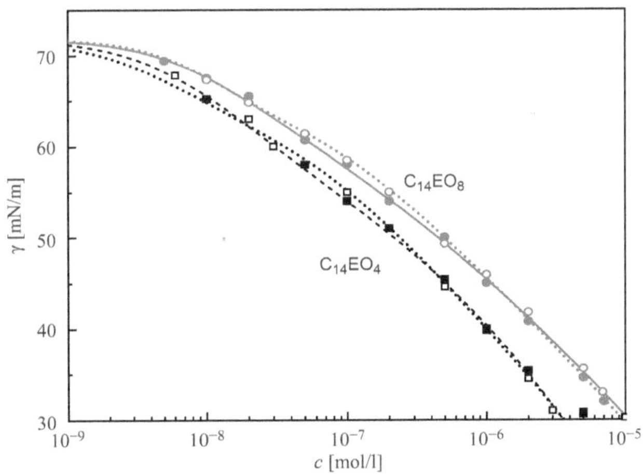

FIGURE 2.5 Surface tension isotherms of aqueous solutions of $C_{14}EO_4$ (squares) and $C_{14}EO_8$ (circles); the lines are calculated with the model given by Equations (2.5) and (2.6) with $s = 2$ (dotted lines), $s = 5$ (dashed line), and $s = 10$ (solid line). Adapted from Fainerman et al.[31] with permission under Creative Common Attribution License 4.0 [CC-BY 4.0].

the respective surface tension isotherm but also by taking into account the dilational viscoelastic surface behavior of the adsorption layer (not shown here) measured in the same concentration interval. Moreover, the studies were performed in a respective interval of rates of surface layer perturbation, which was arranged by tuning the frequency of bubble oscillations during the viscoelasticity measurements. This will be discussed further in Section 2.8.

2.3.2 Diffusion-Controlled Adsorption Kinetics for Surfactants

So far, we have considered adsorption, bulk surfactant concentration, and, as a consequence, surface tension in a mutual thermodynamic equilibrium. However, more generally adsorption is a dynamic process, involving the migration of surfactant molecules to/from the interfacial layer and possible rearrangements of these molecules within the interfacial layer. These processes are triggered by external stimuli, such as formation or loss of interfacial area, increase in concentration or dilution of the surfactant solution, or addition of new chemical species: all conditions that can be typical for many applications involving surfactants. Surface tension and adsorption are therefore 3.5. general dynamic quantities that may, or may not, eventually reach equilibrium values, depending on the boundary conditions imposed by a specific application. The characterization and modeling of adsorption and surface tension under dynamic conditions are clearly of paramount importance for many applications and for understanding several natural phenomena.

For a diffusion-controlled adsorption of surfactant molecules at fluid interfaces, the treatment based on the solution of Fick's diffusion equations is an obvious option. Fick's first law reads,

$$J = -D\nabla c(r,t), \tag{2.7}$$

where **J** is the diffusion flux vector, D is the diffusion coefficient, and the surfactant concentration field $c(\mathbf{r},t)$ is the spatiotemporal distribution of the surfactant. Fick's second law (also known as the diffusion equation) reads,

$$\frac{\partial c(r,t)}{\partial t} = D\Delta c(r,t). \tag{2.8}$$

As applied to the adsorption of a surfactant at the interface, Fick's first law constitutes the boundary condition for the solution of Equation (2.8), and is formulated to account for the obvious fact that the decrease/increase of the adsorbed amount at the interface should correspond to the diffusion flux from/to the interface, i.e.,

$$\frac{\partial \Gamma(t)}{\partial t} = D\nabla_{\perp} c(r,t), \tag{2.9}$$

where $(\nabla_{\perp})$ is the gradient calculated along the normal to the interface. Other boundary conditions are also problem-specific. In particular, for systems with a limited solution reservoir, the boundary condition is a zero flux at the reservoir wall, i.e., Equation (2.7) becomes $\nabla_{\perp} c(\mathbf{r}_{\text{wall}}, t) = 0$, while for an infinite reservoir the boundary condition is $\lim_{\mathbf{r}\to\infty} c(\mathbf{r},t) = c_0$ with c_0 equal to the equilibrium surfactant bulk concentration. The initial conditions are also determined by the problem formulation; a typical initial condition is the homogeneous surfactant distribution in the solution bulk, $c(r,0) = c_0$, and no adsorption at the surface, $\Gamma(0) = 0$. It is worth noting that while the real initial surface coverage is rather ambiguous, in typical cases for any reasonable distribution of a surfactant throughout the interface, the mathematical solution exhibits quite rapid convergence to that obtainable for the initially bare interface.

It must be underlined that, in all cases, the complete solution of the problem requires always appropriate adsorption and surface tension isotherms, in order to describe the relation between $\Gamma(t)$, $c(0,t)$, and $\gamma(t)$, which can be typically assumed to be in an instantaneous mutual equilibrium.

The generalization of the problem onto the cases of surfactant mixtures and/or interfaces penetrable by surfactants is quite straightforward and results in the solution of coupled equations corresponding to the diffusion of each involved species with the isotherm for the mixture as boundary condition at the interface; also, in case of penetrable interfaces the distribution coefficient(s) become involved in the coupling relations.

At present, the solution of Fick's diffusion equations was elaborated mostly for systems, which due to their geometrical symmetry could be simplified to account only for one spatial variable. These systems are either essentially one-dimensional, with one spatial Cartesian coordinate, or spherically symmetrical with a radial spatial coordinate.

Possibly, the first ever analytical solution for the one-dimensional problem was proposed by Sutherland[32]; however, his solution was derived only under the assumption of a linear (Henry) adsorption isotherm Equation (2.1) and reads,

$$\frac{\Gamma(t)-\Gamma_{\text{eq}}}{\Gamma(0)-\Gamma_{\text{eq}}} = \exp\left(\frac{t}{\tau_D}\right)\text{erfc}\left(\sqrt{\frac{t}{\tau_D}}\right), \tag{2.10}$$

where the equilibrium adsorption Γ_{eq} is calculated from the isotherm equation with the surfactant bulk concentration c_0 and the diffusion relaxation time $\tau_D = D^{-1}(d\Gamma/dc)$.

The first solution to describe the dynamics of adsorption and surface tension for a diffusion-controlled adsorption was derived by Ward and Tordai in 1946[33] who considered the accumulation of surfactant at a freshly formed (initially surfactant-free) planar interface. As the experiments are usually performed with drops and/or bubbles, it is more practical to operate with the corresponding mathematical formulation of the diffusion problem for a spherical geometry, for which Equation (2.8) becomes

$$\frac{\partial c(r,t)}{\partial t} = D\left[\frac{\partial^2 c(r,t)}{\partial r^2} + \frac{2}{r}\frac{\partial c(r,t)}{\partial r}\right], \tag{2.11}$$

with corresponding initial and boundary conditions. Here r is the radial coordinate (a scalar quantity). The solution for the case of the diffusion from an infinite bulk phase to the surface of a bubble with radius R immersed into the surfactant solution was obtained by Joos[34] and reads,

$$\Gamma(t) = 2\sqrt{\frac{Dt}{\pi}}c_0 - 2\sqrt{\frac{D}{\pi}}\int_0^{\sqrt{t}} c_s(t-\tau)d\left(\sqrt{\tau}\right) + \frac{D}{R}\left[c_0 t - \int_0^t c_s(t-\tau)d(t)\right], \quad (2.12)$$

where $c_s(t)$ is the subsurface concentration of the surfactant, which defines the adsorption value $\Gamma(t)$ via the corresponding isotherm. The following approximate solution valid for the early stage of the adsorption process is sometimes used,

$$\Gamma(t) \approx 2\sqrt{\frac{Dt}{\pi}}c_0 - 2\sqrt{\frac{D}{\pi}}\int_0^{\sqrt{t}} c_s(t-\tau)d\left(\sqrt{\tau}\right) \pm \frac{c_0 D}{R}t, \quad (2.13)$$

where the + or – sign of the last term on the right-hand side corresponds to a bubble or drop, respectively. Of course, the real shape of drops and bubbles is not spherical, but it can be assumed as a good approximation. More accurate simulations are far more computationally demanding as demonstrated by Dieter-Kissling et al.[35]

Note that the adsorption process of proteins and of other surface-active compounds follows the same physical principle, and its theoretical description is also generally the same, except for the boundary condition at the interface, given by Equation (2.9), where the right-hand term is modified to account for additional or more complicate processes, such as conformational equilibration within the interfacial layer (for instance, folding/unfolding, changes in the stoichiometry, etc.).

The algorithm for the numerical solution of Fick's diffusion problem with boundary conditions, first proposed by Liggieri et al.,[36] involved the discretization of temporal and spatial grids using variable intervals to make the integration nodes network denser in domains with higher concentration gradients, i.e., during the initial stage and at distances close to the interface. This approach was further implemented into a standard software for the solution of a single surfactant[37] and a mixture of two surfactants,[38] respectively. The procedure allows for a non-homogeneous initial surfactant distribution and non-zero initial adsorption at the interface.

An example for the adsorption kinetics studied via measurements of the dynamic surface tension is shown in Figure 2.6 for the nonionic surfactant $C_{14}EO_8$ (the respective surface tension isotherm of this surfactant is shown in Figure 2.5). The curves were calculated using the software described earlier for the case of surfactants adsorbing in two possible adsorption states ($s = 2$). It can be seen that the data $\gamma(t)$ are not described by a single diffusion coefficient but rather by a certain interval of D values. For both concentrations, the adsorption rate decreases, i.e., the apparent diffusion coefficients become smaller during the adsorption process. This deceleration of the adsorption process was explained in the transition of the surfactant molecules between two adsorption states by Fainerman et al.[39].

Although the values of D are physically reasonable in magnitude, such changes in the diffusivity during the adsorption process are non-physical and point at the

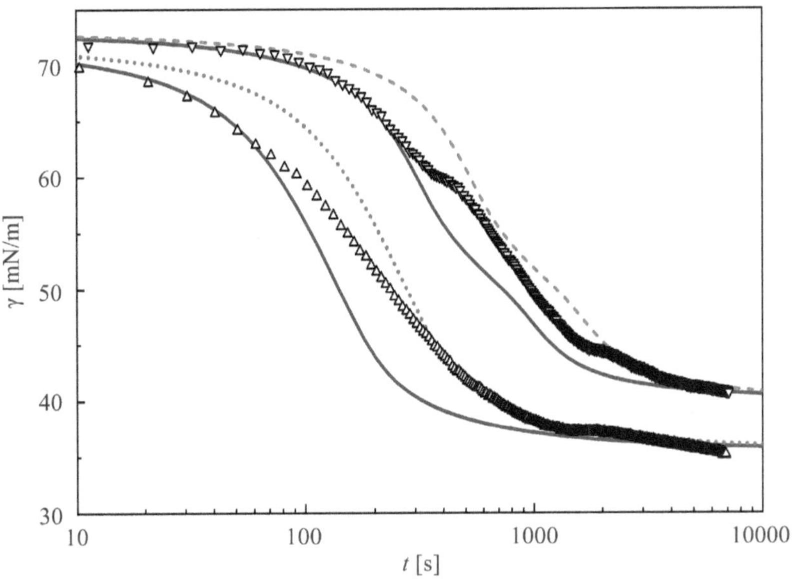

FIGURE 2.6 Dynamic surface tensions (symbols) for two aqueous $C_{14}EO_8$ solutions at surfactant concentrations of 2×10⁻⁶ M and 5×10⁻⁶ M and the corresponding theoretical curves calculated for the reorientation model with different diffusion coefficients: $D = 8\times10^{-10}$ m²/s for both solutions (solid lines), $D = 6\times10^{-10}$ m²/s for 2×10^{-6} M (dashed line), and 4×10^{-10} m²/s for 5×10^{-6} M (dotted line). Adapted from Fainerman et al.[39] with permission under Creative Common Attribution License 4.0 [CC-BY 4.0].

fact that there is an additional process partly influencing the adsorption rate. In the shown case, it is obviously the process of transition at the surface from one adsorption state with a larger molar area into another one with smaller molar area.

2.3.3 Kinetic-Controlled Adsorption Mechanisms

So far, we assumed that the diffusion of surfactants from the bulk to the interface and back is the time-determining step in the adsorption process, as is described in Section 2.3.2. However, in the history of surfactant adsorption modeling, a second group of models was discussed, the so-called kinetic-controlled adsorption. From certain experimental observations, the researchers concluded that what limits the adsorption rate is not the transport by diffusion but rather the adsorption step from the subsurface to the interface. For this purpose, rate equations were used which at infinite time ($d\Gamma/dt = 0$) would lead to one of the known adsorption isotherms.

The simplest is here again the linear dependency of the Henry model,

$$\frac{d\Gamma}{dt} = k_{ads}c - k_{des}\Gamma, \tag{2.14}$$

where k_{ads} and k_{des} are the rate constants of adsorption and desorption, respectively. For $t \rightarrow \infty$ we obtain the Henry Equation (2.1) with $K = k_{ads}/k_{des}$. For the Langmuir adsorption model, the respective rate equation reads,

$$\frac{d\Gamma}{dt} = k_{ads}\left(1 - \frac{\Gamma}{\Gamma_\infty}\right)c - k_{des}\Gamma, \tag{2.15}$$

and at equilibrium $d\Gamma/dt = 0$, it turns into

$$\Gamma = \Gamma_\infty \frac{c}{a+c} \quad \text{with } a = \Gamma_\infty \frac{k_{des}}{k_{ads}} \tag{2.16}$$

Baret proposed to replace the concentration c in Equation (2.15) by $c(0,t)$ while this so-called subsurface concentration can be calculated from the appropriate solution of diffusion kinetic problem described in Section 2.3.2, so that the kinetic process given by Equation (2.15) is linked to the diffusional transport process.[40] In this case, both the diffusional transport and the transfer kinetics are considered, and we speak therefore about a mixed adsorption kinetics.

In the literature, some authors speak about fast or slow surfactants, but such attributes are physically unsound. In fact, adsorption processes faster than expected by diffusion usually mask unrecognized convections in the systems that accelerate the process.

In most cases, however, the adsorption process is slower than expected for a diffusional adsorption mechanism. While in the past, often kinetic mechanisms described by rate equations like Equation (2.15) have been made responsible for the lower adsorption rates, Lunkenheimer and co-workers have shown that it is in many cases the impact of impurities[41] that decelerates the rate of adsorption. More about the history of adsorption models for surfactants at fluid interfaces can be found in publications by Dukhin et al.[2] and Fainerman et al.[25]

2.3.4 Adsorption Kinetics and Equations of State for Mixtures of Surfactants

As mentioned earlier, in many applications, the use of a single surfactant does not provide the required effects so that mixtures of surfactants are required. In publication[42] a broad overview of the theoretical basis is presented and numerous experimental examples are discussed. There are various versions of adsorption models depending on the types of surfactants in the mixtures. For an equimolar mixture of one cationic and one anionic surfactant, we can assume the formation of an ion pair so that in the end only one type of molecules adsorb – a neutralized nonionic molecule, and this can be described adequately by the models for single surfactants. For mixtures of surfactants that adsorb in competition at the interface (for example, two types of nonionic surfactants), however, we have to consider a respective individual space for each component at the interface. For example, for the mixed adsorption layer of two different nonionics of the type C_nEO_m, the following equation of state was proposed by Miller et al.[42]

$$-\frac{\Pi\omega_0^*}{RT} = \ln\left(1-\theta_1-\theta_2\right)+\theta_1\left(1-\frac{\omega_{10}}{\omega_1}\right)+\theta_2\left(1-\frac{\omega_{20}}{\omega_2}\right)+2a\theta_1\theta_2, \quad (2.17)$$

where the subscripts 1 and 2 stand for the respective components, the subscript 0 stands for the minimum molar area required by each component, and $\omega_0^* = \left(\omega_{10}\theta_1+\omega_{20}\theta_2\right)/\left(\theta_1+\theta_2\right)$ is a weighted average molar area. Here, it is assumed that at the interface, there is no interaction between molecules of the same surfactant but an interaction between different ones. In addition, the entropy of adsorption states for both surfactants is accounted for by the second and third terms in the right-hand side of Equation (2.17).

For the description of the adsorption dynamics, the diffusional transport of both components has to be considered. For this purpose, the appropriate procedure should be used, as explained in Section 2.3.2, with coupled Equation (2.11) for each component, the adsorption isotherms for each component, and the joint equation of state (Equation 2.17). As an example, Figure 2.7 shows the dynamic surface tension $\gamma(t)$ for mixed solutions of $C_{12}EO_5$ and $C_{14}EO_8$. The red solid lines are the respective calculations for a diffusion-controlled adsorption mechanism with $D \approx 4.5\times10^{-6}$ cm²/s, a value which is reasonable for both surfactants. As one can see, the agreement between experiment and theory is excellent.

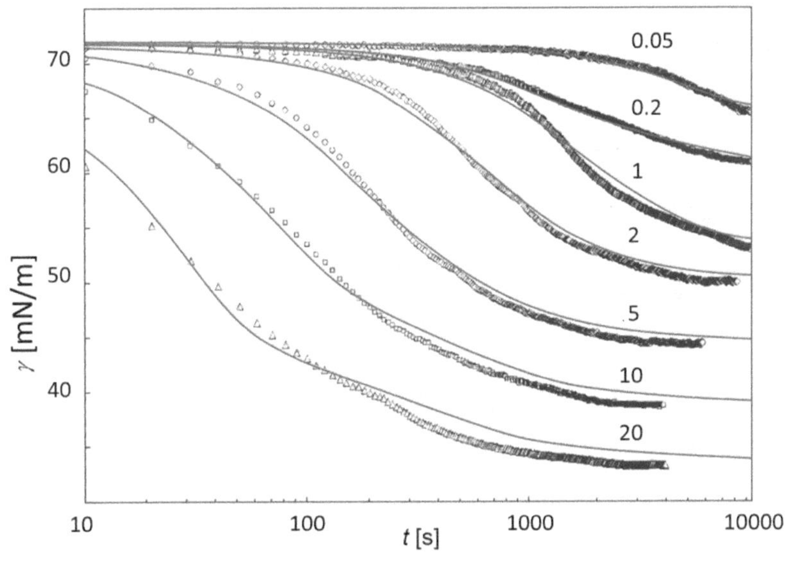

FIGURE 2.7 Dynamic surface tension of mixed solutions of $C_{12}EO_5$ and $C_{14}EO_8$ at a constant mixing ratio of 3:1; symbols are experimental points while the red solid lines are calculated. Labels: curves are labeled by initial total concentrations of the solution in μM. Adapted from Miller et al.,[42] Copyright (2017), with permission from Elsevier.

2.4 PARTITION COEFFICIENTS

In emulsion systems, we typically deal with water/oil interfaces. Since ionic surfactants used as emulsifiers are essentially only soluble in water, they stabilize O/W emulsions according to the Bancroft rule. Nonionic surfactants, however, depending on their partition coefficient, are able to stabilize both types of emulsions. As nonionics are soluble in both adjacent bulk phases, their adsorption process consists not only of the transport of molecules by diffusion in the solution bulk, but it includes also the transition across the interface as long as the distribution equilibrium between the two liquid phases is not reached yet.

Ravera et al.[4] described in detail the theoretical background for the adsorption process of surfactants in water/oil systems for surfactants soluble in both liquid phases. The basis is the same system of equations as that used by Ward and Tordai; however, the transport by diffusion happens in both liquid phases, one in the range $x < 0$ and the other $x > 0$ while the interface is located at $x = 0$. Thus, we have to solve two diffusion equations, one in the water and one in the oil phase, respectively,

$$\frac{\partial c}{\partial t} = D^{(W)} \frac{\partial^2 c}{\partial x^2} \quad \text{for x} < 0\text{, and} \quad \frac{\partial c}{\partial t} = D^{(O)} \frac{\partial^2 c}{\partial x^2} \quad \text{for } x > 0, \tag{2.18}$$

with a boundary condition at the joint interface,

$$\frac{\partial \Gamma}{\partial t} = D^{(O)} \left.\frac{\partial c}{\partial x}\right|_{x=0^-} - D^{(W)} \left.\frac{\partial c}{\partial x}\right|_{x=0^+}. \tag{2.19}$$

At the interface the distribution equation for the subsurface concentrations of the surfactant has to be considered in addition:

$$c^{(O)}(0,t) = k_p c^{(W)}(0,t). \tag{2.20}$$

The result is an integral equation similar to that derived by Ward and Tordai,[33]

$$\Gamma(t) = \frac{2c_0^{(W)}}{\sqrt{\pi}} \left(\sqrt{D^{(W)}} + \frac{c_0^{(O)}}{c_0^{(W)}} \sqrt{D^{(O)}} \right) \sqrt{t} - \frac{1}{\sqrt{\pi}} \int_0^t \left(\sqrt{D^{(W)}} + \frac{c_0^{(O)}}{c_0^{(W)}} \sqrt{D^{(O)}} \right) \frac{c^{(W)}(0,\tau)}{\sqrt{t-\tau}} d\tau \tag{2.21}$$

The superscripts (W) and (O) stand for the water and oil phases, respectively, and the subscript 0 refers to the concentration far from the interface so that it can be assumed that in the solution bulk the surfactant concentration is not changed during the adsorption process. For the almost spherical geometry again the more complicated set of equations equivalent to Equation (2.12) should be used, but this leads to rather complex equations that are beyond the scope of this chapter.

In Figure 2.8, the results obtained via drop profile analysis tensiometry with drops of aqueous C_{13}DMPO solutions of different bulk concentration immersed in

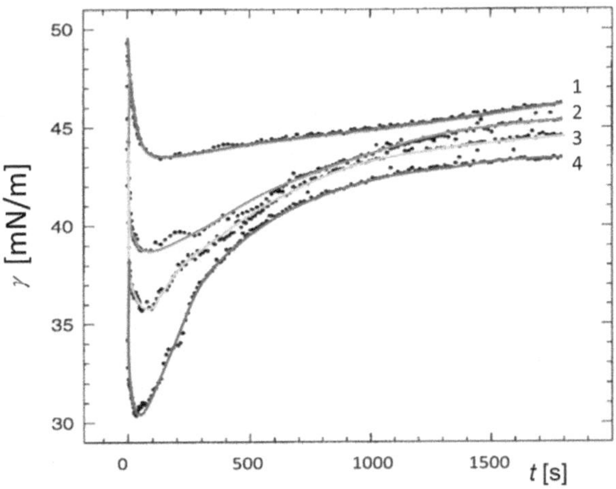

FIGURE 2.8 Dynamic interfacial tensions of $C_{13}DMPO$ at the water/hexane interface using an aqueous solution drop with a radius of 0.2 cm in a cuvette with about 35 mL hexane at four different surfactant concentrations: 1×10^{-5} M (1), 2×10^{-5} M (2), 3×10^{-5} M (3), and 5×10^{-5} M (4). Adapted from Ferrari et al.,[43] Copyright (1997), with permission from Elsevier.

pure hexane are shown.[43] The size of the drop and the cuvette were chosen such that the ratio of the cuvette volume and that of the drop was about one thousand. Water and hexane were mutually saturated prior to the experiments to avoid any transfer of solvent across the interface.

The measured dynamic interfacial tensions first decrease, as expected, but then pass through a minimum and increase again. With increasing bulk concentration, the tension values at the minimum become lower, the interfacial tension increases more steeply, and the adsorption times become shorter. This behavior results from effects of the transport of surfactant from the drop to the external phase across the interface combined with the available number of surfactant molecules within the drop. At the beginning of the process, the influx of surfactant from the drop to the interface overcomes the outflux due to the transfer of surfactant molecules into the external phase, so that there is a net increase of the adsorbed molecules and, therefore, a decrease of the interfacial tension. After a while, however, due to the finiteness of the volume, the remaining amount of surfactant molecules in the drop is no longer sufficient to sustain the above influx, so that the outflux prevails, decreasing the amount of adsorbed molecules and leading to an increase of the interfacial tension. The analysis of the dynamic interfacial tension data, performed with a spherical geometry analogue – a drop of a first liquid surrounded by a shell of the second one – of the model given above by Equation (2.21), describes very well the experimental observations and provides a partition coefficient of $k_p = 34$ as average for the four studied concentrations. In publication by Ravera et al.[4] a table for k_p values for quite a number of surfactants is given, mainly for the systems water/*n*-hexane; water/*n*-heptane, and water/isooctane.

For the nonionic surfactants C_nEO_m, for example, the values for k_p depend strongly on both parameters, n and m.[4] For example, for $C_{10}EO_4$ we have $k_p = 38$ and for $C_{10}EO_8$ only $k_p = 0.8$ (at 20°C), i.e., due to the larger hydrophilic EO group (8 instead of 4 EO units), the k_p value is decreased by a factor of about 50. When we compare $C_{10}EO_5$, however, with $C_{12}EO_5$, we see that for an identical EO headgroup, the k_p values change from about 14 to about 27, i.e., by a factor of about 2 only.

2.5 ADSORPTION MODELS FOR POLYMERS AND PROTEINS

The most frequently cited work on protein adsorption layers at liquid interfaces was published about half a century ago by Graham and Phillips in a series of papers, including the adsorption kinetics,[44] the adsorption isotherms,[45] and also the surface-dilational viscoelasticity.[46] The studies were performed with three important proteins: bovine serum albumin (BSA), β-casein (BCS), and lysozyme as examples for a globular, a random coil, and a very compact small hydrophobic globular protein, respectively. The analysis of the resulting experimental data, measured at water/air and water/*n*-decane interfaces, was still made on the basis of rather simple models, but nonetheless these publications provide a solid basis for understanding the adsorption properties of proteins at fluid interfaces.

The adsorption of proteins was described in the historical work of Graham and Phillips[44] still by models developed for surfactants or for polymers, for example, by the model of Singer.[47] The respective equation of state reads,

$$\Pi = \Pi_0 \left[\frac{x-1}{x} \frac{z}{2} \ln\left(1 - \frac{2\theta}{z}\right) - \ln(1-\theta) \right], \tag{2.22}$$

with $\Pi_0 = RT/\omega_0$, where ω_0 is the molar area of a polymer segment at the interface and θ is again the surface coverage. Moreover, x is the number of statistical units per polymer molecule, and z refers to a lattice correlation number. As one can see, most of the parameters are not relevant to the molecular structure of proteins, so that the development of specific adsorption models was urgently needed.

Only recently, a new and particularly dedicated model for proteins was derived by Fainerman et al.[15] and further refined by Kovalchuk et al.[16] for the analysis of dilational interfacial rheology data. This model (similar, but not identical with the multi-state model described in Section 2.3.1) takes into consideration important details of protein adsorption layers, in contrast to the simple Langmuir model used by Graham and Phillips[45] in their data analysis. The equation of state of the Fainerman model for proteins reads,

$$-\frac{\Pi\omega_0}{RT} = \ln(1-\theta) + \theta\left(1 - \frac{\omega_0}{\omega}\right) + a\theta^2, \tag{2.23}$$

having a structure similar to the Frumkin model (including the interaction between adsorbed molecules expressed by the coefficient a), but with an additional second term on the right-hand side to consider the entropy of molecular conformations of the adsorbed protein molecules. The parameter ω_0 is the molar area difference between two neighboring adsorption configurations and ω is the molar area of a protein at the interface averaged over all possible configurations s, with $\omega_j = \omega_{min} + (j-1)\omega_0$ for $j = 1, ..., s$. The maximum molar area for an almost empty interface ω_{max} is therefore given by $\omega_{max} = \omega_{min} + (s-1)\omega_0$. Note, this equation is very similar to Equation (2.5), but the definitions of molar areas are different.

What has to be considered in describing adsorption layers of proteins at fluid interfaces is the fact that, after complete coverage, a secondary layer, and maybe even a tertiary layer, can be formed. In contrast to adsorption layers of surfactants, where secondary layers are not expected, protein molecules can easily find the possibility to adsorb in a layer underneath the primary adsorption layer via hydrophobic interactions. This feature of protein adsorption layers is demonstrated by the behavior of surface pressure isotherms vs. adsorption (or concentration) of protein above a certain 'critical' concentration c^*, where the steep increase of the surface pressure changes to a relatively low increase (see Figure 2.9).

The dependence for the total surface pressure of protein molecules adsorbed in such multilayer films on the total adsorbed amount Γ_P, in this 'post-critical' range, instead of that defined by Equation (2.23), becomes,

$$\Pi = \Pi^* \left(1 + \frac{1}{n_a} \frac{\Gamma_P - \Gamma_P^*}{\Gamma_P^*}\right), \tag{2.24}$$

where Γ_P^* and Π^* are the adsorption and interfacial pressure, respectively, at the critical concentration c^* at which the interface is completely covered by a monolayer of protein molecules. Beyond this concentration the secondary layer starts to be formed by protein aggregates (via hydrophobic interaction) with an aggregation number n_a.

In Figure 2.9, the surface pressure isotherms for the globular protein human serum albumin (HSA) and the random coil protein β-casein (BCS) are shown, together with the calculated dependencies given by the solid lines. The table of the used values for the model parameters is given in by Kovalchuk et al.[16] As one can see, the isotherms

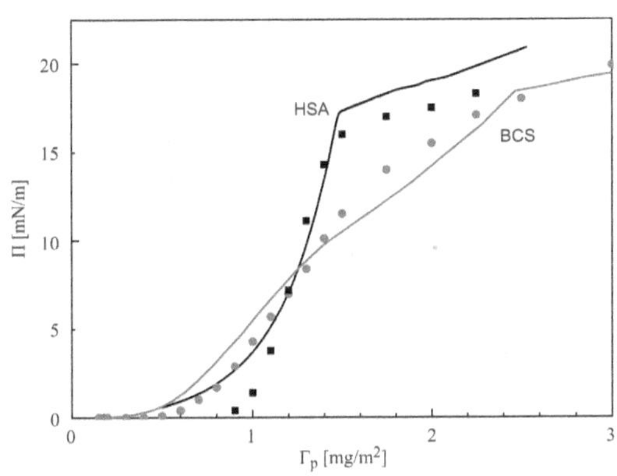

FIGURE 2.9 Surface pressure isotherms for the globular HSA (■) and the random coil BCS (●), the solid lines are calculated from the model explained below. Adapted from Kovalchuk et al.[16] with permission under Creative Common Attribution License 4.0 [CC-BY 4.0].

for the two types of proteins are quite different, but both can be described quite well with the assumption of a secondary adsorption layer, while the onset of the formation of this secondary layer is characterized by the kink at a similar surface pressure of about 18.5 mN/m. The corresponding adsorbed amounts of the proteins, however, are quite different: 1.4 mg/m² for HSA and 2.5 mg/m² for BCS.

In the study by Gochev et al.,[48] the above given model was used to show a good fit between the theoretical predictions and several sets of experimental data, including the surface pressure isotherm, adsorption, and surface-dilational viscoelasticity modulus in a wide range of protein concentrations.

It should be added that in the work of Llamas et al.,[49] the models derived by Fainerman for proteins was successfully applied to describe the adsorption of different copolymers, as they have a structure similar to some proteins. At present, most activities in the field of polymer adsorption modeling deal with the description of the adsorption layer structure via the self-consistent field theory[50–52] rather than to derive equations of state or adsorption isotherms via an interfacial thermodynamics approach.

2.6 MIXED SURFACTANT–PROTEIN ADSORPTION LAYERS

In food technologies, mixtures of low-molecular-weight surfactants and proteins are very often used as stabilizers for foams and emulsions. For ionic surfactants, the primary interaction with proteins is electrostatic interaction, which is true for cationics and anionics. The number of accessible charged sites on the protein molecule depends on the solution pH. For example, for β-lactoglobulin (BLG) with an isoelectric point of about 5.2, the protein has a positive net charge at pH < 5.2 and a negative net charge for pH > 5.2.[53] However, it is important to mention here that, even when the net charge of a protein is for example positive and an interaction with anionic surfactants is preferred, there are still negative charges on the protein, so that even an electrostatically drive interaction with cationic surfactants is possible.

When the added ionic surfactants surpass a certain (critical) concentration, then the available charges on the protein molecules are compensated and hydrophobic interactions start to dominate. This interaction, involving mainly the surfactants' aliphatic chain, leads to a corona around the protein molecule or even protein aggregates, the surface charge of which can change the sign. Note that the surface activity of the protein–surfactant aggregates first increases with the added ionic surfactant, and then decreases again, while the net-uncharged state of the aggregates is the most surface-active one. Above a certain concentration of added surfactants, even precipitation can happen. Similar processes are observed in systems of polyelectrolytes mixed with oppositely charged ionic surfactants.[54]

When a nonionic surfactant is added, electrostatic interactions do not play any role and instead the hydrophobic interactions generally dominate. The resulting protein–surfactant aggregates have a lower surface activity than the original protein, i.e., the nonionic surfactants act like a hydrophilic surface functionalization, which renders the protein less surface-active.

In Fainerman et al.[22] a respective set of equations was derived, providing thermodynamic relations suitable to describe the two very different single components, i.e., surfactant and protein molecules. The approach used in Fainerman et al.[22] assumes multistate properties of protein molecules and a Frumkin-like behavior of surfactant molecules. The result is similar to Equation (2.17) for mixtures of two surfactants; however, we have to consider the entropy only caused by the protein conformations. In addition, the interactions between all possible pairs of adsorbed molecules (a_P, a_S, and a_{PS}) have to be considered with subscripts P and S referring to the protein and surfactant molecules, respectively,

$$-\frac{\Pi\omega_0^*}{RT} = \ln\left(1-\theta_P-\theta_S\right)+\theta_P\left(1-\frac{\omega_0}{\omega_P}\right)+a_P\theta_P^2+a_S\theta_S^2+2a_{PS}\theta_P\theta_S, \tag{2.25}$$

where

$$\omega_0^* = \frac{\omega_0\theta_P+\omega_{S0}\theta_S}{\theta_P+\theta_S}. \tag{2.26}$$

Note that the use of different adsorption protocols, such as simultaneous and sequential adsorption, can lead to different interfacial layer structures. This is caused by the fact that in a simultaneous adsorption, the interaction between the two components happens already in the solution bulk while in a sequential protocol the components adsorb one by one and, therefore, start to interact with each other only at the interface. Hence, we actually cannot deal with the data as if they are describing the equilibrium. Figure 2.10 shows an example for a mixed system of BCS and sodium dodecyl sulfate (SDS), studied at the water/hexane interface. One can see that in a certain concentration range of the added surfactant, a significant difference for the two adsorption protocols is observed (more details are given by Fainerman et al.[22]).

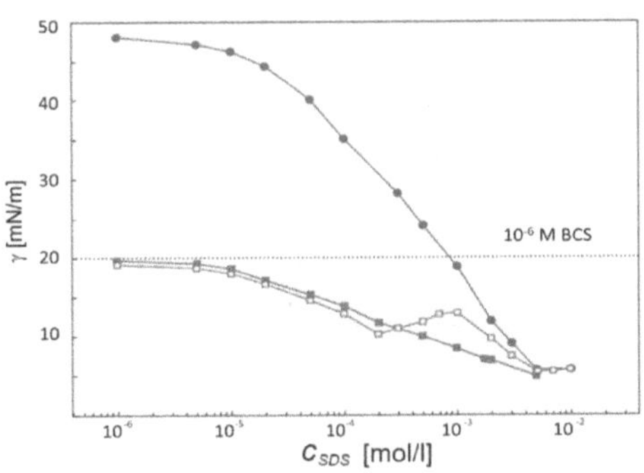

FIGURE 2.10 Surface tension isotherms of pure SDS (●) and mixed BCS/SDS solutions measured by two protocols: sequential adsorption experiments (■) and simultaneous adsorption experiments (□) at the solution/hexane interface; all experiments were performed at a fixed BCS concentration of 10^{-6} M (dotted horizontal line for pure BCS). Adapted from Fainerman et al.,[22] Copyright (2016), with permission from Elsevier.

2.7 COOPERATIVE AND COMPETITIVE ADSORPTION AT LIQUID/LIQUID INTERFACES

Until recently, the physical picture about adsorption layers at water/oil interfaces was simply transferred from the existing knowledge about water/air interfaces: the surface-active molecules adsorb, need some space at the interface, and interact with each other. Depending on the available space at the interface, oil molecules could then penetrate into the adsorbed surfactant layer. Systematic work on the adsorption of surfactants at water/air, water/alkane vapor, and water/liquid alkane interfaces allowed to develop a rather different physical picture for the structure of the respective interfacial layers. The general view of the interface is a mixed adsorption layer comprised of oil molecules together with the surfactant molecules at the water/liquid oil and water/oil vapor interface. Hence, the oil molecules at the interface are not simply replaced by adsorbing surfactants, or the oil molecules penetrate into the hydrophobic part of the surfactant adsorption layer but they are a part of it.

This new physical picture also required a new thermodynamic approach to adequately describe the dynamics and the equilibrium properties of adsorption layers at water/oil vapor and water/liquid oil interfaces. A rather easy way for doing so consisted in the transfer of the adsorption model for surfactant mixtures to the situation at these interfaces, i.e., to consider the adsorption of oil molecules at the interface, although they are not amphiphilic and hence not surface-active in a classical sense. For the aqueous solution/oil vapor systems, we really have mixed adsorption layer in a classical sense, while at the solution/liquid oil interfaces, the oil molecules are embedded into the structure of the interfacial layer.

The experimental origin for this new structure of surfactant adsorption layers at water/oil interfaces is the work by Javadi et al. about the adsorption of SDS and DoTAB in the presence of hexane vapor in the air phase.[55] First of all, even at zero surfactant concentration, a reduced interfacial tension was experimentally observed caused by the hexane molecules adsorbed from the vapor phase. With the addition of traces of one of the two surfactants, the interfacial tension remarkably decreases. This happens at surfactant concentrations at which in the absence of alkane vapor in the air phase, absolutely no changes in the interfacial tension would have been observed. When reducing the partial vapor pressure of the alkane, the effect of the oil molecules on the reduction of the surface tension becomes smaller.[56]

The equation of state of the adsorption model for the present problem is similar to Equation (2.17), except that one component is the surfactant and the other is the oil[57]; also, the entropy contribution is neglected,

$$-\frac{\Pi\omega_0^*}{RT} = \ln\left(1-\theta_1-\theta_2\right)+a_1\theta_1^2+a_2\theta_2^2+2a\theta_1\theta_2, \tag{2.27}$$

where again $\omega_0^* = \left(\omega_{10}\theta_1+\omega_{20}\theta_2\right)/\left(\theta_1+\theta_2\right)$. The respective adsorption isotherm for the oil molecules is given by

$$d_1P_1+k_1\theta_2 = \frac{\theta_1}{1-\theta_1-\theta_2}\exp\left(-2a_1\theta_1-2a_{12}\theta_2\right), \tag{2.28}$$

where P_1 is the partial vapor pressure of the oil and d_1 is the adsorption activity coefficient in the gas phase. The term $k_1\theta_2$ describes the impact of the surfactant molecules on the adsorbed amount of oil molecules. The adsorption isotherm for the surfactant (component 2) reads,

$$b_2c_2 = \frac{\theta_2}{1-\theta_1-\theta_2}\exp\left(-2a_2\theta_2 - 2a_{12}\theta_1\right). \quad (2.29)$$

In the case of liquid oil as the second fluid phase, the isotherm for the oil molecules (component 1) adsorbed at the interface is similar to Equation (2.29).

Also, for protein adsorption layers, it has been shown that oil molecules are part of the adsorption layer at water/oil interfaces.[58] The corresponding thermodynamic model shows a good agreement with the experimental data. The question about a possible cooperativity at very low concentrations also for proteins has not been studied so far because of the generally very high surface activity of proteins and hence the extremely low protein concentration required for such studies.

In Figures 2.11 and 2.12, some results of model calculations are shown using the above discussed thermodynamic model for the adsorption of surfactants at the water/hexane vapor and water/liquid hexane interfaces, respectively. As shown by Fainerman et al.,[57] the model given by the set of Equations (2.27)–(2.29) for the water/hexane vapor, and Equations (2.25) and (2.29) for the water/liquid hexane system, respectively, works quite well. Via the calculations, based on parameter values obtained by best-fit protocols for the comparison of the model and the experimental data, more insight is available about the composition of the interfacial layer. Such information is not easily available by other methods.[59]

In Figure 2.11 we can see the adsorbed amount of C_{16}TAB at the water/hexane vapor interface as a function of the C_{16}TAB bulk concentration c_{CTAB}. The adsorption Γ_{CTAB} increases up to the CMC (8×10^{-4} M) and maybe even further beyond the

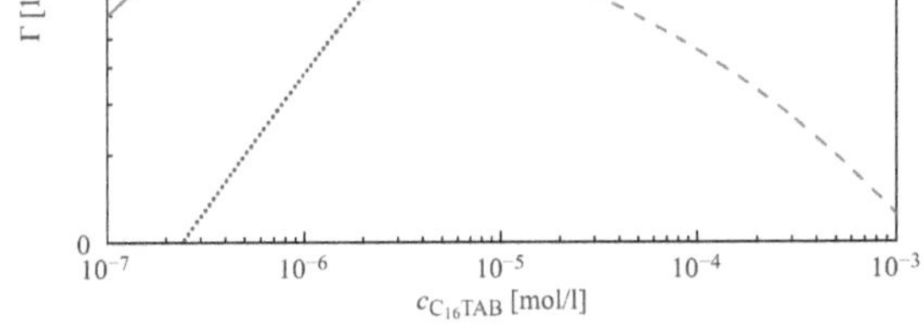

FIGURE 2.11 Adsorbed amounts calculated for the interface between the aqueous solution of C_{16}TAB and hexane saturated air, as calculated via Equations (2.25) and (2.29), adsorption of C_{16}TAB – solid line, adsorption of hexane – dashed line; adsorption of CTAB at the water/air interface as comparison – dotted line. Adapted from Fainerman et al.,[57] Copyright (2020), with permission from Elsevier.

CMC. The dotted curve shows the values for the water/air interface, i.e., the presence of hexane at the interface enhances the adsorption of the surfactant, reaching a similar final value at the CMC. At the same time, the rather high adsorption of the solvent hexane and its decrease with the increasing competition caused by the increasing adsorption of CTAB molecules completes the picture about the mixed surfactant/hexane adsorption layer.

The change of $C_{16}TAB$ adsorption at the water/liquid hexane interface is shown in Figure 2.12. Again, like for the water/hexane vapor interface, the adsorbed amount of hexane is larger than that of $C_{16}TAB$ at low surfactant concentrations and decreases once a sufficient number of $C_{16}TAB$ molecules is adsorbed. The dotted line shows the values for $\Gamma_{CTAB}(c_{CTAB})$ as calculated from the Gibbs fundamental equation given by Equation (1.1) using the experimental data published by Fainerman et al.[60] The agreement with the model given here is quite good.

The described model for the structure of surfactants adsorbed at water/oil interfaces can be best understood from the cartoons shown in Figure 2.13 for the case of $C_{12}TAB$ adsorption at the water/hexane interface. At very low $C_{12}TAB$ concentrations, the water/hexane interface is free of surfactants and the measured interfacial tension is the one known for the pure systems (situation A). Tiny amounts of surfactants, however, lead to a cooperative adsorption of $C_{12}TAB$ and the hexane molecules, i.e., the hexane molecules form a structured layer around the few adsorbed surfactants (situation B), so that changes in the interfacial tension are observed at concentrations several orders of magnitude lower than those leading to a first decrease in surface tension at the water/air interface. The organized hexane molecules at the interface attract further surfactant molecules to adsorb (situation C). Note that this cooperative effect is obviously also the reason for the fact that the purification of oils for professional laboratory studies at water/oil interfaces without impurity effects is extremely difficult.

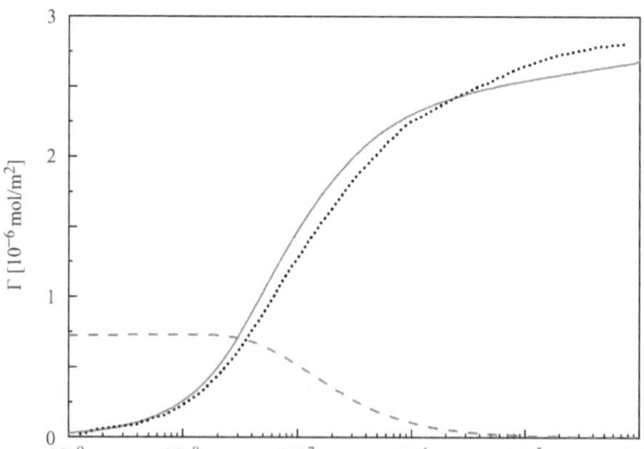

FIGURE 2.12 Adsorbed amount of $C_{16}TAB$ (solid line) and hexane (dashed line) as calculated from Equations (2.25) and (2.29), compared with the dotted line as calculated from the Gibbs equation using the respective interfacial tension data. Adapted from Fainerman et al.[60] with permission under Creative Common Attribution License 4.0 [CC-BY 4.0].

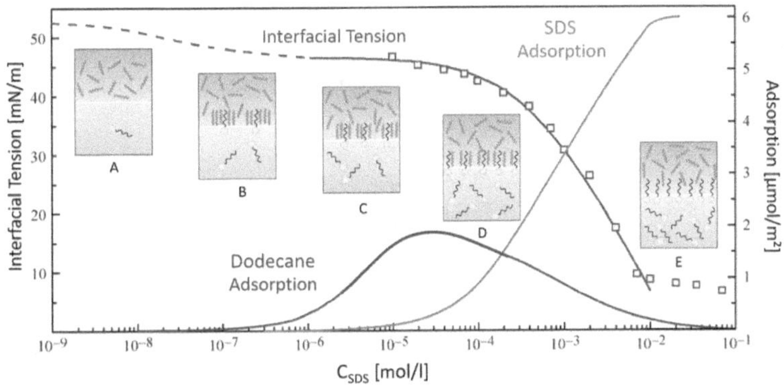

FIGURE 2.13 Schematic of the composition of the interfacial layer between an aqueous SDS solution and liquid dodecane as a function of the SDS concentration. Adapted from Fainerman et al.,[57] Copyright (2020), with permission from Elsevier.

With a further increase in the surfactant concentration, a competition sets in and the hexane molecules are step by step replaced from the interfacial layer (situation D) until the layer finally comprises mainly surfactants with only few interpenetrated hexane molecules (final situation E).

The situation at the water/air interface is similar at higher surface coverages; however, the significant adsorption of the surfactant molecules visualized by measurable changes in the surface pressure starts at much higher bulk concentrations, i.e., when the cooperativity between surfactant and hexane molecules at the water/oil interface crosses over into competition.

2.8 DILATIONAL VISCOELASTICITY OF ADSORBED LAYERS AT LIQUID INTERFACES

The dilational viscoelasticity of interfacial layers is an important characteristic and very important for a tailored use of surfactants and proteins in many applications. In particular, the stability of foams and emulsions depends strongly on this mechanical property. For insoluble (Langmuir-type) monolayers, the elasticity modulus is defined as the change in interfacial tension γ over the relative change in area, either at compression or at expansion,

$$\varepsilon = \frac{d\gamma}{d\ln A} = -\frac{d\gamma}{d\ln\Gamma}, \tag{2.30}$$

In a general case, measuring the interfacial tension as a function of time while the interfacial area is changed allows to determine the viscoelasticity modulus. As approximation, we can use the ratio of the respective differences $\Delta\gamma$ and ΔA to obtain a rough estimation for $\varepsilon = A_0\left(\Delta\gamma/\Delta A\right)$ where A_0 is the interfacial area before the compression/expansion. The rigorously defined viscoelasticity modulus is a frequency-dependent complex function $\varepsilon^*\left(i\varpi\right) = A_0\, F\left\{\Delta\gamma\left(t\right)\right\}/F\left\{\Delta A\left(t\right)\right\}$ calculated by using the Fourier transforms (denoted here as F) of the respective time functions, where i is the imaginary unit and $\varpi = 2\pi f$ is the circular frequency. It is clear that the value of ε depends on the rate of area change. For infinitely fast deformations of surfactant adsorption layers (the so-called high-frequency limit), ε^* approaches the Gibbs viscoelasticity modulus,

$$E_0 = -\left[\frac{d\gamma}{d\ln\Gamma}\right]_{\Gamma\times A=\text{const}}, \tag{2.31}$$

which is an alternative expression of the surface equation of state. The pioneers of the dilational viscoelasticity of interfacial layers, Lucassen and van den Tempel, proposed to handle all theoretical approaches in the domain of complex numbers and defined the parameter ε^* in the following way,

$$\varepsilon^*(i\varpi) = \varepsilon_r + i\varepsilon_i = E + i2\pi f\eta = E + i\varpi\eta, \tag{2.32}$$

with ε_r and ε_i being the real and imaginary part of the complex viscoelasticity, E the dilational elasticity, η the dilational viscosity, and f the frequency of oscillations supposed the area changes are generated by harmonic oscillations. With these definitions, Lucassen and van den Tempel derived a theory for $\varepsilon^*(i\varpi)$ with the assumption that any relaxations induced by the area changes are controlled by diffusion of the adsorbed molecules in the adjacent solution bulk phase (LvdT theory[61]),

$$\varepsilon^*(i\varpi) = E_0\frac{\sqrt{i\varpi}}{\sqrt{i\varpi}+\sqrt{2\varpi_0}}, \tag{2.33}$$

where $\varpi_0 = (D/2)(dc/d\Gamma)^2$ is the characteristic diffusion relaxation frequency, which carries information both on the adsorption isotherm $\Gamma(c)$ and on the diffusive process. Studies of the dilational viscoelasticity are therefore an independent additional source for the determination of the characteristic structural and kinetic parameters concerned with the adsorption phenomena. The real and imaginary parts of the viscoelastic module can be obtained from Equation (2.33),

$$\varepsilon_r = E_0\frac{1+\sqrt{\varpi_0/\varpi}}{1+2\sqrt{\varpi_0/\varpi}+2\varpi_0/\varpi}, \tag{2.34}$$

and

$$\varepsilon_i = E_0\frac{\sqrt{\varpi_0/\varpi}}{1+2\sqrt{\varpi_0/\varpi}+2\varpi_0/\varpi}, \tag{2.35}$$

which are the main relationships of the LvdT theory that are most frequently compared with experimental data. It is worth underlining that, in principle, from a single experiment of measurement of the dilational viscoelasticity, both Equations (2.34) and (2.35) can be utilized as independent relations to cross-check the values of E_0 and ϖ_0. The experimental determination of ε_i is sometimes difficult, in particular for ordinary surfactants, due to the typical sensitivities of the available surface rheology methodologies, which are not sufficient to resolve the usually low values of this parameter.

For the determination of the interfacial dilational viscoelasticity of curved interfaces, two relationships were derived by Joos[34] for the elasticity of drop interfaces by diffusion from inside the drop,

$$E = E_0 \left\{ 1 - i \frac{D}{\varpi R} \frac{dc}{d\Gamma} \left[\nu R \coth(\nu R) - 1 \right] \right\}^{-1}, \tag{2.36}$$

and for the elasticity of bubble or drop interfaces by diffusion from the surrounding surfactant solution,

$$E = E_0 \left\{ 1 - i \frac{D}{\varpi R} \frac{dc}{d\Gamma} \left[\nu R + 1 \right] \right\}^{-1}. \tag{2.37}$$

Here the high-frequency viscoelasticity modulus is given by $E_0 = d\Pi/d\ln\Gamma$ with Γ being the adsorbed amount of surfactant and $\nu^2 = i(\varpi/D)$. In this relation, a spherical drop or bubble with radius R was assumed.

The LvdT approach has been extended to include more time-dependent processes in addition to surfactant diffusion, such as relaxation processes involving the surface-active species within the interfacial layer. This model proposed by Ravera et al.[62] offered the opportunity to describe with great accuracy the surface-dilational rheology behavior of C_nEO_m solutions when considering the response in a large frequency range (10^{-3} to 10^2 Hz) or of complex interfacial layers made of mixtures of surfactants and nanoparticles.[63,64]

The modulus and argument of the dilational elasticity modulus as functions of frequency for an adsorption layer of Span-80 at the water/paraffin oil interface are shown in Figure 2.14. Span-80 was solved in the paraffin oil.[65] The experiments have been performed under low-gravity conditions on the International Space Station. The experiments were performed at different oscillation amplitudes $\Delta A/A_0$ of 5%, 10%, and 20%; however, this had no impact on the measured elasticities, although for the amplitude of 20% we could have expected some first nonlinearity effects.

The analysis of the experimental results in Figure 2.14 shows that the fractional Maxwell model (FMM) can explain the experimental results quite well, while the application of the LvdT model showed quite large differences. However, mixed adsorption layers formed by diffusion from both adjacent liquid bulk phases have a more complex behavior than the one given by the used single-element FMM.[66] A possibility to better understand very complex structures of interfacial layers would consist in a further refined FMM model. Additional experiments at large amplitude oscillations[66] could allow to detect various relaxation processes happening simultaneously in the interfacial layer or between the interface and the adjacent bulk.

2.9 SUMMARY AND OUTLOOK

The adsorption behaviors of surfactants, polymers, proteins, particles, and their mixtures are quite well understood; however, for particular systems, quite a number of

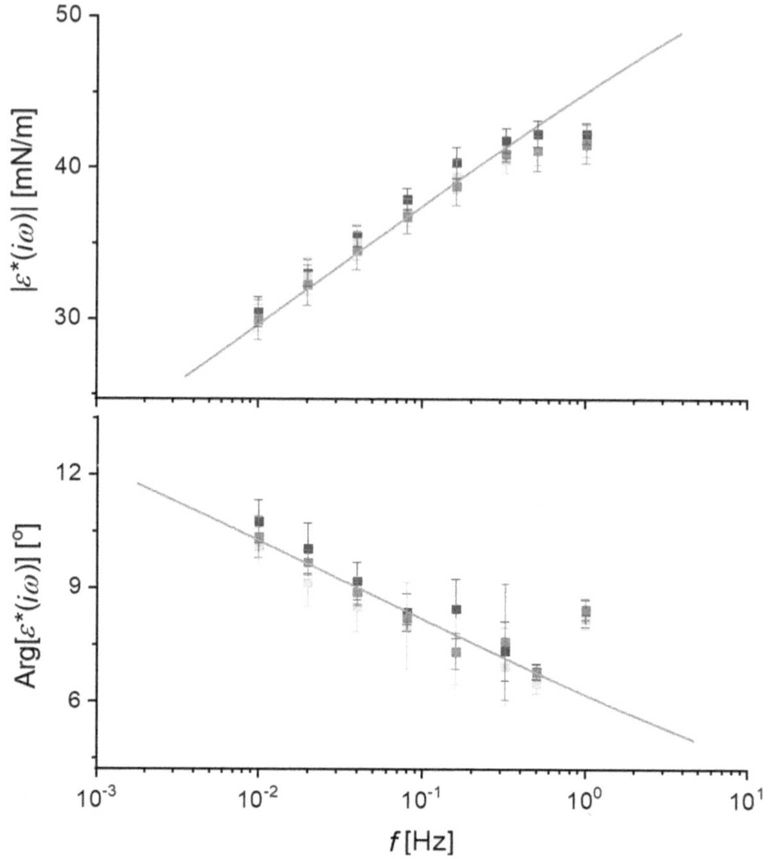

FIGURE 2.14 Modulus and argument of the complex viscoelasticity measured at an amplitude of 20% and temperature 20°C for Span-80 adsorbed at water/paraffin–oil interface (symbols are experimental points for three independent experiments); the line refers to the FMM. Adapted from Loglio et al.[65] with permission under Creative Common Attribution License 4.0 [CC-BY 4.0].

questions are still open. This is true especially when several possible complications occur simultaneously, such as when mixed adsorption layers are formed at liquid/liquid interfaces with a spherical geometry and with a transfer of some components across the interface. Apart from these special cases, many practical situations can be described at least semi-quantitatively. How the interfacial properties determined experimentally and analyzed on the basis of suitable theories can then be applied to practical systems, in particular to predict the stability of corresponding emulsions or even multiple emulsions, is yet far from being solved.

ACKNOWLEDGMENTS

The work was supported by the European Space Agency within the MAP project 'Emulsion Dynamics and Drop Interfaces (EDDI)' (ESA contract n. 4000128643/19).

REFERENCES

1. Aveyard, B. *Surfactants—In Solution, at Interfaces and in Colloidal Dispersions*, Oxford University Press: Oxford, UK, 2019.
2. Dukhin, S.S.; Kretzschmar, G.; Miller, R. Dynamics of adsorption at liquid interfaces: Theory, experiment, application. In *Studies in Interface Science*, Möbius, D., Miller, R. Eds.; Elsevier: Amsterdam, The Netherlands, 1995; Volume 1.
3. Ravera, F.; Dziza, K.; Santini, E.; Cristofolini, L.; Liggieri, L. Emulsification and emulsion stability: The role of the interfacial properties. *Adv. Colloid Interface Sci.* **2021**, *288*, 102344.
4. Ravera, F.; Ferrari, M.; Liggieri, L. Adsorption and partitioning of surfactants in liquid–liquid systems. *Adv. Colloid Interface Sci.* **2000**, *88*, 129–177.
5. Mousavi, N.S.; Romero-Martínez, A.; Miller, R. An empirical model to represent the CMC behavior of aqueous solutions of homologous series of nonionic surfactants, related to its chemical constitution. *J. Mol. Liquid* **2022**, *359*, 119229.
6. Bertsch, H. Sulphuric derivative of higher alcohols. US Patent, **1934**, 1968797.
7. Ropers, M.H.; Czichocki, G.; Brezesinski, G. Counterion effect on the thermodynamics of micellization of alkyl sulfates. *Phys. Chem. B* **2003**, *107*, 22, 5281–5288.
8. Bergeron, V. Disjoining pressures and film stability of alkyl trimethylammonium bromide foam films. *Langmuir* **1997**, *13*, 3474–3482.
9. Mucic, N.; Kovalchuk, N.M.; Aksenenko, E.V.; Fainerman, V.B.; Miller, R. Adsorption layer properties of alkyl trimethyl ammonium bromides at interfaces between water and different alkanes. *J. Colloid Interface Sci.* **2013**, *410*, 181–187.
10. Sarkar, R.; Pal, A.; Rakshit, A.; Saha, B. Properties and applications of amphoteric surfactant: A concise review. *J. Surf. Deterg.* **2021**, *24*, 709–730.
11. Traube, I. Über die Capillaritätsconstanten organischer Stoffe in wässerigen Lösungen. *Liebigs Ann. Chem.* **1891**, *265*, 27–55.
12. Loppinet, B.; Monteux, C. Dynamics of surfactants and polymers at liquid interfaces. In *Soft Matter at Aqueous Interfaces*, Lang, P.R., Liu, Eds.; Springer: Cham, Switzerland, 2016; pp. 137–157
13. Alexandridis P.; Lindman, B. (Eds.). *Amphiphilic Block Copolymers – Self-Assembly and Applications*, Elsevier: Amsterdam, The Netherlands, 2000.
14. Chen, P. (Ed.). *Molecular Interfacial Phenomena of Polymers and Biopolymers*, CRC Press: Boca Raton, FL, USA, 2005.
15. Fainerman, V.B.; Miller, R. Adsorption isotherms of proteins at liquid interfaces. In *Proteins at Liquid Interfaces*, in "Studies in Interface Science", Möbius, D., Miller, R., Eds.; Elsevier: Amsterdam, The Netherlands, 1998; Volume 7, pp. 51–102.
16. Kovalchuk, V.I.; Aksenenko, E.V.; Trukhin, D.V.; Makievski, A.V.; Fainerman, V.B.; Miller, R. Effect of amplitude on the surface dilational visco-elasticity of protein solutions. *Colloids Interfaces* **2018**, *2*, 57.
17. Ramsden, W. Separation of solids in the surface-layers of solutions and 'suspensions' (observations on surface-membranes, bubbles, emulsions, and mechanical coagulation). – preliminary account. *Proc. Roy. Soc. London* **1903**, *72*, 156–164.
18. Pickering, S.U. CXCVI.-Emulsions. *J. Chem. Soc. Trans.* **1907**, *91*, 2001–2021.
19. Iwashita, Y. Pickering–Ramsden emulsions stabilized with chemically and morphologically anisotropic particles. *Curr. Opin. Colloid Interface Sci.* **2020**, *49*, 94–106.
20. Thomas, R.K.; Penfold, J. The adsorption and self-assembly of surfactant mixtures: How the detailed evaluation of adsorption properties provides access to the bulk behaviour. *Adv. Colloid Interface Sci.* **2023**, *319*, 102984.
21. Guzmán, E.; Llamas, S.; Maestro, A.; Fernández-Peña, L.; Akanno, A., Miller, R.; Ortega, F.; Rubio, R.G. Polymer–surfactant systems in bulk and at fluid interfaces. *Adv. Colloid Interface Sci.* **2016**, *233*, 38–64.

22. Fainerman, V.B.; Aksenenko, E.V.; Krägel, J.; Miller, R. Thermodynamics, interfacial pressure isotherms and dilational rheology of mixed protein-surfactant adsorption layers. *Adv. Colloid Interface Sci.* **2016**, *233*, 200–222.
23. Ravera, F.; Ferrari, M.; Liggieri, L.; Loglio, G.; Santini, E.; Zanobini, A. Liquid–liquid interfacial properties of mixed nanoparticle–surfactant systems. *Colloids Surf. A* **2008**, *323*, 99–108.
24. Ravera, F.; Santini, E.; Loglio, G.; Ferrari, M.; Liggieri, L. Effect of nanoparticles on the interfacial properties of liquid/liquid and liquid/air surface layers. *J. Phys. Chem. B* **2006**, *110*, 19543–19551.
25. Fainerman, V.B.; Möbius, D.; Miller, R. (Eds.). *Surfactants: Chemistry, interfacial properties and applications*, in "Studies in Interface Science", Elsevier: Amsterdam, The Netherlands, 2001; Volume 13.
26. Henry, W. Experiments on the quantity of gases absorbed by water, at different temperatures, and under different pressures. *Phil. Trans. Roy. Soc. London* **1803**, *93*, 29–42.
27. Langmuir, I. The constitution and fundamental properties of solids and liquids. II. Liquids. *J. Amer. Chem. Soc.* **1917**, *39*, 1848–1907.
28. von Szyszkowski, B. Experimentelle Studien über kapillare Eigenschaften der wäßrigen Lösungen von Fettsäuren. *Z. Phys. Chem. (Leipzig)* **1908**, *64*, 385–414.
29. Frumkin, A. Die Kapillarkurve der höheren Fettsäuren und die Zustandsgleichung der Oberflächenschicht. *Z. Phys. Chem. (Leipzig)* **1925**, *116*, 466–484.
30. Fainerman, V.B.; Mucic, N.; Pradines, V.; Aksenenko, E.V.; Miller, R. Adsorption of alkyltrimethylammonium bromides at water/alkane interfaces: Competitive adsorption of alkanes and surfactants. *Langmuir* **2013**, *29*, 13783–13789.
31. Fainerman, V.B.; Kovalchuk, V.I.; Aksenenko, E.V.; Ravera, F.; Liggieri, L.; Loglio, G.; Makievski, A.V.; Mishchuk, N.O.; Schneck, E.; Miller, R. A multistate adsorption model for the adsorption of $C_{14}EO_4$ and $C_{14}EO_8$ at the solution/air interface. *Colloids Interfaces* **2021**, *5*, 39.
32. Sutherland, K.L. The kinetics of adsorption at liquid surfaces. *Austr. J. Sci. Res.* **1952**, *A5*, 683–696.
33. Ward, A.F.H.; Tordai, L. Time-dependence of boundary tensions of solutions I. The role of diffusion in time-effects. *J. Chem. Phys.* **1946**, *14*, 453–461.
34. Joos, P. *Dynamic Surface Phenomena*, VSP: Utrecht, The Netherlands, 1999.
35. Dieter-Kissling, K.; Karbaschi, M.; Marschall, H.; Javadi, A.; Miller, R.; Bothe, D. On the applicability of drop profile analysis tensiometry at high flow rates using an interface tracking method. *Colloids Surfaces A* **2014**, *441*, 837–845.
36. Liggieri, L.; Ravera, F.; Ferrari, M.; Passerone, A.; Miller, R. Adsorption kinetics of alkylphosphine oxides at water/hexane interface. 2. Theory of the adsorption with transport across the interface in finite systems. *J. Colloid Interface Sci.* **1997**, *186*, 46–52.
37. Fainerman, V.B.; Lylyk, S.V.; Aksenenko, E.V.; Liggieri, L.; Makievski, A.V.; Petkov, J.T.; Yorke, J.; Miller, R. Adsorption layer characteristics of Triton surfactants. Part 2. Dynamic surface tension and adsorption. *Colloids Surf. A* **2009**, *334*, 8–15.
38. Fainerman, V.B.; Aksenenko, E.V.; Lylyk, S.V.; Petkov, J.T.; Yorke, J.; Miller, R. Adsorption layer characteristics of mixed sodium dodecyl sulfate/C_nEO_m solutions 1. Dynamic and equilibrium surface tension. *Langmuir* **2010**, *26*, 284–294.
39. Fainerman, V.B.; Kovalchuk, V.I.; Aksenenko, E.V.; Ravera, F.; Liggieri, L.; Loglio, G.; Makievski, A.V.; Schneck, E.; Miller, R. Thermodynamics, kinetics and dilational viscoelasticity of adsorbed C_nEO_m layers at the aqueous solution/air interface. *Colloids Interfaces* **2021**, *5*, 16.
40. Baret, J.F. Kinetics of adsorption from a solution. Role of the diffusion and of the adsorption-desorption antagonism. *J. Phys. Chem.* **1968**, *72*, 2755–2758.

41. Lunkenheimer, K.; Haage, K.; Hirte, R. Novel results on the adsorption properties of n-alkyl dimethylphosphine oxides at the air/water interface. *Langmuir* **1999**, *15*, 1052–1058.
42. Miller, R.; Aksenenko, E.V.; Fainerman, V.B. Dynamic interfacial tension of surfactant solutions. *Adv. Colloid Interface Sci.* **2017**, *247*, 115–129.
43. Ferrari, M.; Liggieri, L.; Ravera, F.; Amodio, C.; Miller, R. Adsorption kinetics of alkyl phosphine oxides at the water/hexane interface 1. Pendant drop experiments. *J. Colloid Interface Sci.* **1997**, *186*, 40–45.
44. Graham, D.E.; Phillips, M.C. Proteins at liquid interfaces, I. Kinetics of adsorption and surface denaturation. *J. Colloid Interface Sci.* **1979**, *70*, 403–414.
45. Graham, D.E.; Phillips, M.C. Proteins at liquid interfaces, II. Adsorption isotherms. *J. Colloid Interface Sci.* **1979**, *70*, 415–426.
46. Graham, D.E.; Phillips, M.C. Proteins at liquid interfaces, IV. Dilatational properties. *J. Colloid Interface Sci.* **1980**, *76*, 227–239.
47. Singer, S.J. Note on an equation of state for linear macromolecules in monolayers. *J. Chem. Phys.* **1948**, *16*, 872–876.
48. Gochev, G.G.; Kovalchuk, V.I.; Aksenenko, E.V.; Fainerman, V.B.; Miller, R. β-lactoglobulin adsorption layers at the water/air surface: 5. Adsorption isotherm and equation of state revisited, impact of pH. *Colloids Interfaces* **2021**, *5*, 14.
49. Llamas, S.; Mendoza, A.J.; Guzmán, E.; Ortega, F.; Rubio, R.G. Salt effects on the air/solution interfacial properties of PEO-containing copolymers: Equilibrium, adsorption kinetics and surface rheological behavior. *J. Colloid Interface Sci.* **2013**, *400*, 49–58.
50. Leermakers, F.A.M.; Léonforte, F.; Luengo, G.S. Structure and colloidal stability of adsorption layers of macrocycle, linear, comb, star, and dendritic macromolecules. *Macromolecules* **2020**, *53*, 7322–7334.
51. Jarek, E.; Krasińska-Krawet, Z.; Kruk, T.; Lamch, Ł.; Ronka, S.; Wilk, K.A.; Warszyński, P. Adsorption properties of soft hydrophobically functionalized PSS/MA polyelectrolytes. *Colloids Interfaces* **2021,** *5*, 3.
52. Mu, M.; Leermakers, F.A.M.; Chen, J.; Holmes, M.; Ettelaie, R. Effect of polymer architecture on the adsorption behaviour of amphiphilic copolymers: A theoretical study. *J. Colloid Interface Sci.* **2023**, *644*, 333–345.
53. Ulaganathan, V.; Gochev, G.; Gehin-Delval, C.; Leser, M.E.; Gunes, D.Z.; Miller, R. Effect of pH and electrolyte concentration on rising air bubbles in β-lactoglobulin solutions. *Colloids Surfaces A* **2016**, *505*, 165–170.
54. Uhlig, M.; Miller, R.; von Klitzing, R. Surface adsorption of sulfonated poly(phenylene sulfone)/C_{14}TAB mixtures and its correlation with foam film stability. *Phys. Chem. Chem. Phys.* **2016**, *18*, 18414–18423.
55. Javadi, A.; Moradi, N.; Karbaschi, M.; Fainerman, V.B.; Möhwald, H.; Miller, R. Alkane vapor and surfactants co-adsorption on aqueous solution interfaces. *Colloids Surf. A* **2011**, *391*, 19–24.
56. Mucic, N.; Moradi, N.; Javadi, A.; Aksenenko, E.V.; Fainerman, V.B.; Miller, R. Effect of partial vapor pressure on the co-adsorption of surfactants and hexane at the water/hexane vapor interface. *Colloids Surf. A* **2015**, *480*, 79–84.
57. Fainerman, V.B.; Aksenenko, E.V.; Kovalchuk, V.I.; Mucic, N.; Javadi, A. ; Liggieri, L.; Ravera, F.; Loglio, G.; Makievski, A.V.; Schneck, E.; Miller, R. New view of the adsorption of surfactants at the water/alkane interface – Competitive and cooperative effects of surfactant and alkane molecules. *Adv. Colloid Interface Sci.* **2020**, *279*, 102143.
58. Miller, R.; Aksenenko, E.V.; Zinkovych, I.I.; Fainerman, V.B. Adsorption of proteins at the aqueous solution/alkane interface: Co-adsorption of protein and alkane. *Adv. Colloid Interface Sci.* **2015**, *222*, 509–516.

59. Kovalchuk, V.I.; Aksenenko, E.V.; Schneck, E.; Miller, R. Surfactant adsorption layers – Experiments and modelling. *Langmuir* **2023**, *39*, 3537–3545.
60. Fainerman, V.B.; Aksenenko, E.V.; Makievski, A.V.; Nikolenko, M.V.; Javadi, A.; Schneck, E.; Miller, R. Particular behavior of interfacial tension at the interface between aqueous surfactant solutions and alkane. *Langmuir* **2019**, *35*, 15214–15220.
61. Lucassen, J.; van den Tempel, M. Dynamic measurements of dilational properties of a liquid interface. *Chem. Eng. Sci.* **1972**, *27*, 1283–1291.
62. Ravera, F.; Ferrari, M.; Santini, E., Liggieri, L. Influence of surface processes on the dilational visco-elasticity of surfactant solutions. *Adv. Colloid Interface Sci.* **2005**, *117*, 75–100.
63. Liggieri, L.; Santini, E.; Guzman, E.; Maestro, A.; Ravera, F. Wide-frequency dilational rheology investigation of mixed silica nanoparticle–CTAB interfacial layers. *Soft Matter*, **2011**, *7*, 7699–7709.
64. Guzman, E.; Santini, E.; Ferrari, M.; Liggieri, L.; Ravera, F. Effect of the incorporation of nanosized titanium dioxide on the interfacial properties of 1,2-dipalmitoyl-*sn*-glycerol-3-phosphocholine Langmuir monolayers. *Langmuir* **2017**, *33*, 10715–10725.
65. Loglio, G.; Kovalchuk, V.I.; Bykov, A.G.; Ferrari, M.; Krägel, J.; Liggieri, L.; Miller, R.; Noskov, B.A.; Pandolfini, P.; Ravera, F.; Santini, E. Interfacial dilational viscoelasticity of adsorption layers at the hydrocarbon/ water interface: the fractional Maxwell model. *Colloids Interfaces* **2019**, *3*, 66.
66. de Groot A, Yang J, Sagis LMC. Surface stress decomposition in large amplitude oscillatory interfacial dilatation of complex interfaces. *J. Colloid Interface Sci.* **2023**, *638*, 569–581.

3 Experimental Techniques for the Study of Liquid/Fluid Interfaces

Georgi Gochev, Francesca Ravera, Giuseppe Loglio, Libero Liggieri, Aliyar Javadi, Jan Zawala, Emanuel Schneck, Reinhard Miller, and Valentin B. Fainerman

3.1 INTRODUCTION

Emulsions are liquid dispersions in which one liquid is dispersed dropwise in a second immiscible liquid. Emulsions are thermodynamically instable systems and tend to macroscopically separate with time into the two original liquid phases. Therefore, all processes that can degrade an emulsion have to be hindered or at least decelerated by suitable emulsifiers. Coalescence and Ostwald ripening are the main processes that change the drop size distribution, while flocculation, creaming, or sedimentation promote these processes.[1] The state of the art is characterized by the notion that the stability of emulsions is based on a multiscale scheme due to the many possibilities to destabilize the liquid films between emulsion droplets as the main elements of an emulsion.[2] Also, molecular dynamics simulations help to identify the crucial aspects influenced by the used emulsifiers.

There are various phenomena at liquid interfaces,[3] and for their systematic study, various experimental methods are available. For liquid/gas and liquid/liquid interfaces, measurements of the surface and interfacial tension are most frequently performed. Historically, the ring and plate tensiometry methods were most common; however, they are not easily used for studies at interfaces between two liquids and, quite generally, they suffer from disturbances of the interfacial layer by surface rheological effects. Today, we have a powerful arsenal of modern commercial instruments for interfacial tension measurements based on single drops and bubbles, on planar interfaces, and on whole emulsion samples.

Measurements of interfacial tension γ allow determination of the adsorbed amount Γ of any type of emulsifiers, although this requires the application of the Gibbs fundamental Equation (1.1). Methods for a direct determination of Γ are based mainly on reflectometry measurements, like for example, in ellipsometry as demonstrated by Benjamins et al.[4] or reflectometry with neutrons, X-rays, or visible light. In

DOI: 10.1201/9781032636184-4

Sections 3.5 and 3.6, more details about reflectometric and spectroscopic techniques are presented.

From an experimental point of view, the most suitable for the understanding of emulsions are surface tension measurements with single drops.[5] To get further, surface-sensitive spectroscopic and reflectometric methods can be applied, but their adaptations to water/oil interfaces present technical challenges.[6–11] Therefore, in this chapter, we present first an overview of experimental methods, mainly based on single drops, for investigations of liquid/liquid interfaces at equilibrium and under dynamic conditions. After that, some spectroscopic and reflectometric methods are discussed, which give direct access to the molecular self-assembly at liquid/liquid interfaces in terms of surface excess and structural organization.

3.2 DYNAMIC INTERFACIAL TENSIOMETRY AT SHORT ADSORPTION TIMES

The dynamics of fluid interfaces are the basis for systematic investigations of the behavior of interfacial layers, which in turn are required for a professional application of surface-active materials in practice. Various aspects of the interfacial dynamics were extensively discussed in reference.[12] Although this book was published many years ago, the fundamental principles are of course still correct. In the meantime, however, new knowledge was gained and the current state of the art, essentially with respect to liquid/liquid interfaces and hence emulsions, will therefore be summarized here.

To quantitatively understand the dynamics of surfactant adsorption, tensiometry methods applied at short adsorption times are most suitable.[13] The bubble pressure tensiometry is obviously the method of choice when dynamic surface properties for extremely short adsorption times are required.[14] The disadvantage is the applicability only to liquid/gas surfaces. Anyhow, it is a frequently used method and provides dynamic surface tensions from 1 millisecond surface age and less, and up to several minutes of adsorption times. The principle is based on the registration of the capillary pressure in a bubble during its growth until detachment from a narrow capillary. The short adsorption times cannot be reached only by the bubble formation at high bubble frequencies but also a particular data analysis is required as described in detail in reference.[15]

The equivalent device to the bubble pressure tensiometer for the liquid/gas surface is the capillary pressure tensiometry for liquid/liquid interfaces.[16] This device requires the use of particular pressure sensors suitable for direct contact with liquids. For the measurement of dynamic interfacial tensions at short adsorption times, the experimental mode of growing drops is used, which is a protocol similar to the one applied in the maximum bubble pressure tensiometry. The challenge in reaching short adsorption times consists of the impact of the liquid's hydrodynamics that adds a respective contribution to the measured capillary pressure. The schematic of an instrument for measuring the dynamic interfacial tension is shown in Figure 3.1. This simple design is similar to the first instrument of this kind based on an electronic pressure transducer controlled by a computer that is both standard nowadays, but was completely new when it was first designed by Liggieri et al.[17] in 1995.

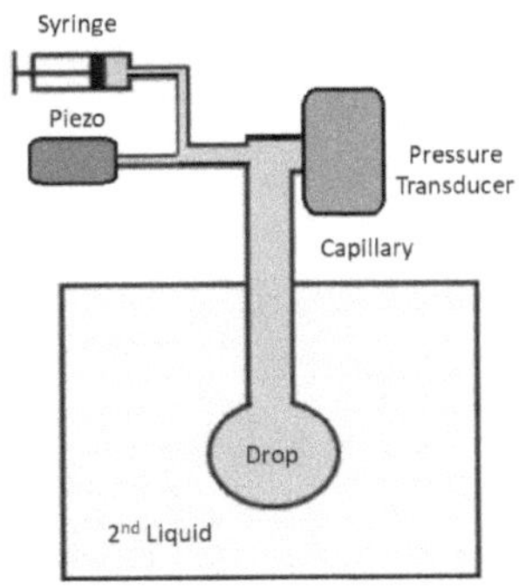

FIGURE 3.1 Main elements of a capillary pressure tensiometer, according to Javadi et al.[16]

The hydrodynamics of the liquid, which according to the experimental protocols is forced to flow through narrow parts of the instrument, produces an additional pressure contribution ΔP_{hyd} to the measured total pressure difference ΔP_{tot}. This contribution has to be separated from the real capillary pressure $\Delta P_{\gamma} = 2\gamma/r$ and the static pressure ΔP_{st} in order to determine the interfacial tension $\gamma(t)$.

The total pressure balance reads,

$$\Delta P_{tot} = \frac{2\gamma}{r} + \Delta P_{st} + \Delta P_{hyd}. \tag{3.1}$$

This impact of the liquid's hydrodynamics makes it impossible to determine values $\gamma(t)$ at adsorption times t shorter than about 10 ms. In addition, when the piezo drive is used to generate harmonic perturbations of the drop size at higher oscillation frequencies, required for measurements of the dilational viscoelasticity of the interfacial layer, again the hydrodynamic contribution has to be separated from the measured total pressure changes. This is even more complicated as the hydrodynamic effect is not an additive term in the determination of the capillary pressure oscillations.[18].

Figure 3.2 shows the dynamic interfacial tensions measured by capillary pressure tensiometry for the water/hexane interface in the absence and presence of the surfactant Span-80 in the hexane phase. The different dependencies $\gamma(t)$ are obtained for different liquid flow rates for the growing water drop in hexane. In the absence of the surfactant (curve 1), the interfacial tension is constant. In the presence of Span-80, the initial tension value at $t = 0$ is much lower. With increasing liquid flow rate Q, a stronger expansion of the drop surface area is produced, leading to higher interfacial tension values.

The analysis of these data requires the consideration of the (approximately radial) liquid flow inside the drop and the expansion of the interfacial area depending on the experimental protocol. Although the experiments provide information at rather short effective adsorption times, their analysis is rather complex.[16]

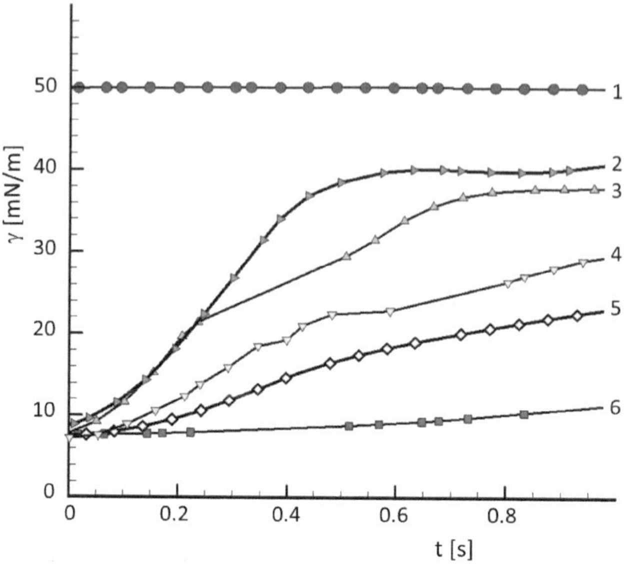

FIGURE 3.2 Dynamic interfacial tension of a water drop growing at a capillary tip with a diameter of about 1 mm in hexane at different liquid flow rates of $Q = 0.85$ mm³/s (1) for a pure water/hexane interface, and $Q = 16.7$ (2), 8.5 (3), 4.8 (4), 2.0 (5), and 0.85 (6) mm³/s for a 10^{-4} M Span-80 solution in hexane formed in pure water. Adapted from Javadi et al.,[16] Copyright (2010), with permission from Elsevier.

3.3 DROP AND BUBBLE PROFILE ANALYSIS TENSIOMETRY

In the late 1980s, Neumann and his team developed a method called axisymmetric drop shape analysis (ADSA) from an experimental principle to a routine method for measuring accurately the surface and interfacial tension of liquids.[19] Historically, photos of drops were used, and the drop profiles were manually measured on the printed photo. Then, these drop profile coordinates were fitted by profiles calculated manually, which was an enormous amount of work. It was only the availability of computers and electronic video cameras that brought the historical step by Neumann to broad applicability. The next milestone was initiated by Benjamins et al.[20] who used a normal personal computer instead of a workstation required before. A comprehensive description of profile analysis tensiometry (PAT) was given by Loglio et al.,[21] and its main elements are shown in Figure 3.3.

The design of all present commercial instruments is still quite similar to the one depicted, although the capacity of the computers and most of all cameras is still constantly increasing. Also, the sensitivity and the number of frames per second of the CCD cameras is increasing continuously. Thus, the overall accuracy is much more governed by the experimental conditions in the lab rather than by the quality of the camera or the controlling computer. However, via statistics or artificial intelligence, the measurement accuracy can be increased from typically ±0.1 mN/m to maybe

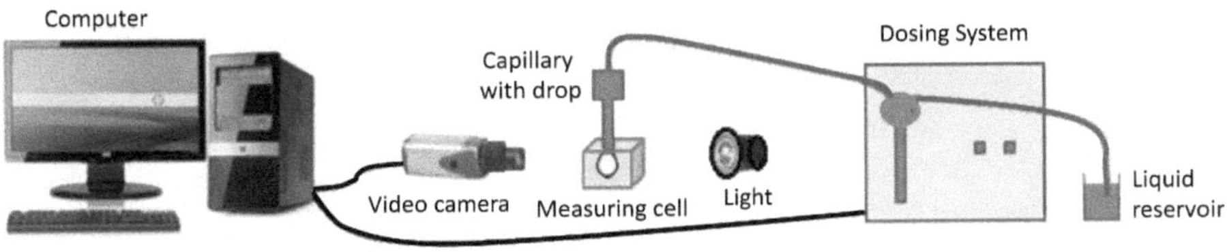

FIGURE 3.3 Main elements of a drop profile analysis tensiometer, according to the design of PAT1 (SINTERFACE Technologies, Berlin, Germany).

±0.01 mN/m in the near future. This would then allow developing experiments to routine protocols like those described by Rivas et al.,[22] i.e., to perform a time-resolved analysis of "near periodic perturbations" of the interfacial layer caused by unknown external perturbation to obtain the time-dependent dilational viscoelasticity of the interfacial layer.

Another application of PAT is the measurement of contact angle in the three-phase contact of a solid surface and two fluids. This situation can be easily arranged with a drop sitting on the solid surface. This was actually the topic that drove Neumann and his team to develop ADSA for measuring not only surface tension but also the contact angle of liquids on solid surfaces in order to determine the surface energy of the solids.[19]

Figure 3.4 shows how drops can be formed for measurements of the contact angle θ. The cartoon in Figure 3.4a shows the formation of the drop from below, which requires that a hole be drilled into the solid sample. However, measurements for the same system can be performed via the arrangement in Figure 3.4b as the presence of the needle in the drop's pole does not disturb the contact angle measurement.

The situation in Figure 3.4c shows that also the contact angle in the three-phase contact between a solid surface and two immiscible liquids can be easily arranged. The definition of all quantities for the measurement of the contact angle θ is given in the scheme of Figure 3.5.

The interfacial tension forces γ_{SV}, γ_{SL} and γ_{LV} are those between the solid and the vapor phase (or the second liquid), solid and liquid, and liquid and the vapor (or a second liquid) phase, respectively. The force balance in the three-phase contact is given by Young's equation:

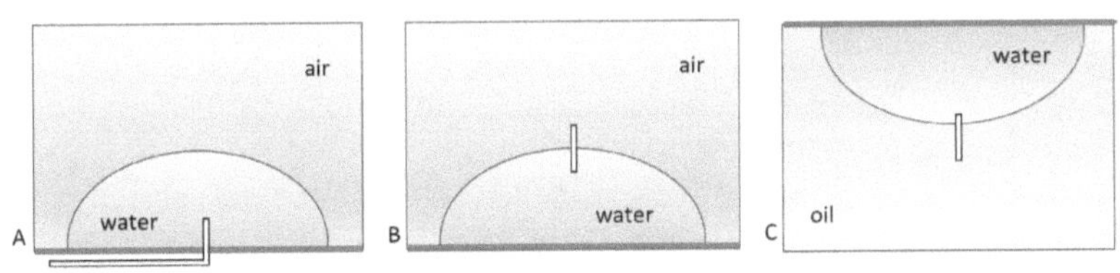

FIGURE 3.4 Contact angle measurements with sessile drops; (A) liquid supply from below a hole in the solid, (B) liquid supply from top, and (C) sessile drop in a solid–water–oil system.

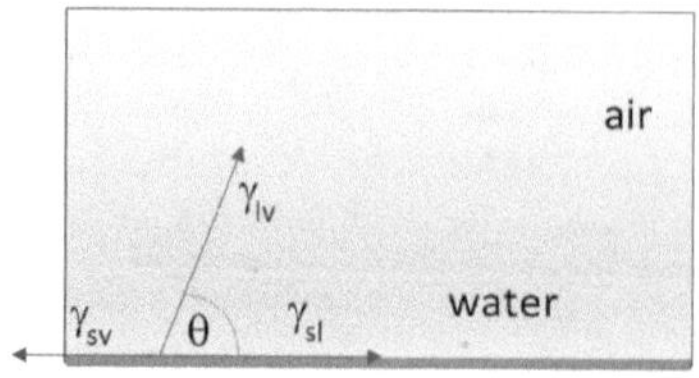

FIGURE 3.5 Definition of the parameters of Young's equation.

$$\gamma_{SV} = \gamma_{SL} + \gamma_{LV} \cos(\theta). \tag{3.2}$$

Measurements of contact angle and the estimation of the surface energy of solid surfaces are relevant when emulsions are stabilized by particles.

In 2017, a systematic surface tension study by Kairaliyeva et al.[23] was published showing that results obtained by experiments with PAT can be quite different when using single drops or bubbles. The origin of these often rather significant differences was explained by the so-called depletion effects. An example is shown in Figure 3.6 for the two nonionic surfactants $C_{10}EO_8$ and $C_{12}EO_5$.[24] The isotherm of $C_{10}EO_8$ is located at concentrations about one order of magnitude higher than that for the surfactant $C_{12}EO_5$. Thus, the surface activity of $C_{12}EO_5$, given by the location of the surface tension isotherm, is about one order of magnitude higher than the one of $C_{10}EO_8$. This difference in surface activity can be expressed by the adsorption activity coefficient b in the Frumkin adsorption isotherm:

$$bc = \Gamma_\infty \frac{\Gamma}{1-\Gamma_\infty} exp\left[-2a\frac{\Gamma}{\Gamma_\infty}\right]. \tag{3.3}$$

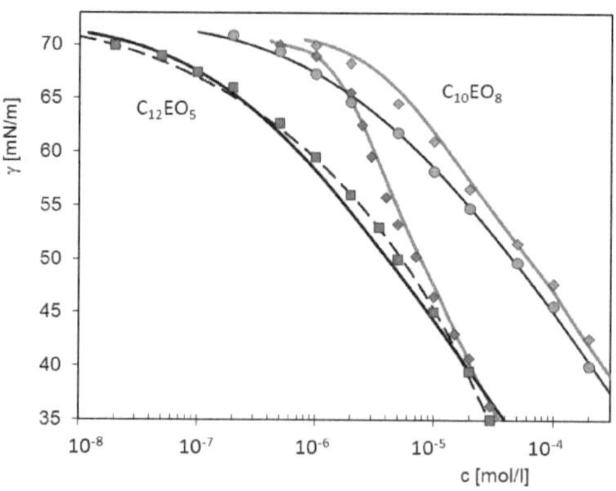

FIGURE 3.6 Interfacial tension isotherms $\gamma(c)$ measured for two nonionic surfactants $C_{10}EO_8$ (green symbols) and $C_{12}EO_5$ (blue symbols) measured by single solution drops (◆) or air bubbles formed in the aqueous solutions (●), respectively; the solid lines are best fits with the Frumkin adsorption model. Adapted from Kairaliyeva et al.[24] with permission under a Creative Commons Attribution [CC-BY 4.0] License.

The smaller the values of b, the larger is the surface activity of a surfactant, i.e., the smaller is the bulk concentration c at which a respective adsorbed amount Γ is reached. Note that the term "surface activity" is not very clearly defined and we use here the value of b as one of the possibilities.

The depletion in experiments with single solution drops is caused by the adsorption of surfactant molecules at the drop surface which decreases the bulk concentration inside the drop. The following mass balance describes this situation,

$$cV + \Gamma A = constant, \tag{3.4}$$

with A and V being the surface area and the volume of the solution drops. At time $t = 0$, the adsorption Γ is assumed to be zero, so that the initial concentration inside the drop is $c = c_{ini}$. As the adsorption proceeds, Γ increases and the actual concentration in the drop gradually decreases,

$$c(t) = c_{ini} - \frac{\Gamma(t)}{V / A}. \tag{3.5}$$

As long as the ratio V/A is large, the second term on the right-hand side of Equation (3.5) is negligible. However, for small drops with a volume of the order of 10 mm^3, as it is typically used in PAT experiments, the ratio V/A is of the order 5×10^{-4} cm (assumed that the drop has a spherical shape). Typical values of Γ are of the order of 10^{-10} mol/cm^2, so that the depletion term $\frac{\Gamma}{V / A}$ has a value of about 2.5×10^{-7} mol/cm^3 = 2.5×10^{-4} M. Hence, for all solutions with a concentration of about 10^{-3} M and less, depletion effects due to adsorption of surfactants at the single drop surface must be expected.

On the contrary, in experiments with single bubbles having a similar size like the drops, the solution volume V in the cuvette is typically of the order of 15 ml or 1.5×10^4 mm^3, i.e., about 1000 times larger so that depletion effects can be expected at concentrations 10^3 times lower than for single drops. Such systems are extremely rare.

Therefore, the data in Figure 3.6 measured with bubbles (●) can be directly fitted by an adsorption model (black solid curves), while those measured with the single drops have to be fitted with corrected concentrations for each point, i.e., at $c = c_{ini} - \Gamma(c) \cdot A/V$. This was done in reference [24] and the corresponding red solid curves were obtained. It was shown that the resulting parameters of the Frumkin model were identical for both sets of data. The process of data analysis, using parallel experiments with drops and bubbles for the same surfactant solutions, is rather time-consuming; however, it has a great advantage: it provides the opportunity to determine the adsorption isotherm experimentally $\Gamma(c)$ as a by-product of the depletion correction.

Drop profile analysis tensiometry is a method well suited for water/oil systems containing surfactants that are soluble in both liquid phases. Thus, when the surfactant is initially solved in one phase, it adsorbs at the interface and then transfers into the second liquid phase. After reaching the adsorption equilibrium, the surfactant will be distributed between the two liquids according to the partition coefficient for

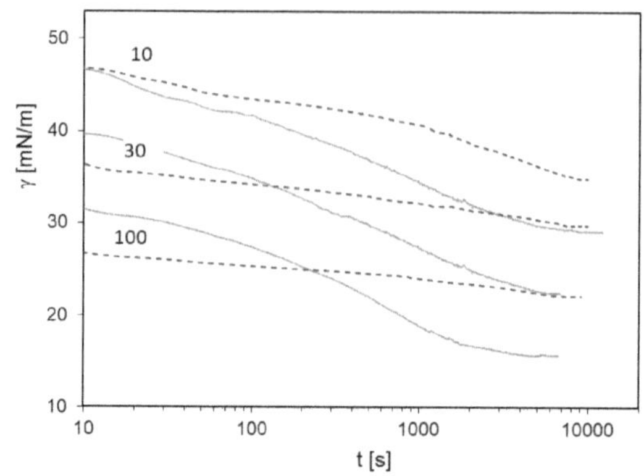

FIGURE 3.7 Dynamic interfacial tension measured for a pure decane drop immersed into aqueous solutions of C_{12}DMPO (green solid curves) and for a pure water drop immersed into solutions of C_{12}DMPO in decane (blue dashed curves) at concentrations of 10, 30, and 100 μmol/L, respectively. Adapted from Fainerman et al.[25] with permission under a Creative Commons Attribution [CC-BY 4.0] License.

this surfactant and the two liquid phases. Figure 3.7 illustrates the results obtained for a hexane drop immersed in an aqueous solution of C_{12}DMPO (green solid curves).[25] The figure also shows the dynamic interfacial tensions for a water drop immersed into solutions of C_{12}DMPO in decane (blue dashed curves) for the same three concentrations (10, 30, and 100 μmol/l).

The equilibrium interfacial tension values for the water drop immersed in the solution of C_{12}DMPO in decane are several mN/m higher than those for the decane drop immersed into the aqueous C_{12}DMPO solutions at the same initial concentrations. Such results can be used to determine the partition coefficient of C_{12}DMPO between the two liquid phases water and decane. As one can see, the time for reaching the adsorption equilibrium in the decane solution is shorter than in the aqueous solution, which is caused by the larger diffusion coefficient of C_{12}DMPO in decane than in water.[26] This is caused by the slightly lower viscosity of decane but also by the higher solubility of C_{12}DMPO in decane. A quantitative analysis of these dynamic processes would however also require the consideration of the real volumes of the two liquid phases.

The adsorption kinetics of ionic surfactants at water/oil interfaces is easier to handle because a transfer of surfactant molecules across the interface can be excluded as they are insoluble in non-polar liquids. As mentioned in a previous chapter, adsorption layers at water/oil interfaces are characterized not only by the surfactants but also by the oil molecules embedded in their structure. This finding was obtained from experiments performed with drops of aqueous surfactant solutions formed in air saturated by alkane vapor. A simple description would impose that the surface tension behavior should be identical to that observed at the water/air surface because alkane molecules are not amphiphilic and hence do not adsorb. The results, however, shown in Figure 3.8 for the adsorption kinetics of an aqueous solution drop of

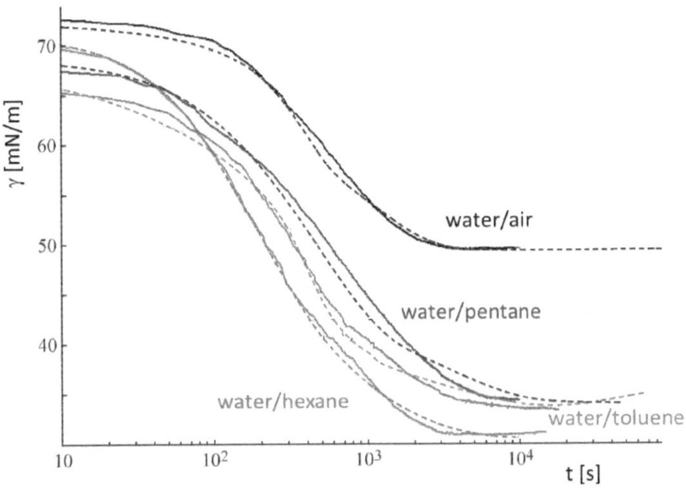

FIGURE 3.8 Dynamic surface tensions of a 5×10^{-6} M $C_{14}EO_8$ in water at the interface to air saturated by different oil vapors, dashed curves are calculated with the diffusion-controlled model for mixed adsorption layers. Adapted from Miller et al.[27] with permission from the PCCP Owner Societies.

5×10^{-6} M $C_{14}EO_8$ at the interface to pure air and air saturated by different oils demonstrate clearly that the oil molecules must be part of the interfacial layer.[27] Moreover, the different oil vapors contribute differently to the decrease in surface tension.

The reason for the frequent use of PAT in interfacial science is its versatile applicability. PAT allows not only measurements of interfacial tension but also of the interfacial dilational viscoelasticity based on harmonic oscillations of the drop volume or drop surface area. In particular, the stability of emulsions is influenced by these mechanical quantities of the interfacial layers.

As an example, in Figure 3.9, the interfacial dilatational elasticity *E'* is shown as a function of the oscillation frequency *f* for silk fibroin adsorbed at the dodecane/water interface.[28] As one can see, the dilational elasticity is almost independent of the frequency and increases with frequency only for the highest concentration of 10^{-4} g/ml.

There are plenty of dilational viscoelasticity data for adsorption layers of surfactants, proteins, and their mixtures in literature, and many of them are measured with the aim to quantify the impact of the adsorption layers on the stabilization of foams and emulsions.

3.4 NEW SINGLE DROP METHODS TO MIMIC EMULSIONS

PAT can be seen as a kind of basis for a number of additional valuable applications. Already the above-discussed capillary pressure tensiometry was presented as a kind of extra equipment for PAT, i.e., the basic instrument was just equipped with a particular measuring cell including a special capillary, a piezo drive, and a pressure sensor

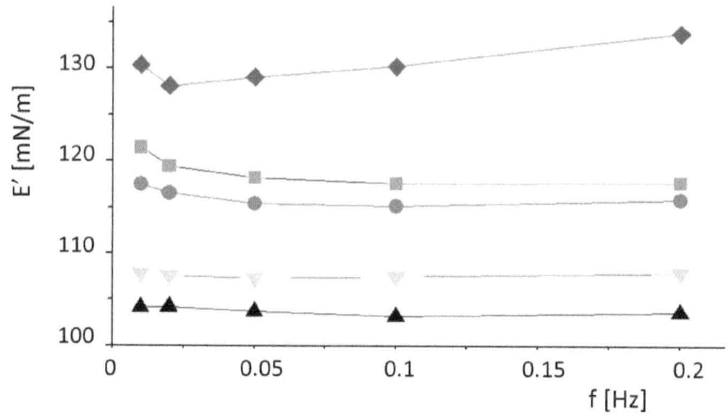

FIGURE 3.9 Dilational interfacial viscoelasticity of silk fibroin at the water/dodecane interface at different concentrations in water: 5×10^{-3} (■); 10^{-2} (●); 2×10^{-2} (▲); 5×10^{-2} (▼); 10^{-1} (◆) mg/ml. Adapted from Qiao et al.,[28] Copyright (2017), with permission from Elsevier.

while the video technique including the light source, the temperature control, and all the required mechanical construction of PAT was available for its use in the new experimental protocol. Moreover, the required capillary pressure software can be made available just as an extra function. This is the way it is provided, for example, by the PAT1 instrument produced by SINTERFACE Technologies.[21]

3.4.1 Coaxial Double Capillaries for Drop Volume Exchange

As a special feature for the drop profile analysis tensiometry, a double capillary was proposed many years ago by Cabrerizo-Vilchez et al.,[29] consisting of a very thin capillary mounted inside a second larger capillary,[30] at the tip of which the drop is formed (see Figure 3.10).

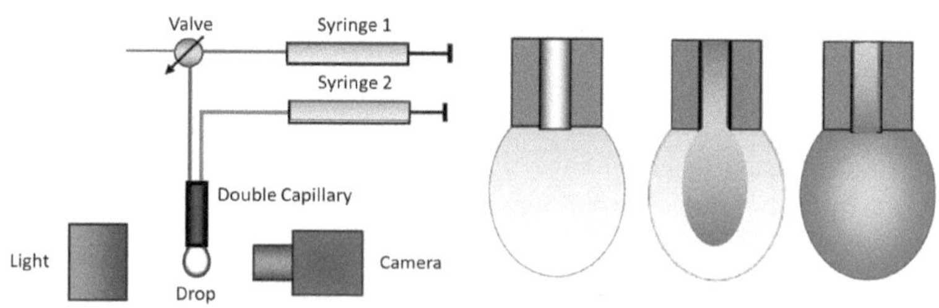

FIGURE 3.10 Design of a double capillary for PAT used for drop volume exchange as extra equipment for PAT. Adapted from Ferri et al.,[30] Copyright (2010), with permission from Elsevier.

Via a controller–responder protocol, the liquid inside a drop can be exchanged in a way that liquid is pumped into the drop via the internal tube and the same amount of liquid is sucked out via the outer tube while the drop size is kept constant. During this drop volume exchange, the process of drop profile analysis can be continued as the disturbances caused by the liquid flow into and out of the drop are sufficiently small.

This exchange process allows not only to change the measurement from one solution to another, which is one interesting protocol.[31] It allows also to create a certain efficient convection inside the drop. Such a situation is sometimes of advantage when the adsorption of surfactants or proteins is investigated at rather low bulk concentrations. Then, the time required to reach the adsorption equilibrium becomes very long. It depends on the surfactant bulk concentration, and each decrease of the concentration by one order of magnitude leads roughly to a 100-times longer adsorption time. This could quickly lead to a situation at which no equilibrium can be reached anymore in a reasonable time.

For an aqueous C_{14}DMPO solution with a concentration of 6×10^{-6} M, the dynamic surface tension changes very slowly, and even after one hour adsorption time, the measured values are still very far from equilibrium,[30] as one can see in Figure 3.11.

When an enhanced convection inside the solution drop is created by a drop exchange using the same solution to be pumped into the drop, the adsorption rate is remarkably increased. With an exchange rate of 0.2 μl/s (for a drop with a volume of typically about 15 μl), already after half an hour the adsorption equilibrium is established. This represents an adsorption rate several orders of magnitude larger than expected by a diffusion-controlled adsorption. Note, due to the large amount of solution used in this experimental protocol, also the depletion effects can be neglected; however, an analysis of the adsorption mechanism becomes very difficult because

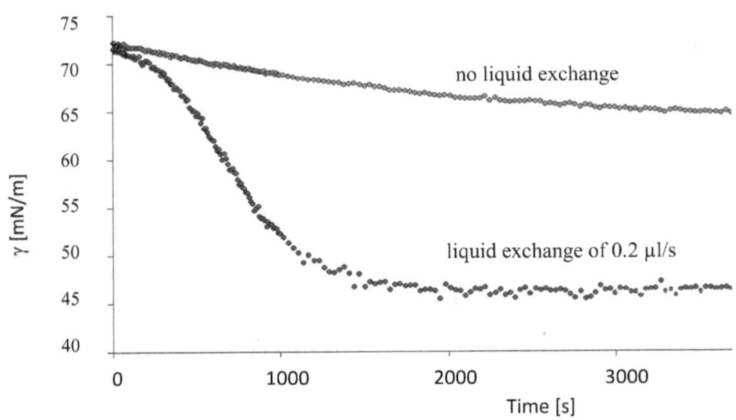

FIGURE 3.11 Dynamic surface tension of C_{14}DMPO at a bulk concentration of $c = 6 \times 10^{-6}$ M for static (◇) adsorption and convection exchange (○) experiments. Adapted from Ferri et al.,[30] with permission from Elsevier.

the impact of the convection on the surfactant transport cannot be easily described quantitatively.

3.4.2 Drop–Drop Micromanipulator to Mimic Emulsions

In addition to the capillary pressure and double capillary modules as extra equipment for PAT, a special design for liquid film manipulations exists. This so-called drop and bubble micromanipulator (DBMM) is more or less equivalent to two modules for capillary pressure measurements, which can be arranged mechanically in a quite simple way. The arrangements for creating films or for producing a drop or a bubble inside another drop is then arranged by respective capillaries. In Figure 3.12, the arrangement is shown for a drop formed inside another drop, i.e., a situation met in multiple emulsions.[32]

The liquid films formed in this way are comparatively thick, but the arrangement allows studies on the transport of surfactant between the different phases of the mimicked multiple emulsion. In a different design, two capillaries can be mounted just opposite to each other so that drops can be formed at either side. With an additional very accurate *xyz*-stage, one of the drops can then be exactly positioned opposite to the other and brought into contact so that the stability of the liquid film between them can be measured exactly, as it was described in reference [32]. This type of experiments is suitable for investigating the stability of emulsions films, even under the condition of an external perturbation. This can be done by generating size oscillations of one the respective drops by the respective piezo drives in the system, which is shown in Figure 3.12.

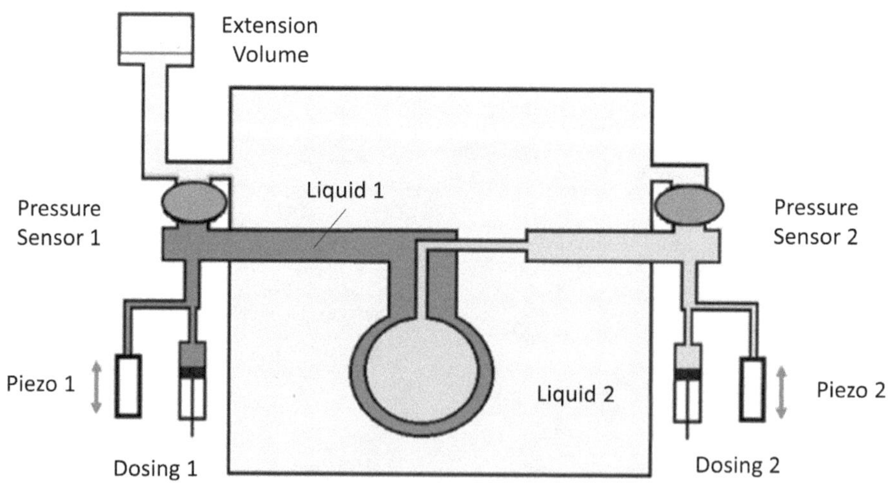

FIGURE 3.12 A drop-in-drop configuration on the basis of the DBMM module used to mimic the situation in emulsions or multiple emulsions. Adapted from Won et al.,[32] Copyright (2014), with permission from Elsevier.

3.4.3 Rising Oil Drops in Aqueous Solutions

Experiments on rising bubbles in solutions of surface-active compounds is a well-established tool to gain information on the adsorption process at the surface of bubbles moving in a liquid flow field. This information was demonstrated to be of great importance for understanding the elementary processes in flotation.[33]

With respect to the formation of emulsions and the corresponding adsorption dynamics of emulsifiers at the water/oil interface, the case of rising oil drops in aqueous surfactant solutions is relevant. This subject is essential for example in various extraction processes at liquid/liquid interfaces.[34] The change in the rising velocity of oil drops due to the formation of internal liquid flow pattern was visualized by Garner and Skelland[35] using alumina particles added to the system. Griffith[36] stated that surfactants form a stagnant cap at the rear pole of drops, equivalent to the situation of rising bubbles. The existing stagnant cap, the extension of which depends on the drop size and the surface activity and concentration of present surfactants, can reduce the rising velocity from a maximum for an absolutely mobile drop surface to a minimum velocity equivalent to that of a solid sphere.

The properties of the so-called dynamic adsorption layers (DAL) established at the surface of bubbles and drops in a liquid flow field were defined by Dukhin.[37] These DAL properties are responsible for the creation of surface tension gradients along the surface layer from the leading pole toward the rear pole of a rising bubble/drop, which in turn lead to Marangoni effects and have a strong impact on the rising behavior of oil drops in aqueous surfactant solutions.[38] The effects of surfactants on the surface properties and rising velocities can now also be modeled with numerical simulations based on the Navier–Stokes equations and corresponding boundary conditions at the surface of rising bubbles and drops.[39,40]

Great progress in studies of rising drops was made first by Kraume et al.[34,38] Following their achievements, Gawel and Zawala[41] developed a set-up for producing drops in a wide range of defined sizes and generation frequencies, i.e., of the initial surface age of a departing drop. The generation of sufficiently small drops with high precision is an important prerequisite for the accurate analysis of experimental data about the impact of adsorbing species like surfactants or proteins[42] on the dynamics of drops in liquid flow fields. In Figure 3.13, the terminal velocity of oil drops of different diameters in aqueous solutions of sodium dodecyl sulfate (SDS) or bovine serum albumin (BSA) is shown. The drops were formed with accurate diameter so that the application of the theory by Schiller and Naumann[43] could be applied.

The terminal velocities were measured with the instrument proposed in reference [42], so that drops in a broad diameter range could be studied. For the dodecane and hexadecane drops with diameters smaller than 1.5 mm, the agreement with the model of Schiller and Naumann[43] is quite good. The measured terminal velocities correspond to the behavior of solid spheres. The same behavior is also observed for dodecane drops rising in SDS or BSA solutions. One would have expected that the terminal velocity of the oil drops for systems without added surfactant or protein is significantly higher; however, the deceleration of the drops was caused by impurities present in water and in particular in the oil phases being most surface active at water/

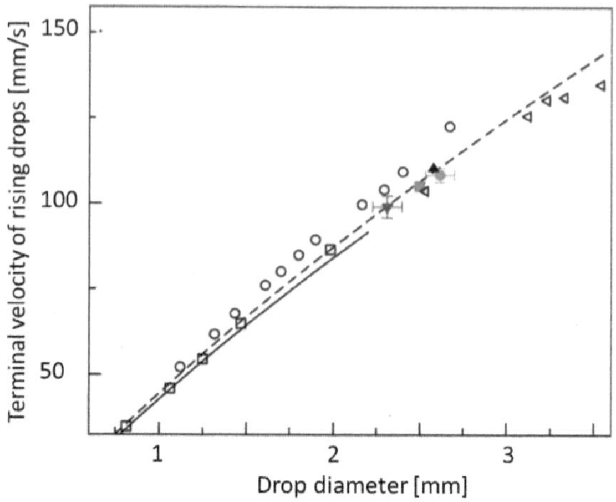

FIGURE 3.13 Terminal velocity of oil drops rising in aqueous solutions of 5 × 10^{-3} M SDS (◁) or 7.5 × 10^{-6} M BSA at pH 9.5 (▲), pH 7 (■), pH 5 (θ), and pH 3.5 (●) all in dodecane; dodecane (○) or hexadecane (□) drops in water; calculations after the model of Schiller and Naumann for dodecane (solid line) or hexadecane (dashed line) drops in water. Adapted from Gawel and Zawala[42] with permission under a Creative Commons Attribution [CC-BY 4.0] License.

oil interfaces.[44] Interestingly, a similar effect of complete interface immobilization was also observed for rising bubbles, but for significantly smaller bubble diameters (<ca. 0.5 mm).[44]

3.5 REFLECTOMETRIC METHODS

Together with the surface tension, the surface excess Γ of adsorbing molecules is another fundamental parameter of pivotal importance for the characterization of interfacial layers as given in the classical interfacial thermodynamics by the Gibbs fundamental Equation (1.1). Historically, the radiotracer technique is perhaps the first method for in situ measurements of the surface excess at a static liquid interface,[11] and it was also employed in adsorption studies at water/oil interfaces,[45,46] but it is somehow less popular in the recent decades as it requires the application of radiolabeled amphiphilic species.

Reflectometry methods for the study of soft interfaces based on polarized light (ellipsometry), neutrons (neutron reflectometry, NR), or X-rays (X-ray reflectometry, XRR) have been developed in the 1970s and 1980s.[11,47] These methods are label-free and provide access to the surface excess Γ and the molecular organization of interfacial layers in terms of layer thickness *d* and volume fraction Φ. Detailed information about the basics of these three reflectometric methods and their applications to interfacial layers of surfactants, polymers, proteins, and polymer nanogels can be found in the recent review articles by Maestro and Gutfreund,[48] Gochev et al.,[11] and

Schneck et al.[49] As in many other cases of surface-sensitive methods, their application to soft interfaces have started with experiments at the water/air interface[11] and only later they were adapted to probe water/oil interfaces.[10,11,47,50–54] In the following, we briefly discuss the experimental basics of ellipsometry, XRR and NR, as well as some examples to demonstrate the achievements at water/oil interfaces.

3.5.1 Ellipsometry

Ellipsometry belongs to the optical methods in which changes in the state of linear polarization of the used light beam are measured after reflection by the studied interface. In reference [55], the description of the light beam properties and the way of its specular reflection by an adsorption layer is given in a comprehensive way. The two main measurable quantities are the ellipsometry angles Ψ and Δ, which present the change of amplitude and phase of the polarized reflected beam. The specular reflection at an interface is described by the corresponding Fresnel coefficients r_p and r_s of the parallel (p) and perpendicular (s) polarization light components, respectively. The ratio ρ of these two coefficients provides the interrelation between the two ellipsometry angles Ψ and Δ,

$$\rho = \frac{r_p}{r_s} = \tan\Psi \cdot e^{i\Delta} = Re(\rho) + Im(\rho), \tag{3.6}$$

ρ is a complex quantity, which reflects variations in the amplitude and the phase shift of the reflected electric field relative to the incident electric field. Equation (3.6) is the basic equation in ellipsometry relating the experimentally accessible ellipsometric angles Ψ and Δ. The Fresnel coefficients are functions of the wavelength λ and the angle of incidence θ of the light beam, and also of the optical properties of the reflecting interface, i.e., the refractive indices n of the two bulk phases and the interfacial layer in between them.

In the special case of measurements at the Brewster angle θ_B, $Re(\rho) = 0$ and

$$Im(\rho) \equiv \bar{\rho} = \frac{\pi}{\lambda} \frac{\sqrt{n_1^2 + n_2^2}}{n_1^2 - n_2^2} \eta, \tag{3.7}$$

where n_1 is the refractive index of the phase, in which the ellipsometry beams travel through, n_2 is the refractive index of the subphase, $\bar{\rho}$ is the coefficient of ellipticity, and η is a surface parameter with the dimension of a length (called ellipsometry thickness) [50].

The thickness of the interfacial layer d and the adsorbed amount Γ cannot be determined independently but require the application of a respective model. The software of commercial ellipsometers provides the necessary routines for fitting a chosen model of the interface to the experimental data for the ellipsometric angles ψ and Δ. It should be stressed that based on (ψ, Δ) measurements, ellipsometry is capable of determining density–distance profiles along the z-direction normal to the interface, i.e., the refractive index n and the thickness d of the interfacial layer,

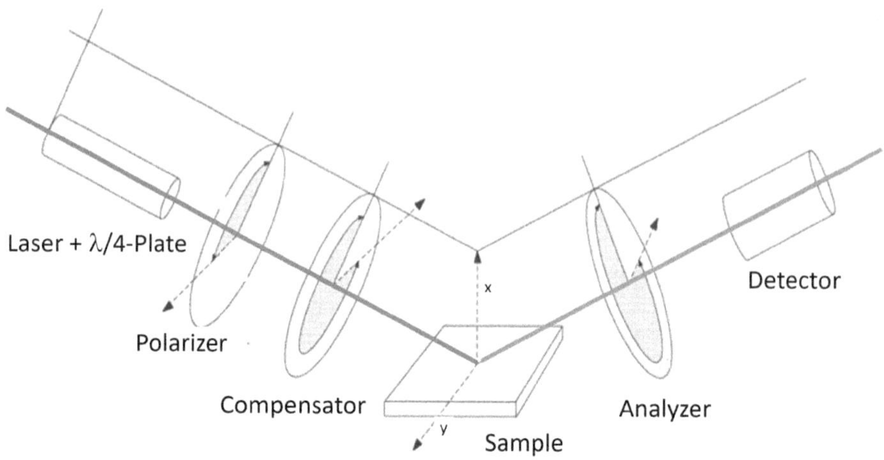

FIGURE 3.14 Scheme of an interfacial ellipsometer in the PCSA configuration. Adapted with permission from Wilkinson et al.[54] Copyright (1993), American Chemical Society.

however, with relatively high uncertainty; hence, decoupling of *n* and *d* is not a trivial task as discussed in more detail in reference [11]. Yet the product (*n,d*) serves to calculate the surface excess Γ(*n,d*) with high reliability.[11]

The typical design of an ellipsometer is shown in Figure 3.14, consisting of the three main optical components polarizer, compensator, and analyzer. While the polarizer produces a linearly polarized light, the compensator provides a defined phase retardation of the light beam sent to the sample, which is then reflected to the analyzer and recorded by the detector.

Based on the development of ellipsometry for probing water/air interfaces,[11,48–51,55] the technique was adapted for experiments with water/oil interfaces.[11,50,51,56–59] The early studies used a 'simple' arrangement,[11,50,56] as schematized in Figure 3.15A.

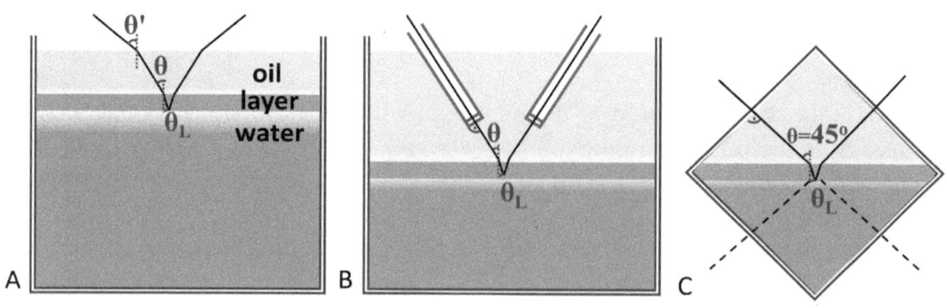

FIGURE 3.15 Schematic illustrations (not to scale) of experimental approaches for ellipsometric measurements at water/oil interfaces (provided the oil is the lighter phase). Legend: θ ' – incident angle (set by the instrument) of the laser beam hitting the air/oil interface, θ – incident angle of the refracted beam hitting the water/oil interface (note in B and C, $\theta' \equiv \theta$), θ_1 – angle of refraction within the adsorption layer.

Benjamins et al.[57] proposed the use of light guides in order to avoid optical effects from the oil/air interface (Figure 3.15B). Figure 3.15C illustrates another experimental approach proposed by Nylander et al.,[58] where a quadratic profile cuvette is used. This experimental arrangement allows for setting the direction of the incident beam to hit the interface by traveling through any of the two liquid phases. Thus, one can perform a coupled experiment with measurements from each side of the interface and then to apply a co-refinement data analysis for estimating n and d with relatively low uncertainty. A limitation of this configuration is that the incident angle is geometrically restricted to a fixed value of $\theta = 45^\circ$. To overcome this restriction, Stocco and coworkers used a cylindrical glass cell, which allowed multiangle scans.[51]

Another approach for decoupling the product (n_L,d) was proposed by Russev et al.,[56] who used the arrangement shown in Figure 3.15A. They performed experiments with a β-casein solution in contact with either air or oil and applied a minimization procedure to analyze the paired measurement data sets for the two interfaces, and reported results for n and d. The authors stressed the need for correction of the actual incident angle θ at the interfacial layer based on the known angle θ ' set by the instrument. This problem is overcome in experiments with the arrangements shown in Figure 3.15B and 3.15C.

As example for the use of the experimental configuration shown in Figure 3.15B, the ellipsometry results by Day et al.[59] for mixed protein/surfactant layers at the water/hexadecane interface are presented in Figure 3.16 in terms of the coefficient

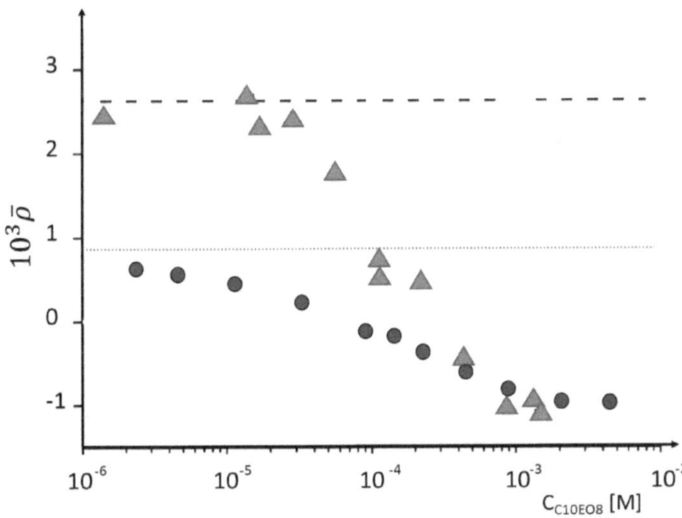

FIGURE 3.16 Ellipticity $\bar{\rho}$ as a function of the $C_{10}EO_8$ concentration in water at the interface to hexadecane in absence (●) and presence of 1 g/L β-casein (▲); the dashed line is the value for the ß-casein solution alone, $\bar{\rho} = +0.93 \times 10^{-3}$ for the bare water/hexadecane interface (green dotted line). Adapted with permission from John Wiley and Sons from Paulsson et al.,[58] Copyright (1999).

of ellipticity $\bar{\rho}$, which may be assumed proportional to the surface excess Γ. The sign of $\bar{\rho}$ is determined by the refractive index of the adsorption layer n_{layer} relative to the refractive indexes of the two macroscopic phases $n_{wat} \approx 1.33$ for water and n_{oil} for oil, as $\bar{\rho} > 0$ when $n_{layer} < n_{wat}$ or $n_{layer} > n_{oil}$, and $\bar{\rho} < 0$ for intermediate values ($n_{wat} < n_{layer} < n_{oil}$). For $n_{hd} = 1.435$ for hexadecane, one obtains $\bar{\rho}_{w/hd} = +0.93 \times 10^{-3}$ for the bare water/hexadecane interface with accounting for surface roughness.[50] Therefore, the decrease of $\bar{\rho}$ below $\bar{\rho}_{w/hd}$ with increasing surfactant concentration and the fact that it adopts negative values (Figure 3.13) means that the saturated surfactant layer has a refractive index $n_{surf} < 1.43$, whereas the increased $\bar{\rho}$ above $\bar{\rho}_{w/o}$ means that the protein layer has a higher refractive index $n_{prot} > 1.43 > n_{surf}$. The stepwise addition of the nonionic surfactant $C_{10}EO_8$ to an aqueous β-casein solution leads to a displacement of the protein from the water/hexadecane interface. At sufficiently high surfactant concentrations, the measured $\bar{\rho}$ values become identical to the values measured in absence of the protein, indicating that all protein molecules are presumably displaced from the interface.

3.5.2 X-ray and Neutron Reflectometry

Methods based on X-rays and neutrons are powerful tools in soft matter research for probing interfacial structures at relevant length scales.[10,11,47,48,60] Concerning water/oil systems, X-ray and neutron scattering, as well as other similar techniques, are used to study real emulsions,[61] while XRR and NR can readily be used to investigate a single macroscopic planar water/oil interface.[10,11,47,52–54,62,63] In XRR or NR experiments, the intensity I of the reflected beam is measured as a function of the grazing angle $\alpha = 90° - \theta$ (θ is the incident angle) and the X-ray or neutron wavelength λ . I is typically presented as a function of the Q_z, the scattering vector component perpendicular to the surface,

$$Q_z = \frac{4\pi}{\lambda}\sin\alpha, \tag{3.8}$$

which depends on the wavelength and the grazing angle. The reflectivity profile $R(Q_z)$ is dictated by the depth profile of the so-called scattering length density (SLD), $r(z)$. For X-rays, the SLD is proportional to the electron density, for neutron it depends on the number densities and scattering lengths of the various atomic nuclei.

The higher the Q_z-range, the higher is the depth resolution. XRR usually provides a wider Q_z-range. The strength of NR is that a strong and defined SLD contrast can be created by the use of isotopic substitutions.[11,47,48,63] In fact, different combinations of isotopic contrasts can be set in order to highlight specific features of the interfacial layers. NR is thus typically more sensitive to the adsorbed amount than XRR. In a recent study,[64] both methods XRR and NR were used to determine the surface excess of adsorbed surfactant impurities at the interface between water and perfluorinated oil. Notably, the study demonstrated that XRR was more sensitive to Γ than NR for this particular liquid/liquid interface, which is not usually the case.[11]

In the 1990s, the first studies appeared on water/oil interfaces employing XRR or NR.[10,11,52–54] In these investigations, the 'simple' experimental arrangement

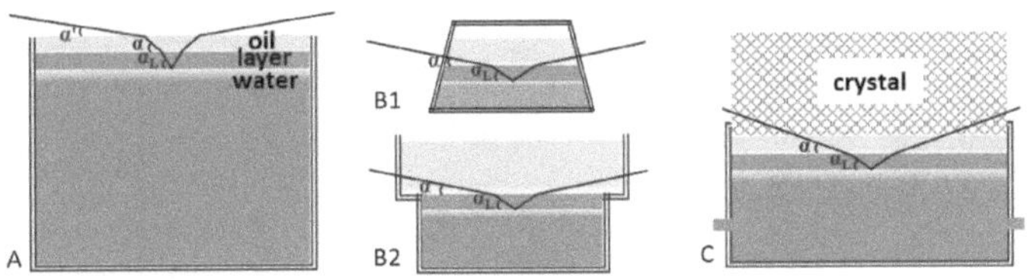

FIGURE 3.17 Schematic illustrations (not to scale) of experimental approaches for XRR and NR measurements at water/oil interfaces (provided the oil is the lighter phase); note that in (C) the oil is contrast-matched to the crystal, and hence, the neutron beam does not refract at the crystal/oil interface (see also in Figure 3.19). Legend: α' – grazing angle (set by the instrument) of the laser beam hitting the air/oil interface, α – grazing angle of the refracted beam hitting the water/oil interface (note in B and C, $\alpha' \equiv \alpha$), α_L – angle of refraction within the adsorption layer. In A and C) a microscopically thin oil layer is used, while in (B) the oil layer is macroscopically thick.

schematized in Figure 3.17a was used, where the oil is deposited on the aqueous subphase in the form of a 'microscopically thin' layer. This way, water/oil interfaces covered by adsorbed surfactants,[52] polymers,[53] or proteins[11,54] were studied. Note that in the case of volatile oils, the oil layer can be evaporated to a nanoscopic thicknesses.[65]

A significant XRR work on water/oil interfaces was made by Schlossman and coworkers using the experimental configuration schematized in Figure 3.17B1.[10,62] An XRR study on bare water/alkane interfaces revealed that the width σ of the interfacial region between water and various hydrocarbons (C_6–C_{16}) of the homologous series of n-alkanes (apolar oils) increases with the C-number within the range 4–6 Å; note, these values are larger than the ones σ_{cap} predicted by capillary wave theory.[62] On the other hand, XRR experiments with nitrobenzene (polar oil) revealed $\sigma < \sigma_{cap}$, which means that the polar oil damps the interfacial capillary wave fluctuations.[62] NR experiments with C_{16} revealing $\sigma \approx 6$ Å[66] are in excellent agreement with the above XRR results[62] while ellipsometry measurements for various polar and apolar oils revealed smaller widths $\sigma < 3$ Å.[50]

Schlossman and coworkers investigated also surfactant layers at the water/hexane interface and as an example, in Figure 3.18a, we present the obtained $\rho_e^*(z)$ profiles (ρ_e^* is the electron density normalized to that of water ρ_e^w) for the nonionic surfactant triacontanol adsorbed either at a water/air or at a water/hexane interface.[62] For these two types of liquid interfaces, the authors performed a systematic XRR study with adsorption layers of various surfactants (alkanols and carboxylic acid) in order to investigate the effects of: (1) the alkyl chain flexibility (flexible hydrocarbon vs. rigid fluorocarbon); (2) the headgroup type (-OH vs. -COOH); and (3) the nature of the interface; on the molecular structure of the interfacial layers.[62] Figure 3.18A shows that the investigated surfactant forms a well-ordered 'solid-like' monolayer at the water/air interface, whereas it forms a disordered 'liquid-like' monolayer at the water/hexane interface. Note that this type of interfacial behavior is valid for all species (C_{20}, C_{22}, C_{24}, and C_{30}) of the hydrocarbon n-alkanol homologous series investigated in reference [62]. In contrast, the two (partially) fluorinated n-alkanols

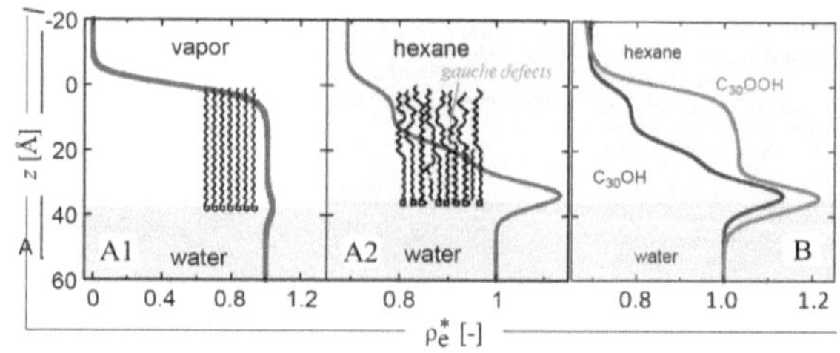

FIGURE 3.18 Schematic illustrations (not to scale) of the structure of adsorption layers of surfactants studied by XRR; interfacial structure and $\rho_e^*(z)$ profiles (ρ_e^* is the electron density normalized to that of water ρ_e^w). (A) $C_{30}OH$ (triacontanol) at the water/air (A1) and the water/hexane (A2) interfaces. (B) $C_{30}OH$ and $C_{30}OOH$ at the water/hexane interfaces. Reprinted from Schlossman,[62] Copyright (2005), with permission from Elsevier.

(C_8F_{17}-C_2H_4-OH and $C_{10}F_{21}$-C_2H_4-OH) display better ordering at the water/hexane interface forming a 'solid-like' monolayer as for the case of the water/air interface. According to the authors, the origin of the different ordering capacity of fluorocarbon and hydrocarbon chains in such surfactant monolayers is the ability for mixing of the surfactant alkyl chain with the solvent phase (hexane). Hydrocarbon chains are prone to mixing with hexane, whereas fluorocarbon chains exclude hexane from the monolayer. Another experiment with $C_{30}OH$ and $C_{30}OOH$ at the water/hexane interface (Figure 3.18B) revealed that unlike the above-discussed alkanols, the carboxylic acid forms a more disordered interfacial monolayer, which was attributed to interactions between the hydrophilic headgroups in the water due to hydrogen bonding.

As a conclusion, it can be summarized that the degree of ordering of surfactants at liquid/fluid interfaces is determined by in-plane intermolecular interactions such as headgroup–headgroup interactions (e.g., electrostatic repulsion, hydrogen bonding, and ion binding) within the 'wet' part of the monolayer immersed in the solution, and van der Waals interactions between alkyl chains in the hydrophobic 'upper' part of the monolayer.[62] The major difference between water/air and water/oil interfaces manifests itself via the modulation of the van der Waals attractions between alkyl chains in the following manner. For water/air interfaces, the 'upper' part of the monolayer is dry and effective van der Waals attractions between alkyl chains promote a high degree of molecular ordering. In contrast, for water/oil interfaces, insertion of solvent molecules and gauche defects (both phenomena being sensitive to the chemical composition, e.g., hydrocarbon vs. fluorocarbon) may disturb the van der Waals attraction, hence leading to a lower degree of molecular ordering (which is, however, entropically favorable). These effects by different types of molecular interactions were also detected by experiments with optical methods, which will be discussed in the next section in the context of spectroscopic studies.

The experimental configuration shown in Figure 3.17B2 was employed in XRR and NR experiments with water/oil interfaces in studies on solvent extraction[67] and on the detection of surface-active impurities in water.[64] Figure 3.17C presents

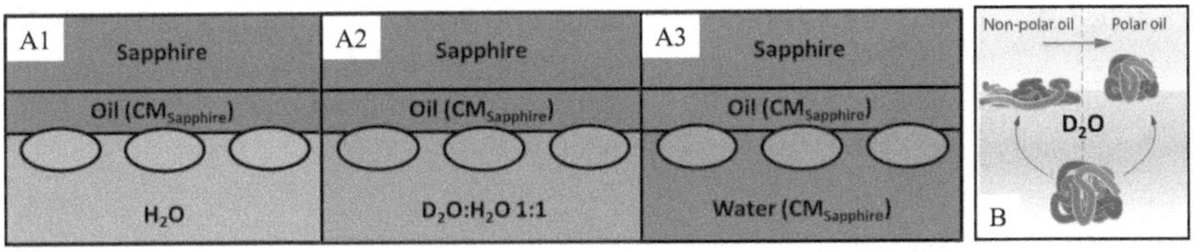

FIGURE 3.19 (A) Schematic illustrations (not to scale) of the structure of adsorption layers of monoclonal antibody (mAb) at the water/hexadecane interface studied by NR for different aqueous isotopic contrasts as denoted in the figures: (A1) H_2O; (A2) protein-matched water (1:1 D_2O:H_2O); and (A3) sapphire-matched water. Reprinted from Hollowell et al.[63] with permission under a Creative Commons Attribution [CC-BY 4.0] License. (B) Schematic illustration of the structure of adsorbed β-lactoglobulin (in D_2O aqueous subphase) at water/oil interface for a polar (methyl *tert*-butyl ether) and a non-polar (n-hexane) oils studied with NR. Adapted from Bergfreund et al.[70] with permission under a Creative Commons Attribution [CC-BY 4.0] License.

the scheme of another experimental arrangement, which we call here the 'crystal' cell. It was introduced by Zarbakhsh et al. [68] and used in NR studies on protein adsorption.[11,63]

Figure 3.19 presents NR experimental approaches using the 'crystal' cell. In all cases the hexadecane used was contrast-matched to the sapphire crystal, while the SLD of the aqueous phase was modulated by varying the H_2O/D_2O ratio in order to match the SLD of the protein layer (Figure 3.19A2) or of the oil/sapphire phase (Figure 3.19A3). The former case provides information about the degree of mixing of the protein molecules into the oil phase. In the latter case, the equal SLDs of both the aqueous and the oil phases render the interface 'invisible' to neutrons; then any detected reflectivity signal originates solely from the presence of adsorbed species and thus such an experiment delivers unambiguously quantitative information about the surface excess. In the case of H_2O (Figure 3.19A1), the protein and each of the two liquid phases have different SLDs, and hence, such an experiment is expected to deliver at once all the structural information about the interfacial region.

In an NR study on polymer adsorption at the water/oil interface, Protat et al.[69] introduced a modified 'crystal' cell. This cell allows for using a macroscopic oil phase with dissolved polymer in contact with an aqueous phase, the pH and salinity of which can be tuned. The authors were able to resolve the structure of the polymer adsorption layers, which showed different degrees of penetration into the aqueous phase depending on the NaCl concentration used. The results were interpreted in the light of various types of emulsions, such as water-in-oil, oil-in-water, or multiple emulsions.

The type of the oil phase and mainly its polarity may have a significant effect on the physicochemical and structural properties of a water/oil interfacial layer being either bare[50,62] or covered with adsorbed species, such as surfactants, proteins, and particles.[70] In the review article by Bergfreund et al.,[70] the authors provide suggestions for experimental studies with water/oil interfaces and for the choice of suitable model oils. Concerning effects of the oil polarity on the interfacial structuring of adsorbed species, Figure 3.19B presents a scheme of the structure of β-lactoglobulin at two different water/oil interfaces – the protein globules adopt a more extended conformation at the interface and penetrate deeper into the oil phase when the polar methyl *tert*-butyl ether is used in comparison to the case of the apolar n-hexane.[65,70]

3.6 SPECTROSCOPIC METHODS

Various spectroscopy and microscopy techniques are available for an in situ characterization of the dynamics and structure of interface-dominated soft matter systems, including foams and emulsions, among which absorption and vibrational spectroscopy (including Raman spectroscopy), light and X-ray scattering, diffusing wave spectroscopy, neutron spin echo, and others are well recognized.[9,71,72] Spectral data in absorption, reflection, or scattering modes are collected from macroscopic bulk samples and analyzed via appropriate optical theories. Such methods provide information about dynamic and structural features in foams and emulsions, including information on the molecular level.

Absorption spectroscopy uses electromagnetic radiation in the ultraviolet-visible (UV-vis) region (190–800 nm). A frequently used related technique is circular dichroism (CD), which could be advanced by the use of synchrotron sources.[9] The application of CD to emulsions was made possible via the so-called refractive index matched emulsion approach.[9,73] CD scans are usually divided into two spectral regions, namely far-UV (170–260 nm) and near-UV (260–320 nm). Applications of vibrational spectroscopy in colloid studies are based mainly on infrared (IR) and Raman spectra. Fourier transform infrared spectroscopy (FTIR) is well established in emulsion science.[9,72] FTIR combined with CD is a powerful tool for investigating changes in protein conformations upon adsorption, and these methods were extensively applied to protein stabilized emulsions.[9,73] Concerning proteins, tryptophan emission fluorescence spectroscopy is also widely used, however, mostly in studies on protein conformations in bulk solution.[9] The application of this technique to emulsions is based on the fact that the maximum of the characteristic emission peak of tryptophan undergoes a red-shift with changes in the environment from hydrophobic (buried in the globules' core in bulk solution) to hydrophilic (exposed to water under interface-induced changes in the tertiary structure).[9]

Indeed, the above-discussed methods deliver valuable information about the adsorption of amphiphiles at droplet interfaces in emulsions; however, the obtained optical signals originate from both interface-adsorbed and bulk-dissolved molecular species. In this respect, nonlinear optical methods such as the spectroscopic techniques of Second Harmonic Generation (SHG) and Sum Frequency Generation (SFG) are inherently surface-sensitive techniques with molecular sensitivity.[6–8,74] These methods are based on second-order nonlinear optical processes, which under the electric dipole approximation are forbidden in centrosymmetric (bulk) media, but manifest at interfaces, where the inversion symmetry is necessarily broken.[6,74] The SHG and SFG techniques working with a macroscopic planar interface in reflection mode[6–8] can readily operate in scattering mode to probe a real emulsion sample, in which case they are called second harmonic scattering (SHS) and sum frequency scattering (SFS).[75–79] The latter two methods are especially efficient in studies on nanoemulsions, where the interfacial properties of the nano-size droplets may differ from those of relevant macroscopic planar interfaces.[75,79]

In a typical SFG experiment at a planar interface, two pulse laser beams – a Vis beam with a fixed frequency (ω_{Vis}) and an IR beam with a tuneable frequency (ω_{IR}) – are spatially and temporally overlapped in a noncollinear manner at the interface under investigation, where the optical fields induce a second-order polarization, thus generating a beam at the sum frequency $\omega_{SF} = \omega_{IR} + \omega_{Vis}$, which is emitted in a well-defined direction.[6–8] While the IR beam matches active vibrational modes, the Vis beam serves to set an easily measurable resulting SFG signal in the visible light range. Contemporary instruments work with broadband femtosecond IR pulses tunable in the wavelength range of 2–20 μm.[80,81] The SFG sensitivity is considerably enhanced under total internal reflection (TIR) geometry, the so-called TIR-SFG.[7,82] Resonance enhancement of the signal is observed when vibrational modes of the adsorbed molecules are excited. Hence, the SFG signal contains a nonlinear spectrum, which encodes vibrational information of the adsorbed species without any contribution from the bulk subphase. In addition, control of the polarizations (⊥ or || to the plane of incidence) of the three coherent SFG, Vis, and IR beams allows for analyzing signals originating from various geometrical configurations of adsorbed molecules.[8] The analysis of the spectral features yields the number density of the adsorbed species, the orientation of certain functional groups as well as information about the chemical nature of the adsorbed material and the local energetics. For the sake of illustration, we present in Figure 3.20 a spectrum of the surfactant C_8AAP. This surfactant is a representative of a group of light-responsive

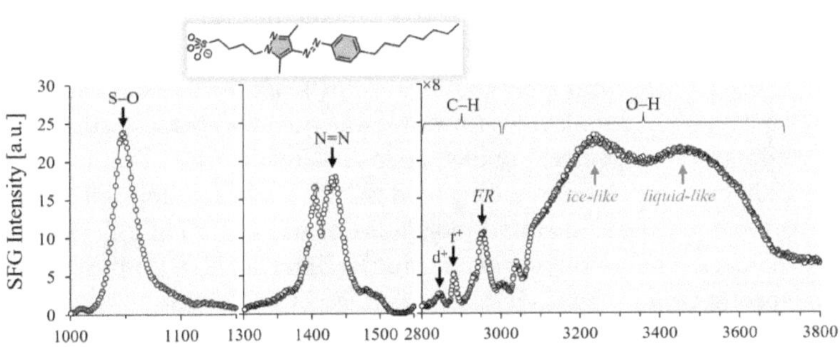

FIGURE 3.20 SFG spectrum of 0.07 mM E-photoisomer of the surfactant C_8AAP (octyl-arylazopyrazole butyl sulfonate) adsorbed at the water/air interface. The ordinate is the measured SFG intensity in arbitrary units, note that the original measurement data for the range 2800–3800 cm^{-1} are multiplied by a factor of 8 for better visibility. The vertical arrows indicate centers of characteristic bands such as S–O stretching vibrations; the peak at around 1440 cm^{-1} is frequently assigned to N=N stretching vibrations; range of C–H stretching vibrations: methylene (d^+) and methyl (r^+) symmetric modes, and Fermi resonance (FR); range of O–H stretching vibrations with peaks for 'ice-like' and 'liquid-like' water. Unpublished data, courtesy of Björn Braunschweig.

low-molecular-weight molecules based on arylazopyrazole, as the respective E- and Z- photo-isomers may have very different surface activity,[83] which fact renders these surfactants successful candidates for development of responsive, adaptive, or smart soft colloids.

Typical spectral ranges for H_2O and hydrogenous alkyl chains are (1) 2800–3000 cm^{-1} for C–H symmetric and asymmetric stretching vibrations in methyl (CH_3) and methylene (CH_2) alkyl chain units; and (2) 3000–3600 cm^{-1} for O-H symmetric stretching vibrations in hydrogen-bonded interfacial water.[7,8,80,84] In the latter case, the ordering of water molecules in the vicinity of the interface can be discriminated by SFG as tetrahedrally coordinated 'ice-like' water and more disordered 'liquid-like' water, with characteristic peaks centered at about 3200–3250 cm^{-1} and 3400–3450 cm^{-1}, respectively. These two water bands can be well pronounced for water/air interfaces depending on the magnitude of the interfacial charge, and on some other factors, which may affect the ordering of interfacial water.[7,80] Dedicated SFG and SFS studies of bare water/oil interfaces revealed the dominance of the 'ice-like' peak over the 'liquid-like' peak at water/CCl_4 and water/alkane interfaces, meaning that interfacial water is somewhat better ordered at those water/oil interfaces than at the neat water/air interface, which was assigned to stronger H-bonding networks.[85] However, for both types of water interfaces, the observed SFG intensity in the 'water' spectral region is determined by the magnitude of the electric field at the interface, and can be used as a qualitative measure of the interfacial charge at both water/air and water/oil interfaces.[7,8,80,83] It is noteworthy that efforts have been put to quantify the variations of the electric double-layer potential at the water/air interface electrified by adsorbed ionic surfactants from combined SFG and thin liquid film measurements.[80]

It was found that the use of D_2O and/or deuterated oil and/or surfactant may enlarge the capacity of the SFG method.[7,8,81–83] This approach is based on the fact that the exchange of H-atoms with D-atoms causes shifts of the characteristic vibrational frequencies in the labeled molecules toward lower wavenumbers. In such cases, SFG signals arise in specific spectral regions: 2025–2125 cm^{-1} for C-D stretching modes and 2400–2700 cm^{-1} for O-D stretching modes.[81] In summary, appropriate combinations of hydrogenous and deuterated molecules of the water and oil phases as well as of the used surfactant allow for examining an interfacial system by SFG via the analysis of spectra containing C–H, C–D, O–H, and O–D vibration signatures.[7,8,81–83]

A significant work on water/oil interfaces was made by Richmond and coworkers employing TIR-SFG and using CCl_4 as a hydrophobic phase.[7,81,84] The operating beams travel through the CCl_4 phase, which has a higher refractive index than aqueous solutions (H_2O or D_2O) and is transparent to the typical IR frequencies used in SFG. The ionic surfactants dodecyl-ammonium chloride and the well-known Na dodecyl sulfate (SDS) were investigated by monitoring the spectral range 2800–3000 cm^{-1} of C–H symmetric stretching vibrations in methyl (CH_3) and methylene (CH_2) alkyl chain units.[7] The authors explained the tendency of disordering of the surfactant monolayers studied at the water/CCl_4 interface in comparison to the cases of ordered monolayers at solid/air and water/air interfaces, with the insertion of oil molecules into the hydrophobic slab (alkyl chains) of the surfactant monolayer, which affects the van der Waals forces between the alkyl chains of the adsorbed molecules. This explanation is in agreement with the above-discussed XRR results[62] and was further supported by other SFG results in reference.[83] However, an increase of the surfactant monolayer density at increasing bulk surfactant concentrations usually leads to a relative increase of the alkyl chain ordering.[7,83] It should be emphasized here that in their SFS study on alkane-in-water nanoemulsions, de Aguiar et al.[75] concluded conceptually different interfacial structuring of the alkane molecules (predominantly oriented parallel to the interface) and the alkyl chains (disordered) of SDS molecules adsorbed at the droplets' interfaces (droplet sizes 80–130 nm). In addition, relatively low SDS surface loads on droplets were found. The authors explain both these findings with nonequilibrium surface state in the studied nanoemulsions.

Richmond and coworkers performed more studies on various water/oil systems containing different oils, surfactants, polymers, as well as amphiphile mixtures, which we will not discuss here, but some of them together with studies by other authors were very recently reviewed in reference [8]. It is worth mentioning the recent study by Altman et al.[81] on the adsorption of mixtures of the nonionic surfactant Span-80 and the ionic surfactant AOT at the water/CCl_4 interface. Span-80 was used in its native hydrogenous form, while the two alkyl chains of the AOT molecule were deuterated. In this way, the authors were able to monitor the stoichiometry of the mixed surfactant adsorption layer with time. Initially, both surfactants adsorb simultaneously and saturate the interface, but with the time Span-80 progressively displaces AOT from the interface. This finding stresses the need of special account on the kinetics of the process of molecular displacement at interfaces in theoretical models on the adsorption dynamics of surfactant mixtures. For further information and topic overviews, the reader is kindly asked to refer for example to the references.[8,76]

3.7 SUMMARY AND OUTLOOK

The experimental methods presented in this chapter, together with the theoretical models discussed in Chapter 2, provide a solid basis for a quantitative characterization of liquid/liquid interfaces relevant for the design of emulsion systems. The group of methods based on single drops, including capillary pressure and drop profile techniques provide access to the formation of adsorption layers and their response to external perturbations. While capillary pressure tensiometry delivers dynamic interfacial tensions in a time interval from about 10 ms up to hours of adsorption time, drop profile tensiometry can provide data for not less than 1 s. With respect to oscillation frequencies for measurements of the interfacial dilational viscoelasticity, the highest frequencies of about 10 Hz can be reached by capillary pressure and of less than 1 Hz for drop profile tensiometry. Using the underlaying theoretical models, the corresponding adsorption mechanisms and equations of state can be determined from the gained experimental data.

These parameters are essential for understanding the formation and stabilization of emulsions. So far, however, there are no direct correlations between these interfacial properties and parameters for describing emulsions, i.e., there are no quantitative relations between the adsorbed amount or the adsorption kinetics mechanism and the stability of a corresponding emulsion. In the near future, we can expect here more insights.

In addition to tensiometry methods, we also described in some details reflectometric and spectroscopic methods suitable to determine the adsorbed amounts, the chemical composition, and the molecular organization of the interfacial layers. While the NR and XRR techniques deliver unambiguous information about the thickness and density of interfacial layers, spectroscopic methods such as CD, FTIR, nonlinear optical techniques, and others provide direct information about the interfacial chemistry and the configurations of adsorbed molecules. However, different surface-sensitive spectroscopic techniques have their advantages and drawbacks. For example, in CD and FTIR, the information about the interfacial characteristics of the investigated emulsion samples is based on the changes in the spectral properties of amphiphilic species upon adsorption at the emulsion droplets' surfaces relative to their bulk phase spectral characteristics. In this respect, the inherently surface-sensitive SHG and SFG techniques appear advantageous, but they are experimentally restricted to the model system of a single macroscopic planar interface. Finally, the novel nonlinear scattering methods SHS and SFS, in fact, probe in situ the droplet interfaces in a real nanoemulsion. To what extent studies on planar interfaces are transferable to properties of real emulsions, for example with regard to the influence of the surface-to-volume ratio and the curvature as well as nonequilibrium effe cts,[75,76,79] still remains to be answered.

ACKNOWLEDGMENTS

The work was supported by the ESA MAP project 'Emulsion Dynamics and Drop Interfaces' (EDDI) with the project number 4000128643/19.

REFERENCES

1. Ravera, F.; Dziza, K.; Santini, E.; Cristofolini, L.; Liggieri, L. Emulsification and emulsion stability: The role of the interfacial properties. *Adv. Colloid Interface Sci.* **2021**, *288*, 102344.
2. Niu, H.; Wang, W.; Dou, Z.; Chen, X.; Chen, X.; Chen, H.; Fu, X. Multiscale combined techniques for evaluating emulsion stability: A critical review. *Adv. Colloid Interface Sci.* **2023**, *311*, 102813.
3. Aveyard, B. *Surfactants—In Solution, at Interfaces and in Colloidal Dispersions,* Oxford University Press: Oxford, UK, 2019.
4. Benjamins, J.-W.; Thuresson, K.; Nylander, T. Ellipsometry studies of nonionic surfactant adsorption at the oil–water interface. *Langmuir* **2005**, *21*, 149–159.
5. Suja, V.C.; Rodríguez-Hakim, M.; Tajuelo, J.; Fuller, G.G. Single bubble and drop techniques for characterizing foams and emulsions. *Adv. Colloid Interface Sci.* **2020**, *286*, 102295.
6. Eisenthal, K.B. Liquid interfaces probed by second-harmonic and sum-frequency spectroscopy. *Chem. Rev.* **1996**, *96*, 1343–1360.
7. Richmond, G.L. Vibrational spectroscopy of molecules at liquid/liquid interfaces. *Anal. Chem.* **1997**, *69*, 536A–543A.

8. El-Zohry, A.M.; Sølling, T. Mini-review on water–oil emulsions through vibrational sum frequency generation. *Energy Fuels* **2024**, *38*, 743–760.
9. Zhai, J.L.; Day, L.; Aguilar, M.-I.; Wooster, T.J. Protein folding at emulsion oil/water interfaces. *Curr. Opin. Colloid Interface Sci.* **2013**, *18*, 257–271.
10. Schlossman, M.L. Liquid–liquid interfaces: studied by x-ray and neutron scattering. *Curr. Opin. Colloid Interface Sci.* **2002**, *7*, 235–243.
11. Gochev, G.G.; Campbell, R.A.; Schneck, E.; Zawala, J.; Warszynski, P. Exploring proteins at soft interfaces and in thin liquid films – from classical methods to advanced applications of reflectometry. *Adv. Colloid Interface Sci.* **2024,** 329, 103187 .
12. Dukhin, S.S.; Kretzschmar, G.; Miller, R. Dynamics of adsorption at liquid interfaces: Theory, experiment, application. In *Studies in Interface Science*, Möbius, D., Miller, R., Eds.; Elsevier: Amsterdam, The Netherlands, 1995; Volume 1.
13. Fainerman, V.B.; Möbius, D.; Miller, R. (Eds.). Surfactants: Chemistry, interfacial properties and applications. In *Studies in Interface Science*, Elsevier: The Netherlands, 2001; Volume 13.
14. Fainerman, V.B.; Miller, R. Maximum bubble pressure tensiometry: Theory, analysis of experimental constrains and applications. In *Bubble and Drop Interfaces*, Miller, R., Liggieri, L., Eds.; Brill Publ.: Leiden, The Netherlands, 2011; pp. 75–118.
15. Fainerman, V.B.; Mys, V.D.; Makievski, A.V.; Petkov, J.T.; Miller, R. Dynamic surface tension of micellar solutions in the millisecond and sub–millisecond time range. *J. Colloid Interface Sci.* **2006**, *302*, 40–46.
16. Javadi, A.; Krägel, J.; Pandolfini, P.; Loglio, G.; Kovalchuk, V.I.; Aksenenko, E.V.; Ravera, F.; Liggieri, L.; Miller, R. Short time dynamic interfacial tension as studied by the growing drop capillary pressure technique. *Colloids Surf. A* **2010**, *365*, 62–69.
17. Liggieri, L.; Ravera, F.; Passerone, A. Dynamic interfacial tension measurements by a capillary pressure method. *J. Colloid Interface Sci.* **1995**, *169*, 226–237.
18. Kovalchuk, V.I.; Ravera, F.; Liggieri, L.; Loglio, G.; Pandolfini, P.; Makievski, A.V.; Vincent-Bonnieu, S.; Krägel, J.; Javadi, A.; Miller, R. Capillary pressure studies under low gravity conditions. *Adv. Colloid Interface Sci.* **2010**, *161*, 102–114.
19. Rotenberg, Y.; Boruvka, L.; Neumann, A.W. Determination of surface tension and contact angle from the shapes of axisymmetric fluid interfaces. *J. Colloid Interface Sci.* **1983**, *93*, 169–183.
20. Benjamins, J.; Cagna, A.; Lucassen-Reynders, E.H. Viscoelastic properties of triacylglycerol/water interfaces covered by proteins. *Colloids Surfaces A* **1996**, *114*, 245–254.
21. Loglio, G.; Pandolfini, P.; Miller, R.; Makievski, A.V.; Ravera, F.; Ferrari, M.; Liggieri, L. Drop and bubble shape analysis as tool for dilational rheology studies of interfacial layers. In *Novel Methods to Study Interfacial Layers*, Möbius, D., Miller, R., Eds.; Elsevier, Amsterdam: The Netherlands, 2001; pp. 439–484.
22. Rivas, J.E.M.; Hussain, S.; Tseng, W.-C.; Noskov, B.; Lin, S.-Y. A study on the dilational modulus of adsorbed globular protein films – under a near periodic area fluctuation and rapid surface perturbation. *J. Taiwan Inst. Chem. Eng.* **2024**, *155*, 105288.
23. Kairaliyeva, T.; Mucic, N.; Spasojevic, Lj.; Bucko, S.; Katona, J.; Aksenenko, E.V.; Aidarova, S.B.; Fainerman, V.B.; Tarasevich, Yu.I.; Miller, R. Surface tension measurements using drop profile analysis tensiometry with consideration of the surfactant mass balance in a single drop. *Colloids Interfaces* **2017**, *1*, 1.
24. Kairaliyeva, T.; Aksenenko, E.V.; Mucic, N.; Makievski, A.V.; Fainerman, V.B.; Miller, R. Surface tension and adsorption studies by drop profile analysis tensiometry. *J. Surfactants Detergents* **2017**, *20*, 1225–1241.
25. Fainerman, V.B.; Sharipova, A.; Aksenenko, E.V.; Makievski, A.V.; Nikolenko, M.V.; Javadi, A.; Aidarova, S.B.; Miller, R. Cooperative effects in surfactant adsorption layers at water/alkane interfaces. *Colloids Interfaces* **2019**, *3*, 67.

26. Fainerman, V.B.; Aksenenko, E.V.; Makievski, A.V.; Nikolenko, M.V.; Javadi, A.; Schneck, E.R.; Miller, R. Particular behavior of surface tension at the interface between aqueous solution of surfactant and alkane. *Langmuir* **2019**, *35*, 15214–15220.
27. Miller, R.; Aksenenko, E.V.; Kovalchuk, V.I.; Fainerman, V.B. Adsorption of $C_{14}EO_8$ at the interface between their aqueous solution drop and air saturated by different alkanes vapour. *Phys. Chem. Chem. Phys.* **2017**, *19*, 2193–2200.
28. Qiao, X.Y.; Miller, R.; Sun, K. Interfacial adsorption, viscoelasticity and recovery of silk fibroin layers at different oil/water interface. *Colloids Surf. A* **2017**, *519*, 179–186.
29. Cabrerizo-Vílchez, M.A.; Wege, H.A.; Holgado-Terriza, J.A.; Neumann, A.W. Axisymmetric drop shape analysis as penetration Langmuir balance. *Rev. Sci. Instrum.* **1999**, *70*, 2438–2444.
30. Ferri, J.K., Kotsmar, Cs., Miller, R. From surfactant adsorption kinetics to asymmetric nanomembrane mechanics: Pendant drop experiments with subphase exchange. *Adv. Colloid Interface Sci.* **2010**, *161*, 29–47.
31. Javadi, A.; Dowlati, S.; Shourni, S.; Miller, R.; Kraume, M.; Kopka, K.; Eckert, K. Experimental techniques to study protein–surfactant interactions: New insights into competitive adsorptions via drop subphase and interface exchange. *Adv. Colloid Interface Sci.* **2022**, *301*, 102601.
32. Won, J.Y.; Krägel, J.; Makievski, A.V.; Javadi, A.; Gochev, G.; Loglio, G.; Pandolfini, P.; Leser, M.E.; Gehin-Delval, C.; Miller, R. Drop and Bubble Micro Manipulator (DBMM) – a unique tool for mimicking processes in foams and emulsions. *Colloids Surf. A* **2014**, *441*, 807–814.
33. Malysa, K.; Krasowska, M.; Krzan, M. Influence of surface active substances on bubble motion and collision with various interfaces. *Adv. Colloid Interface Sci.* **2005**, *114–115*, 205–225.
34. Wegener, M.; Kraume, M.; Paschedag, A.R. Terminal and transient drop rise velocity of single toluene droplets in water. *AIChE J.* **2010**, *56*, 2–10.
35. Garner, F.H.; Skelland, A.H.P. Some factors affecting droplet behaviour in liquid-liquid systems. *Chem. Eng. Sci.* **1955**, *4*, 149–158.
36. Griffith, R.M. The effect of surfactants on the terminal velocity of drops and bubbles. *Chem. Eng. Sci.* **1962**, *17*, 1057–1070.
37. Dukhin, S.S. The dynamic adsorption layer and the analogs of the MARANGONI-GIBBS effect. In *The Modern Theory of Capillarity: To the Centennial of Gibbs' Theory of Capillarity*, Goodrich F.C., Rusanov, AI., Eds.; De Gruyter: Berlin, Germany, 1981; pp. 83–106.
38. Wegener, M.; Fevre, M.; Paschedag, A.R.; Kraume, M. Impact of Marangoni instabilities on the fluid dynamic behaviour of organic droplets. *Int. J. Heat Mass Transfer* **2009**, *52*, 2543–2551.
39. Piedfert, A.; Lalanne, B.; Masbernat, O.; Risso, H. Numerical simulations of a rising drop with shape oscillations in the presence of surfactants. *Phys. Rev. Fluids* **2018**, *3*, 103605.
40. Pesci, C.; Weiner, A.; Marschall, H.; Bothe, D. Computational analysis of single rising bubbles influenced by soluble surfactant. *J. Fluid Mech.* **2018**, *856*, 709–763.
41. Gawel, D.; Zawala, J. Automatic single droplet generator with control over droplet size and detachment frequency. *Colloids Interfaces* **2019**, *3*, 57.
42. Gawel, D.; Zawala, J. Stability of liquid films formed by a single bubble and droplet at liquid/gas and liquid/liquid interfaces in bovine serum albumin solutions. *ACS Omega* **2021**, *6*, 18289–18299.
43. Schiller, L.; Naumann, A. A drag coefficient correlation. *VDI Zeitung* **1933**, *77*, 318–320.

44. Pawliszak, P.; Ulaganathan, V.; Bradshaw-Hajek, B.H.; Manica, R.; Beattie, D.A.; Krasowska, M. Mobile or immobile? Rise velocity of air bubbles in high-purity water. *J. Phys. Chem. C* **2019**, *123*, 15131–15138.
45. Graham, D.E.; Chatergoon, L.; Phillips, M.C. A technique for measuring interfacial concentrations of surfactants at the oil-water interface. *J. Phys. E: Sci. Instrum.* **1975**, *8*, 696–699.
46. Sengupta, T.; Damodaran, S. A new methodology for studying protein adsorption at oil-water interfaces. *J. Colloid Interface Sci.* **1998**, *206*, 407–415.
47. Daillant, J.; Gibaud, A. *X-ray and Neutron Reflectivity: Principles and Applications*, Springer: Berlin, Germany, 1999.
48. Maestro, A.; Gutfreund, P. In situ determination of the structure and composition of Langmuir monolayers at the air/water interface by neutron and X-ray reflectivity and ellipsometry. *Adv. Colloid Interface Sci.* **2021**, *293*, 102434.
49. Schneck, E.; Reed, J.; Seki, T.; Nagata, Y.; Kanduc, M. Experimental and simulation-based characterization of surfactant adsorption layers at fluid interfaces. *Adv. Colloid Interface Sci.* **2024**, *in press*.
50. Day, J.R.P.; Bain, C.D. Ellipsometric study of depletion at oil-water interfaces. *Phys. Rev. E* **2007**, *76*, 041601.
51. Stocco, A.; Tauer, K. High-resolution ellipsometric studies on fluid interfaces. *Eur. Phys. J. E* **2009**, *30*, 431.
52. Lee, L.T.; Langevin, D.; Farnoux, B. Neutron reflectivity of an oil-water interface. *Phys. Rev. Lett.* **1991**, *67*, 2678–2681.
53. Cosgrove, T.; Phipps, J.S.; Richardson, R.M. Neutron reflection from a liquid/liquid interface. *Colloids Surf.* **1992**, *62*, 199–206.
54. Dickinson, E.; Horne, D.S.; Phipps, J.S.; Richardson, R.M. A neutron reflectivity study of the adsorption of β-casein at fluid interfaces. *Langmuir* **1993**, *9*, 242–248.
55. Motschmann, H.; Teppner, R. Ellipsometry in interface science. In *Novel Methods to Study Interfacial Layers*, Möbius, D., Miller R., Eds.; Elsevier: The Netherlands, Amsterdam, 2001; pp. 1–53.
56. Russev, S.C.; Arguirov, T.V.l.; Gurkov, T.D. β-Casein adsorption kinetics on air–water and oil–water interfaces studied by ellipsometry. *Colloids Surf. B* **2000**, *19*, 89–100.
57. Benjamins, J.-W.; Jönsson, B.; Thuresson, K.; Nylander, T. New experimental setup to use ellipsometry to study liquid-liquid and liquid-solid interfaces. *Langmuir* **2002**, *18*, 6437–6444.
58. Nylander, T.; Hamraoui, A.; Paulsson, M. Interfacial properties of whey proteins at air/water and oil/water interfaces studied by dynamic drop tensiometry, ellipsometry and spreading kinetics. *Int. J. Food Sci. Technol.* **1999**, *34*, 573–585.
59. Day, J.P.R.; Pudney, P.D.A.; Bain, C.D. Ellipsometric study of the displacement of milk proteins from the oil–water interface by the non-ionic surfactant $C_{10}E_8$. *Phys. Chem. Chem. Phys.* **2010**, *12*, 4590–4599.
60. Narayanan, T.; Konovalov, O. Synchrotron scattering methods for nanomaterials and soft matter research. *Materials* **2020**, *13*, 752.
61. Heiden-Hecht, T.; Wu, B.; Schwärzer, K.; Förster, S.; Kohlbrecher, J.; Holderer, O.; Frielinghaus, H. New insights into protein stabilized emulsions captured via neutron and X-ray scattering: an approach with β-lactoglobulin at triacylglyceride-oil/water interfaces. *J. Colloid Interface Sci.* **2024**, *655*, 319–326.
62. Schlossman, M.L. X-ray scattering from liquid–liquid interfaces. *Physica B: Condens. Matter* **2005**, *357*, 98–105.
63. Hollowell, P.; Li, Z.; Hu, X.; Ruane, S.; Kalonia, C.; van der Walle, C.F.; Lu, J.R. Recent advances in studying interfacial adsorption of bioengineered monoclonal antibodies. *Molecules* **2020**, *25*, 2047.

64. Scoppola, E.; Micciulla, S.; Kuhrts, L.; Maestro, A.; Campbell, R.A.; Konovalov, O.V.; Fragneto, G.; Schneck, E. Reflectometry reveals accumulation of surfactant impurities at bare oil/water interfaces. *Molecules* **2019**, *24*, 411.
65. Bergfreund, J.; Diener, M.; Geue, T.; Nussbaum, N.; Kummer, N.; Bertsch, P.; Nyström, G.; Fischer, P. Globular protein assembly and network formation at fluid interfaces: Effect of oil. *Soft Matter* **2021**, *17*, 1692–1700.
66. Zarbakhsh, A.; Bowers, J.; Webster, J.R.P. Width of the hexadecane-water interface: A discrepancy resolved. *Langmuir* **2005**, *21*, 11596–11598.
67. Scoppola. E.; Watkins, E.B.; Campbell, R.A.; Konovalov, O.; Girard, L.; Dufrêche, J.F.; Ferru, G.; Fragneto, G.; Diat, O. Solvent extraction: Structure of the liquid-liquid interface containing a diamide ligand. *Angew. Chem. Int. Ed. Engl.* **2016**, *55*, 9326–9330.
68. Zarbakhsh, A., Bowers, J., Webster, J.R.P. A new approach for measuring neutron reflection from a liquid/liquid interface. *Meas. Sci. Technol.* **1999**, *10*, 738–743.
69. Protat, M.; Bodin-Thomazo, N.; Malloggi, F.; Daillant, J.; Campbell, R.A.; Fragneto, G.; Watkins, E.B.; Perrin, P.; Pantoustier, N.; Guenoun, P. Neutron reflectivity measurements at the oil/water interface for the study of stimuli-responsive emulsions. *Eur. Phys. J. E Soft Matter* 2018, *41*, 85.
70. Bergfreund, J.; Bertsch, P.; Fischer, P. Effect of the hydrophobic phase on interfacial phenomena of surfactants, proteins, and particles at fluid interfaces. *Curr. Opin. Colloid Interface Sci.* **2021**, *56*, 101509.
71. Cristofolini, L.; Orsi, D.; Isa, L. Characterization of the dynamics of interfaces and of interface-dominated systems via spectroscopy and microscopy techniques. *Curr. Opin. Colloid Interface Sci.* **2018**, *37*, 13–32.
72. Dinache, A.; Pascu, M.-L.; Smarandache, A. Spectral properties of foams and emulsions. *Molecules* **2021**, *26*, 7704.
73. Husband, F.A.; Garrood, M.J.; Mackie, A.R.; Burnett, G.R.; Wilde, P.J. Adsorbed protein secondary and tertiary structures by circular dichroism and infrared spectroscopy with refractive index matched emulsions. *J. Agric. Food Chem.* **2001**, *49*, 859–866
74. Shen Y.R. Surface properties probed by second-harmonic and sum-frequency generation. *Nature* **1989**, *337*, 519–525.
75. de Aguiar, H.B.; Strader, M.L.; de Beer, A.G.; Roke, S. Surface structure of sodium dodecyl sulfate surfactant and oil at the oil-in-water droplet liquid/liquid interface: A manifestation of a nonequilibrium surface state. *J. Phys. Chem. B* **2011**, *115*, 2970–2978.
76. Roke, S.; Gonella, G. Nonlinear light scattering and spectroscopy of particles and droplets in liquids. *Annu. Rev. Phys. Chem.* **2012**, *63*, 353–378.
77. Wang, J.; Arrachart, G.; Giusti, F.; Martin-Gassin, G.; Gassin, P-M.; Jonchere, A.; Diat, O.; Girard, L. Synthesis and characterization of a chromo-extractant to the probe liquid–liquid interface in a solvent extraction process. *J. Phys. Chem. C* **2020**, *124*, 10916–10923.
78. Tran, E.; Richmond, G.L. Interfacial steric and molecular bonding effects contributing to the stability of neutrally charged nanoemulsions. *Langmuir* **2021**, *37*, 12643–12653.
79. Glikman, D.; Wyszynski, L.; Lindfeld, V.; Hochstädt, S.; Hansen, M.R.; Neugebauer, J.; Schönhoff, M.; Braunschweig, B. Charge regulation at the nanoscale as evidenced from light-responsive nanoemulsions. *J. Am. Chem. Soc.* **2024**, *146*, 8362–8371.
80. García Rey, N.; Weißenborn, E.; Schulze-Zachau, F.; Gochev, G.; Braunschweig, B. Quantifying double-layer potentials at liquid–gas interfaces from vibrational sum-frequency generation. *J. Phys. Chem. C* **2019**, *123*, 1279–1286.
81. Altman, R.M.; Christoffersen, E.L.; Jones, K.K.; Krause, V.M.; Richmond, G.L. Playing favorites: Preferential adsorption of nonionic over anionic surfactants at the liquid/liquid interface. *Langmuir* **2021**, *37*, 12213–12222.

82. Knock, M.M.; Bell, G.R.; Hill, E.K.; Turner, H.J.; Bain C.D. Sum-frequency spectroscopy of surfactant monolayers at the oil–water interface. *J. Phys. Chem. B* **2003**, *107*, 10801–10814.
83. Hardt, M.; Busse, F.; Raschke, S.; Honnigfort, C.; Carrascosa-Tejedor, J.; Wenk, P.; Gutfreund, P.; Campbell, R.A.; Heuer, A.; Braunschweig, B. Photo-responsive control of adsorption and structure formation at the air-water interface with arylazopyrazoles. *Langmuir* **2023**, *39*, 5861–5871.
84. Hosseinpour, S.; Götz, V.; Peukert, W. Effect of surfactants on the molecular structure of the buried oil/water interface. *Angew. Chem. Int. Ed. Engl.* **2021**, *60*, 25143–25150.
85. Scatena, L.F.; Brown, M.G.; Richmond, G.L. Water at hydrophobic surfaces: weak hydrogen bonding and strong orientation effects. *Sci.* **2001**, *292*, 908–912.

4 Particle-Laden Interfaces and Their Role in the Stabilization of Emulsions

Carlo Carbone, Armando Maestro, Francisco Ortega, Ramón G. Rubio and Eduardo Guzmán

4.1 INTRODUCTION

The early 20th century marked a significant milestone in particle-laden fluid interface research with the seminal contributions of Ramsden and Pickering. Their pioneering work on the use of particle adsorption to control the stability of emulsions inaugurated an exploration of a research area that has since flourished remarkably. Over the past decades, this field has experienced rapid advances that have shaped our current understanding of particle-laden fluid interfaces.[1–3] Indeed, the extensive research activity in this type of systems has led to a profound appreciation of particle-laden fluid interfaces as compelling models for investigating various aspects of quasi-2D system physics. In particular, these interfaces serve as valuable tools for studying phase diagrams and the potentials governing interparticle interactions.[4] At the same time, particle-laden fluid interfaces have emerged as versatile platforms that facilitate the creation of a wide range of soft materials governed primarily by interfacial dynamics. Examples include colloidosomes, capsules, photonic crystals, integrated circuits, and a host of other innovations.[5] In addition, the reliance on particle-stabilized interfaces is paramount in the quest for environmentally friendly emulsions and foams, a necessity in today's industrial landscape. By bypassing the need for traditional surfactants, this approach addresses environmental concerns while providing significant benefits.[6–8]

The diverse and rich perspectives inherent in the study of particle-laden interfaces stem, in part, from the wide range of systems that can be finely engineered by assembling particles at fluid interfaces. This phenomenon has been fueled by impressive advances in synthetic pathways, bringing in an era of precision where particles of different shapes, sizes (nanometers to micrometers), and surface chemistries can be manufactured. This groundbreaking advance has opened the door to tailoring the physicochemical and structural properties of particles with remarkable precision, thereby dictating their behavior both in bulk and when trapped at fluid interfaces.[9, 10] Consequently, this development significantly expands the range of phenomena that can be explored when particles are adsorbed at fluid interfaces, providing the

 DOI: 10.1201/9781032636184-5

opportunity to engineer particle-laden fluid interfaces with distinct equilibrium and dynamic properties that are finely tuned across a spectrum of interaction potentials.[4, 11] In essence, the domain of particle-laden fluid interfaces is proving to be fertile ground for the creation of all-liquid materials in which particles are confined within specific geometries to generate novel functionalities and behaviors.[12, 13] This is possible because the accumulation and confinement of particles at fluid interfaces lead to novel behaviors and properties. Examples include intriguing 2D phase transitions and unconventional rheological responses. These phenomena cannot be easily explained using traditional physicochemical concepts used to describe their 3D counterparts. Instead, their elucidation requires the incorporation of molecular-scale thermodynamic aspects along with mechanical considerations that manifest at the microscale or even greater distances. These influences are strongly dependent on the particle properties and the nature of the fluids forming the confining interface.[9, 14]

This chapter briefly outlines the current understanding of the fundamental physicochemical aspects that govern the assembly of particles at fluid interfaces, in particular at liquid/liquid interfaces, and their potential impact on the stabilization of so-called Pickering emulsions. This requires the consideration of various aspects related to the adsorption of particles at the interfaces, as well as the response of such interfaces to mechanical perturbations. Understanding these aspects is crucial for modulating the stability of emulsions as a result of the assembly of particles at a liquid/liquid interface. However, it should be emphasized that this chapter does not only provide a critical view to delve deeper into the understanding of the phenomena exhibited by particles trapped at fluid interfaces but is also crucial for expanding the practical use of such systems beyond academia to produce stable emulsions.

4.2 FACTORS INVOLVED IN THE CONTROL OF EMULSION STABILITY FOR PARTICLES

Nanoparticles are a type of additive used to occupy the liquid/liquid fluid interfaces formed during homogenization. By adsorbing onto the oil/water interface, nanoparticles reduce the energy required for interface formation, effectively decreasing the area of the bare interface. The free-energy change associated with the transfer of an isolated spherical particle from the bulk water or oil phase to a planar interface is maximal for a particle that is moderately wetted by both fluid phases. Such particles are effectively irreversibly adsorbed at the liquid–fluid interface and, when coverage of the drops is sufficient, hinder drop coalescence through steric hindrance, thus stabilizing the dispersed system. In the case of emulsions, the type of drops formed (oil or water) is primarily determined by particle wettability. Similar to the behavior of surfactant layers at liquid/liquid interfaces, it is thought that the particle layer bends such that the majority of particle surfaces reside on the external surface of the drop in the liquid that wets the particles more favorably. It is possible to reverse the emulsion type by, for example, gradually increasing the volume fraction of the dispersed phase, leading to a phenomenon known as catastrophic phase inversion. It should be noted that there are numerous factors influencing the degree of droplet coalescence during emulsion. This includes particle concentration, emulsification time, and dispersed phase volume fraction. Key properties of particles-stabilized emulsions related to particle involvement can be summarized as follows[15]:

1) Particle concentration directly affects disperse system stability; higher concentrations lead to greater stability. This implies that a sufficient particle layer density around droplets is necessary for effective stabilization. Additionally, as the average droplet size decreases with increased solid concentration, the total droplet surface area increases, allowing for additional particles to be accommodated at the interface. The relationship between solid concentration and disperse system stability is typically nonlinear; there is an upper limit beyond which system stability does not further increase.
2) The particle size is a crucial parameter for colloidal particle effectiveness in stabilizing emulsions, as it determines their ability to reside at the interface. Stability increases with decreasing particle size until a critical size is reached. Below this threshold, Brownian effects significantly influence particle partitioning at the oil/water interface. Consequently, system stability decreases as particle sizes fall below this critical value. Moreover, particle size influences the initial deformation of droplet menisci, impacting system instability via the coalescence process. The meniscus deformation when large particles are considered results from gravitational effects, whereas particles smaller than 5 μm exhibit behavior unrelated to gravity.
3) Interactions between particles are vital for system stabilization. Effective stabilization requires particles to be in a state of incipient flocculation. This observation emphasizes the importance of interparticle interactions in the stabilization process, impacting the properties of interfacial films on droplets and the rheological characteristics of colloid-laden oil/water interfaces. These interactions also contribute to the non-Newtonian behavior seen in concentrated suspensions, playing a role in the mechanical stability of the particle layer at the oil–water interface.
4) The presence of inorganic ions in the aqueous phase significantly affects particle ability to stabilize dispersed systems. At high electrolyte concentrations, particles tend to coagulate, leading to large aggregates that cannot remain at the oil/water interface. These aggregates have a propensity to settle out of the system.
5) Rheological characteristics of the interfacial region, critical for governing the drainage rate of thin liquid films trapped between coalescing droplets, play a role in emulsion stability.

Furthermore, the presence of particles at the fluid interface can prevent droplet coalescence. Interparticle interactions contribute to system stability and hinder the displacement of particles from the interface, which promotes coalescence. The reduction in particle interaction magnitude can induce droplet coalescence, achieved by either complete particle flocculation or deflocculation. Complete deflocculation eliminates the elastic component of surface viscoelasticity, leading to faster film thinning and reduced energy required to displace particles from the droplet interface contact region (steric hindrance). Complete flocculation, on the other hand, results in larger particle aggregates, reducing their interface residence ability. Lowering interfacial tension can also promote instability.

The effectiveness of colloidal particles in stabilizing emulsions requires the formation of a sufficiently dense particle layer at the interface. As particle concentration increases and complete surface coverage is attained, the rheological properties of interfacial layers change. At high particle concentrations, interfaces exhibit viscoelastic behavior, with viscous properties dominating at low concentrations and elastic properties becoming more prominent as particle concentrations rise. The elastic contribution to viscoelastic behavior primarily stems from interparticle interactions, which become significant at high concentrations and influence film-thinning rates between coalescing droplets.

A specific low particle concentration interaction known as 'particle bridging' is a stability mechanism. In cases where droplets have incomplete coatings, particles adsorbed onto one droplet can directly interact with the interface of neighboring droplets, leading to stable droplet aggregates. Under certain conditions, closely linked particle-laden interfaces can undergo long-range particle bridging, referred to as 'particle zips'. When two networked particle-laden interfaces come into contact, they merge to form a tightly networked monolayer, with this monolayer continuously extending outward from the initial contact point, analogous to crystal growth.[16, 17]

4.3 A BRIEF INTRODUCTION TO THE ASSEMBLY OF PARTICLES AT FLUID INTERFACES

The formation of any particle-laden fluid interface requires consideration of the role of two essential processes: (i) the transport of colloidal particles to the interface, and (ii) the perturbation of the interface by particle insertion. Once particles are in contact with the interface, their incorporation into the interface results primarily from the energetic cost associated with the reduction of the contact area between the two fluid phases. This reduction in contact area serves to minimize unfavorable fluid–fluid interactions and consequently reduce the overall free energy of the system.[6] This resulting decrease in energy plays a pivotal role in ensuring interfacial stabilization.

4.3.1 Interaction of Particles with the Interface

Particle–interface interactions become of paramount importance once particles approach the vicinity of the fluid interface. Given that many types of particles carry surface charges that yield electrical double-layer forces, and both the water/vapor and water/non-polar fluid interfaces hold negative effective charges,[18] adsorption of negatively charged particles at fluid interfaces generally occurs slowly or not at all.[19] In contrast, the adsorption of positively charged particles is generally considerably favored, albeit with a potentially slow diffusion process. Furthermore, as particles draw near to an interface comprised of fluid with differing dielectric constants, repulsive image charge repulsions, and other confinement effects come into play, distorting the electrical double layer of the particles near the interface.[19–21] In fact, when charged colloidal particles approach an interface between fluids with markedly distinct dielectric constant values, a novel form of electrostatic interaction emerges. This can be rationalized by envisioning a charged particle positioned close to the interface (at a distance d) that experiences a force akin to what it would experience when interacting with a similar particle placed in the opposite phase, at an equivalent distance from the interface. In essence, the particle and the image particle are separated by a distance of $2d$.[22] When charged particles are suspended in water near an interface with air or oil, the image charge interactions invariably result in repulsion, irrespective of the sign of the charge. This allows modulating adsorption by manipulating the ionic strength of the dispersion, as increasing the electrolyte concentration screens the electrostatic repulsion between particles and the interface. Conversely, interactions turn attractive when the sign of the image charges is the opposite of that of the particle.[23]

Furthermore, short-range van der Waals interactions between particles and interfaces are the result of the combination of three distinct contributions: (i) interactions involving permanent dipoles (Keesom interactions); (ii) interactions between permanent and induced dipoles (Debye interactions), and (iii) interactions encompassing fluctuating dipoles (London interactions). The resulting interaction can be quantified through the Hamaker constants of the three materials, i.e., those constituting the particle and the two fluids. For a smooth, chemically homogeneous spherical particle with a radius R entirely immersed in one fluid (fluid 1), positioned close to a fluid interface, the Hamaker constant for the interaction between solid particles and the second fluid (fluid 2), mediated through a thin layer of fluid 1 (A_{P12}), can be expressed according to the following expression,[23]

$$A_{P12} \approx \left(\sqrt{A_{PP}} - \sqrt{A_{11}}\right)\left(\sqrt{A_{22}} - \sqrt{A_{11}}\right), \tag{4.1}$$

where A_{PP}, A_{11}, and A_{22} represent the Hamaker constants for the particles, fluid 1, and fluid 2, respectively. It is worth highlighting that A_{P12} assumes a positive value, indicating that van der Waals interactions between particles and interfaces are inherently attractive.[24] The different contributions involved in the Hamaker constant A_{P12} can be associated with the dielectric permittivity ε and refractive indices of both fluids and the particles, as delineated by Lifshitz's theory [49]. When dealing with dielectric media having identical absorption frequencies (ν_e), an approximate expression can be derived as follows:

$$A_{P12} \approx \frac{3}{4} k_B T \left(\frac{\varepsilon_P - \varepsilon_1}{\varepsilon_P + \varepsilon_1}\right)\left(\frac{\varepsilon_2 - \varepsilon_1}{\varepsilon_2 + \varepsilon_1}\right) + \frac{3h\nu_e}{8\sqrt{2}} \frac{\left(n_P^2 - n_1^2\right)\left(n_2^2 - n_1^2\right)}{\sqrt{n_P^2 + n_1^2}\sqrt{n_2^2 + n_1^2}\left(\sqrt{n_P^2 + n_1^2} + \sqrt{n_2^2 + n_1^2}\right)}, \tag{4.2}$$

where h denotes Planck's constant and ε_i and n_i represent the static (zero-frequency) dielectric constants and refractive indexes within the visible range, while k_B and T stand for the Boltzmann constant and absolute temperature, respectively. The subscripts P, 1, and 2 correspond to the particle, and fluids 1 and 2, respectively. In Equation (4.3), the initial term encompasses the Keesom and Debye contributions, while the subsequent term encompasses the influence of London interactions, frequently the most important one.[24, 25]

4.3.2 Kinetics of Particle Adsorption to the Fluid Interface

There are various transport mechanisms facilitating the migration of colloidal particles from a bulk phase toward the fluid interface. These can be categorized as either (i) spontaneous, which includes Brownian diffusion, sedimentation, or flotation, or (ii) externally triggered mechanisms involving field-induced or hydrodynamically guided processes.[26–30] If diffusion serves as the predominant mechanism, gravitational effects often hold significance due to particle density. Within the dilute regime, where particle–particle interactions can be disregarded, the terminal velocity of particle settling arises from the equilibrium between gravitational and viscous forces. According to Stokes' law,[31] the terminal velocity of descending particles is expressed as.

$$v_{St} = \frac{2R(\rho_p - \rho_1)g}{9\eta}, \tag{4.3}$$

where ρ_0 denotes the density of the particles, ρ_1 represents the density of the fluid phase, g signifies the gravitational acceleration, and η symbolizes the dynamic viscosity. The interplay between gravity and diffusion is assessed using the Péclet number, a dimensionless metric quantifying the ratio of convective to diffusive transport, which can be expressed as follows:

$$Pe = \frac{Rv_{St}}{D}, \tag{4.4}$$

where D represents the bulk diffusion coefficient of the particles. For instance, in the case of 1 µm silica particles suspended in water, the Péclet number typically assumes values around 0.1. However, particles of 10 µm exhibit a Péclet number nearly four orders of magnitude higher. Hence, it becomes evident that increasing particle size tends to promote ballistic particle movement over diffusive transport.[32] When gravitational forces can be disregarded ($Pe << 1$) and there are no adsorption barriers or external flow fields influencing diffusion, the transportation of particles toward the interface can be described assuming Fick's law. This approach resembles the behavior of molecular surfactant adsorption as proposed by Ward-Tordai.[21] In this scenario, the rate of mass transport is expressed as

$$\frac{\partial c(x,t)}{\partial t} = D\frac{\partial^2 c(x,t)}{\partial x^2}, \tag{4.5}$$

where $c(t, x)$ accounts for the bulk concentration, x represents the distance to the fluid interface, and t denotes the diffusion time. If it is assumed that an initially uniform particle bulk concentration is denoted as $c(x, 0)=c_{\infty}$, then the boundary condition governing adsorption kinetics can be expressed as follows:

$$\frac{\partial \Gamma}{\partial t} = D\left[\frac{\partial c(x,t)}{\partial x}\right]_{x=0}, \tag{4.6}$$

where $\Gamma(t)$ is the time-dependent interfacial excess of particles adsorbed at the interface, with an initial value of $\Gamma(0)=0$. By applying the Ward–Tordai framework assuming an irreversible adsorption and in the absence of any adsorption barriers, and considering the complete depletion of particles within the sublayer while accounting for a minor alteration in the bulk concentration (i.e., $c(\infty, t)\to c_{\infty}$), the combination of Equations (4.5) and (4.6) yields the following expression[21]:

$$\Gamma(t) = 2c_{\infty}\sqrt{\frac{Dt}{\pi}}. \tag{4.7}$$

Nonetheless, it is quite common the existence of an electrostatic barrier between particles in the bulk phase and the fluid interface, necessitating an expansion of the aforementioned framework. The influence of this electrostatic barrier on the temporal development of surface concentration can be estimated by incorporating an effective diffusivity,[33]

$$\Gamma(t) = -2c_{\infty}\Delta E_{s}\sqrt{\frac{D_{eff}t}{\pi}}, \tag{4.8}$$

with ΔE_s denoting the decrease in interfacial energy linked to the shielding of the fluid–fluid interface, and D_{eff} representing the effective diffusion coefficient, formulated as

$$D_{eff} = D\exp\left(-\frac{\Delta E_{barrier}}{k_B T}\right), \tag{4.9}$$

where $\Delta E_{\text{barrier}}$ represents the height of the energy barrier within the energetic landscape that arises during the adsorption of a particle to a fluid interface.[33]

4.3.3 Driving Forces Controlling the Particle Trapping from the Bulk to the Interface

The accumulation of particles at fluid interfaces is controlled by the decrease in Gibbs energy. This energy shift can be attributed to the interfacial tension, γ_{12}, between the two immiscible fluids, as well as the interactions between each fluid and the solid surface of the particles. This phenomenon can be understood by recognizing that the accumulation of particles at the interface corresponds to a reduction in the contact area between the two fluids due to the disruption of the interfacial continuity. This is the particle penetration at the interface and therefore depends on the particle position in relation to the interface, which can be evaluated in terms of the contact angle (θ). As the contact area decreases, the total free energy of the system also decreases. However, because particle sizes typically exceed the interfacial thickness, the conventional microscopic definition of interfacial tension is not readily applicable to describe its change caused by the formation of particle-laden fluid interfaces.[6] Consequently, the standard thermodynamic description of interfacial tension becomes infeasible for such interfaces, forcing us to consider the interfacial tension as an effective quantity.[6,13] This limitation significantly hinders the development of a physically reliable thermodynamic framework for describing particle-laden interfaces, which is a notable challenge. To date, most of the models developed provide only a phenomenological insight into the thermodynamic behavior of particle-laden interfaces, often falling short of comprehensive coverage.[10]

Considering the simplest situation, characterized by a particle composed of a material that has an identical interfacial tension with both fluid phases, it is expected that its adsorption at the interface primarily result in a reduction of direct contact between the two fluids. This would occur without introducing any additional energy cost due to particles interacting with the fluids. The above scenario requires an explanation of the interfacial tension from a macroscopic point of view, considering its effective character in a particle-laden interface. Thus, the decrease of the effective interfacial tension results from the appearance of a 2D lateral pressure Π. This pressure results from a complex equilibrium involving entropy and interparticle interactions, counteracting the contraction of the interfacial area associated with the interfacial tension. Consequently, this dynamic leads to a reduction of the interfacial tension $\gamma = \gamma_{12} - \Pi$, as the interfacial packing of the particle-laden fluid interface is increased. This interfacial tension change is quantifiable by several common methodological approaches used to evaluate the interfacial tension.[6, 13, 34–37] It should be noted that the lateral pressure contribution emerges from the interplay of the interface coverage (Γ) and the energy associated with the entrapment of a single particle in the fluid interface (ΔE_p), the so-called trapping energy, which can be expressed according to the following expression, $\Pi(\Gamma) = \Gamma|\Delta E_p|$.[38]

In addition to the reduction in interfacial tension associated with the particle trapping to the fluid interface, the change in energy associated with the trapping of particles at the fluid interface plays a key role in determining the final equilibrium state of a particle-laden interface. This energy change is expressed as $\Delta E_p = E_{interface} - E_{bulk}$, where $E_{interface}$ and E_{bulk} are the energies of the particle at the fluid interface and when they are dispersed in the continuous phase, respectively. Therefore, ΔE_p accounts for the energy change that occurs when a particle is transported from a dispersed state to its stable position at the fluid interface.[39, 40] The magnitude of this energy change defines the mechanical equilibrium conditions of the particle-laden fluid interface, which can be framed in the context of Young's law according to the geometric representation shown in Figure 4.1a.[4, 41] In fact, when the sum of the various forces (interfacial tensions) acting within the interfaces equals zero, i.e., the Young–Dupre equation is fulfilled ($0 = \gamma_{P1} - \gamma_{P2} - \gamma_{12} \cos\theta$, where γ_{P1} and γ_{P2}

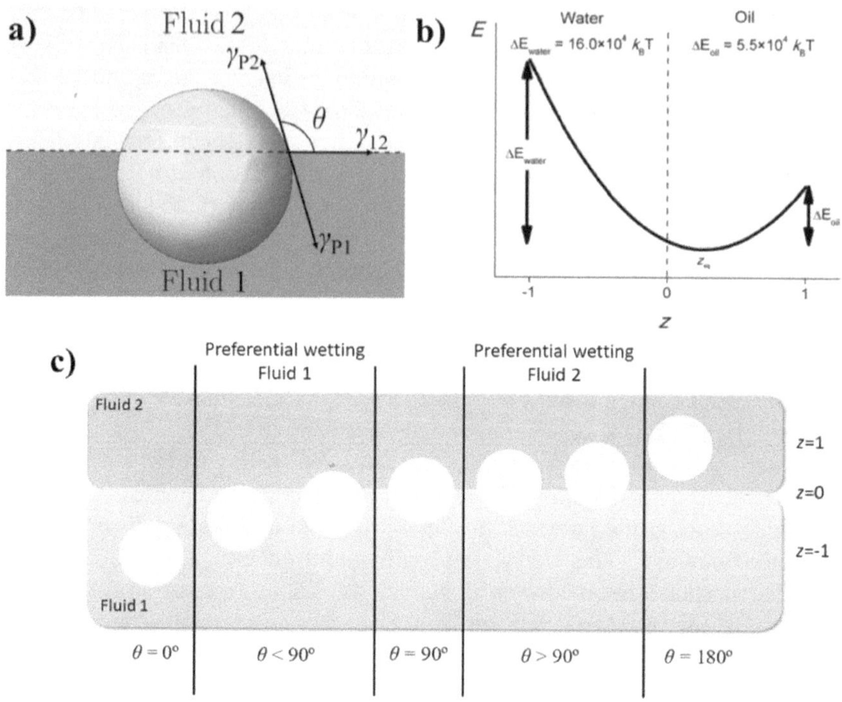

FIGURE 4.1 (a) Sketch of the mechanical equilibrium condition for a particle adsorbed at a fluid interface. Here, θ symbolizes the contact angle of the particles, γ_{P1} and γ_{P2} stand for the interfacial tensions between the particle and the two fluid phases, and γ_{12} corresponds to the interfacial tension of the fluid interface. (b) Energy evolution of a particle concerning its position relative to the interfacial plane (z). The minimum in the potential energy curve at z_{eq} corresponds to the equilibrium position of the particles upon their trapping at the fluid interface. In the panel, the energies required for detaching the particle from the interface to the oil (ΔE_{oil}) or water (ΔE_{water}) phases are indicated. Adapted with permission from Ballard et al.[43] Copyright (2019) Royal Society of Chemistry. (c) Evolution of the particle position in relation to the interfacial plane as a function of its contact angle. Reprinted from Guzmán et al.[3] under a Creative Commons Attribution [CC-BY 4.0] License.

are the solid/fluid interfacial tensions between the particle and fluid 1 and fluid 2, respectively), it becomes possible to formulate a definition of ΔE_p that is relevant to the entrapment of a spherical particle at an interface between two immiscible fluids. This relationship is given by the following expression[42]:

$$\Delta E_p = 2\pi R^2 \gamma_{12} cos\theta_\infty \left(1 - cos\theta\right) + 2\pi R \tau sin\theta - \pi R^2 \gamma_{12} sin^2\theta, \qquad (4.10)$$

where the contact angle, denoted as θ, is the angle formed between the tangent plane of the particle surface and the interface at the precise point where the interface intersects the solid material (see Figure 4.1a).[41] τ stands for the line tension. θ_∞ denotes the macroscopic contact angle, establishing the mechanical equilibrium condition for a particle of infinite size trapped within the fluid interface. This concept is equivalent to the contact angle exhibited by a liquid droplet resting on a completely flat surface possessing similar surface chemistry to the particle. Consequently, it becomes feasible to formulate an extended version of Young's equation as follows:

$$cos\theta = cos\theta_{\infty}\left[1-\frac{\tau}{\gamma_{12}R sin\theta}\right]^{-1}, \tag{4.11}$$

In most cases, unless dealing with extremely small particles or those with exceedingly high surface roughness, the influence of line tension is negligible, leading to $\cos\theta$ being approximately equal to $\cos\theta_{\infty}$. This simplification streamlines the characterization of trapping energy, which can be expressed as,

$$\Delta E_p = -\pi R^2\gamma_{12}\left(1 \pm cos\theta\right)^2, \tag{4.12}$$

The presence of the ± sign accounts for the fact that particles can exhibit quite different affinities for the fluid phases. As a result, the mechanical equilibrium condition is commonly not determined by an equal penetration of particles within both fluid phases. Thus, the equilibrium position of the particle center relative to the interfacial plane, denoted as z_{eq}, deviates from a null value (where $z = 0$ corresponds to the interfacial plane, and $z = -1$ and 1 signify the lower and upper phases, respectively), resulting in two distinct boundary scenarios. The first scenario arises when the particle center is positioned above the interfacial plane, indicating that the particles are primarily immersed in the upper phase (typically the less polar one), leading to the utilization of the + sign. The second scenario arises when the particle center is situated below the interfacial plane, indicating that the particles are predominantly immersed in the lower phase (generally the more polar one), necessitating the use of the – sign.[43] This distinction results in two clearly defined regions for the relative wettability of particles at the fluid interface. A pivotal delineation between these regions is marked by $\theta = 90°$, which indicates an equilibrium state where the particle is equally immersed in both fluid phases ($z_{eq} = 0$).

Figure 4.1b shows an idealized energy landscape, denoted as E, illustrating the total energy involved in the entrapment of a hydrophobic particle within an arbitrary interface that separates polar and non-polar phases (such as a water/oil interface). This energy landscape is presented as a function of the vertical coordinate (z). Notably, the detachment energies of a particle from each fluid phase ($-\Delta E_p$, specifically defined as ΔE_{oil} and ΔE_{water} in Figure 4.1b) inversely correlate with the affinity of the particle for the respective phase. In other words, when $-\Delta E_p$ is lower, the particle exhibits a stronger preference for that phase, causing a non-zero equilibrium position ($z_{eq} \neq 0$) relative to the interfacial plane. Consequently, in the specific case illustrated in Figure 4.1b where $\Delta E_{oil} < \Delta E_{water}$, it should anticipate a greater extent of particle penetration into the oil phase. The equilibrium position, z_{eq}, corresponds to the height at which the free energy is minimized $(\partial E/\partial z) = 0$, dependent on factors such as the interfacial tensions, particle size (expressed as $z_{eq} = \frac{\gamma_{P1}-\gamma_{P2}}{\gamma_{12}}R$), and the particle's hydrophilic–lipophilic balance (HLB) defined in terms of the contact angle θ. For instances where two fluid phases possess substantially divergent dielectric constants, hydrophilic particles tend to position their centers below $z = 0$, characterized by $\theta < 90°$, while hydrophobic particles tend to be primarily immersed in the upper phase with $\theta > 90°$. In Figure 4.1c, the change in the position of particles in relation to the interface is depicted as a function of its contact angle. It is important to emphasize that the position of the particle in relation to the interfacial plane significantly influences interface stabilization, a critical aspect for the stabilization of Pickering emulsions and foams.[44] This position can be altered by additives (such as surfactants, polymers, or small molecules) capable of modifying the chemical nature of the particle surface, either reversibly or irreversibly. Additionally, these additives can impact the interfacial tension, thereby modifying the relative wettability of the particles for the interface, essentially altering their contact angle.[45]

The framework discussed above holds when the attachment of the particles to the fluid interface via capillarity is sufficiently strong to prevent detachment processes as a result of the thermal fluctuations. This implies that the trapping of the particles at the fluid interface can be guaranteed if the trapping energy (ΔE_p) overcomes several times the thermal energy (k_BT). If ΔE_p falls below this threshold ($\Delta E_p < k_BT$), particle trapping does not occur. In such cases, particles are either expelled to one of the fluid phases (with z_{eq} being -1 for the aqueous phase and 1 for the oil phase) or if trapping does occur, it might be linked to an equilibrium between adsorption and desorption, similar to what is observed in molecular surfactants (ΔE_p approximately equal to k_BT or only slightly higher, as in cases where ΔE_p is similar to k_BT).[46] Figure 4.2a displays three schematic scenarios that depict different possible outcomes for particle trapping at a fluid interface, contingent upon the value of ΔE_p.

The effectiveness of particle trapping at the fluid interface is governed by three distinct parameters: θ, R, and γ_{12}. The first parameter, θ, accounting for the relative wettability of the particle for the interface can be altered through changes in the chemical composition of the particles, as well as its physicochemical attributes such as surface roughness, size, and shape. On the other hand, γ_{12} relies solely on the physicochemical characteristics of the fluids defining the interface. Consequently, if a specific scenario in which spherical particles with identical surface roughness and wettability for an interface formed by the same immiscible fluids adsorb to the interface is assumed, the propensity of these particles to remain trapped at the fluid interface will be solely dependent on their size (as indicated by Equation (4.10)).[4]

Figure 4.2b shows for the sake of example how the trapping energy (ΔE_p) varies in relation to the contact angle when spherical particles of different sizes (characterized by parameter R) are trapped at an arbitrary interface with a γ_{12} value of 50 mN/m. For particles ranging from several micrometers to 1 nm in diameter, the trapping energy assumes negative values. However, the plots in Figure 4.2b make it evident that the strength of particle adhesion to the interface is strongly dependent on the size of the considered particle size. This becomes clearer when examining the dependency of trapping energy for particles with a contact angle of 90° on their dimensions, which is displayed in Figure 4.2c. When micron-sized particles adhere

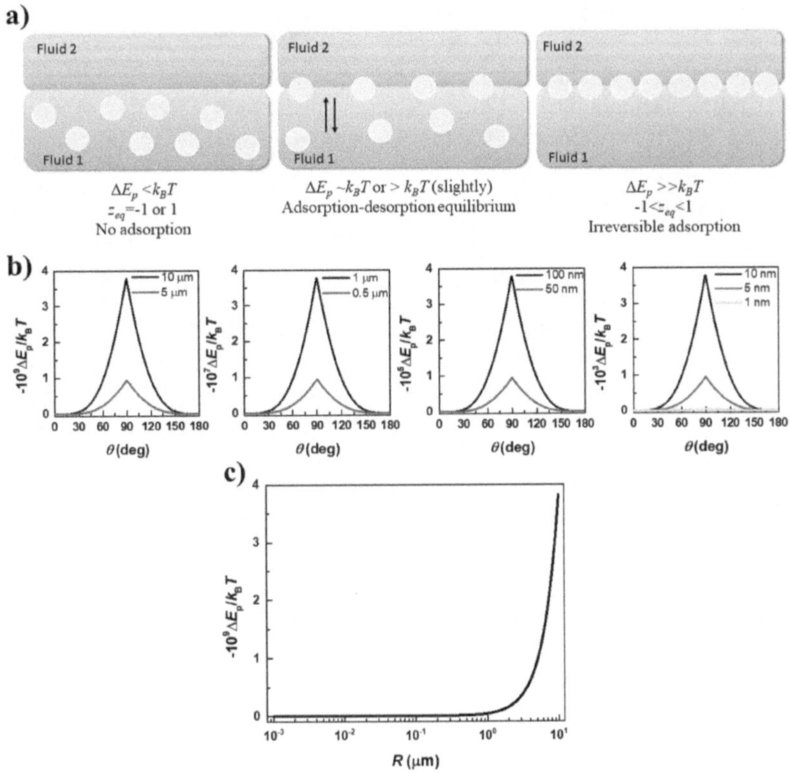

FIGURE 4.2 (a) Different scenarios emerging from particle–fluid interface interactions, depending on the magnitude of ΔE_p. (b) Influence of contact angles on the trapping energy of colloidal particles, possessing different R values. The different curves correspond to the adsorption at an arbitrary fluid interface characterized by $\gamma_{12} = 50$ mN/m under a temperature of 25°C. The scales vary across different panels. (c) Relationship between particle radius and trapping energy for particles adsorbed with a contact angle of 90° for the same conditions considered in panel (b). Reprinted from Guzmán et al.[3] under a Creative Commons Attribution [CC-BY 4.0] License.

to a fluid interface, they incur trapping energies significantly surpassing the thermal energy k_BT. Consequently, the binding of these micron-sized particles is generally irreversible, except for scenarios where γ_{12} values are very low and/or when the $\cos\theta$ approaches 1. It is worth noting that particle trapping at the fluid interface is a dynamic process, characterized by ongoing changes over time.[47] This dynamic nature is evident because particles that are initially trapped with a very small contact angle can experience desorption. This is due to the fact that during their transition to equilibrium positions, they might go through configurations characterized by binding energies comparable to the thermal energy. As a result, a small proportion of micron-sized particles might spontaneously desorb from the interface.[4] However, this pattern is not commonly observed in most experimental systems.[39] Conversely, as particle size approaches the nanometric scale, the trapping energy diminishes to magnitudes approximately on the order of several k_BT. This decrease in trapping energy can lead to thermally driven desorption of particles from the interface, akin to what occurs in situations involving molecular surfactants, polymers, and proteins.[4]

The above framework describing the attachment of particles at fluid interfaces is only valid when particles are small enough to ensure that the action of the interfacial tension can oppose to gravity, avoiding sedimentation.[41] The complex interplay between both types of forces can be defined in terms of the Bond or Eötvos number, $Bo = {}^{2R\Delta\rho g}\!/\!{}_{\gamma_{12}}$, with $\Delta\rho$ representing the difference between the densities of the solid particle and the fluid, and g the gravity acceleration. In general, when the particle size is smaller than 10 μm, the contribution of the interfacial tension forces becomes the dominant contribution, except for those cases in which the particle density is high in comparison with the liquid phases where the role of the gravity may introduce a very important constrain to the trapping of particles at the fluid interface.[41]

The above framework provides a simple description of the attachment of particles to fluid interfaces in terms of three key parameters: R, γ_{12}, and θ. The former two parameters are readily accessible through experimental measurements, while determining the contact angle, θ, can pose challenges due to the diverse range of techniques available for its measurement, which often leads to discrepancies in the obtained values, making the accurate determination of the particle wettability a very important challenge.[4, 40] In addition, there are diverse factors that can alter the energetic landscape associated with particle trapping at fluid interfaces. These factors encompass attributes such as particle shape, size, surface roughness, and the pinning–depinning dynamics of the contact line. Typically, these factors are accounted for within the framework of line tension contribution, which in turn modifies the trapping energy, as represented by Equation (4.10).[39] The curvature of the interface can also introduce an additional level of complexity. Curvature influences three distinct aspects that impact trapping energy: (i) the Laplace pressure, (ii) the intricate interfacial profile, and (iii) the contact line behavior. These combined effects lead to a lateral force generated by curvature, aiming to expel particles from the interface and thereby reducing the trapping energy.[48] Similarly, particle anisotropy plays a role in particle–interface binding. The presence of anisotropy can result in particles adsorbing with a certain tilt angle, altering their penetration into the interface. This, in turn, influences the trapping energy dynamics.[49]

It is important to highlight that whether the attachment of particles to fluid interfaces is reversible or irreversible, along with the specific value of the contact angle, holds major influence over various physicochemical aspects relevant to potential applications involving interfaces laden with particles. These aspects encompass the arrangement of particles at the interface, the type and strength of interactions between particles, the friction coefficient values derived from microrheology experiments, and how particle-laden fluid interfaces react to external mechanical deformations. These factors collectively govern the capability of particles to effectively stabilize emulsions and foams.[50–55]

4.3.3.1 Tuning Particle Contact Angle at a Fluid Interface

Tuning the wettability of particles offers a means to control the arrangement of particles at the interface, thereby influencing interparticle interactions. There are currently numerous physical and chemical approaches to manipulate the wettability of colloidal particles and thereby adjust their arrangement at fluid interfaces. A common strategy to modify the ability of particles to remain trapped at fluid interfaces involves the addition of various chemical additives, such as surfactants, polymers, salts, or low molecular weight compounds (e.g., alcohols), to the particle dispersion. These molecular entities can adhere to the particle surface through diverse interactions including non-electrostatic mechanisms like hydrogen bonds, electrostatic forces, or van der Waals interactions.[8, 29, 36, 45, 56–61] For instance, a complex interplay between hydrophobic and non-covalent electrostatic interactions enables the in situ modification of particle wettability by varying the surfactant concentration in the dispersion containing colloidal particles or in the secondary fluid.[62] The progressive increase of surfactant concentration can lead to the emergence of various surfactant structures adsorbed on the particle surface, which provide different degrees of modification on the particle hydrophobicity. Maximum hydrophobization of particles typically occurs when surfactant concentrations are sufficiently high to neutralize the surface charge of particles when charged colloidal particles are concerned. Beyond the neutralization point, further increases in surfactant concentration tend to an increase in the hydrophilicity of the particles. Reduction in the contact angle is often a result of the formation of a surfactant bilayer on the surface of the particles, driven by hydrophobic interactions among the alkyl tails of the surfactant molecules. This approach was explored by Maestro et al. [53] and Binks et al.,[63] who introduced different alkyltrimethylammonium bromide surfactants (hexadecyltrimethylammonium bromide, CTAB, and dodecyltrimethylammonium bromide, DTAB) into dispersions containing silica nanoparticles to modulate their ability to be trapped at the water/vapor interface. Binks et al.[64] extended this approach using a two-tailed surfactant (didecyldimethylammonium bromide) to tune the wettability of silica particles. It is worth noting that the effect of surfactant addition on particle wettability is subject to the specific chemical composition of the considered particles. Deleurence et al.[65] investigated the modification of the wettability of anionic hydrophobic polystyrene and hydrophilic silica particles with a cationic surfactant. While polystyrene particles displayed significant changes in contact angle, silica particles exhibited minimal modification despite significant changes in particle charge. Another example highlighting the substantial impact of additives on particle interfacial properties is shown in the work by Perrin et al.[66] Their findings illustrated that incorporating diethylene glycol monobutyl ether into silver nanoparticles led to a substantial alteration in the ability of particles to absorb at both fluid/fluid and fluid/solid interfaces.

Alternatively, a well-controlled adjustment of particle wettability can be achieved by irreversibly bonding specific coating ligands to particle surfaces.[67] For instance, the modification of silicon dioxide particles by silanization, resulting in a reduction of surface-free silanol groups from 34% to 20%, led to a twofold increase in the particle contact angle at the water/vapor interface.[68] Consequently, particles with higher hydrophobicity (with 20% free silanol groups) exhibited a contact angle of 113°, while hydrophilic particles (with 34% free silanol groups) displayed a contact angle of 55°. Similarly, grafting poly(glycerol-monomethacrylate) chains onto the surface of polystyrene latex particles makes it possible to modify their contact angle at the fluid interface.[69]

Recent advancements in particle wettability control have spurred substantial research endeavors toward the systematic design of wettability-manipulated particles, particularly in the realm of colloidal particles with asymmetrical wetting attributes, such as Janus particles and patchy colloids.[32, 70, 71]

4.3.3.2 Effect of the Particle Wettability on the Stabilization of Emulsions

Extensive research has demonstrated that partially hydrophobic micro- and nanoparticles play a significant role in the stabilization of air bubbles or droplets, leading to the well-known Ramsden–Pickering emulsions or foams.[1, 2] The stability of foams and emulsions, being inherently thermodynamically metastable, poses a crucial challenge across various applications. In these systems, instabilities primarily arise from film rupture (coalescence) and coarsening due to fluid diffusion between bubbles or droplets stemming from the substantial energy associated with the air/liquid and liquid/liquid interfaces, respectively.[72] Notably, the elevated energy required for particle attachment at fluid interfaces, often reaching hundreds of k_BT, marks a distinct contrast from standard amphiphiles, lipids, or proteins that typically undergo reversible adsorption. Considering the above framework, the wetting characteristics of solid particles stand as a pivotal determinant in discerning the distinct attributes of particle dispersion.[73] For particles trapped at a liquid/liquid interface, two scenarios can arise. First, when the particle displays relatively hydrophilic behavior with an angle $\theta < 90°$, an oil/water dispersion becomes more favorable in the case of equal volumes of these two fluids. Conversely, for particles demonstrating relatively hydrophobic tendencies with $\theta > 90°$, a water-in-oil dispersion becomes the preferred state. A θ value of 90° signifies the point of inversion, where a particle is equally wetted by both phases.[2] Analogous to surfactants, the attachment of particles reduces interfacial energy by curving the interface to incorporate the dispersed phase, typically the phase they preferentially wet. Nonetheless, unlike surfactants, particles do not significantly reduce surface tension, thereby not affecting dispersion stability to the same extent.[53] In fact, it is the network of particles at the interface that assumes the role of bolstering dispersion stability through the formation of a steric barrier against destabilization processes.

A transitional inversion of the curvature in oil-in-water interfaces laden with adsorbed particles can be achieved by tuning θ while maintaining a constant oil/water ratio.[74] Binks et al.[64] investigated a double-phase inversion in an emulsion stabilized by bicatenary cationic surfactant mixed interfacial layers with silica nanoparticles. They observed that as surfactant concentration increased for a fixed particle quantity, stable emulsions underwent an inversion from oil-in-water to water-in-oil, subsequently recovering the oil-in-water configuration at higher surfactant concentrations.[44] This double-phase inversion was strongly correlated to the contact angle of the nanoparticles. At low surfactant concentrations, $\theta < 90°$, giving rise to oil-in-water emulsions. In the intermediate concentration range, θ reaches its maximum value around 105°, signifying the transition of the particles to a hydrophobic state and the consequent formation of water-in-oil emulsions. Lastly, at higher surfactant concentrations, θ decreases to values below 90°, rendering the particles hydrophilic again and leading to the formation of oil-in-water emulsions.[75] It should be noted that the modification of particle wettability can also be exploited for controlling the stability of foams.[76]

It is important to highlight that any parameter influencing the wettability of particles trapped at fluid interfaces can contribute to controlling the stability of emulsions. In fact, according to the study by San-Miguel et al.,[77] the increase of the particle roughness can contribute to enhancing the stability of Pickering emulsions. This may probably be ascribed to a non-uniform wetting of the particles for the interface, which leads to the formation of an irregular contact line, thereby resulting in long-range attraction.[78]

4.4 INTERPARTICLE INTERACTIONS AT THE INTERFACE

Upon the interface becoming covered by various particles, a diverse set of interactions, emerging from different sources, start to exert influence between them.[14, 55, 79] Figure 4.3 illustrates several interactions that can emerge when particles are trapped at fluid interfaces. It is important to note that these schemes account for particles with smooth surfaces and uniform charge distribution, and the obtained representation may diverge from the actual scenario, thereby introducing adjustments to the significance and intensity of the interparticle interactions occurring within the interface.[14]

A comprehensive study of the interparticle interactions between pairs needs the consideration of two primary categories of interactions: (i) direct interactions and (ii) interface-mediated interactions. The former type also manifests in bulk systems, although their presence is slightly altered by the interface. In contrast, the latter category of forces emerges due to confinement at the interface (comprising capillary and hydrodynamic interactions) and is absent in the absence of a fluid interface. The interface-mediated interactions are notably influenced by the specific chemical and morphological attributes of the particles, as well as the nature of the fluid phases present at the interface.[14] Similar to the approach used for analyzing trapping energy, the equilibrium of interparticle interactions driving the arrangement of particles at a fluid interface can be defined by comparing the energy associated with the particles at the interface, denoted as E_{int}, and the corresponding energy of the particle dispersed within the bulk, E_{bulk},

$$\Delta E = E_{int} - E_{bulk}. \tag{4.13}$$

As stated above, a comprehensive examination of the alteration in free energy (ΔE) linked to the creation of an interface laden with particles requires the consideration of several factors. These encompass the trapping energy (ΔE_{p}), the interplay of interparticle forces (encompassing van der Waals, electrostatic, capillary, density fluctuation, and hydrophobic forces) within the interface denoted as μ_{i}, as well as the adverse entropic contribution denoted as μ_{e}. This entropic contribution emerges

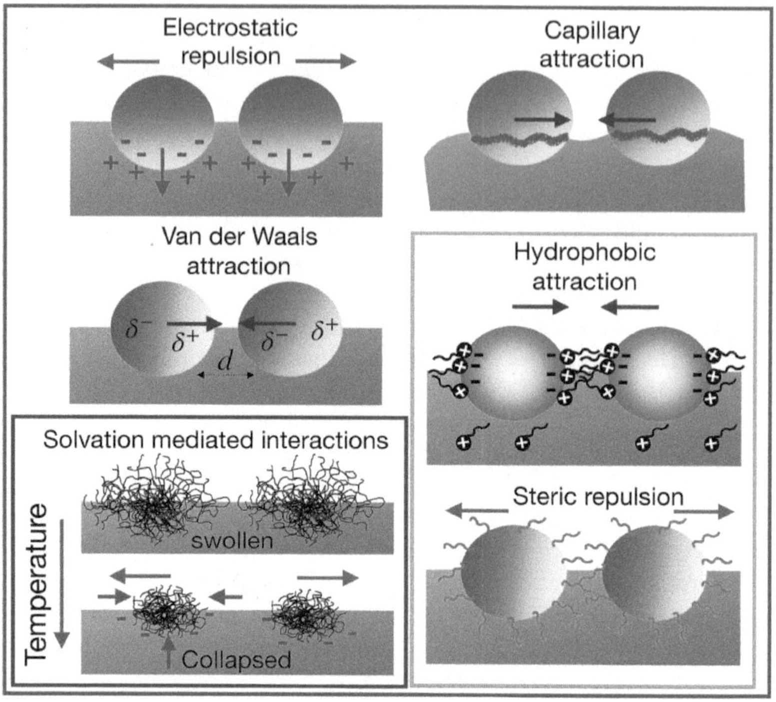

FIGURE 4.3 Illustration depicting the most prevalent interparticle interactions found in interfaces laden with particles. Adapted from Maestro et al.[55], Copyright (2019), with permission from Elsevier.

from the reduction in degrees of translational freedom of particles upon their attachment to the interface. Hence, it becomes feasible to formulate the change in free energy connected to the formation of a particle-laden interface using the following expression:

$$\Delta E = \Delta E_p + \mu_e + \mu_i. \tag{4.14}$$

It is important to highlight that the component linked to trapping energy considers individual particles, whereas the inputs attributed to entropy and interparticle interactions arise from several particles. Consequently, these elements are dependent on the separation between particles, thereby being influenced by the coverage of the interface and the particle concentration within the bulk. This significance cannot be overstated, as the characteristics of particle assemblies at fluid interfaces have the potential to be tuned through modifications in interparticle interaction potentials, particle attributes (such as surface texture and geometry), temperature, or the particle contact angle for the interface.[12]

4.4.1 Direct Interactions at Fluid Interfaces

The first type of direct interactions is the van der Waals forces, which depend on the complex interplay among the various chemical and structural constituents of the particles. These interactions also manifest in particle dispersion, although their effect is less pronounced. Despite the pivotal role of van der Waals forces in shaping the energy landscape of interfaces laden with particles, their precise quantification is complex. Only an effective van der Waals contribution can be approximated by introducing an effective Hamaker constant for the particle across the two fluid phases and the fraction of particle submerged in fluid 2 (denoted as z_{eq}). Consequently, the effective van der Waals contribution between a pair of particles adsorbed at an arbitrary fluid interface is dependent on the specific characteristics of the particle and fluid phases, alongside the relative wettability of the particles for the interface, thereby linking it with the trapping energy. On the other hand, when electrostatic is considered, noticeable differences are found between interfacial systems and bulk counterparts. In fact, the presence of an interface induces a long-range dipole–dipole interaction accompanying the short-range screened Coulomb effect, reminiscent of bulk behavior. The former interaction is heavily influenced by the characteristics of the fluid phases. It is noteworthy that the screened Coulomb component can be neglected when the interfacial coverage is extremely low.

The confinement inherent to the interface imparts an asymmetric character to electrostatic interparticle interactions. Specifically, the dipolar contribution for particles at a water/non-polar fluid interface differs substantially between the water and non-polar phases due to the different electrical permittivity of these phases. In the case of immersed particle fractions, interactions occur between dipoles formed by the dissociated surface groups of particles and the free counterions within the bulk when the polar fluid is considered. Conversely, interactions across the non-polar phase occur between particle surface charges and the image charges that emerge in water.[14] As a result, the position of the particles relative to the interfacial plane (z_{eq}) also significantly influences the strength of dipole–dipole repulsion. Additionally, the strength of the electrostatic contribution is affected by the extent of screening of interactions across the aqueous phase. Notably, when dealing with particles having rough surfaces, ions from the aqueous phase might be transferred to the non-polar one, consequently altering dipole–dipole interactions.[74] Finally, the role of the double-layer contribution to the familiar DLVO (Derjaguin-Landau-Verwey-Overbeek) theory, which holds strictly for two extensive, flat layers surpassing the dimensions of ions in the solution, might not be applicable when considering particles trapped at fluid interfaces.[80]

Additional direct interactions that influence the energy landscape of particle-laden interfaces, albeit with less significance than van der Waals and electrostatic forces, include hydrophobic and steric interactions. The former stems from the necessity to minimize unfavorable interactions between hydrophobic surfaces and water and depends on the molecular intricacies of both particles and fluid phases. However, there is no standardized quantification of the role of hydrophobic interactions in interfaces laden with particles.

Steric interactions emerge due to additives in particle dispersions that alter stability by forming a capping layer on particle surfaces. This capping layer induces osmotic pressure as particles draw closer, leading to an adverse entropic contribution associated with capping layer compression. This, in turn, introduces a repulsive contribution to the overall energy balance of the system.

The existence of direct interactions introduces a contribution to the energy landscape of particle-laden interfaces, which can be envisioned as a combination of a DLVO component, encompassing van der Waals forces and long-range Coulombic interactions, along with a short-range dipolar interaction (E_{dipolar}). These interactions are intimately influenced by the equilibrium location of particles at the fluid interface. Altering these direct interactions with changing interfacial coverage holds profound implications for materials, as it guides the formation of particle assemblies with diverse structures that govern the mechanical performance of the interface, and therefore its stability.[81]

4.4.2 Interactions Mediated by the Presence of a Fluid Interface

The confinement within the interface introduces novel interparticle forces that are not present in bulk systems. Among these, capillary forces are the most significant, originating from interface deformation due to particle adsorption.[82] This phenomenon is markedly influenced by particle properties, particularly size and density, which dictate interface deformation. Larger particles, for instance, induce substantial interface deformation due to gravity effects, leading to the emergence of flotation forces. The parameters governing these forces encompass particle shape, buoyancy, and particle trapping energy at the fluid interface. As particle size diminishes, flotation force strength decreases, becoming negligible for particles smaller than the capillary length of the fluid interface. Even though flotation forces are reduced with the increase of the particle size, the capillary contribution remains important within the energy balance. While particle weight might not be enough to deform the interface, local deformations of the particle contact line can arise, resulting in long-range capillary immersion forces. The intensity of this type of forces is influenced by particle surface roughness and/or chemical heterogeneity, contributing to the formation of non-uniform, undulating, or irregular three-phase contact lines.[14, 83]

The presence of an interface can also induce hydrodynamic interactions that emerge as a result of the particle motion through a viscous medium. This particle motion generates a flow that can influence other particles at the interface, resulting in interface-mediated interactions. However, the significance of hydrodynamic interactions is generally less pronounced within the energy landscape compared to capillary interactions.

Taking the above into account, a working approximation for interparticle interactions within particle-laden interfaces can be reached by combining direct forces with the capillary contribution ($E_{\text{capillary}}$), summarized in the following expression:

$$\mu_i = E_{DLVO} + E_{dipolar} + E_{capillary}. \tag{4.15}$$

The evaluation of the exact magnitudes and ratios of these distinct contributions necessitates meticulous examination, potentially using optical tweezers or related techniques.[84] This exploration is crucial due to its significant role in dictating particle assembly at fluid interfaces. The relative importance of each contribution shifts significantly depending on the interfacial coverage. At minimal coverages, particles predominantly interact through dipolar forces. However, increasing coverage elevates the role of the attractive component, with capillary forces exerting substantial influence on the energy landscape of fluid interfaces laden with particles. Indeed, the prevalence of capillary interactions significantly influences the ordering transition of particles at fluid interfaces as coverage increases.[85] This enhanced attractive contribution as density increases also impacts the dynamics of particles at the fluid interface.[86]

Controlling interparticle interactions is pivotal for modulating particle organization at the interface, which, in turn, can modify interface mechanical properties. This consideration is particularly crucial in scenarios involving particle assembly on oil droplets or gas bubbles within emulsions and foams.[87]

4.5 INTERFACIAL RHEOLOGY OF PARTICLE-LADEN INTERFACES

Rheology involves the examination of how a material deforms and flows when subjected to external stress. Basic solids store energy and exhibit a spring-like, elastic reaction, while simple liquids disperse energy through viscous flow. In the case of intricate viscoelastic materials, rheological assessments unveil responses that resemble both solids and fluids, depending on the timescale of the probing. Interfacial rheology, akin to its 3D counterpart, establishes the connection between stress and the rate of deformation for substances present either by adsorption or by deposition at fluid interfaces.[88]

The behavior of fluid interfaces containing particles is determined by the surface pressure, which opposes surface tension. This pressure contributes to *in-plane* deformations, changes in material area due to dilations, and shear stresses inducing shape changes. These changes emerge from the interaction between interfacial and bulk flows. Conversely, complex fluid interfaces, subject to specific stress conditions, also manifest *out-of-plane* deformations that displace monolayers or portions thereof from the interface plane, leading to effects like monolayer buckling, material expulsion into the bulk, or the formation of multilayers.

The examination of the rheological response of particle-laden interfaces necessitates accounting for the interplay between thermodynamic and hydrodynamic interactions among particles and surrounding fluids. Solving this intricate issue becomes straightforward solely when the mechanical response of bulk fluids is decoupled from that of the interface. In other words, this is true when the mechanical reaction of the fluid phases is negligible compared to that of the interfacial layer.[89, 90] If this condition holds, the deformation of a particle-laden fluid interface can be attributed to two concurrent processes: (i) changes in size and/or shape of fluid interfaces, and (ii) deformation of the interfacial layer. Consequently, the energy needed for interfacial deformation can be expressed as follows:

$$\sigma_{ij} = \gamma\delta_{ij} + T_{ij}. \tag{4.16}$$

Here, δ_{ij} represents the Kronecker delta, and σ_{ij} stands for the stress tensor at the interface. This tensor comprises two elements: (i) the interfacial energy, which corresponds to the energy required for the presence of a fluid interface with a constant area. This element also encompasses any adsorption/desorption processes that alter the interfacial tension, and (ii) the Marangoni stresses that emerge from spatial gradients in interfacial tension. The second contribution, referred to as the anisotropic tensor (T_{ij}), includes the energy required to distort the particle layer.[13, 91]

4.5.1 Interfacial Dilational Rheology

The attachment of particles at fluid interfaces not only affects interfacial tension but also results in enhanced resistance to compression, denoted by the interfacial dilational viscoelastic modulus ε.[55] This magnitude provides insights into alterations in interfacial tension arising from changes in the available area for particle distribution. These changes can be evaluated from the temporal variations of surface pressure $\delta\Pi(t)$ due to alterations in interfacial area $\delta A(t)$ caused by the application of uniaxial stress,[92]

$$\delta\Pi(t) = \Pi(t) - \Pi_0 = \frac{\partial\Pi}{\partial A}\delta A = -\varepsilon(t)u(t), \tag{4.17}$$

where $u(t)$ and $\varepsilon(t)$ represent the time evolution of compressional strain and dilational viscoelastic modulus, respectively. Similarly, $\Pi(t)$ and Π_0 denote the time evolution of surface pressure and initial surface pressure, respectively. The time dependence of the viscoelastic dilational modulus is governed by Equation (4.18), while the time dependence of the compression strain is described by Equation (4.19):

$$\varepsilon(t) = -A_0\left(\frac{\partial\Pi}{\partial A}\right)_T \tag{4.18}$$

and

$$u(t) = \frac{\delta A}{A_0}. \tag{4.19}$$

Under equilibrium conditions of a fluid layer, the limit value at zero frequency of the dynamic modulus can be defined as the static modulus in terms of Gibbs elasticity ε_0, as follows:

$$\varepsilon(t) \rightarrow \varepsilon_0 = \Gamma\left(\frac{\partial\Pi}{\partial\Gamma}\right)_{eq}, \tag{4.20}$$

where $\Gamma = 1/A$ signifies interfacial concentration. For small-amplitude oscillatory deformation at characteristic frequency ω, the complex dilational viscoelastic modulus, described in the following equation:

$$\varepsilon(\omega) = \varepsilon^{'(\omega)} + i\omega\kappa(\omega), \tag{4.21}$$

consists of the storage modulus ($\varepsilon'(\omega)$) representing interfacial dilational elasticity, and the loss modulus ($\varepsilon''(\omega) = \omega\kappa(\omega)$). Here, κ accounts for interfacial dilational viscosity. This definition of the viscoelastic interfacial dilational modulus considers that stress related to interfacial deformation can impact both particle adsorption states and interfacial structure, leading to various relaxation processes with distinct characteristic timescales.[27, 93–95]

The response of particle-laden fluid interfaces to dilational deformations was initially explored by Miller et al.,[95] who provided a theoretical framework for relaxation mechanisms in such interfaces under dilational stress. This approach draws parallels to prior studies on the dilational rheology of proteins and protein–surfactant systems.[37, 96] Consequently, the dilational response of particle-laden interfaces can be elucidated using adsorption isotherms. Despite the desire for models integrating equilibrium and dynamic information, the model was limited due to challenges in determining an authentic equilibrium isotherm.

Particle wettability significantly influences particle trapping at fluid interfaces, as demonstrated by Safouane et al.[97] They investigated the rheological response of monolayers of fumed silica nanoparticles at the water/vapor interface and identified correlations between interfacial morphology and rheological behavior. Storage modulus increased with particle hydrophobicity and interfacial coverage, while the loss modulus remained nearly constant. Hydrophilic particles exhibited low compressibility modulus at the water/vapor interface, gradually increasing to approximately 200 mN/m with increased particle hydrophobicity. This stiffening due to hydrophobicity might be attributed to interfacial layer densification and enhanced particle interactions. Interparticle interactions were also found to be crucial in controlling interfacial mechanical response, introducing strong hysteresis and incompressible films. Similar findings were reported by Zang et al.,[98] who expanded the explored frequency range, revealing relaxation processes related to dilational stress application. These processes increase is more significant with decreasing particle hydrophobicity, influencing interfacial coverage and interfacial particle reorganization.

Kobayashi and Kawaguchi[99] presented different observations for polymer particle-laden interfaces compared to silica nanoparticles. The partially deformable nature of polymer particles influenced the mostly viscoelastic dilational response of latex particles at the water/vapor interface. Elastic and viscous dilational moduli increased with interfacial coverage up to a critical value, after which they steeply decreased. This change was attributed to *out-of-plane* deformations (buckling) of the monolayer due to dilational stress, disrupting the quasi-2D particle-laden interface organization. The viscoelastic moduli transitioned from fluid-like to solid-like behavior with increasing interfacial packing density, reflecting the phase diagram region under analysis.

The interplay between interfacial organization and rheological response was further explored by del Río et al.,[100] who underscored the influence of the monolayer phase on particle-laden interface dilational viscoelastic modulus. They found mainly elastic monolayers, with storage modulus exceeding loss modulus, and storage modulus increasing with interfacial coverage up to a threshold, where it sharply increased. This was attributed to close-packed particle film formation. Particle size was also influential, inversely correlating with Young's modulus of monolayers of monodisperse particles at fluid interfaces, as reported by Vella et al.[101] Hilles et al.[102] revealed that particle-laden interfaces exhibit a limited linear region of low strain under typical laboratory conditions, highlighting the disparity between laboratory-scale dilational flows and those in industrial processes.

Commonly, interactions between oppositely charged particles and surfactants result in various synergistic effects. This implies that combining particles and surfactants typically induces a more pronounced alteration of interfacial tension than expected from individual components. However, this does not hold when silica particles are capped with a cationic surfactant like CTAB (cetyltrimethylammonium bromide). Ravera et al.[29] showed that the dilational viscoelastic response of silica particle layers capped with CTAB displayed significant frequency-dependent behavior at low frequencies (0.005–0.2 Hz) using an oscillating drop tensiometer. Elasticity of the particle-laden interface was also observed to surpass that of surfactant-laden interfaces, aligning with results by Tambe and Sharma[89] for graphite particle–stearic acid mixtures at the water/decane interface. Furthermore, the effect of forming particle-laden interfaces with surfactant-capped particles was more prominent in water/oil interfaces than in water/vapor interfaces. Notably, the interfacial dilational viscoelastic modulus of surfactant-decorated colloidal particle layers at fluid interfaces increased under compression, attributed to the increase of the interfacial particle excess.[35, 103, 104]

The seminal work by Ravera et al.[29] was expanded by broadening the frequency range (from 10^{-3} Hz to 10^{3} Hz) through a combination of three experimental methods (oscillatory drop tensiometer, capillary pressure tensiometer, and electrocapillary wave devices). This revealed two distinct relaxation processes in response to dilational stress, stemming from (i) diffusion-controlled particle adsorption from bulk to interface, and (ii) reorganization of adsorbed material at the interface.[59, 103] The former process predominated at lower frequencies, while the latter governed the dilation response at higher frequencies. These results were effectively described using a theoretical model that incorporated diffusion-controlled adsorption and arbitrary kinetics processes within the interface.[103, 105, 106] The interfacial dilational response of CTAB-decorated silica particles highlighted a complex interplay of interactions, mainly electrostatic and hydrophobic, governing interfacial layer equilibration. This involved CTAB molecule exchange from particle surfaces to the water/vapor interface, influencing particle wettability and interparticle interactions in bulk and upon interface adsorption.[52, 53]

Interfacial aging is a crucial factor in considering the dilational rheological response of layers formed by surfactant-capped particles. Interfacial aging often leads to significant increases in dilational elasticity for layers at water/air or water/oil interfaces.[35, 59, 107] However, this aging-induced rigidification does not affect interfacial tension. Consequently, once maximum interfacial coverage is reached, particle exchange between bulk and interface becomes negligible or occurs at very slow rates, indicating irreversible particle trapping at the fluid interface.[35] This rigidification process results in the formation of solid-like films at the interface, confirmed by Brewster angle microscopy (BAM) imaging,[35] in line with findings by Santini et al.[36, 52] and Whitby et al.[104] in other surfactant-decorated particle systems. Notably, the elasticity of solid-like films formed during interfacial aging matches the levels identified by Zang et al.[108, 109]

4.5.2 Interfacial Shear Rheology

The presence of bulk flow can induce the emergence of interfacial shear deformations at particle-laden interfaces, involving mass and momentum transport.[110] In the context of shear deformations, the amplitude and temporal evolution are decoupled from other interfacial modes.[111] Consequently, a definition of shear elasticity in the context of *in-plane* shear deformation can be established as a constant that relates applied strain (u_{xy}) to stress (σ_{xy}). In cases of solid-like films with Hookean behavior, shear elasticity is expressed as $\sigma_{xy} = Gu_{xy}$. Conversely, fluid-like layers are influenced by the viscous nature of the layer, and shear stress can be formulated as

$$\sigma_{xy} = \eta^S \frac{du_{xy}}{dt}, \tag{4.22}$$

where du_{xy}/dt represents strain rate and η^S is the interfacial shear viscosity. Generally, viscoelastic parameters (G, η^S) tend to be higher when strong particle interactions are present within the system, reflecting the flow induced in surface elements due to the energy required for overcoming interfacial interactions.[55] In the case of oscillatory deformation, the complex shear modulus G^* is defined as

$$G^*(\omega) = G'(\omega) + iG''(\omega) \equiv G' + i\omega\eta^S, \tag{4.23}$$

where G' and G'' signify the storage or shear elastic modulus and the loss modulus, respectively. For experiments where the amplitude of oscillatory deformations at a fixed frequency ω remains small, shear viscosity can be approximated as $G'' = \omega\eta^S$.

Investigations into the response of particle-laden fluid interfaces to shear stress have gained significance, particularly due to the implications of shear processes for emulsion.[27, 95, 112] This has driven considerable research to understand the connections between interfacial microstructure and the shear response of particle-laden fluid interfaces, with interparticle interactions playing a crucial role in modulating these relationships.[113] It is worth noting that differences in interparticle interactions between particles dispersed in bulk and at fluid interfaces give rise to notable distinctions in the shear behavior of three-dimensional particle suspensions and their interfacial counterparts.[114]

One of the first studies characterizing particle-laden interfaces under shear stresses was conducted by Cicuta et al.[115] They compared the behavior of polystyrene latex particles (3 μm in diameter) at the water/decane interface with that of β-lactoglobulin layers. The interfacial rheological response of particle monolayers predominantly exhibited a viscous character ($G''>G'$), whereas β-lactoglobulin-laden interfaces showed primarily elastic behavior. This discrepancy can be attributed to the possibility that proteins undergo deformation and compression processes under shear. At low coverage, the shear response was primarily governed by the viscous contribution, akin to liquid-like films. In contrast, particle-laden interfaces displayed a sharp viscoelasticity increase for surface coverage exceeding 75–80%, attributed to particle jamming. This qualitative alignment was with the results by Yu et al.[116] for silica particle monolayers. At the upper limits of coverage, the dominant response shifts toward elastic character, leading to monolayers with solid-like films. It is worth highlighting that the shift from a viscous-like to a solid-like behavior can be influenced by altering the ionic strength, which in turn modifies the interactions between particles. While increasing the ionic strength does not notably alter the relationship between viscoelastic moduli and interfacial coverage, the transition from a viscous to an elastic-like response takes place at lower coverage values with higher ionic strength. This phenomenon can be attributed to the screening of electrostatic repulsions among particles, promoting the aggregation of particles at the interface.[116]

The interplay between interfacial coverage and the response of particle-laden fluid interfaces to shear stresses was also explored by Imperiali et al.[117] They emphasized the crucial role of particle reorganization in influencing the response of the interface to shear, a conclusion in line with the findings by Barman and Christopher[118] concerning polystyrene latex at the water/vapor interface. Using a double-walled interfacial shear rheometer, they assessed the microstructure of the interface and observed a transition from shear-thinning behavior, linked to the breakup of particle clusters, to yielding behavior as interfacial coverage increased. This transition was attributed to distinct mechanisms of viscous stress dissipation. Interestingly, yielding was detected at interfacial coverages significantly below those related to interfacial jamming, implying the potential for solid-like behavior before close-packing or jamming conditions are reached. Additionally, the researchers noted that high interfacial coverage could lead to a shear-induced ordering phenomenon. Barman and Christopher[118] showed that the increase in interfacial coverage limits the linear range of the rheological response, with the coverage-dependent viscoelastic shear modulus following a power-law trend. These investigations culminated in the understanding that the viscoelastic response of particle-laden interfaces is governed by constraints on particle motion, arising from interparticle capillary attraction and the local microstructure's caging effects. Furthermore, Barman and Christopher[119] emphasized the significance of meso-structural organization in dictating elasticity and yield. They identified that the formation of aligned domains of hexagonally packed particles leads to elastic-like particle films while diminishing domain size results in interfacial flows and viscous-like behavior. Reynaert et al.[120] further extended the analysis of the effect of interfacial packing on particle-laden interface response to shear stresses. They provided additional evidence of the role of interparticle interactions in shaping the interfacial rheological response, observing that particle aggregation triggers an interfacial shear response reminiscent of bulk counterparts. This conclusion aligns with the results by Wijmans and Dickinson,[121] derived from Brownian dynamics simulations. The role of interparticle interactions in particle-laden interfaces' reaction to shear stresses was also investigated by Beltramo et al.[95] Their research demonstrated that strong lateral interactions between particles lead to the emergence of interfacial yield stress, with the elastic modulus and yield stress increasing significantly as interfacial coverage rises from 0.47 to 0.88. For interfacial coverages below 0.64, the viscoelastic modulus was nearly negligible, while above this threshold, the viscoelastic modulus of particle films showed an increase. Within the interfacial coverage range of 0.75 to 0.80, the viscoelastic modulus reached a plateau due to the onset of interfacial jamming. In recent work, Muntz and Thijssen[122] deepened the correlation between particle-laden interface structure and rheological behavior. Their study focused on poly(methyl methacrylate) particles at water/dodecane interfaces, utilizing a shear rheometer coupled with a confocal microscope. Findings indicated that at low surface coverage, the particle-laden interface behaved as a two-dimensional Newtonian fluid, undergoing aggregation when subjected to shear stress surpassing a certain threshold. An increase in interfacial coverage resulted in the dominance of the elastic contribution, with the interface displaying plastic deformation beyond the yield stress.

Additionally, the influence of particle roughness in the presence of shear stresses must be acknowledged, as it plays a critical role in the entrapment of particles at fluid interfaces and their organization. The relationship between particle roughness and shear response at particle-laden interfaces is closely tied to interfacial coverage[123] and can be understood in terms of interparticle friction.[124] The emergence of contact line pinning, due to the trapping of particles at metastable positions, can alter particle behavior and render hydrophilic particles behave like hydrophobic ones and vice versa.[125] This phenomenon is influenced by the hydrophilic–lipophilic balance of the particles.[97] Such studies provide insights into the intricate interplay between interparticle interactions, wettability, coverage, and shear response in particle-laden fluid interfaces.

Another crucial aspect to consider in understanding the response of particle-laden fluid interfaces to shear stresses relates to the examination of its temporal evolution. Shedding light on this matter, Krishnaswamy et al.[126] investigated the adsorption kinetics of silver nanoparticles (ranging from 10 to 50 nm) at toluene/water interfaces using shear rheology. Their study revealed an increase in both the storage and loss moduli over time, attributed to the densification of the monolayer during the adsorption process. Furthermore, following interfacial equilibration, the particle-laden interface displayed behavior reminiscent of a 2D glassy material, with the shear response highly contingent on the strain amplitude. In contrast, strain sweep measurements unveiled shear thickening behavior in G'' under significant strain amplitudes, while no dependence of G' on the strain amplitude was observed up to the point of shear thickening. The shear thickening phenomenon was found to conform to a power law, with exponents of 2 and 1 describing the storage and loss moduli, respectively. This behavior exhibited by silver nanoparticles notably differs from that of gold nanoparticles at the water/vapor interface,[127] with the latter demonstrating behavior akin to a gel-like system experiencing strain-induced softening. Furthermore, the rheological response of gold nanoparticles showed no dependence on applied strain for values below a critical threshold of 0.1%, after which the storage modulus diminished. Moreover, the viscoelastic moduli were well described as a function of interfacial coverage through a power law with an exponent around 0.65, which bears resemblance to observations in percolating systems.[93]

The exploration of the shear response of monolayers formed by particles capped with various ligands at fluid interfaces has received relatively less attention compared to their uncapped counterparts. Despite this limited focus, several investigations have highlighted the intricate connections between monolayer shear response, interfacial microstructure, and interparticle interactions. A pioneering study addressing this aspect was conducted by Orsi et al.,[127] who delved into the shear response of monolayers of gold nanoparticles capped with dodecanethiol at the water/vapor interface. Their findings revealed a predominant elastic response characterized by moduli scaling with interfacial concentration, indicating a percolated network – a correlation that resonates with their earlier results using X-ray photon correlation spectroscopy.[128] Moreover, the frequency-dependent complex modulus aligned with the rheological behavior of soft glass,[129] indicative of a fractal nature in the film and in accordance with microscopic observations.

Maestro et al.[58] investigated the impact of surfactant concentration on the shear response of CTAB-decorated silica particle monolayers at water/vapor interfaces through a combination of strain- and frequency-sweep oscillatory shear experiments. High surfactant concentrations led to solid-like behavior beneath a yield point analogous to that of a 2D glass, with shear modulus ranging from 0.1 to 1 N/m, similar to silica particle monolayers.[108, 109] The comprehensive analysis of the shear response unveiled power-law behavior of both storage and loss moduli with respect to frequency – a characteristic of a soft glassy material.[129] Employing a strain-rate frequency superposition approach, they revealed a gradual relaxation process, nearly independent of strain-rate amplitude. This phenomenon was attributed to a transition from a fluid-like to a solid-like interface, in agreement with the results by Zang et al.[109] on monolayers of hydrophobic silica with varying degrees of hydrophobicity.

The above discussions underscore how capping ligands increase the rigidity and resistance of particle-laden fluid interfaces against deformations.[55] Notably, solid-like particle-laden interfaces undergo deformation beyond a yielding point upon a shear strain ramp, resulting in a plastic flow.[130] The yield stress τ^* obeys a power-law relationship with interfacial coverage ($\tau^* \sim \Gamma^{\alpha}$). Elucidating the molecular mechanism behind this dynamic response necessitates accounting for particle-level and many-body physics, as proposed by Maestro and Zaccone.[131] Their model introduces a description of shear deformation response, encompassing local particle connectivity, cage-breaking temporal evolution, and the dynamics of elastic response microstructural heterogeneity. The fusion of these contributions yields non-affine deformations, as evidenced by analyzing data from oscillatory shear measurements, requiring only two fitting parameters: the 'cage' relaxation time and viscous relaxation time. Furthermore, the model integrates an interparticle spring constant reflecting interaction strength, modifiable through particle hydrophobicity – governed by surfactant quantity. This model establishes a connection between non-affine deformation, local cage rearrangements, and emergent plastic flow, offering insight into the mechanical response of colloidal particles (both capped and uncapped) ensnared at the water/vapor interface.

Interfacial shear rheology has proven valuable for assessing particle adsorption kinetics at fluid interfaces, as demonstrated by Maas et al.[132] Their study of lipid-decorated silica particle adsorption at the water/oil interface unveiled bimodal kinetics, with the first process occurring over a timescale near 1 h. During this initial stage, G' increased due to strong interparticle interaction layer formation. Subsequently, a second phase lasting several hours led to slight increases in both G' and G'', attributed to interface material accumulation. This accumulation continued until the steady state was achieved, characterized by storage and loss moduli values orders of magnitude higher than those typical of pristine surfactant monolayers.[97]

4.5.3 Correlations between Interfacial Rheology and the Emulsion Stabilization

The mechanical behavior of interfaces holds a pivotal role in governing the stability and deformation of coated droplets. This influences their ability to stabilize emulsions by minimizing coalescence, Ostwald ripening, and gravitational destabilization through drainage or creaming processes.[133, 134] A significant impact on stabilizing dispersed systems is exerted by the attachment of particles to the interface, forming a self-assembled robust armor that hinders deformations, preventing droplets from colliding, and thereby maintaining the stability of the dispersed systems. The stabilization of particular interest involves jammed layers, occurring when the available particle distribution area approaches close-packing.[57, 135] Such jamming counteracts interfacial tension-driven coarsening, conferring notable stability to biphasic liquid–liquid systems. The rheological properties, encompassing shear and dilational characteristics, contribute to regulating emulsion stability by influencing film drainage.[136]

Stancick et al.[137] corroborated the role of interfacial coverage in emulsion stabilization by nanoparticles. They emphasized that the particle-laden interface's role in stabilizing dispersed systems extends beyond mechanical property modification, highlighting the steric hindrance induced by nanoparticles to prevent coalescence. The work by Frith et al.[138] underscored the importance of forming close-packed silica layers, conferring rigidity that bolsters dispersed system stability. Sullivan and Kilpatrick[139] demonstrated the use of emulsion-stabilized particles to enhance oil-recovery emulsion stabilization, forming rigid, stagnant films around drops to hinder coalescence. Recent studies, however, have stirred controversy regarding the relationship between interfacial layer mechanical properties and their ability to stabilize dispersed systems. Santini et al.[60, 140] explored the possibility of carbon particles stabilizing emulsions but found no definitive correlation between interfacial viscoelastic properties and stabilization ability. Similarly, Santini et al.[8, 141]

studies on silica nanoparticles and palmitic acid raised questions about the relationship between particle-laden interface mechanical properties and stabilization of dispersed systems. High dilational viscoelastic moduli (>100 mN/m) were observed for silica nanoparticles + palmitic acid monolayers at air/water interfaces, yet this system could not stabilize foams. Conversely, it proved effective for forming highly stable water-in-oil emulsions, with limited dependence on rheological properties for emulsion stabilization.

In contrast, the role of shear modulus appears clearer due to its direct connection with interfacial structure and coverage. Madivala et al.[142] highlighted the high interfacial shear modulus of ellipsoidal particle monolayers and their unique packing determining interfacial gel formation. These gel-like structures enveloping droplets significantly contribute to dispersed system stabilization, particularly ultra-stable emulsions, whether water-in-oil or oil-in-water. This stabilization by ellipsoidal particles surpassed that of spherical counterparts.[142] Kim et al.[143] confirmed this scenario using clays to stabilize oil-in-water emulsions, showcasing the superior performance of hydrophobic clays forming close-packed, high-shear modulus layers for emulsion stabilization. Notably, the addition of surfactants to hydrophilic clays induced hydrophobization, enhancing emulsion stabilization. Brugger et al.[144] revealed that higher G' values corresponded to improved emulsion stability due to increased shear modulus. However, high G'' values reduced emulsion stability due to the formation of easily breakable soft layers.

Another factor to be considered in assessing the effectiveness of particles for stabilizing dispersed systems is the influence of particle roughness.[77] Emulsion stability is positively impacted by increased particle roughness, as it contributes to elevated shear viscosity, consequently impeding drainage processes. Moreover, comprehending interfacial rheology concerning particle-laden interfaces is vital for controlling bulk rheology in the resultant foams or emulsions.

4.6 CONCLUSIONS

Over the past decades, the investigation of particle-laden fluid interfaces has witnessed remarkable expansion, primarily owing to their acknowledged significance in various technological domains centered around particle–interface interactions, notably the stabilization of dispersed systems such as emulsions. Nevertheless, a comprehensive understanding of the control mechanisms underlying the behavior of these systems remains elusive, largely due to the intricate nature of the theoretical and experimental methodologies required for their study. This chapter has provided an updated perspective on the current understanding of the physics behind particle-laden interfaces, paying special attention to the most fundamental aspects contributing to emulsion stabilization. This has required to encompass both static and dynamic facets, and various theoretical and experimental considerations devised to ascertain the science behind Pickering emulsion stabilization.

The characteristics of particle arrangements within fluid interfaces are intricately intertwined with numerous factors encompassing particles, fluids, and their interactions. This gives rise to diverse scenarios for the properties of particle-laden fluid interfaces, intricately linked with the structural attributes of these interfaces. Through a thorough investigation of these facets, it becomes feasible to identify optimal conditions for effectively stabilizing emulsions. Nonetheless, there remains a significant knowledge gap concerning the interplay between interfacial rheological properties, morphologies, and the stabilization of such emulsions.

Attaining a comprehensive understanding of the physicochemical intricacies associated with the behavior of particles trapped at fluid interfaces holds the potential to unlock novel perspectives in interfacial studies and the utilization of these systems for both research and technological advancements in the field of the stabilization of emulsions.

ACKNOWLEDGMENTS

The team in Madrid was funded by grants PID2023-147156NB-I00 funded by MCIN/AEI/10.13039/501100011033 (Spain) and PR12/24-31566 (Ayudas para la Financiación de Proyectos de Investigación UCM 2023), and the European Innovative Training Network-Marie Sklodowska-Curie Action NanoPaInt (grant agreement 955612), funded by the E.U. A.M. was funded by MCIN/AEI/10.13039/501100011033 (Spain) under grant PID2021-129054NA-I00 and the IKUR Strategy under the collaboration agreement between the Ikerbasque Foundation and Materials Physics Center on behalf of the Department of Education of the Basque Government.

REFERENCES

1. Ramsden, W. Separation of solids in the surface-layer of solutions and "suspension" (Observations on Surface-membranes, Bubbles, Emulsions, and Mechanical Coagulation). Preliminary account. *Proc. R. Soc. Lond.* **1903**, *72*, 156–164.
2. Pickering, S.U. CXCVI. – Emulsions. *J. Chem: Soc. Trans.* **1907**, *91*, 2001–2021.
3. Guzmán, E.; Ortega, F.; Rubio, R.G. Forces controlling the assembly of particles at fluid interfaces. *Langmuir* **2022**, *38*, 13313–13321.
4. Guzmán, E.; Martínez-Pedrero, F.; Calero, C.; Maestro, A.; Ortega, F.; Rubio, R.G. A broad perspective to particle-laden fluid interfaces systems: from chemically homogeneous particles to active colloids. *Adv. Colloid Interface Sci.* **2022**, *302*, 102620.
5. Lotito, V.; Zambelli, T. Approaches to self-assembly of colloidal monolayers: A guide for nanotechnologists. *Adv. Colloid Interface Sci.* **2017**, *246*, 217–274.
6. Garbin, V. Colloidal particles: Surfactants with a difference. *Phys. Today* **2013**, *66*, 68.
7. González-González, A.; Sánchez-Arribas, N.; Santini, E.; Rodríguez-Villafuerte, J.L.; Carbone, C.; Ravera, F.; Ortega, F.; Liggieri, L.; Rubio, R.G.; Guzmán, E. Effects of oil phase on the inversion of Pickering emulsions stabilized by palmitic acid decorated silica nanoparticles. *Colloids Interfaces* **2022**, *6*, 27.
8. Santini, E.; Guzmán, E.; Ferrari, M.; Liggieri, L. Emulsions stabilized by the interaction of silica nanoparticles and palmitic acid at the water–hexane interface. *Colloids Surf. A* **2014**, *460*, 333–341.
9. Ji, X.; Wang, X.; Zhang, Y.; Zang, D. Interfacial viscoelasticity and jamming of colloidal particles at fluid–fluid interfaces: a review. *Rep. Progr. Phys.* **2020**, *83*, 126601.
10. Guzmán, E.; Abelenda-Núñez, I.; Maestro, A.; Ortega, F.; Santamaria, A. Particle-laden fluid/fluid interfaces: Physico-chemical foundations. *J. Phys.: Condens. Matter* **2021**, *33*, 333001.

11. Vogel, N.; Retsch, M.; Fustin, C.-A.; del Campo, A.; Jonas, U. Advances in colloidal assembly: The design of structure and hierarchy in two and three dimensions. *Chem. Rev.* **2015**, *115*, 6265–6311.
12. Bayles, A.V.; Vermant, J. Divide, conquer, and stabilize: Engineering strong fluid–fluid interfaces. *Langmuir* **2022**, *38*, 6499–6505.
13. Forth, J.; Kim, P.Y.; Xie, G.; Liu, X.; Helms, B.A.; Russell, T.P. Building reconfigurable devices using complex liquid–fluid interfaces. *Adv. Mat.* **2019**, *31*, 1806370.
14. Bresme, F.; Oettel, M. Nanoparticles at fluid interfaces. *J. Phys.: Condens. Matter* **2007**, *19*, 413101.
15. Arditty, S.; Whitby, C.P.; Binks, B.P.; Schmitt, V.; Leal-Calderon, F. Some general features of limited coalescence in solid-stabilized emulsions. *Eur. Phys. J. E* **2003**, *11*, 273–281.
16. Stancik, E.J.; Fuller, G.G. Connect the drops: Using solids as adhesives for liquids. *Langmuir* **2004**, *20*, 4805–4808.
17. Horozov, T.S.; Aveyard, R.; Clint, J.H.; Neumann, B. Particle zips: Vertical emulsion films with particle monolayers at their surfaces. *Langmuir* **2005**, *21*, 2330–2341.
18. Jena, K.C.; Scheu, R.; Roke, S. Surface impurities are not responsible for the charge on the oil/water interface: A comment. *Angew. Chem. Int. Ed.* **2012**, *51*, 12938–12940.
19. Dugyala, V.R.; Muthukuru, J.S.; Mani, E.; Basavaraj, M.G. Role of electrostatic interactions in the adsorption kinetics of nanoparticles at fluid–fluid interfaces. *Phys. Chem. Chem. Phys.* **2016**, *18*, 5499–5508.
20. McGorty, R.; Fung, J.; Kaz, D.; Manoharan, V.N. Colloidal self-assembly at an interface. *Mat. Today* **2010**, *13*, 34–42.
21. Ward, A.F.H.; Tordai, L. Time-dependence of boundary tensions of solutions I. The role of diffusion in time-effects. *J. Chem. Phys.* **1946**, *14*, 453–461.
22. Wang, H.; Singh, V.; Behrens, S.H. Image charge effects on the formation of Pickering emulsions. *J. Phys. Chem. Lett.* **2012**, *3*, 2986–2990.
23. Kettlewell, S.L.; Schmid, A.; Fujii, S.; Dupin, D.; Armes, S.P. Is latex surface charge an important parameter for foam stabilization? *Langmuir* **2007**, *23*, 11381–11386.
24. Van Oss, C.J.; Chaudhury, M.K.; Good, R.J. Interfacial Lifshitz-van der Waals and polar interactions in macroscopic systems. *Chem. Rev.* **1988**, *88*, 927–941.
25. Israelachvili, J. *Intermolecular and Surface Forces*, Academic Press: Waltham, MA, USA, 1985.
26. Deshmukh, O.S.; van den Ende, D.; Stuart, M.C.; Mugele, F.; Duits, M.H.G. Hard and soft colloids at fluid interfaces: Adsorption, interactions, assembly & rheology. *Adv. Colloid Interface Sci.* **2015**, *222*, 215–227.
27. Mendoza, A.J.; Guzmán, E.; Martínez-Pedrero, F.; Ritacco, H.; Rubio, R.G.; Ortega, F.; Starov, V.M.; Miller, R. Particle laden fluid interfaces: Dynamics and interfacial rheology. *Adv. Colloid Interface Sci.* **2014**, *206*, 303–319.
28. Schwenke, K.; Isa, L.; Del Gado, E. Assembly of nanoparticles at liquid interfaces: Crowding and ordering. *Langmuir* **2014**, *30*, 3069–3074.
29. Ravera, F.; Santini, E.; Loglio, G.; Ferrari, M.; Liggieri, L. Effect of nanoparticles on the interfacial properties of liquid/liquid and liquid/air surface layers. *J. Phys. Chem. B* **2006**, *110*, 19543–19551.
30. Garbin, V.; Crocker, J.C.; Stebe, K.J. Forced desorption of nanoparticles from an oil–water interface. *Langmuir* **2012**, *28*, 1663–1667.
31. Berg, J.C. *An introduction to Interfaces and Colloids: The Bridge to Nanoscience*, World Scientific: Singapore, Singapore, 2010.
32. Correia, E.L.; Brown, N.; Razavi, S. Janus particles at fluid interfaces: Stability and interfacial rheology. *Nanomaterials* **2021**, *11*, 374.
33. Joos, P. *Dynamic Surface Phenomena*, CRC Press: Boca Ratón, FL, USA, 1999.

34. Noskov, B.A.; Bykov, A.G. Dilational rheology of monolayers of nano- and microparticles at the liquid-fluid interfaces. *Curr. Opin. Colloid Interface Sci.* **2018**, *37*, 1–12.
35. Yazhgur, P.A.; Noskov, B.A.; Liggieri, L.; Lin, S.Y.; Loglio, G.; Miller, R.; Ravera, F. Dynamic properties of mixed nanoparticle/surfactant adsorption layers. *Soft Matter* **2013**, *9*, 3305–3314.
36. Santini, E.; Guzmán, E.; Ravera, F.; Ciajolo, A.; Alfè, M.; Liggieri, L.; Ferrari, M. Soot particles at the aqueous interface and effects on foams stability. *Colloids Surf. A* **2012**, *413*, 216–223.
37. Akanno, A.; Guzmán, E.; Fernández-Peña, L.; Llamas, S.; Ortega, F.; Rubio, R.G. Equilibration of a polycation–anionic surfactant mixture at the water/vapor interface. *Langmuir* **2018**, *34*, 7455–7464.
38. Du, K.; Glogowski, E.; Emrick, T.; Russell, T.P.; Dinsmore, A.D. Adsorption energy of nano- and microparticles at liquid–liquid interfaces. *Langmuir* **2010**, *26*, 12518–12522.
39. Maestro, A.; Guzman, E.; Ortega, F.; Rubio, R.G. Contact angle of micro- and nanoparticles at fluid interfaces. *Curr. Opin. Colloid Interface Sci.* **2014**, *19*, 355–367.
40. Zanini, M.; Isa, L. Particle contact angles at fluid interfaces: Pushing the boundary beyond hard uniform spherical colloids. *J. Phys.: Condens. Matter* **2016**, *28*, 313002.
41. Davies, G.B.; Krüger, T.; Coveney, P.V.; Harting, J. Detachment energies of spheroidal particles from fluid-fluid interfaces. *J. Chem. Phys.* **2014**, *141*, 154902.
42. McBride, S.P.; Law, B.M. Influence of line tension on spherical colloidal particles at liquid-vapor interfaces. *Phys. Rev. Lett.* **2012**, *109*, 196101.
43. Ballard, N.; Law, A.D.; Bon, S.A.F. Colloidal particles at fluid interfaces: Behaviour of isolated particles. *Soft Matter* **2019**, *15*, 1186–1199
44. Binks, B.P.; Rodrigues, J.A. Double inversion of emulsions by using nanoparticles and a di-chain surfactant *Angew. Chem. Int. Ed.* **2007**, *46*, 5389–5392.
45. Vialetto, J.; Anyfantakis, M. Exploiting additives for directing the adsorption and organization of colloid particles at fluid interfaces. *Langmuir* **2021**, *31*, 9302–9335.
46. Wi, H.S.; Cingarapu, S.; Klabunde, K.J.; Law, B.M. Nanoparticle adsorption at liquid–vapor surfaces: Influence of nanoparticle thermodynamics, wettability, and line tension. *Langmuir* **2011**, *27*, 9979–9984.
47. Kaz, D.; McGorty, R.; Mani, M.; Manohara, V.N. Physical ageing of the contact line on colloidal particles at liquid interfaces. *Nature Mater.* **2012**, *11*, 138–142.
48. Léandri, J.; Würger, A. Trapping energy of a spherical particle on a curved liquid interface. *J. Colloid Interface Sci.* **2013**, *405*, 249–255.
49. Coertjens, S.; Dier, R.D.; Moldenaers, P.; Isa, L.; Vermant, J. Adsorption of ellipsoidal particles at liquid–liquid interfaces. *Langmuir* **2017**, *33*, 2689–2697.
50. Oettel, M.; Dietrich, S. Colloidal interactions at fluid interfaces. *Langmuir* **2008**, *24*, 1425–1441.
51. Zang, D.Y.; Rio, E.; Delon, G.; Langevin, D.; Wei, B.; Binks, B.P. Influence of the contact angle of silica nanoparticles at the air–water interface on the mechanical properties of the layers composed of these particles. *Mol. Phys.* **2011**, *109*, 1057–1066.
52. Santini, E.; Krägel, J.; Ravera, F.; Liggieri, L.; Miller, R. Study of the monolayer structure and wettability properties of silica nanoparticles and CTAB using the Langmuir trough technique. *Colloids Surf. A* **2011**, *382*, 186–191.
53. Maestro, A.; Guzmán, E.; Santini, E.; Ravera, F.; Liggieri, L.; Ortega, F.; Rubio, R.G. Wettability of silica nanoparticle–surfactant nanocomposite interfacial layers. *Soft Matter* **2012**, *8*, 837–843.
54. Ortega, F.; Ritacco, H.; Rubio, R.G. Interfacial microrheology: Particle tracking and related techniques. *Curr, Opin. Colloid Interface Sci.* **2010**, *15*, 237–245.
55. Maestro, A. Tailoring the interfacial assembly of colloidal particles by engineering the mechanical properties of the interface. *Curr. Opin. Colloid Interface Sci.* **2019**, *39*, 232–250.

56. Maestro, A.; Bonales, L.J.; Ritacco, H.; Rubio, R.G.; Ortega, F. Effect of the spreading solvent on the three-phase contact angle of microparticles attached at fluid interfaces. *Phys. Chem. Chem. Phys.* **2010**, *12*, 14115–14120.
57. Maestro, A.; Rio, E.; Drenckhan, W.; Langevin, D.; Salonen, A. Foams stabilised by mixtures of nanoparticles and oppositely charged surfactants: Relationship between bubble shrinkage and foam coarsening. *Soft Matter* **2014**, *10*, 6975–6983.
58. Maestro, A.; Deshmukh, O.S.; Mugele, F.; Langevin, D. Interfacial assembly of surfactant-decorated nanoparticles: On the rheological description of a colloidal 2D glass. *Langmuir* **2015**, *31*, 6289–6297.
59. Ravera, F.; Ferrari, M.; Liggieri, L.; Loglio, G.; Santini, E.; Zanobini, A. Liquid–liquid interfacial properties of mixed nanoparticle–surfactant systems. *Colloids Surf. A* **2008**, *323*, 99–108.
60. Santini, E.; Ravera, F.; Ferrari, M.; Alfè, M.; Ciajolo, A.; Liggieri, L. Interfacial properties of carbon particulate-laden liquid interfaces and stability of related foams and emulsions. *Colloids Surf. A* **2010**, *365*, 189–198.
61. Zabiegaj, D.; Santini, E.; Guzmán, E.; Ferrari, M.; Liggieri, L.; Ravera, F. Carbon soot-ionic surfactant mixed layers at water/air interfaces. *J. Nanosci. Nanotechnol.* **2015**, *15*, 3618–3625.
62. Cappelli, S.; de Jong, A.M.; Baudry, J.; Prins, M.W.J. Interparticle capillary forces at a fluid–fluid interface with strong polymer-induced aging. *Langmuir* **2017**, *33*, 696–705.
63. Binks, B.P.; Rodrigues, J.A.; Frith, W.J. Synergistic interaction in emulsions stabilized by a mixture of silica nanoparticles and cationic surfactant. *Langmuir* **2007**, *23*, 3626–3636.
64. Binks, B.P.; Isa, L.; Tyowua, A.T. Direct measurement of contact angles of silica particles in relation to double inversion of Pickering emulsions. *Langmuir* **2013**, *29*, 4923–4927.
65. Deleurence, R.; Parneix, C.; Monteux, C. Mixtures of latex particles and the surfactant of opposite charge used as interface stabilizers – influence of particle contact angle, zeta potential, flocculation and shear energy. *Soft Matter* **2014**, *10*, 7088–7095.
66. Perrin, L.; Pajor-Swierzy, A.; Magdassi, S.; Kamyshny, A.; Ortega, F.; Rubio, R.G. Evaporation of nanosuspensions on substrates with different hydrophobicity. *ACS Appl. Mat. Interfaces* **2018**, *10*, 3082–3093.
67. Howes, P.D.; Chandrawati, R.; Stevens, M.M. Colloidal nanoparticles as advanced biological sensors. *Science* **2014**, *346*, 1247390.
68. Zang, D.; Stocco, A.; Langevin, D.; Wei, B.; Binks, B.P. An ellipsometry study of silica nanoparticle layers at the water surface. *Phys. Chem. Chem. Phys.* **2009**, *11*, 9522–9529.
69. Reed, K.M.; Borovicka, J.; Horozov, T.S.; Paunov, V.N.; Thompson, K.L.; Walsh, A.; Armes, S.P. Adsorption of sterically stabilized latex particles at liquid surfaces: effects of steric stabilizer surface coverage, particle size, and chain length on particle wettability. *Langmuir* **2012**, *28*, 7291–7298.
70. Bradley, L.C.; Chen, W.-H.; Stebe, K.J.; Lee, D. Janus and patchy colloids at fluid interfaces. *Curr, Opin. Colloid Interface Sci.* **2017**, *30*, 25–33.
71. Fernandez-Rodriguez, M.A.; Chen, L.; Deming, C.P.; Rodriguez-Valverde, M.A.; Chen, S.; Cabrerizo-Vilchez, M.A.; Hidalgo-Alvarez, R. A simple strategy to improve the interfacial activity of true Janus gold nanoparticles: A shorter hydrophilic capping ligand. *Soft Matter* **2016**, *12*, 31–34.
72. Weaire, D.; Hutzler, S. *The Physics of Foams*, Clarendon Press: Oxford-United Kingdom, 1999.
73. Binks, B.P.; Murakami, R. Phase inversion of particle-stabilized materials from foams to dry water. *Nature Mat.* **2006**, *5*, 865–869.

74. Aveyard, R.; Binks, B.P.; Clint, J.H. Emulsions stabilised solely by colloidal particles. *Adv. Colloid Interface Sci.* **2003**, *100 –102*, 503–546.
75. Binks, B.P.; Isa, L.; Tyowua, A.T. Direct measurement of contact angles of silica particles in relation to double inversion of Pickering emulsions. *Langmuir* **2013**, *29*, 4923–4927.
76. Binks, B.P.; Kirkland, M.; Rodrigues, J.A. Origin of stabilisation of aqueous foams in nanoparticle–surfactant mixtures. *Soft Matter* **2008**, *4*, 2373–2382.
77. San-Miguel, A.; Behrens, S.H. Influence of nanoscale particle roughness on the stability of Pickering emulsions. *Langmuir* **2012**, *28*, 12038–12043.
78. Stamou, D.; Duschl, C.; Johannsmann, D. Long-range attraction between colloidal spheres at the air–water interface: The consequence of an irregular meniscus. *Phys. Rev. E* **2000**, *62*, 5263–5272.
79. Maestro, A.; Santini, E.; Guzmán, E. Physico-chemical foundations of particle-laden fluid interfaces. *Eur. Phys. J. E* **2018**, *41*, 97.
80. Kralchevsky, P.A.; Danov, K.D. Interactions between particles at a fluid interface. In *Nanoscience: Colloidal and Interfacial Aspects,* Starov, V.M., Ed.; CRC Press: New York, NY, USA, 2010; pp 397–435.
81. Pepicelli, M.; Jaensson, N.; Tregouët, C.; Schroyen, B.; Alicke, A.; Tervoort, T.; Monteux, C.; Vermant, J. Surface viscoelasticity in model polymer multilayers: From planar interfaces to rising bubbles. *J. Rheol.* **2019**, *63*, 815–828.
82. Kralchevsky, P.A.; Denkov, N.D.; Danov, K.D. Particles with an undulated contact line at a fluid interface: Interaction between capillary quadrupoles and rheology of particulate monolayers. *Langmuir* **2001**, *17*, 7694–7705.
83. Dasgupta, S.; Auth, T.; Gompper, G. Nano- and microparticles at fluid and biological interfaces. *J. Phys.: Condens. Matter* **2017**, *29*, 373003.
84. Pradhan, S.; Whitby, C.P.; Williams, M.A.K.; Chen, J.L.Y.; Avci, E. Interfacial colloidal assembly guided by optical tweezers and tuned via surface charge. *J. Colloid Interface Sci.* **2022**, *621*, 101–109.
85. Bonales, L.J.; Rubio, J.E.F.; Ritacco, H.; Vega, C.; Rubio, R.G.; Ortega, F. Freezing transition and interaction potential in monolayers of microparticles at fluid interfaces. *Langmuir* **2011**, *27*, 3391–3400.
86. Parolini, L.; Law, A.D.; Maestro, A.; Buzza, D.M.A.; Cicuta, P. Interaction between colloidal particles on an oil–water interface in dilute and dense phases. *J. Phys.: Condens. Matter* **2015**, *27*, 194119.
87. Rahman, S.E.; Laal-Dehghani, N.; Barman, S.; Christopher, G.F. Modifying interfacial interparticle forces to alter microstructure and viscoelasticity of densely packed particle laden interfaces. *J. Colloid Interface Sci.* **2019**, *536*, 30–41.
88. Larsson, R.G. *The Structure and Rheology of Complex Fluids*. Oxford University Press: Oxford, UK, 1999.
89. Tambe, D.E.; Sharma, M.M. The effect of colloidal particles on fluid-fluid interfacial properties and emulsion stability. *Adv, Colloid Interface Sci.* **1994**, *52*, 1–63.
90. Langevin, D. Influence of interfacial rheology on foam and emulsion properties. *Adv, Colloid Interface Sci.* **2000**, *88*, 209–222.
91. Fuller, G.G.; Vermant, J. Complex fluid-fluid interfaces: Rheology and structure. *Ann. Rev. Chem. Biomol. Eng.* **2012**, *3*, 519–543.
92. Noskov, B.A. Fast adsorption at the liquid-gas interface. *Adv. Colloid Interface Sci.* **1996**, *69*, 63–129.
93. Maestro, A.; Guzmán, E.; Chuliá, R.; Ortega, F.; Rubio, R.G.; Miller, R. Fluid to soft-glass transition in a quasi-2D system: Thermodynamic and rheological evidences for a Langmuir monolayer. *Phys. Chem. Chem. Phys.* **2011**, *13*, 9534–9539.

94. Guzmán, E.; Fernández-Peña, L.; Akanno, A.; Llamas, S.; Ortega, F.; G. Rubio, R. Two different scenarios for the equilibration of polycation—anionic solutions at water–vapor interfaces. *Coatings* **2019**, *9*, 438.
95. Beltramo, P.J.; Gupta, M.; Alicke, A.; Liascukiene, I.; Gunes, D.Z.; Baroud, C.N.; Vermant, J. Arresting dissolution by interfacial rheology design. *Proc. Nat. Acad. Sci. U.S.A.* **2017**, *114*, 10373–10378.
96. Kotsmar, C.; Pradines, V.; Alahverdjieva, V.S.; Aksenenko, E.V.; Fainerman, V.B.; Kovalchuk, V.I.; Krägel, J.; Leser, M.E.; Noskov, B.A.; Miller, R. Thermodynamics, adsorption kinetics and rheology of mixed protein–surfactant interfacial layers. *Adv. Colloid Interface Sci.* **2009**, *150*, 41–54.
97. Safouane, M.; Langevin, D.; Binks, B.P. Effect of particle hydrophobicity on the properties of silica particle layers at the air-water interface. *Langmuir* **2007**, *23*, 11546–11553.
98. Kirby, S.M.; Anna, S.L.; Walker, L.M. Effect of surfactant tail length and ionic strength on the interfacial properties of nanoparticle–surfactant complexes. *Soft Matter* **2018**, *14*, 112–123.
99. Kobayashi, T.; Kawaguchi, M. Surface dilational moduli of latex-particle monolayers spread at air–water interface. *J. Colloid Interface Sci.* **2013**, *390*, 147–150.
100. del Río, O.I.; Kwok, D.Y.; R.Wu; Alvarez, J.M.; Neumann, A.W. Contact angle measurements by axisymmetric drop shape analysis and an automated polynomial fit program. *Colloids Surf. A* **1998**, *143*, 197–210.
101. Vella, D.; Aussillous, P.; Mahadevan, L. Elasticity of an interfacial particle raft. *Europhys. Lett.* **2004**, *68*, 212–218.
102. Hilles, H.; Monroy, F.; Bonales, L.J.; Ortega, F.; Rubio, R.G. Fourier-transform rheology of polymer Langmuir monolayers: Analysis of the non-linear and plastic behaviors. *Adv, Colloid Interface Sci.* **2006**, *122*, 67–77.
103. Liggieri, L.; Santini, E.; Guzmán, E.; Maestro, A.; Ravera, F. Wide-frequency dilational rheology investigation of mixed silica nanoparticle–CTAB interfacial layers. *Soft Matter* **2011**, *7*, 7699–7709.
104. Whitby, C.P.; Fornasiero, D.; Ralston, J.; Liggieri, L.; Ravera, F. Properties of fatty amine–silica nanoparticle interfacial layers at the hexane–water interface. *J. Phys. Chem. C* **2012**, *116*, 3050–3058.
105. Guzman, E.; Liggieri, L.; Santini, E.; Ferrari, M.; Ravera, F. Mixed DPPC-cholesterol Langmuir monolayers in presence of hydrophilic silica nanoparticles. *Colloids Surf. B* **2013**, *105*, 284–293.
106. Guzman, E.; Liggieri, L.; Santini, E.; Ferrari, M.; Ravera, F. Influence of silica nanoparticles on dilational rheology of DPPC-palmitic acid Langmuir monolayers. *Soft Matter* **2012**, *8*, 3938–3948.
107. Bykov, A.G.; Noskov, B.A.; Loglio, G.; Lyadinskaya, V.V.; Miller, R. Dilational surface elasticity of spread monolayers of polystyrene microparticles. *Soft Matter* **2014**, *10*, 6499–6505.
108. Zang, D.Y.; Rio, E.; Langevin, D.; Wei, B.; Binks, B.P. Viscoelastic properties of silica nanoparticle monolayers at the air-water interface. *Eur. Phys. J. E* **2010**, *31*, 125–134.
109. Zang, D.; Langevin, D.; Binks, B.P.; Wei, B. Shearing particle monolayers: Strain-rate frequency superposition. *Phys. Rev. E* **2010**, *81*, 011604.
110. Erni, P. Deformation modes of complex fluid interfaces. *Soft Matter* **2011**, *7*, 7586–7600.
111. Langevin, D. *Light Scattering by Liquid Surfaces and Complementary Techniques*, Marcel Dekker, Inc.: New York, N.Y., U.S.A., 1992.
112. Thijssen, J.H.J.; Vermant, J. Interfacial rheology of model particles at liquid interfaces and its relation to (bicontinuous) Pickering emulsions. *J. Phys.: Cond. Matt.* **2017**, *30*, 023002.

113. Gray, J.J.; Bonnecaze, R.T. Rheology and dynamics of sheared arrays of colloidal particles. *J. Rheol.* **1998**, *42*, 1121–1151.
114. Trappe, V.; Weitz, D.A. Scaling of the viscoelasticity of weakly attractive Particles. *Phys. Rev. Lett.* **2000**, *85*, 449–453.
115. Cicuta, P.; Stancik, E.J.; Fuller, G.G. Shearing or compressing a soft glass in 2D: Time-concentration superposition. *Phys. Rev. Lett.* **2003**, *90*, 236101.
116. Yu, K.; Zhang, H.; Biggs, S.; Xu, Z.; Cayre, O.J.; Harbottle, D. The rheology of polyvinylpyrrolidone-coated silica nanoparticles positioned at an air-aqueous interface. *J. Colloid Interface Sci.* **2018**, *527*, 346–355.
117. Imperiali, L.; Liao, K.-H.; Clasen, C.; Fransaer, J.; Macosko, C.W.; Vermant, J. Interfacial rheology and structure of tiled graphene oxide sheets. *Langmuir* **2012**, *28*, 7990–8000.
118. Barman, S.; Christopher, G.F. Simultaneous interfacial rheology and microstructure measurement of densely aggregated particle laden interfaces using a modified double wall ring interfacial rheometer. *Langmuir* **2014**, *30*, 9752–9760.
119. Barman, S.; Christopher, G.F. Role of capillarity and microstructure on interfacial viscoelasticity of particle laden interfaces. *J. Rheol.* **2015**, *60*, 35–45.
120. Reynaert, S.; Moldenaers, P.; Vermant, J. Interfacial rheology of stable and weakly aggregated two-dimensional suspensions. *Phys. Chem. Chem. Phys.* **2007**, *9*, 6463–6475.
121. Wijmans, C.M.; Dickinson, E. Simulation of interfacial shear and dilatational rheology of an adsorbed protein monolayer modeled as a network of spherical particles. *Langmuir* **1998**, *14*, 7278–7286.
122. Muntz, I.; Richards, J.A.; Brown, S.; Schofield, A.B.; Rey, M.; Thijssenar, J.H.J. Contactless interfacial rheology: Probing shear at liquid–liquid interfaces without an interfacial geometry via fluorescence microscopy. *J. Rheol.* **2023**, *67*, 67–80.
123. Wilson, H.J.; Davis, R.H. Shear stress of a monolayer of rough spheres. *J. Fluid Mech.* **2002**, *452*, 425–441.
124. Brown, E.; Zhang, H.; Forman, N.A.; Maynor, B.W.; Betts, D.E.; DeSimone, J.M.; Jaeger, H.M. Shear thickening and jamming in densely packed suspensions of different particle shapes. *Phys. Rev. E* **2011**, *84*, 031408.
125. Zanini, M.; Marschelke, C.; Anachkov, S.E.; Marini, E.; Synytska, A.; Isa, L. Universal emulsion stabilization from the arrested adsorption of rough particles at liquid-liquid interfaces. *Nat. Commun.* **2017**, *8*, 15701.
126. Krishnaswamy, R.; Majumdar, S.; Ganapathy, R.; Agarwal, V.V.; Sood, A.K.; Rao, C.N.R. Interfacial rheology of an ultrathin nanocrystalline film formed at the liquid/liquid interface. *Langmuir* **2007**, *23*, 3084–3087.
127. Orsi, D.; Baldi, G.; Cicuta, P.; Cristofolini, L. On the relation between hierarchical morphology and mechanical properties of a colloidal 2D gel system. *Colloids Surf. A* **2012**, *413*, 71–77.
128. Orsi, D.; Cristofolini, L.; Baldi, G.; Madsen, A. Heterogeneous and anisotropic dynamics of a 2D gel. *Phys. Rev. Lett.* **2012**, *108*, 105701.
129. Sollich, P.; Lequeux, F.; Hébraud, P.; Cates, M.E. Rheology of soft glassy materials. *Phys. Rev. Lett.* **1997**, *78*, 2020–2023.
130. Vandebril, S.; Vermant, J.; Moldenaers, P. Efficiently suppressing coalescence in polymer blends using nanoparticles: Role of interfacial rheology. *Soft Matter* **2010**, *6*, 3353–3362.
131. Maestro, A.; Zaccone, A. Nonaffine deformation and tunable yielding of colloidal assemblies at the air–water interface. *Nanoscale* **2017**, *9*, 18343–18351.
132. Maas, M.; Ooi, C.C.; Fuller, G.G. Thin film formation of silica nanoparticle/lipid composite films at the fluid-fluid interface. *Langmuir* **2010**, *26*, 17867–17873.

133. Pelipenko, J.; Kristl, J.; Rosic, R.; Baumgartner, S.; Kocbek, P. Interfacial rheology: An overview of measuring techniques and its role in dispersions and electrospinning. *Acta Pharm.* **2012**, *62*, 123–140.
134. Salonen, A.; Lhermerout, R.; Rio, E.; Langevin, D.; Saint-Jalmes, A. Dual gas and oil dispersions in water: Production and stability of foamulsion. *Soft Matter* **2011**, *8*, 699–706.
135. Cheng, H.-L.; Velankar, S.S. Controlled jamming of particle-laden interfaces using a spinning drop tensiometer. *Langmuir* **2009**, *25*, 4412–4420.
136. Bhamla, M.S.; Giacomin, C.E.; Balemans, C.; Fuller, G.G. Influence of interfacial rheology on drainage from curved surfaces. *Soft Matter* **2014**, *10*, 6917–6925.
137. Stancik, E.J.; Widenbrant, M.J.O.; Laschitsch, A.T.; Vermant, J.; Fuller, G.G. Structure and dynamics of particle monolayers at a liquid-liquid interface subjected to extensional flow. *Langmuir* **2002**, *18*, 4372–4375.
138. Frith, W.J.; Pichot, R.; Kirkland, M.; Wolf, B. Formation, stability, and rheology of particle stabilized emulsions: Influence of multivalent cations. *Ind. Eng. Chem. Res.* **2008**, *47*, 6434–6444.
139. Sullivan, A.P.; Kilpatrick, P.K. The effects of inorganic solid particles on water and crude oil emulsion stability. *Ind. Eng. Chem. Res.* **2002**, *41*, 3389–3404.
140. Santini, E.; Guzmán, E.; Ravera, F.; Ciajolo, A.; Alfè, M.; Liggieri, L.; Ferrari, M. Soot particles at the aqueous interface and effects on foams stability. *Colloids Surf. A* **2012**, *413*, 216– 223.
141. Santini, E.; Guzmán, E.; Ravera, F.; Ferrari, M.; Liggieri, L. Properties and structure of interfacial layers formed by hydrophilic silica dispersions and palmitic acid. *Phys. Chem. Chem. Phys.* **2012**, *14*, 607–615.
142. Madivala, B.; Vandebril, S.; Fransaer, J.; Vermant, J. Exploiting particle shape in solid stabilized emulsions. *Soft Matter* **2009**, *5*, 1717–1727.
143. Kim, J.K.; Rühs, P.A.; Fischer, P.; Hong, J.S. Interfacial localization of nanoclay particles in oil-in-water emulsions and its reflection in interfacial moduli. *Rheol Acta* **2013**, *52*, 327–335.
144. Brugger, B.; Rosen, B.A.; Richtering, W. Microgels as stimuli-responsive stabilizers for emulsions. *Langmuir* **2008**, *24*, 12202–12208.

5 Emulsions and Interfacial Properties

Francesca Ravera, Eva Santini, Libero Liggieri, Luigi Cristofolini and Davide Orsi

5.1 INTRODUCTION

Emulsions are liquid-in-liquid dispersions of sub-micro/micro-sized droplets widespread in nature, technologies, and products.[1] They are of great interest in many applied fields[2–6] spanning from pharmaceuticals, cosmetics, and foods to other practical applications like oil recovery or development of new nanostructured soft materials. Ordinary emulsions are thermodynamically unstable, and the constituting liquids tend to separate under the effect of external forces or perturbations. The characteristic time of this process may differ by several orders of magnitude, depending on the chemical composition of the system and the environmental conditions. For this reason, the prediction of the lifetime of emulsions and their microstructure is essential for their utilization in diverse applications.

The control of emulsification and emulsion stability is obtained by employing appropriate surface-active components such as low-weight surfactants, biomolecules, polymers, proteins, solid nanoparticles, or more complex associations of surfactants and nanoparticles. These amphiphilic components, or emulsifiers, tend to accumulate at the interfaces between the two liquid phases (typically aqueous and oily phases), modifying their physicochemical and mechanical properties. In this way, they favor emulsification lowering interfacial tension and, at once, contrast the mechanisms at the bases of the emulsion destabilization. As explained in more detail in the following sections, besides gravity-driven creaming/sedimentation, the main mechanisms governing the emulsion evolution, related to droplet interactions, are flocculation, coalescence, and Ostwald ripening (OR).[7,8]

The mechanisms involved in emulsification and emulsion destabilization have been widely investigated for many years and the basic concepts are well consolidated. Moreover, several studies have been recently carried out dealing with the utilization of classical and emerging emulsifiers in various application fields,[8–14] some of these focusing on how the emulsion controlling mechanisms are influenced by the adsorption properties of the emulsifiers. For example, many studies[7, 11, 12, 14] have underlined the importance of dynamic and structural features of the interfacial layers on stabilization against coalescence and Ostwald ripening.

Despite the huge number of studies dedicated to emulsions, several aspects are still worthy of investigation because of their relevance to today's emulsion-based technologies and products. One of the challenging topics in this field concerns, for

DOI: 10.1201/9781032636184-6

example, the development of efficient formulations of emulsions with a low impact on the environment and health.

In fact, the mainly used emulsifiers present potential negative effects on health and, at once, contribute in a large extent to environmental pollution when spread as domestic and industrial waste, especially for rivers and sea waters. Moreover, despite the advances in the direction of a better understanding of emulsions behavior, most of the formulations adopted by the industry still rely on semi-empirical approaches using an overabundant amount of surface-active components.

In this contest, many studies on emulsions nowadays aim to promote a more efficient and safe use of emulsifiers in emulsion formulations via a deeper understanding of their role in emulsification and stabilization. This should contribute to improving the biocompatibility and biodegradability of products, using molecules with lower toxic/pollutant effects and/or significantly lower concentrations. The use of natural surfactants, derived from plants or vegetable wastes, also goes in the direction of more sustainable solutions, even if these compounds are typically complex (macro) molecules with variable and often ill-defined structures, requiring a rational utilization of the development of new more suitable interpretative models.

On the basis of these considerations, it is clear that the way to develop a new formulation minimizing the impact on health and the environment is necessary through a deeper understanding of the relation between emulsion behavior and interfacial properties, mainly governed by the equilibrium and dynamic adsorption of emulsifiers. In the following, the basic well-consolidated concepts on emulsions are recalled together with an overview of the principal results obtained during the last years, focusing on the role of the interfacial properties in emulsification and emulsion evolution, the specificity of some emulsifiers, and the classical and emerging experimental methods recently adopted.

5.2 THE ROLE OF INTERFACIAL PROPERTIES IN EMULSION FORMATION AND MORPHOLOGY

The process of emulsification is of fundamental importance for all emulsion-based technologies and especially for industry because it is strictly related to the tailoring of the morphology of the products and their long-term stability.

The methods adopted for emulsification in industrial and research fields may be very different depending on the application and the kind of emulsifier employed.[6, 15] An important aspect is the level of energy transferred to the systems, related to the different kinds of employed mechanical devices that may be, for example, shear mixers, high-pressure valve or membrane homogenizers, micro-fluidizers, and sonicators.

The different factors influencing the emulsification process and the resulting characteristics of the obtained emulsions have been widely investigated in relation to the different application fields.[6, 9, 11, 15, 16] The basic principles on the role of the adsorption properties of emulsifiers and the dynamic processes occurring during emulsion formation are, however, common to all the applications.

In all cases, in fact, important requirements for efficient emulsifiers are strong ability to lower the interfacial tension, short characteristic times of adsorption, and high surface coverage degree once adsorbed at water–oil interfaces. The initial morphology of the produced emulsions, i.e., the droplet average size and the size distribution, is strongly dependent on these characteristics, which, in turn, also affect the long-term stability.

To go into more detail, it is important to remember that the emulsification process, which is essentially the fragmentation of one liquid phase into the other one with the production of a huge number of droplets, is associated with an important increase of the total interfacial area, ΔA. From a thermodynamic point of view, this corresponds to the work done on the system proportional to the interfacial tension, γ, that is $W = \gamma \, \Delta A$.

This means that for a given energy transferred to the liquid–liquid system, the lower the interfacial tension, the greater the increase of the interfacial area corresponding to the production of small droplets. In other words, emulsification is favored when interfacial tension is low. Moreover, the efficiency of the process is warranted if the interfacial tension tends to remain low and also under the dynamic conditions typical of the emulsification process, that is water–oil interfaces continuously renewed and submitted to stretching. This can be obtained only if the characteristic times of the adsorption of emulsifiers are short enough to balance the area enlarging inducing droplet generation.

The correlation between the average droplet size of the produced emulsions and the value of the interfacial tension has been demonstrated in many experimental works[16, 17] as well as it has been shown that the reduction of the droplet size is favored if emulsifiers presenting short characteristic time of adsorption are used. This has been shown in experimental studies concerning water-in-oil and oil-in-water emulsions.[18, 19]

Another important aspect to be considered regarding the efficiency of the emulsifiers is the capability to prevent the phenomenon of coalescence, which is especially relevant during the process of emulsification. For this aspect, appropriate emulsifiers must be present, not only strong amphiphilicity and short adsorption characteristic time but also the capability to provide a high degree of surface coverage. This property depends on their molecular structure and their arrangement at water–oil interfaces once adsorbed.[17, 20]

Results on the relation between interfacial and emulsification properties have also been reported in recent works on natural surfactants[17, 21] employed in foods, cosmetics, and pharmaceutical products. In reference [21], for example, the ability as emulsifiers of saponins of different origins is investigated in relation to their effects on interfacial tension.

Another important effect to be taken into account concerns the depletion of the surfactant solution during the emulsification, due to the huge increase of the interfacial area, associated with the fragmentation of the disperse phase. The consequent adsorption of emulsifiers at these freshly formed water–oil interfaces may result in an important impoverishment of the matrix phase of the produced emulsions, which must be considered for formulation.

This phenomenon has been taken into account in many studies on emulsions.[9, 17, 20] In reference [17] in particular, it has been shown that the knowledge of the adsorption of thermodynamic properties of the emulsifiers and, in particular, some isotherm parameters such as the surface activity and molecular area, makes it possible to estimate the concentration of the depleted continuous phase from the drop sizes of the formed emulsions. In practice, a relation among the initial concentration of the solution, the concentration of the depleted matrix phase, and the droplet size distribution, in a quasi-equilibrium condition achieved just after emulsification, can be expressed as the solution of the following set of equations:

$$\begin{cases} c = c_0 - 3\dfrac{\sum_i f_i r_i^2}{\sum_i f_i r_i^3}\phi\Gamma, \\ c = c(\Gamma) \end{cases} \tag{5.1}$$

where c_0 is the concentration of the continuous phase before emulsification, ϕ is the volume ratio of the disperse phase, and f_i are the frequencies corresponding to the radii r_i of the distribution. The second equation of the set (Equation 5.1) expresses the adsorption isotherm, which can be evaluated for the specific system through equilibrium interfacial tension versus concentration measurements, adopting appropriate thermodynamic models. This relation will be written in terms of key isotherm parameters such as the surface activity and molecular area.

For narrow-size distributions, the first equation of the set (Equation 5.1) can be approximated by introducing the average radius, r_{avg}, as

$$c = c_0 - 3\frac{\phi\Gamma}{r_{avg}}. \tag{5.2}$$

Notice that this is important for the prediction of the emulsion morphology because it allows the minimum amount of emulsifiers necessary for a given droplet size distribution to be estimated.

Another effect that can contribute to the depletion of the emulsion matrix phases is the partitioning of the emulsifier between the two liquid phases, which is particularly relevant when nonionic surfactants without negligible solubility in both water and oil phases are used as emulsifiers. The transfer of the surfactant across the water–oil interfaces and the consequent impoverishment of one phase with respect to the other one is an important aspect that must be taken into account because it can affect the dynamics of the interfacial layers and modify the morphology and stability of the corresponding emulsions.

The approach above reported to calculate the depletion due to adsorption can be adapted to take into account also the depletion due to the transfer of surfactant in the dispersed phase. This can be done by adding the term $-cK\phi$ in the mass balance (first in the set of Equation (5.1)), where K is the partition coefficient or the equilibrium ratio between the concentration of the disperse and continuous phases.

Partitioning of nonionic surfactants and, in general, their solubility in both liquid phases has been widely investigated,[22] and methods to quantify this property through the measurement of the partition coefficient have been developed.[23, 24]

The solubility of emulsifiers in both liquid phases can also be responsible for the formation of emulsions with a complex morphology such as, for example, double emulsions where the disperse phase is an emulsion in turn.

Double emulsions are commonly obtained in some application fields, like cosmetics or foods, where the presence of mixtures of lipophilic and hydrophilic surfactants[25, 26] results in the coexistence of water-in-oil and oil-in-water emulsions. According to the empirical Bancroft rule, in fact, the continuous phase in emulsions is the liquid where the emulsifier is predominantly soluble. This condition for the onset of double emulsification can be encountered also with emulsifiers presenting comparable solubility in water and in oil phases, as just observed in some experimental studies on emulsions stabilized by nonionic surfactants.[17]

5.3 THE ROLE OF INTERFACIAL PROPERTIES IN EMULSION EVOLUTION

The principal mechanisms governing the emulsion evolution are gravity-driven separations, such as creaming and sedimentation, flocculation, coalescence, and Ostwald ripening. While creaming and sedimentation are essentially governed by the bulk properties of the two liquids involved, mainly the density difference and the droplet size but also the viscosity, the other phenomena are related, more or less directly, to the properties of the liquid–liquid interface.[27] Figure 5.1 reports a sketch of these mechanisms.

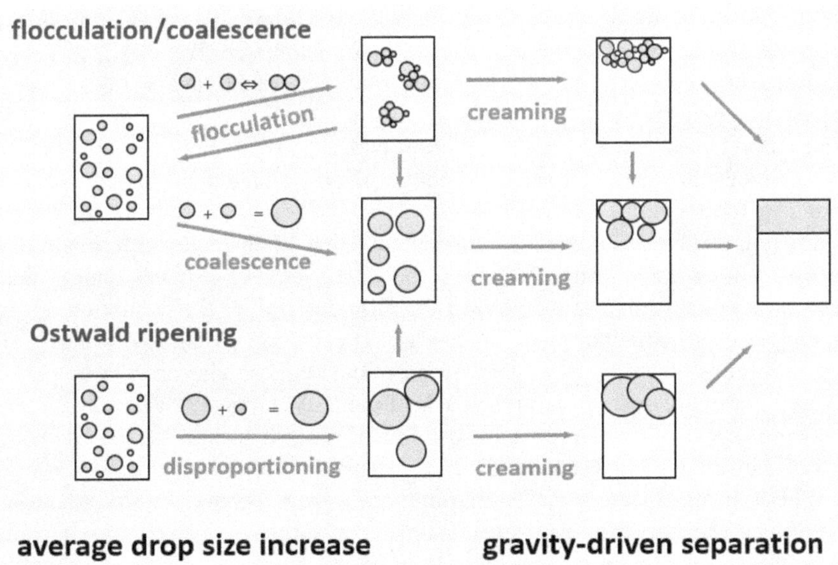

FIGURE 5.1 Schematic description of the main mechanisms governing the emulsion evolution. Adapted from Ravera et al.[27] Copyright (2020) with permission from Elsevier.

Flocculation consists in the formation of droplet aggregates, driven by the interaction of the adsorbed layers at the surface of the emulsion droplets. Coalescence is the merging of two droplets into a single one, occurring when moving droplets get in contact or consequent to their flocculation. It is essentially related to the thinning and rupture of the liquid film between droplets and, for that, strongly influenced by the interfacial properties of the adsorbed layers. Ostwald ripening (OR) consists of a partial dissolution of the dispersed liquid phase induced by the capillary pressure, resulting in a mass transfer from small to big droplets. Though mainly driven by the solubility of the disperse liquid, the surfactant adsorption at droplet interfaces and the rheological properties of the interfacial layers can influence, in principle, this phenomenon.

It is important to underline that, in real systems, these destabilizing mechanisms are all interconnected and they influence each other during the emulsion evolution. As an example, the gravity-driven creaming/flocculation, even if it does not modify the drop size distribution, which is instead affected by coalescence and Ostwald ripening, tends to make more compact the droplet population characterized by higher size and, consequently, may favor the onset of flocculation and coalescence. On the other side, the increase of the average droplet sizes induced by coalescence and Ostwald ripening tends to quicken the creaming process and eventually the destabilization of the emulsion. This is the principal motivation for investigating emulsions under weightless conditions where, as better explained in the following section, the effects of the single mechanisms related to the interfacial properties on the kinetics and stability can be assessed more in detail because gravity-driven effects are excluded.

In the following, these mechanisms are discussed singularly in order to describe in more detail the role that the adsorption of emulsion stabilizers and, in general, the thermodynamic and dynamic properties of the interfacial layers at the water–oil interface may play on them.

The theoretical treatment of flocculation and coalescence phenomena as destabilizing processes of emulsions has been well consolidated for many years and reported in fundamental reference literature.[28–30] When two emulsion droplets are forced to interact, because of coming in contact during their random motion or under the effect of long-range attractive forces, their deformable interfaces tend to flatten generating a liquid film. The stability of this film determines the formation of droplet aggregates, or flocculation, which can also involve more than two droplets. The eventual rupture of the film induces droplet coalescence. The properties of these films are commonly dealt with on the basis of the DLVO theory,[30] where the interaction potential between two approaching colloidal spheres is expressed as the combination of the van der Waals attraction and electrostatic repulsion.

On the basis of this theory, a key parameter for the experimental investigation of liquid films is the disjoining pressure, V_d, or surface force between the two interfaces limiting the liquid film, as a function of their distance, or film thickness. According to the DLVO theory, the disjoining pressure is the sum of long-range repulsive electrostatic forces and attractive van der Waals forces, while in the presence of adsorbed interfacial layers, further terms must be considered, depending on the surface-active species like, for example, short-range repulsive steric forces. Thus, the feature of the disjoining pressure versus the film thickness, or disjoining pressure isotherm, presents a minimum in correspondence to the irreversible aggregation, associated with the formation of a thin film between the droplets, termed 'Newton Black films', with a typical thickness of 10 nm or less. The reversible aggregation, or flocculation, is instead associated with a secondary minimum, corresponding to the formation of thicker films (around 50–100 nm), termed 'common black films'.

The relation between the feature of the disjoining pressure isotherm and the properties of the adsorption layers at the single interface has been widely investigated, especially for ionic and nonionic surfactants[12] and proteins,[13, 31] and the results have been put in relation with the stability of the respective emulsions.

From these studies, it is evident that the principal characteristics of the adsorption layers improving the stability against flocculation and coalescence are related to the electrostatic and steric repulsion between droplets, obtained by an adequate surface charge. This condition is typically obtained by using ionic surfactants. Nonionic surfactants may have a role in the stabilization against coalescence and Ostwald ripening, related to the rheological properties of the adsorption layers, but are not effective against flocculation because they do not provide sufficient electrostatic droplet repulsion.

As concerns in particular coalescence, one has to consider that, even if ionic surfactants, inducing repulsive interactions between the adsorbed layers, are surely effective stabilizers, the thinning and rupture of the liquid film between the emulsion droplets is a complex hydrodynamic process, not only influenced by the electrical and steric repulsion but, in general, by the adsorption properties of the emulsifiers. These determine, in fact, the degree of the interfacial coverage and the key characteristics of the interfaces, like dynamic interfacial tension and dilational rheology, which are the basis of the mechanical response of the film to perturbation.

In the case of ionic surfactants, the electrical repulsion is surely the most relevant stabilizing effect, and a small amount of adsorbed molecules is sufficient to improve remarkably the stability of emulsions, while for nonionic surfactants, a more important role is played by other equilibrium and dynamic adsorption properties.[27]

The interfacial rheology of the adsorbed layers plays a fundamental role in the fluid dynamic behavior of the liquid films between droplets in an emulsion. The contrast to the thinning and rupture of these films depends on how the adsorbed emulsifiers respond to the perturbation of the interfacial layer. The key rheological parameter mainly responsible for such response is the dilational viscoelasticity E, defined as the variation of the interfacial tension, γ, associated with compression/extension of the interfacial layer, that is

$$E = \frac{\delta\gamma}{\delta \ln A_0}. \tag{5.3}$$

According to classical approaches,[32, 33] E can be expressed in the domain of frequency, ν, as a complex quantity,

$$E(\nu) = E_r + iE_i = E_0\kappa(\nu), \tag{5.4}$$

where $E_0 = -\partial\gamma / \partial ln\Gamma$ is the Gibbs elasticity, i.e., a thermodynamic quantity that can be calculated through the adsorption isotherm of the system. The complex character and the frequency dependence of the term κ express the viscoelastic behavior of the adsorbed layer strictly related to the adsorption dynamics. Theoretical expressions for this factor can be found assuming different models for the adsorption dynamics.[34] In particular cases where the adsorption from the film bulk is negligible, like with insoluble surfactant or very depleted matrix phases, and if the relaxation processes within the adsorption layer are excluded, κ is equal to 1 and the interfaces behave as purely elastic.

It is clear from its definition that high values of the dilational elasticity correspond to an increase in interfacial rigidity under hydrodynamic conditions. This implies a reduction of the rate of drainage in the film between the emulsion droplets, which is a stabilizing effect against coalescence.

Another important effect concerns the behavior of thin films under stable or meta-stable conditions. This is the case, for example, of the films between flocculated droplets. A small mechanical perturbation of the adsorbed layer may induce a local adsorption dilution associated with a surface gradient of the interfacial tension. Such gradient, essentially expressed by the Gibbs elasticity, induces a surface flow, consequent to the restoring of the interfacial tension uniformity (Marangoni effect). This drags a bulk flow in the film which contrasts its thinning and, consequently, the drop coalescence. This effect is schematically described in Figure 5.2.

In the case of soluble surfactants and relevant interfacial relaxation processes, the interfacial tension gradient induced by the local depletion of the adsorption layer is not simply expressed by the Gibbs elasticity but by an extended expression of Equation (5.4). In fact, adsorption kinetics is described by more complicate models that take into account the involved relaxation processes.[33] In the case of soluble surfactants, for example, the adsorption from the continuous phase tends to dampen the interfacial tension gradient induced by the area perturbation, depending on the adsorption–diffusion properties of surfactants. The effect above described, stabilizing the film against thinning, is in this case weaker in comparison to the case of insoluble surfactant. In most cases, however, even for soluble surfactants, the

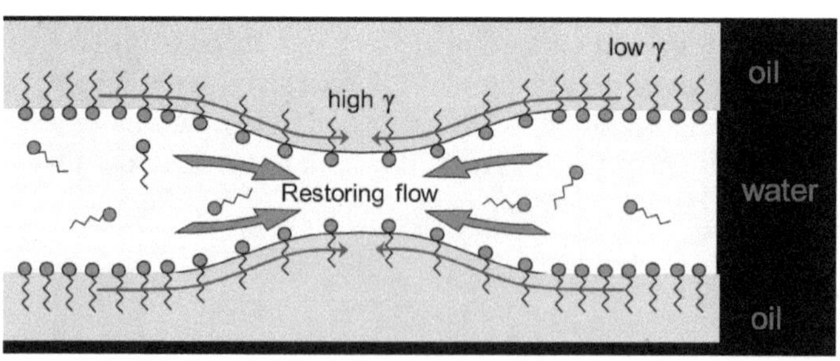

FIGURE 5.2 Schematic description of the mechanism of stabilization of the liquid film between two oil drops by Marangoni effect, contrasting drop coalescence. Adapted from Ravera et al.[27] Copyright (2020) with permission from Elsevier.

finiteness of the liquid film volume makes it impossible to restore the adsorption gradient by surfactant transfer from the matrix phase to the interface. The restoring is then also governed in this case by the Gibbs elasticity. This leads to the conclusion that the systems characterized by high value of Gibbs elasticity present better stability against coalescence.

Moreover, in the presence of kinetic processes within the adsorbed layers, such as molecular reorientation or conformational changes, with characteristic times comparable with those of the perturbation, the dilational elasticity may assume values larger than the Gibbs elasticity,[34] resulting in an increased efficiency to contrast the film thinning.

Ostwald ripening (OR) is a mechanism according to which the larger droplets of an emulsion, or the larger bubbles in a foam, grow at the expense of the smaller ones also referred to as disproportionation. OR is a non-equilibrium process triggered by the difference of Laplace pressure in the emulsion droplets and driven by the transport of molecules between the droplets.[35] The larger Laplace pressure, in fact, results in the increase of the solvability of the molecules of the smaller droplets,[36] compared to those of the larger ones. In a system of droplets, that results statistically in the increase of the size of larger droplets at the expenses of the small ones, mediated by the transport of the corresponding molecules in the matrix phase. Due to thermal fluctuations, the process eventually evolves up to the complete separation of the emulsion.

OR is typically a dominant mechanism of destabilization and evolution of the bubble size distribution in foams, due to the typically large solubility of gases in water. In emulsion, the oil–water solubility is typically some order of magnitude lower, and OR is correspondingly slowed down. Nevertheless, its comprehension and effects can be important to understand the long-term stability of some emulsified products, or definitely needed, to model the behavior of geological reservoirs of oils.[14]

The current models predict growth of the droplet size following time dependence with exponents ranging between 1/2, for concentrate emulsions, and 1/3, for diluted emulsions.[37] These two values correspond to two limit situations that are the exchange of molecules between the droplets occurs through diffusion in the thin films between droplets[38, 39] or through diffusion in the bulk[36] of the matrix phase. In the cases of flocculated emulsions, the evolution of the process is, however, more complicated due to the coexistence of both types of transport.

The problem of OR in a system of droplets or bubbles, characterized by a size distribution, was first challenged by Lifshitz, Slyozov, and Wagner (LSW). Their theory[40–42] was developed in simplified conditions set for a very dilute system of droplets/bubbles, for pure liquids, and for purely diffusive transport. As evidenced in reference [35], this simplified theory predicts a critical radius, r_c, discriminating between the growing and the shrinking of droplets. At sufficiently large time, such critical radius grows accordingly to

$$\frac{d(r_c^3)}{dt} = \frac{8D\gamma Mc}{9RT\rho^2} = \Omega, \tag{5.5}$$

where c, γ, and D, are respectively the solubility of the disperse liquid, the interfacial tension, and the diffusion coefficient with respect to the matrix liquid; M and ρ are the molar mass and the density of the disperse phase; R is the gas constant, and T is the absolute temperature.

The achievement of a stationary frequency distribution, $\phi(u)$, of the scaled radii, $u=r/r_c$ is also predicted

$$\begin{cases} \phi(u)=\dfrac{81e}{\sqrt{32}}u^2(3+u)^{-7/3}\left(\dfrac{3}{2}-u\right)^{-11/3} exp\left[\left(\dfrac{2}{3}u-1\right)^{-1}\right] \quad for \ u<\dfrac{3}{2} \\ \phi(u)=0 \ for \ u\geq\dfrac{3}{2} \end{cases}. \tag{5.6}$$

Since r_c grows with time, it means that the unscaled frequency distribution $\phi(r)$ has a self-similar evolution with time. The average u in the above distribution is about 1.14, so that the cube of the average radius grows linearly with a rate of about $(1.14)^3\Omega$.

In practice, the LSW theory has been shown to be accurate in the prediction of the OR growth only for emulsions very dilute – that is, with the disperse-phase content below 5% – and even in the presence of surfactants, provided to use the correct interfacial tension.

In most cases, a linear growth of the cube of the average radius is also observed for larger disperse-phase fractions. In these cases, however, the values of Ω are typically larger than the LSW estimation, which has been interpreted as an enhancement of the dissolved molecules transport caused by the hydrodynamic effect of the presence of a large number of droplets. An example of this linear dependence found in the literature for a dodecane-in-water emulsion stabilized by SDS[20] is given in Figure 5.3.

Being proportional to the value of the Laplace pressure in the droplets, the interfacial tension plays a fundamental role in setting the driving force of the process. Thus, based on the definition of Ω, the presence of surfactant causes a slowing down of the OR. The role of surfactant appears however far more important. For dilute oil-in-water emulsions formulated with surfactant concentrations below or slightly above the micellar concentration, the experiments report[20, 43–47] Ω decreasing with interfacial tension and with the oil solubility, in agreement with the LSW theory, but with values that are variously different from those expected by the LSW theory. In fact, based on its simplified assumption, this theory does not consider additional mass transport that may arise from the presence of surfactant, in particular, the possible role of micelles as carriers for an oily disperse phase,[48] as it appears from the common understanding in detergency. This role appears however still controversial, in particular, for ionic surfactants[43] where the observed growth rates resulted to be much smaller than expected, possibly due to effects related to repulsive interactions between micelles and droplet interfaces.

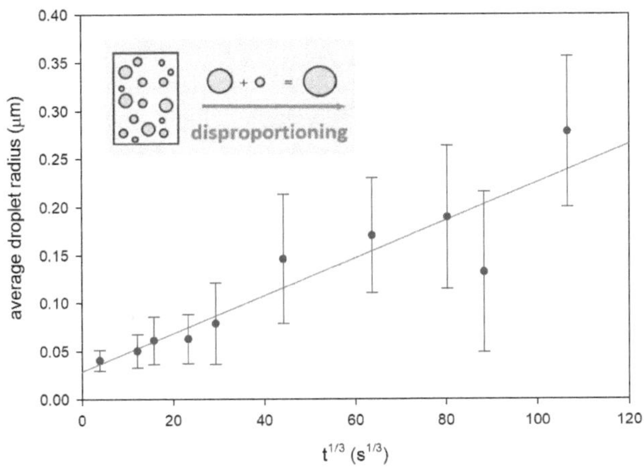

FIGURE 5.3 Example of the linear dependence of the cube of the average droplet radius with time, obtained in a water–dodecane emulsion stabilized by SDS. Adapted with permission from Llamas et al.[20] Copyright (2018) with American Chemical Society.

A stabilizing effect of adsorption interfacial layers against OR has been speculated on the basis of the dilation viscoelasticity, E. The treatment has been presented for foams,[49] but it has been extended to the cases of emulsions.[20]

According to its definition, Equation (5.4), the dilational viscoelasticity E can also be expressed in terms of the variation of the adsorption Γ, caused by relative variations of the interfacial area, A_0, i.e.,

$$E = -E_0 \frac{\delta ln\Gamma}{\delta lnA_0}. \tag{5.7}$$

The second factor in the r.h.s. represents the kinetic response of the adsorption layer and can be expressed through the mere variation of the interfacial area and the surfactant exchange between the interface and the bulk phases. This exchange tends to restore the adsorption equilibrium between bulk and interface during the dilation/compression of the interfacial layer. The values of this factor range between 0 and 1, so that E_0 is the maximum value for E.

Since the driving force for OR disappears when $dP / dr \geq 0$, using the above definition, OR is suppressed when $E \geq \gamma / 2$, which is known as the Gibbs criterion.

Besides predicting the possibility of preventing the onset of OR using proper surfactants, this criterion also states the possibility that OR could spontaneously stop when drops become sufficiently large. In fact, with the progress of OR the total interfacial area of droplets decreases, shifting the equilibrium of the interfacial layer toward larger adsorptions and, therefore, lower surface tensions.

Elaborations based on such criteria have been proposed, based on simulations and models accounting for the variation of interfacial tension and dilation elasticity/viscosity caused by the above area reduction process.[35,49–52] In most cases, however, only situations where the adsorption layer behaves purely elastic are considered, that is, the number of molecules at the interface maintains constant and, therefore, E coincides with E_0. These are typical situations referring to irreversibly adsorbed amphiphiles, which in fact maximize the value of E and, therefore the possibility to contrast OR in emulsions, using species such as proteins, polymers, or amphiphilic particles.[16, 53]

While likely possible for foams, for typical emulsions, stabilized by soluble surfactants the exchange with the bulk cannot be neglected, so that E must be considered in its general form. This is particularly true when considering that OR in emulsions is generally a slow process due to the typically limited solubility of the disperse liquid. In this case, most likely, the exchange of molecules with the bulk is able to restore the adsorption equilibrium between bulk and interface, within the timescale of the area shrinking caused by OR. In these conditions, E can be in practice vanishing.

Recently, the applicability of the Gibbs criterion in emulsions stabilized by soluble surfactants has been rationalized and quantified,[27] by comparing the characteristic time of area shrinking due to OR and that of the diffusional exchange of surfactant molecules between the bulk and the interface. The latter can be estimated as

$$\tau_D = \frac{1}{D_s}\left(\frac{d\Gamma}{dc_s}\right)^2, \tag{5.8}$$

where D_s and c_s are the diffusion coefficient and the concentration of the surfactant in the bulk.

After introducing the growth rate, Ω, the characteristic time for the area variation of a droplet subject to OR is estimated as

$$\tau_A^{OR} = \left(\frac{d\ln A}{dt}\right)^{-1} = \frac{3r^3}{2\copyright}. \tag{5.9}$$

The necessary condition to get values of E different from zero is that τ_A^{OR} is comparable to or smaller than τ_D. Equating these characteristic times results in the condition that E is vanishing for $r>r^*$, where

$$r^* = \left(\frac{2\copyright}{3D_s}\left(\frac{d``}{dc_s}\right)^2\right)^{\frac{1}{3}}, \tag{5.10}$$

and therefore ineffective in stopping OR.

Estimating $d\Gamma/dc_s$ for low-molecular-weight surfactants, within the most effective models describing the adsorption equilibrium, one realizes that the values of r^* are of the order of 1 nm, which is an unrealistic droplet size for an ordinary emulsion. It is therefore clear that for these types of surfactants, there is little chance to stabilize an emulsion against OR.

It is however important to notice that while the values of D_s remain more or less of the same order of magnitude, $d\Gamma/dc_s$ may increase even of order of magnitude with the degree of amphiphilicity of the surfactant. According to the above analysis, for these surfactants r^* can be significantly larger, providing a more realistic possibility to stabilize OR in drops of the order of 20–30 nm. Due to its weaker dependence on Ω, it appears instead less realistic to find situations where its value may provide larger values of r^*.

It appears instead less realistic, for emulsions, to find situations where larger values of r^* can be obtained varying Ω, contrary to foams for which Ω may become several orders of magnitude larger, because of the large solubility and diffusion coefficients in the water of gas molecules. In any case, in emulsions, the above analysis predicts that only for the population of droplets smaller than r^*. In that situation, one can expect that for a long time, the d.s.d. split in a superposition of the initial population of small droplets stabilized against OR and of a population of large droplets evolving according to the prediction of the LSW theory.

5.4 PARTICLE-STABILIZED EMULSIONS

Particle-stabilized emulsions, or Pickering emulsions, present unique properties regarding the control of stability, morphology, and droplet size distribution.[6, 15,53] For this reason, they are of great interest to many applications spanning from foods and cosmetics to biomedical applications and development of new nanomaterials.[54–59].

The interaction of solid particles with liquid interfaces and their effects on the interfacial properties have been widely investigated in the last decades.[60]

The driving force for the adsorption of nanoparticles to liquid–air interfaces is mainly the partial wettability, as expressed by the well-known expression of the energy variation associated with the transfer of one spherical particle of radius R, from bulk to interface,[61]

$$\Delta E = -\pi R^2 \gamma \left(1 - \cos\theta\right)^2 \tag{5.11}$$

where γ is the surface tension and θ is the equilibrium contact angle. This latter, together with the particle size, is a key parameter in determining the degree of the particle amphiphilicity. From this expression, it follows that, while micro-sized particles already at a very low contact angle remain irreversibly attached to the interface, as the energy needed to detach them is several orders of magnitude higher than the thermal energy k_BT ($\sim 10^6\ k_BT$), nano-sized particles partially wetted by water present a much weaker adsorption energy. Thus, nanoparticles (NPs) may behave as common surfactants generating, in some cases, an interfacial phase similar to a Gibbs adsorption layer with the achievement of a dynamic equilibrium for the partitioning between bulk and interface.[62–66].

In many cases, the affinity of NPs with the liquid interface is modified by adding appropriate surfactants to the systems that, adsorbing at the NP surface, induce the formation of amphiphilic complexes. The NP–surfactant interaction may be

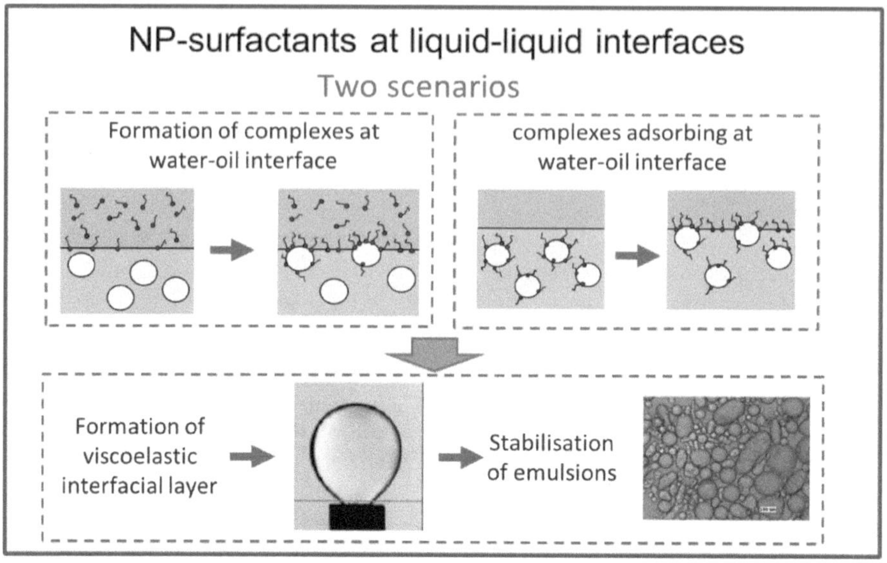

FIGURE 5.4 Schematic description of the NP–surfactant layer formation.

electrostatic, like in the case of ionic surfactants associated with hydrophilic charged NPs, or originate from hydrogen-bond or hydrophobic interactions.

For these systems, depending on the solubility in water and oil phases of the involved components, two different scenarios are in general applicable (Figure 5.4). In the first one, NPs are captured by the surfactant adsorbed at the fluid interface. This occurs, especially, in liquid–liquid systems where surfactants adsorb from a phase other than the NP dispersion.[67, 68] In these cases, stable composite interfacial layers are usually formed similar to membranes with specific structural and rheological characteristics. In the second scenario, short-chain soluble surfactants, added to the dispersions, adsorb on the surface of NPs, modifying their wettability and consequently their affinity with the liquid–liquid interface according to Equation (5.11).[69] In this case, the partial hydrophobization of the complexes results from the hydrophobic chains of the surfactant adsorbed on the NPs. Therefore, the degree of amphiphilicity of such surfactant-decorated NPs can be tuned by changing the amount of surfactant in the dispersion. Notice that the screening of the NP surface charge due to the adsorption of oppositely charged surfactants may also play a role in determining such degree of amphiphilicity. However, this second effect is, in many cases, negligible with respect to the first one.

Many experimental studies on particle-stabilized emulsions rely on nanoparticles associated with ionic surfactants, allowing the tuning of their partial hydrophilicity and, consequently, their degree of adsorption at water–oil interfaces.[69–71]

The effect of the interfacial properties on the coalescence mechanism is particularly important for this kind of emulsion. The principal reason for the stabilizing effect against the coalescence of nanoparticles or composite nanoparticle–surfactant complexes is the fact that such systems form rigid quasi-solid membranes at the drop surfaces, which hinders coalescence mainly for steric or electrostatic repulsion.[72] The surface coverage is very important for these systems, in fact, as in most cases nanoparticles adsorb irreversibly to the liquid–liquid interface, coalescence is stopped with the achievement of the surface saturation. This mechanism is at the origin of the monodisperse character typical of Pickering emulsions.

On the other hand, experimental studies on the rheological properties of composite nanoparticle–surfactant layers[67, 69, 73] also suggest that for these systems when they are rather far from saturation, a mechanical stabilizing effect may arise from the dilational viscoelastic properties of the interfacial layers, as discussed above for common surfactants stabilized emulsion. Characterizing these systems with measurements of dilational viscoelasticity versus frequency is an effective method to assess the dynamic processes occurring in the interfacial layer and to correlate these properties with the stability of emulsions. This kind of investigation is promising in view of a new formulation of Pickering emulsions at low content of solid emulsifier but still requires further investigations.

Studies on the stabilization against Ostwald ripening in Pickering emulsions can be found in recent literature[74, 75] and this process is rather important for the application of this kind of emulsions, especially because they present a rather high surface area, associated with the very low average droplet radius. Despite this aspect, the high values that the elasticity modulus of the particle-laden interfacial layers can reach ensure the Gibbs criterion for the arrest of the Ostwald ripening to be satisfied. For Pickering emulsions, however, the stabilization against OR is also related to the reduction of the droplet interfacial area effectively subjected to molecule dissolution, which is expected to be reduced by the presence of particle layers at the droplet surfaces.

It is important to highlight that particle-stabilized emulsions may offer the possibility to significantly reduce the use of surfactants. Degrees of amphiphilicity of NP–surfactant complexes, providing the required adsorption at the droplet interfaces for the emulsion stabilization, can be obtained using a rather low amount of surfactants, typically 1–2 orders of magnitude below *cmc*.[15, 69, 70, 71] Above a given hydrophobicity degree, which in many cases can be obtained with such low surfactant concentrations, the complexes adsorb irreversibly at the liquid–liquid interface and tend to saturate it, forming compact membranes, hindering coalescence by steric effects or electrostatic repulsion.[72] Far from saturation, instead, the complexes behave as soluble surfactants[67,72,73] and coalescence hindering arises from the rheological properties of the interfacial layers.

From these considerations, one can conclude that the stabilization mechanisms against coalescence and OR, typical of these composite NP–surfactant systems, may be very promising especially for the development of new formulations more biocompatible and at low environment impact, utilizing, for example, NP–surfactant systems like silica NP or biocompatible crystalline cellulose,[76–78] associated with small amounts of ionic or nonionic surfactants.

5.5 EXPERIMENTAL METHODS TO INVESTIGATE EMULSIONS

As evidenced in previous sections, the properties of the layers at the liquid–liquid interfaces have a fundamental role in determining the emulsion morphology and stability. Many experimental approaches adopted to investigate emulsions are then aimed at determining the possible correlation between interfacial properties and emulsions contributing, in this way, to the engineering of emulsion features and stability.

In particular, investigating the equilibrium and dynamic adsorption properties of the various surfactant species used as emulsifiers or emulsion stabilizers provides information on the mechanical properties of liquid interfaces and films that, as reported in previous sections, play a role during emulsification and affect drop coalescence and Ostwald ripening.

The effect of these properties may be only marginal, when electrical and steric repulsions are predominant, like in the cases of certain macromolecules, charged solid particles, or concentrated ionic surfactants. For other systems, instead, they are expected to play a fundamental role because governing the hydrodynamic response to mechanical perturbation. This is the case, for example, of emulsions stabilized by nonionic surfactants, but also more complicated systems like mixtures of nanoparticles and surfactants at moderate or low concentrations.

Thus, for these systems of particular interest are the studies on adsorption kinetics and thermodynamics of emulsifiers, allowing the adsorption properties relevant for emulsion stabilization, such as the characteristic time of adsorption, the interfacial coverage, and the rheological response of the adsorption layers, to be assessed. These studies are commonly based on the measurement of the interfacial tension against concentration, as well as rheological investigation such as measurements of the dilational viscoelasticity versus frequency, which also allow the intrinsic dynamic properties of the adsorbed layers to be accessed.

In recent literature,[17, 20, 79–82] complementary techniques are applied to compare the adsorption properties on a single liquid–liquid interface with the feature of the corresponding emulsions. In these studies, multi-technique methodologies are applied paying particular attention to the effects of partitioning and other specific characteristics typical of liquid–liquid systems, such as the low values of the interfacial tension and, in some cases, the more complex adsorption mechanisms.[16, 20]

In most of these studies, the adsorption properties of the surfactant system at water–oil interface are carried out by adopting drop tensiometers to measure the dynamic and equilibrium interfacial tension and dilational viscoelasticity and, in particular those relying on the acquisition of the drop profile, or drop shape tensiometer.[83] The measurement of the dynamic interfacial tension, from the formation of a fresh drop, until the achievement of the adsorption equilibrium allows the adsorption characteristic time to be assessed, which is a key parameter especially relevant for the process of emulsification.

The evaluation of the adsorption isotherm, or the equilibrium interfacial tension versus the surfactant concentration, allows the thermodynamic parameters characterizing the adsorption properties to be assessed. Among these are the surface activity, the interaction parameters between the adsorbed molecules and their occupational area, which give information about the coverage degree obtainable with the employed surfactants.

To assess these characteristics, it is important to adopt appropriate theoretical models for the interpretation of the experimental data. In recent years, innovative adsorption models have been developed,[84] which are in many cases extensions of the classical Langmuir or Frumkin adsorption models. These new theoretical approaches take into account multiple states of the adsorbed layers characterized by different molecular reorientation and molar area varying with the advancement of the adsorption and particularly relevant for nonionic surfactants with larger polyoxylenated chains and some macromolecules like proteins.[85]

Drop tensiometers are also commonly employed for the measurement of the dilational viscoelastic modulus versus frequency, according to the oscillating drop method.[86] Due to the limitation in the frequency range related to the violation of the quasi-equilibrium mechanical condition required for the drop shape technique,[87] capillary pressure drop tensiometers are, in some cases, more appropriate, as explained in detail in reference [88].

Also in this case, the optimization of the information from the acquired dilational rheology data is obtained adopting appropriate theoretical models to analyze the experimental data. Such models must take into account both the thermodynamic adsorption properties of the surfactants and the possible relaxation processes occurring in the interfacial layers.[33, 34]

A deeper knowledge of the dynamic response of the interface, as well as of the equilibrium parameters describing the thermodynamic adsorption properties, is important because it allows the prediction of the surfactant/emulsifier efficiency in emulsion formation and stabilization, especially as concerns the hindering of coalescence.

Examples of these kinds of studies are reported in references [17] and [20] where the above approach has been applied to investigate oil-in-water emulsions stabilized by ionic (SDS) and nonionic synthetic surfactants – Tween-80, and SDS, respectively, and by surfactants derived from vegetable sources, such as saponin and citronella glucoside (CG). Dynamic and equilibrium interfacial tension and dilational viscoelasticity were measured by drop shape tensiometry. The integrated analysis of these data allowed for assessing the appropriate interfacial model for the description of the adsorption layer.

In order to investigate the correlation between the results on the interfacial properties to the behavior of the corresponding emulsions, different approaches are usually adopted. As shown in references [17] and [20], the monitoring of the emulsion volume during its aging till the complete separation of the two liquid phases is a very simple method allowing comparative results on the behavior of different emulsifiers and/or different concentrations of the same emulsifier to be obtained. Other methods to characterize the emulsion stability and also the efficiency of emulsification rely on optical microscopy techniques and/or optical scattering techniques like classical dynamic light scattering (DLS).[89] These can provide, even if with some limitations, the drop size distributions of the emulsions at different stages of their evolution. By optical microscopy in dense emulsions, in fact, one often observes drops close to the container walls, assuming that these are representative of the population in bulk. On the contrary, by DLS, drops are characterized in the bulk of the volume, but the technique is only applicable in extremely diluted emulsions fulfilling the assumption of optical single scattering. Due to these limitations, in order to obtain reliable results, a combination of more than one technique is often employed,[94, 90] including also the diffusing wave scattering (DWS),[91] which, as explained in detail in the following, allows analyzing optically dense emulsions and obtaining information on the evolution of the average drop size and the droplet dynamics.

DWS is a correlation spectroscopy technique ideally suited for the non-invasive study of turbid systems, such as concentrated colloidal suspensions and foams, and of emulsions in particular.[92–98] From the point of view of the instrumentation, it bears striking similarities to dynamic light scattering (DLS). Coherent laser light, either expanded into a planar wave or focused, impinges on a sample cell; light scattered by the emulsion is collected by detectors placed typically in backscattering and transmission geometries. The scattered light that reaches the detectors is analyzed using the time autocorrelation function; their analysis is used to characterize the tiny relative displacements of the emulsion drops, easily reaching sub-nanometric sensitivity. As such, DWS yields information averaged on the whole bulk of the emulsion. This is a significant advantage with respect to optical techniques accessing the regions of the sample closer to the container walls, where dynamics and aging processes are influenced by the presence of the wall itself, or to investigations where the emulsion is diluted in concentrated surfactant solutions to 'freeze' emulsion aging.[90, 96]

These advantages are counterbalanced by a certain degree of difficulty in the DWS data interpretation, compared for instance to DLS. In its path of length s through the emulsion, the direction of each photon is randomized on average after a transport length l^*, roughly proportional to the number density of drops, then it can be assumed that on average light experiences s / l^* randomizing scattering events on its path to the detector. The drop motions affect the light phase, hence resulting in a dynamics speckle pattern. Detailed analysis of these fluctuations provides info on the dynamics of drops. In fact, the temporal autocorrelation of the scattered electric field $g^1(t)$ is given by a convolution of the exponential de-correlation contribution connected to a path of length s, with the path length distribution $P(s, l^*)$,[90]

$$g^1(t) = \int_{-\infty}^{\infty} P(s, l^*) \exp\left(-\frac{s}{3l^*} k_0^2 \Delta r^2(t)\right) ds, \tag{5.12}$$

$g^1(t)$ is related to the detected autocorrelation function of the scattered intensity by the Siegert relation $g^2(t) = A + \beta\left(g^1(t)\right)^2$, where the parameters A and β depend on the characteristics of the experimental geometry. Equation (5.12) makes evident the need to use more than one detection geometries (e.g., backscattering and transmission), each characterized by its own set of $P(s, l^*)$, to take into account the details of sample cell geometry and then to decouple the information related to average drop size and drop density, i.e., the light transport length l^*, from information on the drop

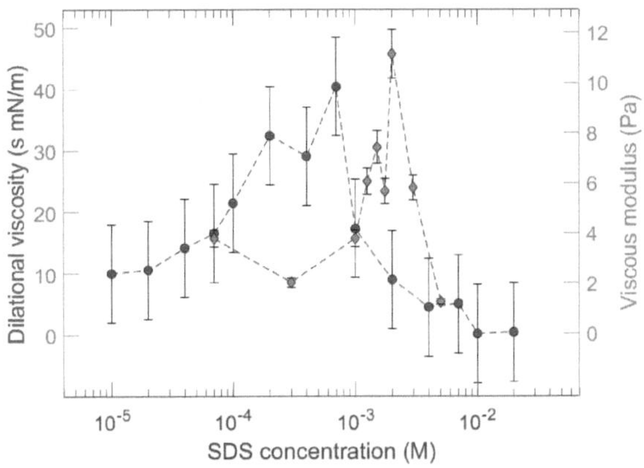

FIGURE 5.5 Comparison between shear modulus of a concentrate dodecane-in-water emulsion stabilized by SDS and the dilational viscosity of the corresponding interfacial layer. Adapted from Salerni et al.[95] Copyright (2019) with permission from Elsevier.

dynamics, contained within the temporal decay of the DWS correlation functions *A* scheme of a typical DWS experiment is reported in Figure 5.5.

Discriminating the type of drop dynamics is particularly important in emulsions, as it relates to the de-emulsification mechanism in play during aging. In concentrated emulsions, droplet dynamics and the corresponding mechanical properties on the macroscale are determined by the coupling between interfacial energy and long-range molecular interactions, which result in jamming solid-like dense packing.[96] In diluted emulsions on the contrary, the stationary dynamics is expected to be diffusive, with possible effects of confinement on the longer timescale and/or at the highest volume fractions of dispersed phase; ballistic motion, however, may occur either as a consequence of creaming or sedimentation,[92] under standard gravity conditions, or appears as a transient acceleration of the dynamics in microgravity.

For drop sizes in the colloidal regime, the analysis of Brownian diffusion yields the average drop size, in an analogy with the more common DLS sizing technique. In the case of strongly polydisperse emulsions, containing drops as large as tens and hundreds of microns in size, accounting for a large fraction of oil volume but for a small number density, do not contribute significantly to the decorrelation of the scattered intensity. Therefore, the Brownian relaxation time can be linked to the average size of the smaller drops within the emulsion.[92, 94, 95,99]

The analysis of 'ballistic' regimes observed by DWS in the early stages of emulsion aging, due to constant velocity relative motion of drops, yields information on the onset of de-emulsification mechanisms such as coalescence. In fact, a possible source of these motions is the occurrence of drop–drop coalescence, with subsequent hydrodynamic flows affecting neighboring drops, or because of Marangoni-like forces generated by surfactant gradients connected to the release of surfactant from the larger – but with less surface – resulting drop.

An example of multi-technique experimental work reporting the comparison between results on interfacial properties of an emulsion analysis through DWS is shown in reference [94]. When an interesting correlation was evidenced between the shear modulus obtained by DWS in a 50:50 water/dodecane emulsion stabilized by SDS and the dilational viscosity of the corresponding interfaces measured by tensiometric techniques.

As shown in Figure 5.5, both quantities show a surprising resemblance of the dependence on surfactant concentration, with a maximum at about the same concentration. The concentration at the maximum becomes nearly coincident if we apply the correction to obtain the actual concentration of surfactant in the emulsion matrix phase, as discussed in Section 5.2. This suggests that, at least in concentrate emulsions, the dynamic of caged droplets could be conditioned by the dilational viscoelasticity, though Marangoni-like effects, generated as in reaction to dilations of the interfacial layer caused by the confined Brownian translation of the droplets.

5.6 A SPECIAL ENVIRONMENT TO STUDY EMULSIONS: THE MICROGRAVITY CONDITIONS

As already underlined, under terrestrial laboratory conditions, the single destabilization mechanisms active in emulsions are strongly interconnected and their effects on the emulsion separation are enhanced gravity-driven droplet creaming or sedimentation. Once the size of the droplets becomes larger than a few microns, where not contrasted by large viscosities of the continuous phase, these effects become actually preponderant and lead to separation in times of the order of a few minutes. Since the mentioned single destabilization mechanisms are all increasing the average droplet size, unless making them ineffective, or, at least very slow, ordinary emulsions, even generated with sub-micrometric droplet size, will eventually separate in a short time. The study of emulsions in the absence of buoyancy, where droplets move only under the effect of Brownian forces, offers therefore the possibility to disentangle the different effects of mechanisms underlying the emulsion separation. Buoyancy can be avoided by studying emulsions made of couples of density-matched liquids, but of course that strongly limits the number of systems that can be investigated. Alternatively, nowadays emulsion can be investigated using platforms simulating the absence of gravity. These are typically free-fall laboratories. Among them, considering the typical timescales involving the destabilization process, the most suitable are orbital platforms, such as satellites or orbital vehicles. The International Space Station (ISS) offers, in particular, an ideal environment for these types of experiments that can be performed, using properly designed instrumentation for times of the order of several days or even months.

Besides providing a simplified environment for acquiring high-quality data, useful to test and develop models for the mentioned mechanisms, these experiments are also mandatory to better understand the behavior of emulsified systems under reduced gravity conditions, in order to help the development of technologies for space vehicles and satellites, or in view of designing efficient strategies in support to planetary exploration. Such technologies are especially concerned with waste recycling, separative processes, exploitation and processing of in situ resources, and heath transport technologies.

Based on that, a few experiments, specifically aimed at investigating the destabilization process on board the ISS, have been proposed and performed. Hexane/water and paraffin oil/water, stabilized either by the nonionic polyoxyethylene alkyl ether surfactant $C_{10}EO_5$ or by the sorbitan surfactant Span-80, have been investigated in an experimental campaign named FASES (Fundamental and Applied Studies in Emulsion Stability). The experiments made use of an instrument specifically developed by the European Space Agency, installed inside the 'Columbus' module of the ISS, as a scientific container of the Fluid Science Laboratory rack. The latter instrument was designed to investigate the evolution of the droplet size distribution in dilute (transparent) emulsions, with volume ratio below 5%, based on a microtomography technique providing the 3D reconstruction of an image of the emulsion volume (1 mm^3) from a series of 2D pictures of the corresponding volume taken at different deep of field.[100]

A different diagnostics housed in the same experimental module allowed following the evolution of the average droplet size in concentrate (opaque) emulsion, with a volume ratio above 20%, based on a calorimetric technique. The latter exploits the undercooling phenomena to which microscopic droplets are subjected in emulsions. Based on the current nucleation theory, the undercooling extent to which emulsion droplets remain liquid depends, in fact, on the droplet size. Thus, identifying such temperature by the heat releases during the solidification process in a calorimeter allows for estimating the average size of the droplets in the emulsion, by the use of proper nucleation modeling.[101] Based on this approach, freezing/heating cycles performed at different ages of the emulsion created and stored under microgravity conditions were then used to resolve the evolution of the average droplet size for several days.[102]

In general, both types of experiments evidenced that the process of destabilization of emulsions in reduced gravity conditions proceeds on timescales that can be even orders of magnitude larger than under normal gravity conditions. While this result is not surprising for small volume ratio emulsions, where droplets moving under Brownian effects have little probability to meet and interact, less clear is the situation with the larger volume ratios, where drops may be strongly confined, with a probability for mutual interaction of the drop layers comparable with that of drops separated by gravity effects.

The use of the DWS technique to investigate emulsions in weightlessness conditions is recently of increasing interest because can provide significant contributions to the understanding of the mechanisms behind the temporal evolution of drop populations during aging.

As reported in the previous section, the information obtained by DWS analysis on average drop size, number density, and dynamical regimes can be linked to the role of surfactants, and to the interactions between drops. Moreover, DWS could facilitate the development of rational formulation models based on concepts such as energy efficiency, preservation of resources, and improved predictability of formulation aging and shelf-life. This not only brings evident benefits for terrestrial applications but also opens up new avenues for technological advancements in space and planetary exploration.

DWS experiments performed on the International Space Station,[97,103] enabling the study of aging over prolonged times, make it possible to follow the evolution of the main descriptors of the drop size distribution during aging, as well as to disentangle different destabilization mechanisms, isolating intrinsic evolution processes. However, this complex operation has to take into account several peculiarities of the microgravity environment.

Even minor degradations of the weightlessness condition cause drop motions that, in the DWS correlation functions, are indistinguishable from dynamical features that could be attributed, for instance, to coalescence events. These degradations are typically caused by ISS orbit correction maneuvers, docking and undocking of space vehicles and cargos, and astronaut activity. To account for these, avoiding possible misinterpretation of dynamical features in the DWS correlation functions, the DWS apparatus must include a dedicated set of micro-machined accelerometers (MMA) that continuously monitor the microgravity conditions with high temporal resolution, at least 10 Hz.

Additionally, the design of the sample environment and of the DWS apparatus must meet additional constraints to adapt to the microgravity conditions. Analytical solutions of the diffusion equation exist for simplified geometries, like cells made by a wide rectangular slab with thickness $L \gg l^*$, while real cells may have different shapes because they serve other needs, e.g., to accommodate an integrated emulsification device. Therefore, sets of $P\left(s, l^*\right)$ must be computed numerically by optical Monte Carlo simulations, or by solving the diffusion equation of light, or by ray-tracing algorithms. To validate this, careful calibration procedures must be planned, as well as repeating calibration measurements on standard DWS instruments.

Finally, several sample cells must be arranged into a positioning mechanism that allows for an easy, automatic sample exchange. The accelerations of this positioning mechanism during sample exchange affect the position of drops within the emulsion. This implies that parallel investigations of several sample cells at very long aging times are not feasible.

This expertise comes from DWS campaigns in microgravity conditions performed on the 'Soft Matter Dynamics' (SMD) instrument, designed and produced by Airbus for the European Space Agency.[102] Figure 5.6 reports a scheme of the experimental apparatus used in SMD. DWS experiments on emulsion using SMD have been validated by preparatory experiments exploiting a standard DWS setup, as well as using a replica of SMD on the ground. These validation campaigns led to scrutinizing existing methods and determining the accuracy by which the DWS correlation function is determined. Emulsion aging was studied continuously for the first two hours, and then periodically for up to five days, obtaining information on the temporal evolution of the transport mean free path l^*, related to the number density of drops and to their size. In polydisperse samples, additional information from microscopy experiments can be used to derive the number density of drops from this parameter. Moreover, DWS characterized the modifications over time of the internal

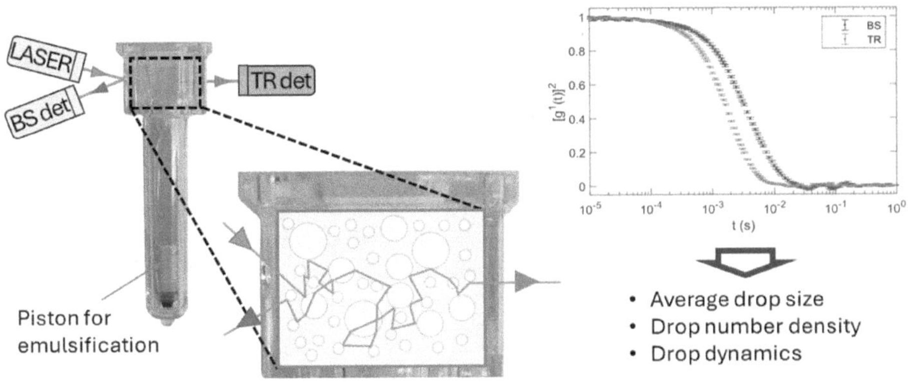

FIGURE 5.6 Scheme of a DWS experiment performed on an emulsion in backscattering (BS) and transmission (TR) geometries, using a sample cell developed by Airbus and ESA for microgravity investigation of emulsions by the 'Soft Matter Dynamics' instrument.

dynamics of the emulsion, decoupling Brownian diffusion of drops from ballistic motion that can be related to coalescence events.

5.7 CONCLUSIONS

The overview of the recent studies and adopted experimental methods presented in previous sections proves that the role played by the interfacial properties in the different processes concerned with the production, processing, and stability of emulsions has gained in the last years a good understanding.

On the other side, the development of emulsion-based technologies in the various production fields requires important challenges to be faced mainly related to sustainable development and the low impact on the environment and health.

As concerns more specifically the emulsion-related technologies and products, to answer these nowadays needs it is necessary to advance in the direction of more efficient, energy and resource-saving processes, in emulsification and emulsion breaking; the reduction of emulsifiers – in particular of surface-active species – and their substitution with compounds with a milder impact for environment and health; the increase of shelf-life; and the performance of emulsified products for consumers and industry.

To achieve these objectives, it is very important to push our knowledge to a new level of detail, in order to foster the development of methodologies and modeling tools for an on-design approach to emulsions and emulsifiers formulation.

ACKNOWLEDGMENTS

The work was supported by the ESA MAP project 'Emulsion Dynamics and Drop Interfaces (EDDI)' with the project number 4000128643/19.

REFERENCES

1. Becher, P. *Encyclopedia of Emulsion Technology*, Marcel Dekker: New York, NY, USA, 1985.
2. Wang, D.; Yang, D.; Huang, C.; Huang, Y.; et al. Stabilization mechanism and chemical demulsification of water-in-oil and oil-in-water emulsions in petroleum industry: A review. *Fuel* **2021,** *286*, 119–390.
3. Sarkar, A.; Dickinson, E. Sustainable food-grade Pickering emulsions stabilized by plant-based particles. *Curr. Opin. Colloid Interface Sci.* **2020,** *49*, 69–81.
4. Sagalowicz L.; Leser M.E. Delivery systems for liquid food products. *Curr. Opin. Colloid Interface Sci.* **2010**, *15*, 61–72.
5. Cai, Z.; Wei, Y.; Shi, A.; Zhong, J.; Rao, P.; Wang, Q.; Zhang, H. Correlation between interfacial layer properties and physical stability of food emulsions: Current trends, challenges, strategies, and further perspectives. *Adv. Colloid Interface Sci.* **2023**, *313*, 102863.
6. McClements, D.J.; Gumus, C.E. Natural emulsifiers — Biosurfactants, phospholipids, biopolymers, and colloidal particles: Molecular and physicochemical basis of functional performance. *Adv. Colloid Interface Sci.* **2016**, *234*, 3–26.
7. Meinders, M.B.J., Van Vliet. T. The role of interfacial rheological properties on Ostwald ripening in emulsions. *Adv. Colloid Interface Sci.* **2004**, *108–109*, 119–126.
8. Schmitt, V.; Cattelet, C.; Leal-Calderon, F. Coarsening of alkane-in-water emulsions stabilized by nonionic poly(oxyethylene) surfactants: The role of molecular permeation and coalescence. *Langmuir* **2004**, *20*, 46–52.
9. Li, J.; Li, L. Preparation of a novel nanoparticle with extruded soy protein isolate-oat β-glucan: Interfacial properties and mechanism of emulsion stability. *Food Hydrocolloids* **2024,** *150*, 109686.
10. Tcholakova, S.; Denkov, N.D.; Banner, T. Role of surfactant type and concentration for the mean drop size during emulsification in turbulent flow. *Langmuir* **2004,** *20*, 7444–7458.
11. McClements, D.J.; Jafari, S.M. Improving emulsion formation, stability and performance using mixed emulsifiers: A review. *Adv. Colloid Interface Sci.* **2018,** *251*, 55–79.
12. Aveyard, R.; Binks, B.P.; Esquena, J.; Fletcher, P.D.I.; Bault, P.; Villa, P. Flocculation transitions of weakly charged oil-in-water emulsions stabilized by different surfactants. *Langmuir* **2002,** *18*, 3487–3494.
13. Dickinson, E. Strategies to control and inhibit the flocculation of protein-stabilized oil-in-water emulsions. *Food Hydrocoll.* **2019,** *96*, 209–223.
14. Zhang, Q.; Li, Y.; Cao, L.; Li, L.; Huang, K.; Li, W.; Yang, C. Dilational viscoelastic properties of water-fuel interfaces in single and binary surfactant systems. *Energy Fuels* **2019**, *33*, 9055–9066.
15. Tcholakova, S.; Denkov, N.D.; Lips, A. Comparison of solid particles, globular proteins and surfactants as emulsifiers. *Phys. Chem. Chem. Phys.* **2008,** *10*, 1608–1627.
16. Goloub, T.P.; Pugh, R.J. The role of the surfactant head group in the emulsification process: Binary (nonionic-ionic) surfactant mixtures. *J. Colloid Interface Sci.* **2005**, *291*, 256–262.
17. Dziza, K.; Santini, E..; Liggieri, L.; Jarek, E.; Krzan, M.; Fischer, T.; Ravera, F. Interfacial properties and emulsification of biocompatible liquid-liquid systems. *Coatings.* **2020**, *10*, 397.
18. Santini, E.; Liggieri, L.; Sacca, L.; Clausse, D.; Ravera, F. Interfacial rheology of Span 80 adsorbed layers at paraffin oil-water interface and correlation with the corresponding emulsion properties. *Colloids Surf. A* **2007,** *309*, 270–279.

19. Kalogianni, E.P.; Varka, E.M.; Karapantsios, T.D.; Kostoglou, M.; Santini, E.; Liggieri, L.; Ravera, F. A multi-probe non-intrusive electrical technique for monitoring emulsification of hexane-in-water with the emulsifier C10E5 soluble in both phases. *Colloids Surf. A* **2010**, *354*, 353–363.
20. Llamas, S.; Santini, E.; Liggieri, L.; Salerni, F.; Orsi, D.; Cristofolini, L.; Ravera, F. Adsorption of sodium dodecyl sulfate at water-dodecane interface in relation to the oil in water emulsion properties. *Langmuir* **2018,** *34*, 5978–5989.
21. Zhu, Z.; Wen, Y.; Yi, J.; Cao, Y.; Liu, F.; McClements, D.J. Comparison of natural and synthetic surfactants at forming and stabilizing nanoemulsions: Tea saponin, Quillaja saponin, and Tween 80. *J. Colloid Interface Sci.* **2019,** *536*, 80–87.
22. Ravera, F.; Ferrari, M.; Liggieri, L. Adsorption and partitioning of surfactants in liquid-liquid systems. *Adv. Colloid Interface Sci.* **2000,** *88*, 129–177.
23. Fainerman, V.; Sharipova, A.; Aidarova, S.; Kovalchuk, V.; Aksenenko, E.; Makievski, A.; Miller, R. Direct determination of the distribution coefficient of tridecyl dimethyl phosphine oxide between water and hexane. *Colloids and Interfaces* **2018,** *2*, 28–39.
24. Ravera, F.; Ferrari, M.; Liggieri, L.; Miller, R.; Passerone, A. Measurement of the partition coefficient of surfactants in water/oil systems. *Langmuir* **1997,** *13*, 4817–4820.
25. Dickinson, E. Double emulsions stabilized by food biopolymers. *Food Biophys* **2011,** *6,* 1–11.
26. Silva, B.F.B.; Rodríguez-Abreu, C.; Vilanova, N. Recent advances in multiple emulsions and their application as templates. *Curr. Opin. Colloid Interface Sci.* **2016**, *25*, 98–108.
27. Ravera, F.; Dziza, K.; Santini, E.; Cristofolini, L.; Liggieri, L. Emulsification and emulsion stability: The role of the interfacial properties. *Adv Colloid Interf Sci* **2021,** *288*, 102344–102355.
28. Ivanov, I.B.; Kralchevsky, P.A. Stability of emulsions under equilibrium and dynamic conditions. *Colloids Surf. A* **1997**, *128*, 155–175.
29. Petsev, D.N.; Denkov, N.D.; Kralchevsky, P.A. Flocculation of deformable emulsion droplets. ii. interaction energy. *J. Colloid Interface Sci.* **1995**, *176*, 201–213.
30. Israelachvili, J.N. *Intermolecular and Surface Forces,* Academic Press: New York, NY, USA, 1992.
31. McClements, D.J. Protein-stabilized emulsions. *Curr. Opin. Colloid Interface Sci.* **2004**, *9*, 305–313.
32. Lucassen, J.; van den Tempel, M. Dynamic measurements of dilational properties of a liquid interface. *Chemical Engineering Sci.* **1972**, *27*, 1283–1291.
33. Ravera, F.; Ferrari, M.; Santini, E.; Liggieri, L. Influence of surface processes on the dilational visco-elasticity of surfactant solutions. *Adv. Colloid Interface Sci.* **2005**, *117*, 75–100.
34. Ravera, F.; Ferrari, M.; Liggieri, L. Modelling of dilational visco-elasticity of adsorbed layers with multiple kinetic processes. *Colloids Surf. A* **2006,** *282–283*, 210–216.
35. Taylor, P. Ostwald ripening in emulsions. *Adv. Colloid Interface Sci.* **1998**, *75*, 107–631
36. Voorhees, P.W. The theory of Ostwald ripening. *J. Stat. Phys.* **1985**, *38*, 231–252.
37. Cantat, I.; Cohen-Addad, S.; Elias, F.; Graner, F.; Hohler, R.; Pitois, O.; Rouyer, F.; Saint-Jalmes, A. *Foams - Structure and Dynamics,* Oxford University Press: Oxford, UK, 2013.
38. Farajzadeh, R.; Krastev, R.; Zitha, P.L.J. Foam film permeability: Theory and experiment. *Adv. Colloid Interface Sci.* **2008**, *137*, 27–44, 2.
39. Krustev, R.; Platikanov, D.; Nedyalkov, M. Permeability of foam common black films to gas. Part 2. *Colloids Surf. A* **1997**, *383*, 383–390.
40. Lifshits, I.M.; Slezov, V.V. Kinetics of the diffusion decomposition of supersaturated solid solutions. *Zhurnal. Eksp. i Teor. Fiz.* **1958,** *35*, 479–492.

41. Lifshits, I.M.; Slezov, V.V. The kinetics of precipitation from supersaturated solid solutions. *J Phys. Chem. Solids* **1961**, *19*, 35–50.
42. Wagner, C. Theorie der Alterung von Niederschlägen durch Umlösen (Ostwald-Reifung). *Ber. Bunsenges Phys. Chem.* **1961,** *65*, 581–591.
43. Hoang, T.K.N.; La, V.B.; Deriemaeker, L.; Finsy, R. Ostwald ripening of alkane in water emulsions stabilized by sodium dodecyl benzene sulfonate. *Langmuir* **2002,** *18*, 10086–10090.
44. Weiss, J.; Herrmann, N.; McClements, D.J. Ostwald ripening of hydrocarbon emulsion droplets in surfactant solutions. *Langmuir* **1999,** *15*, 6652–6657.
45. De Smet, Y.; Deriemaeker, L.; Finsy, R. Ostwald ripening of alkane emulsions in the presence of surfactant micelles. *Langmuir* **1999,** *15*, 6745–6754.
46. Soma, J.; Papadopoulos, K.D. Ostwald ripening in sodium dodecyl sulfate-stabilized decane-in-water emulsions. *J. Colloid Interface Sci.* **1996**, *181*, 225–231.
47. Kabalnov, A.S.; Makarov, K.N.; Pertzov, A.V.; Shchukin, E.D. Ostwald ripening in emulsions. 2. Ostwald ripening in hydrocarbon emulsions: Experimental verification of equation for absolute rates. *J. Colloid Interface Sci.* **1990**, *138*, 98–104.
48. Ayaprakai, S.; Dungan, S.R. Influence of surfactant structure on the contribution of micelles to Ostwald ripening in oil-in-water emulsions. *J. Colloid Interface Sci.* **2010**, *343*, 102–108.
49. Kloek, W.; Van Vliet, T.; Meinders, M. Effect of bulk and interfacial rheological properties on bubble dissolution. *J. Colloid Interface Sci.* **2001**, *237*, 158–166.
50. van Meinders, M.B.J. The role of interfacial rheological properties on Ostwald ripening in emulsions. *Adv. Colloid Interface Sci.* **2004**, *108*, 19–26.
51. Hu, X.; Miller, R.; Guo, L. Experimental study on interfacial characteristics during bubble dissolution. *Colloids Surf. A* **2016**, *505*, 179–185.
52. Salonen, A.; Gay, C.; Maestro, A.; Drenckhan, W.; Rio, E. Arresting bubble coarsening: A two-bubble experiment to investigate grain growth in the presence of surface elasticity. *Europhys. Lett.* **2016**, *116*, 46005–46011.
53. Hunter, T.N.; Pugh, R.J.; Franks, G.V.; Jameson, G.J. The role of particles in stabilising foams and emulsions. *Adv. Colloid Interface Sci.* **2007,** *137*, 57–81.
54. Wei, Y.; Tong, Z.; Dai, L.; Wang, D.; Lv, P.; Liu, J.; Mao, L.; Yuan, F.; Gao, Y. Influence of interfacial compositions on the microstructure, physiochemical stability, lipid digestion and β-carotene bioaccessibility of Pickering emulsions. *Food Hydrocolloids* **2020**, *104*, 105738–105754.
55. Binks, B.P. Colloidal particles at a range of fluid-fluid interfaces. *Langmuir* **2017**, *33*, 6947–6963.
56. Destribats, M.; Rouvet, M.; Gehin-Delval, C.; Schmitt, C.; Binks, B. Emulsions stabilised by whey protein microgel particles: Towards food-grade Pickering emulsions. *Soft Matter* **2014**, *10*, 6941–6954.
57. Bai, L.; Lv, S.; Xiang, W.; Huan, S.; McClements, D.J.; Rojas, O.J. Oil-in-water Pickering emulsions via microfluidization with cellulose nanocrystals: 1. Formation and stability. *Food Hydrocoll* **2019**, *96*, 699–708.
58. Harman, C.L.G.; Patel, M.A.; Guldin, S.; Davies, G.L. Recent developments in Pickering emulsions for biomedical applications. *Curr. Opin. Colloid Interface Sci.* **2019**, *39*, 173–189.
59. Arditty, S.; Schmitt, V.; Giermanska-Kahn, J.; Leal-Calderon, F. Materials based on solid-stabilized emulsions. *J. Colloid Interface Sci.* **2004**, *275*, 659–664.
60. Maestro, A.; Santini, E.; Guzmán, E. Physico-chemical foundations of particle-laden fluid interfaces. *Eur. Phys. J. E* **2018**, *41*, 97–118.
61. Levine, S.; Bowen, B.D.; Partridge, S.J. Stabilization of emulsions by fine particles I. Partitioning of particles between continuous phase and oil/water interface. *Colloids Surf* **1989**, *38*, 325–343.

62. Fujii, S.; Mouri, E.; Akiyama, K.; Nakayama, S.; Uda, K.; Nakamura, Y.; Matsuoka. H. pH-Sensitive adsorption behavior of polymer particles at the air– water interface. *Langmuir* **2017**, *33*, 1451–1459.
63. Schwenke, K., Isa, L., Del Gado, E. Assembly of nanoparticles at liquid interfaces: Crowding and ordering. *Langmuir* **2014**, *30*, 3069–3074.
64. Hua, X.; Bevan, M.A.; Frechette, J. Reversible partitioning of nanoparticles at an oil–water interface. *Langmuir* **2016**, *32*, 11341–11352.
65. Garbin, V.; Crocker, J.C.; Stebe, K.J. Nanoparticles at fluid interfaces: Exploiting capping ligands to control adsorption, stability and dynamics. *J. Colloid Interface Sci.* **2012**, *387*, 1–11.
66. Miller, R.; Fainerman, V.B.; Kovalchuk, V.I.; Grigoriev, D.O.; Leser, M.E.; Michel, M. Composite interfacial layers containing micro-size and nano-size particles. *Adv. Colloid Interface Sci* **2006**, *128–130*, 17–26.
67. Santini, E.; Guzmán, E.; Ravera, F.; Ferrari, M.; Liggieri, L. Properties and structure of interfacial layers formed by hydrophilic silica dispersions and palmitic acid. *Phys. Chem. Chem. Phys* **2012**, *14*, 607–615.
68. Whitby, C.P.; Fornasiero, D.; Ralston, J.; Liggieri, L.; Ravera, F. Properties of fatty amine–silica nanoparticle interfacial layers at the hexane–water interface. *J. Phys. Chem. C* **2012**, *116*, 3050–3058.
69. Ravera, F.; Ferrari, M.; Liggieri, L.; Loglio, G.; Santini, E.; Zanobini, A. Liquid-liquid interfacial properties of mixed nanoparticle-surfactant systems. *Colloids Surf. A* **2008**, *323*, 99–108.
70. Arab, D.; Kantzas, A.; Bryant, S.L. Nanoparticle stabilized oil in water emulsions: A critical review. *J. Pet. Sci. Eng.* **2018**, *163*, 217–242.
71. Vatanparast, H.; Eftekhari, M.; Javadi, A.; Miller, R.; Bahramian, A. Influence of hydrophilic silica nanoparticles on the adsorption layer properties of non-ionic surfactants at water/heptane interface. *J. Colloid Interface Sci.* **2019**, *545*, 242–250.
72. Arditty, S.; Whitby, C.P.; Binks, B.P.; Schmitt, V.; Leal-Calderon, F. Some general features of limited coalescence in solid-stabilized emulsions. *Eur. Phys. J. E* **2003**, *11*, 273–281.
73. Liggieri, L.; Santini, E.; Guzmán, E.; Maestro, A.; Ravera, F. Wide-frequency dilational rheology investigation of mixed silica nanoparticle-CTAB interfacial layers. *Soft Matter* **2011**, *7*, 7699–7709.
74. Thompson, K.L.; Derry, M.J.; Hatton, F.L.; Armes, S.P. Long-term stability of n-alkane-in-water Pickering nanoemulsions: Effect of aqueous solubility of droplet phase on ostwald ripening. *Langmuir* **2018**, *34*, 9289–9297.
75. Hunter, S.J.; Chohan, P.; Varlas, S.; Armes, S.P. Effect of temperature, oil type, and copolymer concentration on the long-term stability of oil-in-water Pickering nanoemulsions prepared using diblock copolymer nanoparticles. *Langmuir* **2024,** *40*, 3702–3714.
76. Liu, L.; Hu, Z.; Sui, X.; Guo, J.; Cranston, E.D.; Mao, Z. Effect of counterion choice on the stability of cellulose nanocrystal Pickering emulsions. *Ind. Eng. Chem. Res.* **2018**, *57*, 7169–7180.
77. Dai, H.; Wu, J.; Zhang, H.; Chen, Y.; Ma, L.; Huang, H.; Huang, Y.; Zhang, Y. Recent advances on cellulose nanocrystals for Pickering emulsions: Development and challenge. *Trends Food Sci. Technol.* **2020**, *102*, 16–29.
78. Bazazi P.; Hejazi S.H. Cellulose nanocrystal laden oil–water interfaces: Interfacial viscoelasticity, emulsion stability, and the dynamics of three-phase contact-lines. *Ind. Eng. Chem. Res.* **2021**, *60*, 4892–4902.
79. Zhang, X.; Qi, B.; Xie, F.; Hu, M.; Sun, Y.; Han, L.; Li, L.: Zhang, S.; Li, Y. Emulsion stability and dilatational rheological properties of soy/whey protein isolate complexes at the oil-water interface: Influence of pH. *Food Hydrocolloids* **2021,** *113*, 106391.

80. Ho, T.M.; Razzaghi, A.; Ramachandran, A.; Mikkonen K.S. Emulsion characterization via microfluidic devices: A review on interfacial tension and stability to coalescence. *Adv. Colloid Interface Sci.* **2022,** *299,* 102541.
81. Krstono, V.; Pavlovic, N.; Nikolic, I.; Milutinov, J.; Cirin, D. Physicochemical properties and stability of oil-in-water emulsions stabilized by soy protein isolate and xanthan gum. *Int. J. Biol. Macromol.* **2024,** *260*, 129610.
82. Lei, J.; Gao, Y.; Ma, Y.; Zhao, K.; Du, F. Improving the emulsion stability by regulation of dilational rheology properties. *Colloids Surf. A* **2019**, *583*, 123906–123917.
83. Loglio, G.; Pandolfini, P.; Liggieri, L.; Makievski, A.V.; Ravera, F. Determination of interfacial properties by the pendant drop tensiometry: Optimisation of experimental and calculation procedures. In *Bubble Drop Interfaces*, Miller, R., Liggieri, L., Eds.; Bril: Leiden, The Netherlands, 2011; 385–400.
84. Fainerman, V.B.; Kovalchuk, V.I.; Aksenenko, E.V.; Ravera, F.; Liggieri, L.; Loglio, G.; Makievski, A.V.; Mishchuk, N.O.; Schneck, E.; Miller, R. A multistate adsorption model for the characterization of C_{13}DMPO adsorption layers at the aqueous solution/air interface. *Langmuir* **2022**, *38*, *16*, 4913–4920.
85. Fainerman, V.B.; Aksenenko, E.V.; Krägel, J.; Miller, R. Thermodynamics, interfacial pressure isotherms and dilational rheology of mixed protein-surfactant adsorption layers. *Adv. Colloid Interface Sci* **2016**, *233*, 200–222.
86. Ravera, F.; Loglio G.; Kovalchuk, V.I. Interfacial dilational rheology by oscillating bubble/drop methods. *Curr. Opin. Colloid Interface Sci.* **2010**, *15*, 217–228.
87. Leser, M.E.; Acquistapace, S.; Cagna, A.; Makievski, A.V.; Miller, R. Limits of oscillation frequencies in drop and bubble shape tensiometry. *Colloids Surf. A* **2005**, *261*, 25–33.
88. Ravera, F.; Loglio, G.; Pandolfini, P.; Santini, E.; Liggieri, L. Determination of the dilational viscoelasticity by the oscillating drop/bubble method in a capillary pressure tensiometer. *Colloids Surf. A* **2010**, *365*, 2–13.
89. Berne, B.J.; Pecora, R. *Dynamic Light Scattering*, Dover: New York, NY, USA, 2000.
90. Orsi, D.; Salerni, F.; Macaluso, E.; Santini, E.; Ravera, F.; Liggieri, L.; Cristofolini, L. Diffusing wave spectroscopy for investigating emulsions: I. Instrumental aspects. *Colloids Surf. A* **2019**, *580*, 123574–123584.
91. Weitz, D.A.; Pine, D.J. Diffusing-wave spectroscopy. In *Dynamic Light Scattering: The Method and Some Applications,* Brown, W., Ed.; Clarendon Press: Oxford, UK, 1993.
92. Höhler, R.; Cohen-Addad, S.; Durian, D.J. Multiple light scattering as a probe of foams and emulsions. *Curr. Opin. Colloid Interface Sci.* **2014**, *19*, 242–252.
93. Lorusso, V.; Orsi, D.; Salerni, F.; Liggieri, L.; Ravera, F.; McMillin, R.; Ferri, J.;. Cristofolini, L. Recent developments in emulsion characterization: Diffusing Wave Spectroscopy beyond average values. *Adv. Colloid Interface Sci.* **2021**, *288*, 102341–102356.
94. Pasquet, M.; Galvani, N.; Requier, A.; Cohen-Addad, S.; Höhler, R.; Pitois, O.; Rio, E.; Salonen, A.; Langevin, D. Coarsening transitions of wet liquid foams under microgravity conditions. *Soft Matter* **2023**, *19*, 6267–6279.
95. Salerni, F.; Orsi, D.; Santini, E; Liggieri, L..; Ravera, F.; Cristofolini, L. Diffusing wave spectroscopy for investigating emulsions: II. Characterization of a paradigmatic oil-in-water emulsion. *Colloids Surf. A* **2019**, *580*, 123724–123731.
96. McMillan, R.E.; Orsi, D.; Cristofolini, L.; Ferri, J.K. Particle sizing in non-dilute dispersions using diffusing wave spectroscopy with multiple optical path lengths. *Colloids Interface Sci. Comm.* **2022,** *49*, 100641–100649.
97. Cohen-Addad, S.; Höhler, R. Rheology of foams and highly concentrated emulsions. *Curr. Opin. Colloid Interface Sci.* **2014**, *19*, 536–548.

98. Carpy, R.; Picker, G.; Amann B.; Ranebo, H.; Vincent-Bonnieu, S.; Minster, O.; Winter, J.; Dettmann, J.; Castiglione, L.; Höhler, R.; Langevin D. Foam generation and sample composition optimization for the FOAM-C experiment of the ISS. *J Phys Conf Ser* **2011**, *327*, 012025–012031.
99. Scheffold, F. Particle sizing with diffusing wave spectroscopy. *J. Dispers. Sci. Technol.* **2002**, *23*, 591–599.
100. Antoni, M.; Krägel J.; Liggieri, L.; Miller, R.; Sanfeld, A.; Sylvain, J.D. Binary emulsion investigation by optical tomographic microscopy for FASES experiments. *Colloids Surf. A* **2007**, *309*, 280–285.
101. Clausse, D.; Gomex, F.; Dalmazzone, C.; Noik, C. A method for the characterization of emulsions, thermogranulometry: Application to water-in-crude oil emulsion. *J. Colloid Interface Sci.* **2005,** *287*, 694–703.
102. Drelich, A.; Dalmazzone, C.; Pezron, I.; Liggieri, L.; Clausse, D. DSC (Differential Scanning Calorimetry) used to follow the evolution of W/O emulsions versus time on ground and in space in the ISS. *Oil & Gas Science and Technology – Rev. IFP Energies nouvelles* **2018**, *73*, 16–25.
103. Born, P.; Braibanti, M.; Cristofolini, L.; Cohen-Addad, S.; Durian, D.J.; Egelhaaf, S.U.; Escobedo-Sánchez, M.A.; Höhler, R.; Karapantsios, T.D.; Langevin, D.; Liggieri, L.; Pasquet, M.; Rio, E.; Salonen, A.; Schröter, M.; Sperl, M.; Sütterlin, R.; Zuccolotto-Bernez, A.B. Soft matter dynamics: A versatile microgravity platform to study dynamics in soft matter. *Rev. Sci. Instr.* **2021**, *92*, 124503–124518.

6 Thin Liquid Emulsion Films

Georgi Gochev, Jan Zawala, Plamen Tchoukov, Ljubomir Nikolov and Elena Mileva

6.1 INTRODUCTION

Thin liquid films (TLF) are ubiquitous building blocks of soft colloids and play the lead role in the stabilization mechanisms in foams[1] and emulsions.[2–8] Formation, properties and stability of emulsions and the corresponding thin liquid emulsion films formed between interacting emulsion droplets have been studied for more than a century and were reviewed in many articles in scientific journals and books,[2–8] as well as in the current book, where various aspects of the state-of-the-art of the emulsion science are discussed. Concerning stability of emulsions in terms of inter-droplet surface forces, the early works by Sonntag and coworkers[5] and Bibette and coworkers.[7] are noteworthy. Before starting our overview on emulsion films, we recall that there are several types of emulsion systems. Generally, two-phase water-oil systems can be divided into two types according to their continuous and dispersed liquid phases: (1) *simple* emulsions consisting of dispersed monophasic droplets in a continuous liquid matrix, such as water droplets in oil (W/O emulsions) or oil droplets in water (O/W emulsions); and (2) *multiple* (or *double*) emulsions consisting of dispersed biphasic droplets in a continuous liquid matrix. Such biphasic droplets (here called *secondary* droplets) contain dispersed droplets of the same type as of the continuous phase (here called *primary* droplets). Multiple emulsions are typically categorized into two types, namely W/O/W emulsions (*primary* water droplets – in *secondary* oil droplets – in water) and O/W/O emulsions (*primary* oil droplets – in *secondary* water droplets – in oil). Figure 6.1a and b present schemes of various interactions of droplets in emulsions and the formation of *symmetric* emulsion films: water-oil-water (W/O/W) or oil-water-oil (O/W/O) films. An interesting case are three-phase oil-water-air (O/W/A) dispersed systems, such as aerated emulsions (*foamulsions*)[9] or foams containing small emulsified oil droplets.[10] In such three-phase systems, *asymmetric* O-W-A films, so-called *pseudo-emulsion* films, are formed (Figure 6.1c).[10,11] In this chapter, we are not considering in detail this exotic type of TLFs, but it is noteworthy that comparison of the disjoining pressure isotherms for a foam film and a pseudo-emulsion film obtained from equivalent aqueous solutions of the well-known surfactant sodium dodecyl sulfate (SDS) revealed that the pseudo-emulsion film is considerably less stable than the corresponding foam film.[11]

A substantial difference between O/W/O and W/O/W films is that the latter are characterized by low permittivity and electroconductivity. Thus, in W/O/W films,

 DOI: 10.1201/9781032636184-7

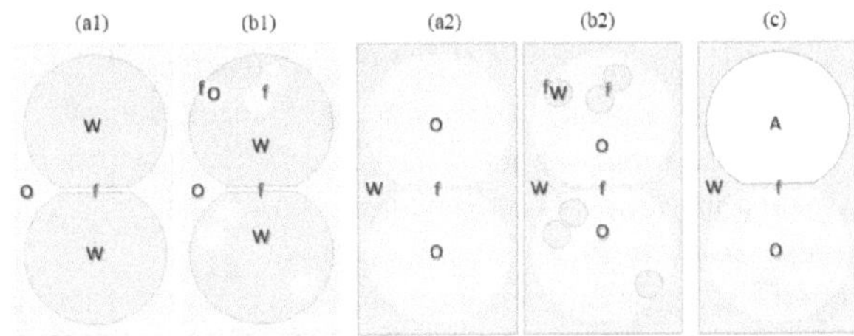

FIGURE 6.1 Schemes of droplet interactions in emulsions (not to scale; simplified with respect to droplets sizes). (a) *Simple* emulsions; W/O/W film in W/O emulsion (a1) and O/W/O film in O/W emulsion (a2). (b) *Multiple* emulsions; interactions of *primary* droplets or *secondary* droplets as well as of a *primary* droplet with the continuous phase; both types of O/W/O and W/O/W *symmetric* emulsion films can be formed either in a O/W/O (b1) or in a W/O/W (b2) emulsions. (c) *Asymmetric* 'pseudo-emulsion' O/W/A film. Letter code: W – water, O – oil, A – air, f – film.

electrostatic interaction surface forces are inefficient.[12–15] On the contrary, O/W/O films can be stabilized by electrostatic forces as the equilibrium film thickness is determined by the magnitude of the electrical double-layer potential φ_0 at the film's surfaces and the solvent conditions (ionic strength I, and pH); however, when the electrostatic forces are suppressed, O/W/O films can be stabilized by other types of surface forces,[15–18] including short-range interactions in Newton black films (NBF) and bilayer lipid membranes (BLM).[1,19] The equilibrium W/O/W films can be stabilized by steric and structural types of surface forces.[12–14,20] All these interactions in symmetric emulsion films are discussed in the following sections.

In relevance to emulsion stability, various classes of amphiphilic molecules have been and still are under careful examination by researchers, yet the hunt for the perfect emulsifier is continuously running.[21] The emulsifying tendency of a surface-active agent to form O/W or W/O emulsion is usually related to the Bancroft's rule '*the phase in which the stabilizing agent is more soluble will be the continuous phase*' and can be quantified by the hydrophile-lipophile balance (HLB) number.[15,22] Both types of O/W/O or W/O/W films can be stabilized by appropriate emulsifiers.[16–20] However, the majority of known amphiphilic molecules are well soluble in water and O/W/O film stabilized by various types of low-molecular-weight (LMW) surfactants,[16] polymers,[17,18] and proteins[23–25] have been studied in the literature for several decades.

The process of coalescence of emulsion droplets can be investigated by various experimental approaches.[26,27] Notably, the formation, drainage, and rupture of the relevant emulsion films are investigated by the model system of a single microscopic emulsion film via different experimental approaches,[6–8,16,23] which will be considered in Section 6.3. Molecular dynamics (MD) simulations were also involved in studies on emulsion films. Rekvig et al.[28,29] used a hybrid *dissipative particle dynamics/Monte Carlo* simulation scheme to predict the interaction forces in both types of O/W/O and W/O/W emulsion films stabilized by model hypothetical LMW surfactants and to describe the mechanism of film rupture (in agreement with a channel nucleation theory). The results revealed the action of steric forces between overlapping hydrophobic chains in W/O/W films. Aarøen et al.[30] reported combined results from optical tweezers experiments and MD simulations.

It should be emphasized that the advances in our current understanding of TLFs were based mostly on a large body of studies on foam films,[1,16,25,31] whereas emulsion films are studied less extensively.[16–18,20,23–25] As this chapter is dedicated to emulsion films, we try to keep the focus on the theme and thus refer to foam film studies only in cases needed to guide the reader. Note that, on the one hand, there are many common features in the physical chemistry of foam films and O/W/O films, and on the other hand, the nature of the condensed oil phase inevitably implies certain differences between these two types of thin aqueous films.[32–34]

6.2 THEORETICAL BASICS OF THIN LIQUID FILMS

The thin liquid film problem has a long history – beginning in the 17th century with curious experiments by prominent scientists and their attempts for measurement of the film thickness as the most intriguing TLF parameter, through the introduction of the concept of the *disjoining pressure* by Derjaguin in the 1930s, and the appearance of the celebrated DLVO (Derjaguin-Landau-Verwey-Overbeek) theory, and up to date.[35] In this section, some elements of the basic theoretical aspects of TLF dynamics are briefly outlined. For further details on the topic one can look, for instance, in Refs 1,15,27,32, 36, and 37.

6.2.1 Elements of Thin Liquid Film Thermodynamics

The coalescence stability of the emulsion systems is closely related to the structural details of the liquid phases in the thin-liquid-layer domain. It is well-known that the configuration of the liquid elements in the vicinity of a fluid boundary is different in comparison to the bulk medium.[1,32,38] These peculiarities are consequences of the essential reorganizations in the interfacial area between the adjacent bulk liquids and lay the basis of the concept of interfacial (surface) tension at a liquid-fluid interface. According to Bakker, the interfacial tension σ between two isotropic bulk phases is defined as[1,32,38]

$$\sigma = \int_{-\infty}^{+\infty} \left[P^{n} - P^{\tau}(z) \right] \mathrm{d}z, \tag{6.1}$$

where P is the pressure and the superscripts (^{n}) and ($^{\tau}$) denote normal and tangential coordinates in a local coordinate system. In the case of a liquid phase A situated in-between two bulk liquid phases B (droplets), a 'thick' liquid layer is formed if the thickness of the gap between the approaching droplets is significantly smaller compared to the droplets' sizes (Figure 6.2a1) and the definition in Equation (6.1) is valid for either interface (B/A and A/B). However, when both interfacial layers are overlapping, the properties of the thin liquid layer are different from the initial bulk of the A phase and an emulsion TLF is formed (Figure 6.2a2).[1,32,36,39]

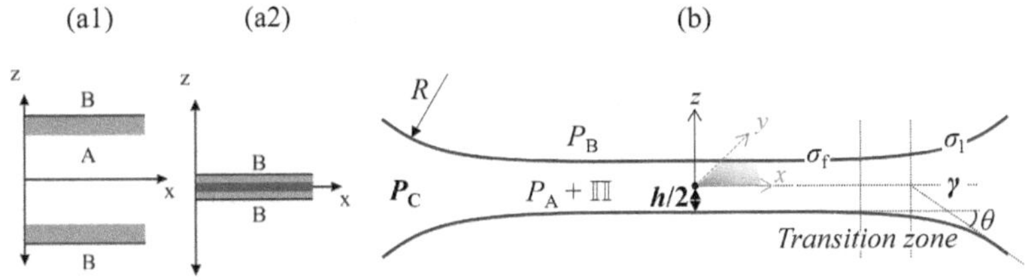

FIGURE 6.2 (a) Schematic presentation of a symmetric plane-parallel liquid layer (phase A) between emulsion droplets (phase B): a thick liquid layer of phase A with bulk properties (a1) and a TLF, where interfacial regions overlap (a2). (b) Symmetric plane-parallel TLF in equilibrium with a bulk meniscus. Denotations: R – radius of the meniscus curvature; P_A – pressure in the film phase A; P_B – pressure in the droplet phase B; P_C – capillary pressure in the meniscus; Π – disjoining pressure; h – film thickness; σ_l – surface tension of the bulk liquid; σ_f – surface tension of the film's surfaces; γ – film tension; θ – contact angle.

In line with the Bakker's formula (Equation (6.1)) and postulating $P^n = \text{const}$, the interfacial tension σ_f at each film's interface may be presented as (see also Figure 6.2b)[1,32]

$$\sigma_f = \int_0^{h/2} \left[P_A - P^\tau(z)\right] dz + \int_{h/2}^{\infty} \left[P_B - P^\tau(z)\right] dz. \tag{6.2}$$

This constitutes the so-called TLF model with two Gibbs dividing planes. Due to the overlap of the transition interfacial regions (Figure 6.2a2), there is anisotropy of the pressure tensor in the film bulk. This has been accounted for by the introduction of the concept of thickness-dependent *disjoining pressure* $\Pi(h)$.[35,36] In the case of a microscopic horizontal TLF, the disjoining pressure counterbalances the capillary pressure P_C (the pressure difference in the film and the bulk phase in the meniscus):

$$\Delta P = P_B - P_A = P_C = \Pi(h). \tag{6.3}$$

Insofar, as no real bulk liquid core exists in the TLF, the film tension may also be defined as[1,32]

$$\begin{aligned} \gamma &= 2\int_0^{\infty}\left[P_B - P^\tau(z)\right] dz = 2\int_0^{\frac{h}{2}}\left[P_B - P^\tau(z)\right] dz + 2\int_{\frac{h}{2}}^{\infty}\left[P_B - P^\tau(z)\right] dz \\ &= 2\int_0^{\frac{h}{2}}\left[P_B - P_A(z)\right] dz + 2\int_0^{\frac{h}{2}}\left[P_A - P^\tau(z)\right] dz + 2\int_{\frac{h}{2}}^{\infty}\left[P_B - P^\tau(z)\right] dz. \end{aligned} \tag{6.4}$$

Therefore,

$$\gamma = 2\sigma_f + \Pi(h)h. \tag{6.5}$$

The key peculiarity of the tensions γ and σ_f is their dependence on the (equilibrium) TLF thickness h. Thus, the properties of the TLF are related to the interactions of the interfaces across its thickness and are linked to the concept of film disjoining pressure. The disjoining pressure may be presented as a sum of different components referring to various (DLVO and non-DLVO) surface forces:[1,32,40]

$$\Pi(h) = \Pi_{VW}^{\text{DLVO}}(h) + \Pi_{el}^{\text{DLVO}}(h) + \Pi_{st}^{\text{non-DLVO}}(h) + \Pi_{osc}^{\text{non-DLVO}}(h) + \Pi_{im}^{\text{non-DLVO}}(h) + \cdots, \tag{6.6}$$

where the subscripts VW denotes van der Waals attraction, el denotes electrostatic repulsion, st denotes steric interactions (overlap repulsion, intersegment, or bridging attraction), osc denotes oscillatory forces (structural, H-bond ordering of water, presence of colloidal particles, micelles, depletion attraction), and im denotes image forces (charge reversal phenomena).

The DLVO forces were at the foundations of the initial development of the disjoining pressure concept.[1,36,41] For the interpretation of the experimental results, the most often used expressions are for $\Pi_{VW}(h)$ and $\Pi_{el}(h)$, the superposition of which is supposed to be the most important for the stability of the TLF:[39,42]

$$\Pi_{VW}(h) = -K_{VW} / h^n, \; K_{VW} = \frac{A(h)}{6\pi}, \tag{6.7}$$

where K_{VW} is the van der Waals-Hamaker constant[39] and A is the Hamaker constant for symmetric films (however, a weak function of h); $n = 3$ or $n = 4$ (without or with correction for the electromagnetic retardation of the dispersion forces).[1,32] This component of the disjoining pressure refers to the so-called van der Waals molecular interactions (dispersion molecular forces). The electrostatic component may be expressed as follows:

$$\Pi_{\text{el}}(h) = 2c_{\text{el}}RT\left(\cosh zY^m - 1\right), \; Y^m = F\psi^m / RT, \; m \equiv h/2, \tag{6.8}$$

where c_{el} is the concentration of a symmetric electrolyte with valence z, F is the Faraday constant, RT is the product of the molar gas constant R and the temperature T ψ^m is the potential at the central plane of the symmetric film at $h/2$, and it is related to the potential ψ^0 of the double electric layer at the film/adjacent interface as follows:

$$-\frac{\kappa h}{2} = \int_{Y=Y^0}^{Y=Y^m} \frac{z}{\sqrt{2\left[\cosh(zY) - \cosh(zY^m)\right]}} \mathrm{d}Y,$$
$$Y = \frac{F\psi}{RT}, \quad Y^0 = \frac{F\psi^0}{RT}, \tag{6.9}$$
$$\kappa = \sqrt{F^2 z^2 c_{\text{el}} / \varepsilon_0 \varepsilon RT},$$

where $1/\kappa$ is the Debye length, ε and ε_0 are the relative dielectric permittivities of the fluid and of vacuum, respectively.

In a TLF with adsorbed polymers at the film's surfaces, an effective (non-DLVO) steric interaction operates when the two polymer layers come into contact, i.e., $h \leq 2\delta$, where δ is the polymer layer thickness. Note that this type of surface force is different from the short-range surface force acting in NBFs from LMW surfactants.[1,19,32]

Such 'polymer-induced' steric surface forces are well described by the scaling theory of de Gennes for the interaction of two polymer 'brushes',[43] where the steric disjoining pressure scales with the film thickness as

$$\Pi_{st} \sim \left[\left(\frac{h}{2\delta} \right)^{-9/4} - \left(\frac{h}{2\delta} \right)^{3/4} \right]. \tag{6.10}$$

The two terms in the brackets reflect osmotic pressure and elastic (restoring) force, respectively, as the osmotic term becomes dominant at high compressions.

The action of these three major Π components in terms of the disjoining pressure isotherm are illustrated in Figure 6.3. The figure shows also the shape of the classical DLVO disjoining pressure isotherm ($\Pi_{VW}(h) + \Pi_{el}(h)$) and the conditions for obtaining the three main types of TLFs:[1,16,32]

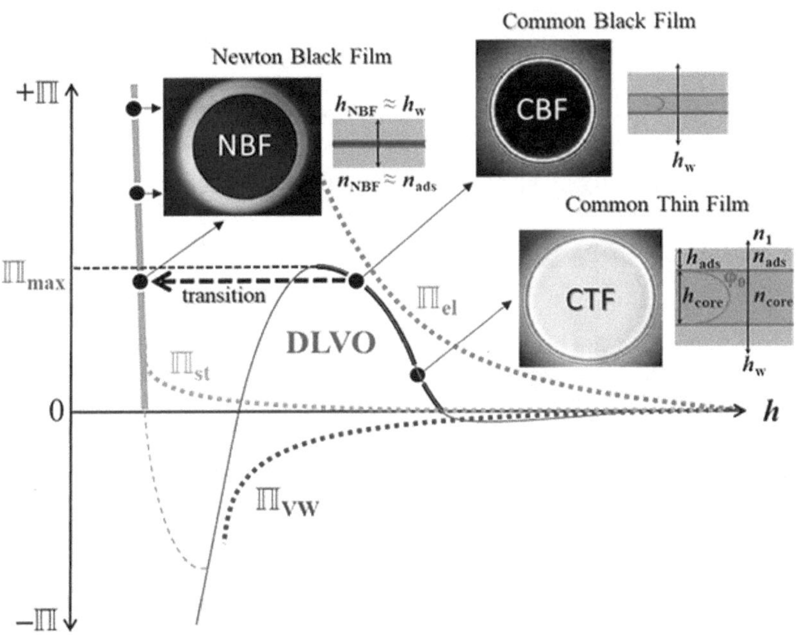

FIGURE 6.3 TLF disjoining pressure isotherm $\Pi(h)$. An equilibrium TLF is formed only at $\Pi > 0$; dotted lines follow the course of the individual components of Π and the solid black line follows the DLVO theory ($\Pi = \Pi_{VW} + \Pi_{el}$) – after a shallow minimum, the disjoining pressure increases up to a critical value Π_{max} and after that it decreases infinitely; however, experimental results have shown the existence of a second minimum, after which Π sharply ascends, providing formation of an equilibrium NBF stabilized by non-DLVO short-range forces; the thick dashed arrow indicates CBF→NBF transition; microphotographs and three-layer model schemes for different types of TLF are also shown. Denotations: h_w – equivalent film thickness, h_{core} and n_{core} – thickness and refractive index of the film's core, h_{ads} – thickness and refractive index of the adsorption layer at either film's interface (further explanations are given throughout the main text).

1. *Common thin film* (CTF) with thickness above ca. 30 nm, stabilized by electrostatic repulsion;
2. *Common black film* (CBF) with thickness between ca. 10 nm and 30 nm, stabilized by electrostatic repulsion;
3. *Newton black film* (NBF) with thickness below 10 nm, stabilized by short-range surface forces.

Apart from the above-considered Π_{VW}, Π_{el}, and Π_{st}, other (non-DLVO) contributions to the total disjoining pressure in a TLF, such as Π_{osc} and Π_{im}, were identified and described in the literature, e.g., in Refs 15, 40, and 42.

One substantial property of the emulsion systems is their coalescence stability. The stability of the thin liquid layer, most generally presented by the so-called thickness of rupture of the film, plays a decisive role. The theory of film rupture was first formulated for the case of foam films.[32,39] The studies show that the true stability of emulsion films is related to their thermodynamic properties (surface tension, disjoining pressure). As for the intrinsic coupling of hydrodynamics and mass transfer of surfactants, it concerns the so-called kinetic stability related to the time evolution of the draining TLF. It may be expressed by the interfacial mobility of thin liquid layer alone (see in Section 6.2.2). The basic result is that while for some characteristics (such as the lifetime of films) interfacial mobility is of utmost importance, the thickness of rupture is actually independent of mobility factors.

6.2.2 Elements of Thin Liquid Film Hydrodynamics

The basic characteristics of thin emulsion films are described through several parameters which link the key film drainage outcomes and the structural peculiarities of the fluid media. They are summarized here for the axisymmetric case of the TLF formed between two closely approaching droplets.[27,32,44]

Unlike the case of a foam film (with liquid-gas interfaces), for an emulsion film there is an essential coupling of the fluid flows in the contiguous liquid phases. This is due to the innate mobility of the liquid-liquid interfaces. Because of the viscosities of the adjoining liquids, the flow in one phase is transferred through the boundary interface into the neighboring phase, and these flows mutually influence each other. Generally, the flows outside and inside the approaching droplets obey the Navier-Stokes and the continuity equations

$$\rho_i \left(v_i \cdot \nabla\right) v_i = \nabla P_i + \mu_i \nabla^2 v_i \ ; \ \nabla \cdot v_i = 0, \tag{6.11}$$

where ρ is the density and μ is the dynamic viscosity. The following boundary conditions at the droplet-film interface represent the coupling of the flows in the contiguous phases,[37,45] namely, continuity of the velocity vectors and the tangential stress components at the droplet-continuous phase interfaces:

$$\mathbf{v}_{\mathrm{f}} = \mathbf{v}_{\mathrm{d}}; \ P_{\mathrm{f}}^{n\tau} = P_{\mathrm{d}}^{n\tau}. \tag{6.12}$$

Here and in what follows, all quantities referring to the film (continuous phase) are marked with a subscript ($_f$), and those for the droplets are supplied with a subscript ($_d$).

Important dynamic characteristics are the film thinning velocity V and the time of thinning τ until critical film thickness h_{cr} is reached. For a plane-parallel horizontal TLF with thickness h, these are defined (in cylindrical coordinates) as[27,32,39]

$$V = -\frac{dh}{dt}, \quad \tau = \int_{h_o}^{h_{cr}} \frac{dh}{V}, \tag{6.13}$$

where t is the time. The two-droplet model presents a 'primitive model' of emulsion film hydrodynamics and gives the basic relationships that govern the peculiarities of the flows in the adjacent phases. When the distance between the approaching interfaces becomes smaller than the droplets' dimensions, the liquid motion outside the liquid particles is localized in the narrowest domain between them. An anizodiametric 'effective' flow region of major dissipation of energy is formed there, namely a TLF (Figure 6.4).[37,45,46] The TLF generates the major term of the hydrodynamic interaction force between the two droplets during the film drainage process. When the fluid flow is mostly localized in this anizodiametric gap region, the 'degeneration' of the flow pattern in the thin layer leads to the formation of a respective 'effective' zones of interaction inside the droplets, occupying regions adjacent to the thin layer outside.[45,46] The mutual deformation of the hydrodynamic fields results in additional forces acting on the fluid particles.

The idea about the existence of 'effective' zones of interaction is specified by the above-defined characteristic film dimension parameters H_f and R_f with a dimensionless anisotropy factor $\epsilon = H_f/R_f << 1$, and by introducing the following scaling parameters (illustrated in Figure 6.4):[37,46] H_d and $R_d \sim R_f$ – characteristic dimensions of the 'effective' zone inside the droplets, adjacent to the film; U_f – characteristic radial velocity of the film outflow; U_d – characteristic radial velocity of the film-droplet interface; $V_f \sim \varepsilon^2 U_f$ – characteristic approach velocity of the droplets along the z-axis. The Reynolds numbers Re for the emulsion film and inside the

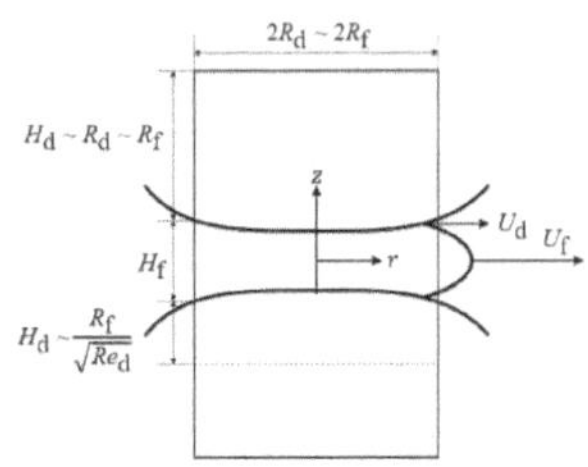

FIGURE 6.4 Two emulsion droplets of radii R at small gap widths. H_f, R_f – characteristic dimensions of the effective zone between the droplets (liquid film); H_d, R_d – characteristic dimensions of the effective zone inside the droplets, adjacent to the TLF; U_f, U_d – characteristic radial velocities of the TLF and at the film-droplet interface; h – distance of the closest approach between the droplets; (r, φ, z) – cylindrical coordinate system.

droplets are measures for relative significance of the viscous and convective terms in Equation (6.11), and are defined as

$$Re_{\mathrm{f}} = \frac{\rho_{\mathrm{f}} V_{\mathrm{f}} H_{\mathrm{f}}}{\mu_{\mathrm{f}}}; \quad Re_{\mathrm{d}} = \frac{\rho_{\mathrm{d}} U_{\mathrm{d}} R_{\mathrm{d}}}{\mu_{\mathrm{d}}}. \tag{6.14}$$

The relative interfacial mobility at the film-droplet interface is estimated against the TLF drainage outflow by the hydrodynamic factor f^{h}, which reflects the coupling of flows in the contiguous phases. In the case of pure liquids, the latter depends on the viscosities of both phases, the introduced geometric scaling parameters of the TLF, and the 'effective' zone of interaction inside the droplets:[37]

$$\Delta U = U_{\mathrm{f}} - U_{\mathrm{d}}, \frac{\Delta U}{U_{\mathrm{f}}} \sim \frac{f^{\mathrm{h}}}{1+f^{\mathrm{h}}}, f^{\mathrm{h}} \sim \left(\frac{\mu_{\mathrm{d}}}{\mu_{\mathrm{f}}}\right)\left(\frac{H_{\mathrm{f}}}{H_{\mathrm{d}}}\right). \tag{6.15}$$

The scaling concepts allow the application of dimensional and similarity theories,[47] so that the hydrodynamic equations and the respective boundary conditions (Equations. (6.11) and (6.12)) are rendered dimensionless. As a result, a range of physically reasonable simplifications is outlined. In view of the overall delay of the drainage kinetics at small gap widths between the approaching droplets $Re_{\mathrm{f}} < 1$, and the governing drainage equations for the TLF reduce to a creeping-flow Stokes model.[44,37,46] These equations may be further reduced to a lubrication-flow model (Reynolds model)[39,48] only if $(\Delta U / U_{\mathrm{f}}) > \varepsilon^2$. However, the Reynolds model does not represent correctly the fluid motion in the TLF when $(\Delta U / U_{\mathrm{f}}) \leq \varepsilon^2$ and additional members of the Stokes equations have to be accounted for Refs 37, 45 and 46. These ideas have been further developed in recent studies.[42,49]

Insofar as the flow inside the droplets is concerned, the general Navier-Stokes equations are usually applied. Depending on the specific system's properties, the latter may be further simplified based on the following scaling estimates:[37,45,46]

$$H_{\mathrm{d}} \sim \frac{R_{\mathrm{d}}}{\sqrt{1+Re_{\mathrm{d}}}} \sim \frac{R_{\mathrm{f}}}{\sqrt{1+Re_{\mathrm{d}}}}. \tag{6.16}$$

So, the flow equations reduce to either Stokes equations for $H_{\mathrm{d}}(Re_{\mathrm{d}} < 1) \sim R_{\mathrm{d}} \sim R_{\mathrm{f}}$ or to hydrodynamic boundary-layer equations[50] if $H_{\mathrm{d}}(Re_{\mathrm{d}} > 1) \sim \dfrac{R_{\mathrm{f}}}{\sqrt{Re_{\mathrm{d}}}}$.

Another important factor for the drainage performance of emulsion systems is the presence of emulsifiers (LMW surfactants, macromolecular surfactants, nanoparticles, etc.). In the simplest case of a LMW surfactant, one common (empirical) observation is the above-mentioned Bancroft's rule. Usually, the presence of an emulsifier was interpreted in view of its impact on the interfacial mobility of the fluid interfaces. This point of view stems from the physicochemical hydrodynamics of Levich[51] and was further developed for the specific interactions in emulsion films by Lee and Hodgson.[52] The key idea is that at small separations the coupling of the surfactant mass transfer and the fluid motion is additionally modified by the formation of thin emulsion layers between the approaching droplets. In certain cases, this results in an enhanced suppression of the tangential mobility of the film surfaces. This effect may be accounted for as a specification of the interfacial mobility scaling by introducing the so-called Marangoni factor.[37,45,53–55] So, the boundary conditions (Equation (6.12)) at the droplet-TLF interface (in the regime of diffusion-controlled mass transfer of the surfactant) are modified as

$$v_{\mathrm{f}} = v_{\mathrm{d}},\ P_{\mathrm{f}}^{n\tau} = P_{\mathrm{d}}^{n\tau} + \nabla^{\tau}\sigma,\ \nabla^{\tau}\left(v^{\tau}\Gamma\right) - D_{\mathrm{s}}\left(\nabla^{\tau}\right)^{2}\Gamma = -D_{\mathrm{f}}\frac{\partial c_{\mathrm{f}}}{\partial n} + D_{\mathrm{d}}\frac{\partial c_{\mathrm{d}}}{\partial n}. \quad (6.17)$$

Here, Γ is the interfacial concentration of the adsorbed surfactant; c_{f} and c_{d} are the bulk concentrations in the film and inside the droplets, respectively; D_{f}, D_{d}, and D_{s} are bulk and surface diffusion coefficients; σ is the interfacial tension of the droplet-continuous phase interface; (n,τ) are the above-defined normal and the tangential coordinates in a local coordinate system bound to the droplet's surface.

The leading implication is that during the film formation and drainage the surfactant mass transfer couples with the emulsion film hydrodynamics. The geometrical film anizodiametry modifies both the flow in the contiguous phases and the surfactant mass transfer regimes. To analyze this option, bulk Peclet numbers Pe are introduced as a measure of the surface convection and diffusion at the interface, and then the characteristic lengths δ_i normal to the film-droplet interface from the film side (δ_{f}), and from the side of the droplets (δ_{d}), within which the major concentration changes appear (see in Figure 6.5), are defined as [37,53–55]

$$Pe_{\mathrm{f}} = \frac{U_{\mathrm{f}}R_{\mathrm{f}}}{D_{\mathrm{f}}};\ \delta_{\mathrm{f}} \sim \frac{R_{\mathrm{f}}}{\sqrt{1+Pe_{\mathrm{f}}}};\ Pe_{\mathrm{d}} = \frac{U_{\mathrm{d}}R_{\mathrm{d}}}{D_{\mathrm{d}}};\ \delta_{\mathrm{d}} \sim \frac{R_{\mathrm{d}}}{\sqrt{1+Pe_{\mathrm{d}}}} \sim \frac{R_{\mathrm{f}}}{\sqrt{1+Pe_{\mathrm{d}}}}. \quad (6.18)$$

Therefore, one arrives at the following estimation of the tangential mobility of the film-droplet interface:

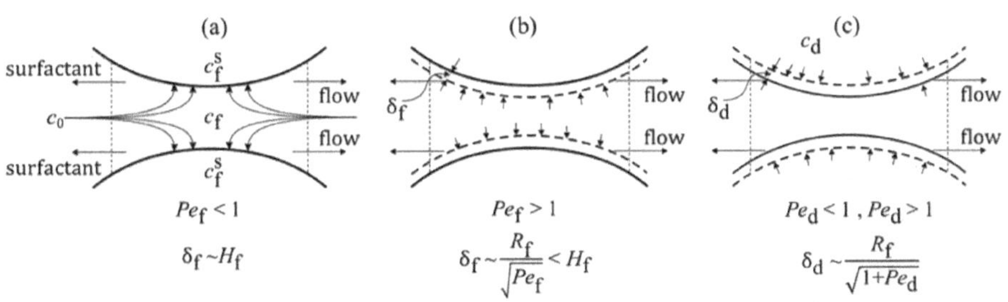

FIGURE 6.5 Mass transfer fluxes: (a) and (b) for surfactant, soluble in the TLF; (c) for surfactant, soluble in the droplet phase. c_0, c_{f}, c_{d}, and $c_{\mathrm{f}}^{\mathrm{s}}$ are the surfactant concentrations in the film-forming solution (meniscus region), in the film bulk, in the droplet bulk, and in the subsurface region of the film-droplet interface, respectively; $\delta_{\mathrm{f}}, \delta_{\mathrm{d}}$ are the characteristic lengths normal to the interface (depths of the subsurface diffusion layers), in the film and in the droplet, within which the major concentration changes occur.

$$\frac{\Delta U}{U_{\mathrm{f}}} \sim \frac{f^{\mathrm{h}} + f_i^{\mathrm{c}}}{1 + f^{\mathrm{h}} + f_i^{\mathrm{c}}}. \tag{6.19}$$

Here, along with the above-defined hydrodynamic factor f^{h}, the so-called concentration (Marangoni) factors f_i^{c} (subscript i = f or d) are defined as follows:

$$f_{\mathrm{f}}^{\mathrm{c}} \sim \frac{\dfrac{\Gamma^0}{\mu_{\mathrm{f}} D_{\mathrm{f}}} \left|\dfrac{\partial \sigma_{\mathrm{f}}}{\partial c_{\mathrm{f}}}\right|_0}{1 + \dfrac{D_{\mathrm{s}}}{H_{\mathrm{f}} D_{\mathrm{f}}} \left|\dfrac{\partial \Gamma}{\partial c_{\mathrm{f}}}\right|_0} \text{ for } Pe_{\mathrm{f}} < 1; \quad f_{\mathrm{f}}^{\mathrm{c}} \sim \frac{\dfrac{\Gamma^0}{\mu_{\mathrm{f}} D_{\mathrm{f}}} \left|\dfrac{\partial \sigma_{\mathrm{f}}}{\partial c_{\mathrm{f}}}\right|_0 \dfrac{\varepsilon}{\sqrt{Pe_{\mathrm{f}}}}}{1 + \dfrac{D_{\mathrm{s}}}{H_{\mathrm{f}} D_{\mathrm{f}}} \left|\dfrac{\partial \Gamma}{\partial c_{\mathrm{f}}}\right|_0 \dfrac{\varepsilon}{\sqrt{Pe_{\mathrm{f}}}}} \text{ for } Pe_{\mathrm{f}} > 1, \tag{6.20a}$$

and

$$f_{\mathrm{d}}^{\mathrm{c}} \sim \frac{\dfrac{\Gamma^0}{\mu_{\mathrm{f}} D_{\mathrm{d}}} \left|\dfrac{\partial \sigma_{\mathrm{f}}}{\partial c_{\mathrm{d}}}\right|_0 \varepsilon}{1 + \dfrac{D_{\mathrm{s}}}{H_{\mathrm{f}} D_{\mathrm{d}}} \left|\dfrac{\partial \Gamma}{\partial c_{\mathrm{d}}}\right|_0 \varepsilon} \text{ for } Pe_{\mathrm{d}} < 1; \quad f_{\mathrm{d}}^{\mathrm{c}} \sim \frac{\dfrac{\Gamma^0}{\mu_{\mathrm{f}} D_{\mathrm{d}}} \left|\dfrac{\partial \sigma_{\mathrm{f}}}{\partial c_{\mathrm{d}}}\right|_0 \dfrac{\varepsilon}{\sqrt{Pe_{\mathrm{d}}}}}{1 + \dfrac{D_{\mathrm{s}}}{H_{\mathrm{f}} D_{\mathrm{d}}} \left|\dfrac{\partial \Gamma}{\partial c_{\mathrm{d}}}\right|_0 \dfrac{\varepsilon}{\sqrt{Pe_{\mathrm{d}}}}} \text{ for } Pe_d > 1, \tag{6.20b}$$

where Γ^0 is the interfacial concentration of the adsorbed surfactant in the absence of fluid motion, all quantities marked as $|\ |_0$ refer to the respective equilibrium properties. From these scaling relationships, it clearly follows that a surfactant, soluble in the droplet, would always have a less pronounced immobilizing effect on the interfacial mobility as compared to a surfactant soluble in the emulsion TLF, and this has also been proven experimentally.[56] This influence is further diminished in thinner films, irrespective of the particular mass transfer feeding pattern to the interface.

The above scaling considerations show that the coupling of the film flow conditions with the surfactant flux peculiarities may be added to the appropriate hydrodynamic background in the understanding of the experimentally observed Bancroft's rule. Only in the specific case of a diffusion-controlled mass transfer of the surfactant, soluble in the emulsion film, it is possible to ensure the necessary conditions (the so-called 'controlled slip'), so that the tangentially mobile interfaces can be immobilized.[52] The mechanism is as follows: the radial outflow in the TLF during drainage carries the adsorbed surfactant out of the film domain. This results in the increase of the additional tangential force $\nabla^{\tau}\sigma$ (Equation (6.17)) and in the onset of bulk surfactant flux toward the film-droplet interface. The TLF bulk region, however, quickly becomes depleted of surfactant molecules, and the compensating flux has to come from the meniscus region (Figure 6.5a).

In all the rest of the cases (Figure 6.5b and c), the surfactant feed-up flux toward the film surfaces is very powerful and cannot impose a lower tangential mobility of the interfaces related to the presence of LMW surfactants. Besides, the formation of the diffusion boundary layer alone, on either side of the mobile interface, is of no particular importance: it has none special effect on interfacial mobility, and consequently on the flow regime in the system. The important issue for a surfactant, soluble in the droplets, is that the onset of a powerful flux from inside the droplet toward the film surface masks all the rest effects (see Figure 6.5c). Thus, the film drains almost like in the surfactant-free case and even for a surfactant soluble in both phases, the effect would be exactly as by the preferential solubility inside the droplets.

In more sophisticated cases of surfactant structures in the bulk of the TLF, e.g., (pre)micellar species,[41,57] or if the adsorbed emulsifier is reorganized in a specific manner in the interfacial layer, e.g., conformational changes and aggregation process in systems containing macromolecules,[58] and proteins, in particular,[25,59,60] the interfacial properties have to be correlated with the film drainage hydrodynamics. This is now a 'hot topic' in the emulsion science.

6.3 EXPERIMENTAL METHODS

Emulsion films have been investigated already in the early/middle 20th century. Rehbinder and Wenström (1930),[3] and later on other authors,[26,61–63] used simple experimental arrangements and visual observations to investigate the stability of emulsion films in terms of the lifetime of a film formed between a freestanding emulsion droplet (water or oil) attached to a planar water-oil interface (Figure 6.6a). In more recent studies, such experiments were supported by video cameras and various technical accessories, which allow for investigating in detail the act of drop collision with the surface and the evolution of the thus created TLF until its rupture (including film drainage).[64–66] Current setups have also the option of simultaneous opto-spectroscopic measurements of the film thickness[66] and such results are discussed in Section 6.4.1.

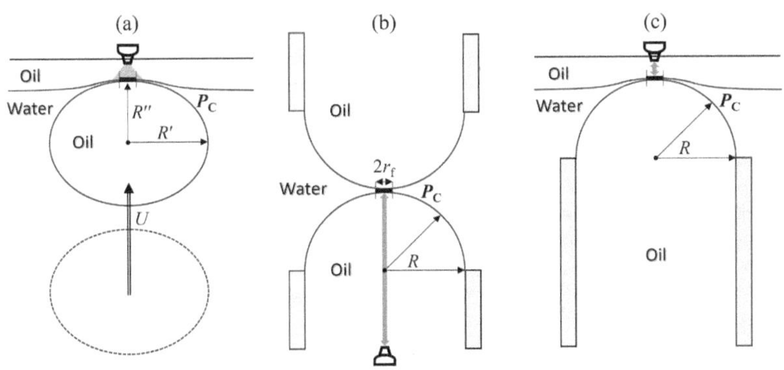

FIGURE 6.6 Schemes of experimental configurations for investigating thin emulsion films (here illustrated for the case of O/W/O films); R is the drop radius, P_C is the capillary pressure, r_f is the film radius. (a) The colliding drop experiment with opto-spectroscopic measurements; the vertical arrow indicates the velocity U of a freely rising drop in solution. (b) A drop-drop experimental arrangement. (c) The LSFA technique. In parts (b) and (c), the grey vertical double arrows indicate the monochromatic beam path in interferometry measurements of the film thickness.

In relation to O/W emulsions, van den Tempel (1958)[67] used an experimental arrangement where two oil droplets emerging from opposing vertical capillaries residing in an aqueous SDS solution are brought into contact and thus an O/W/O film is formed (Figure 6.6b). The drainage of the obtained emulsion film was observed under microscope until equilibrium film thickness was reached, which was measured by light interferometry. The same experimental approach was employed by Sonntag coworkers[5]. in the 1960-1970s to study O/W/O films, obtained from surfactant[68] or polymer[69] solutions, as well as W/O/W films.[70] On a similar basis, recently Won et al.[71] introduced an automated instrument called *Drop-Bubble-Micro-Manipulator.* The instrument allows for monitoring droplet-droplet or bubble-bubble coalescence as well as droplet-bubble interactions with a high-speed camera and precise measurements of the drop/bubble capillary pressure.

The so-called *liquid surface forces apparatus* (LSFA), was introduced by Aveyard et al.,[72] where a stationary drop on a capillary is brought into contact with a planar water-oil interface (Figure 6.6c). The thickness is measured interferometrically. The technique has the advantage to exploit emulsion films of radius of a few μm, but however, it is restricted to a relatively narrow pressure range of up to 2 kPa,[72,73] hence, the obtained $\Pi(h)$ isotherms may be incomplete, when the critical pressure of film rupture is higher than this experimental pressure limit.[16,72]

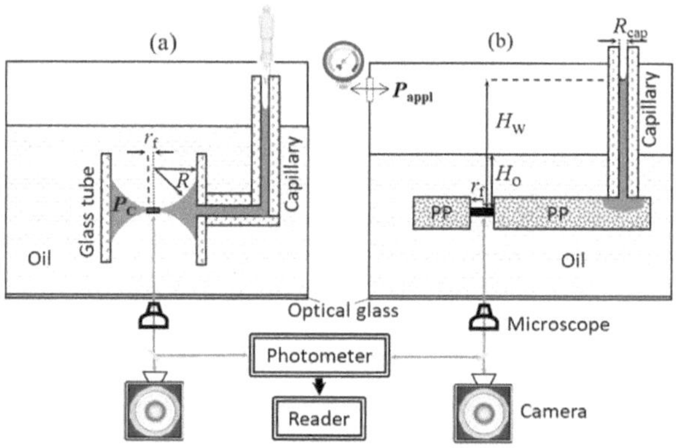

FIGURE 6.7 Scheme of the setup for emulsion film investigations (here illustrated for the case of O/W/O films). (a) Scheludko tube cell. (b) Exerowa-Scheludko porous plate cell. 'PP' stands for porous plate and further denotations are explained in the main text.

The *Microinterferometric* technique[1,16] allows for simultaneous measurements of the thickness h and the disjoining pressure Π of a TLF, i.e., the disjoining pressure isotherm $\Pi(h)$, which is the equation of state of an equilibrium TLF. Two types of measuring cells are shown in Figure 6.7 for the case of experiments with O/W/O emulsion films.[74]

In the *Scheludko tube cell*, a microscopic freestanding plane-parallel film is formed in the center of a biconcave solution drop entrapped in a glass tube holder. In such an arrangement, the film is connected with the drop's meniscus and at equilibrium the film disjoining pressure Π is equal to the capillary pressure in the meniscus P_C:

$$\Pi = P_C = 2\sigma_l R/(R^2 - r_f^2) \approx 2\sigma_l/R, \tag{6.21}$$

where R is the radius of curvature of the meniscus (which under good wetting conditions can be approximated to the radius of the tube) and r_f is film radius. Typically, R is on the order of 1–2 mm, while r_f is on the order of 0.1 mm, hence, the term $R/(R^2 - r_f^2)$ in Equation (6.21) is frequently reduced to $1/R$. For the sake of illustration, Figure 6.8 presents microphotographs of O/W/O films obtained in a *Scheludko cell.* This cell was used in studies on emulsion films of either O/W/O[16-19,24,25,74] or W/O/W[14,20] type.

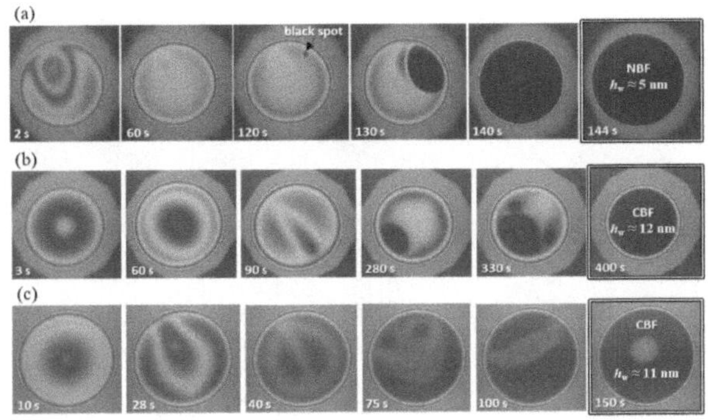

FIGURE 6.8 Microphotographs following the drainage of O/W/O films with $r_f \approx 100$ μm obtained in a *Scheludko cell* at P_C of *ca.* 20–40 Pa; in each row, the last microphotographs (in frames) are of equilibrium films with equivalent thickness h_w. Oil phase was isoparafinic oil Isopar™ M. Aqueous phase. (a) 25 mM C_{12}TAB + 0.4 M NaCl; NBF, $h_w \approx 5$ nm. (b) 5 mM C_{12}TAB + 0.1 M NaCl; CBF, $h_w \approx 12$ nm. (c) 20 μM INUTEC® SP1 + 0.1 M NaCl; CBF, $h_w \approx$ 11 nm. Unpublished results by Gochev.

It is noteworthy that some modifications of the *Scheludko cell* were elaborated. In the so-called *miniaturized Scheludko cell* the film holder has a radius of $R \approx 0.3$ mm, allowing for measurements at higher capillary pressures, including emulsion films studies.[23] Wierenga et al.[75] elaborated a cell, which allows for exchange of the film-forming solution with minor disturbances of pre-formed adsorption layers. Another modification is the so-called *bike-wheel cell*, which allows for control on the capillary pressure and thus for obtaining $\Pi(h)$ isotherms.[16] However, to our best knowledge, neither the *exchange cell* nor the *bike-wheel cell* were used in emulsion films studies.

In the *Exerowa-Scheludko porous plate cell*, additional pressure P_{appl} can be applied on the film's surfaces (Figure 6.7b). This technique is frequently referred to in the literature as the *Thin Film Pressure Balance* (TFPB). The film is formed in the center of a solution drop entrapped in a small hole (radius of 0.5–1 mm). In such an arrangement, the film may be set to occupy the full area of the hole, remaining connected with the solution soaked in the porous material. Typically, an experiment consists of three major steps: (1) Initially, the film is formed at a low P_{appl}. (2) Subsequently, P_{appl} is gradually increased until the film occupies the full area of the hole. (3) P_{appl} is further increased in a stepwise manner as at each step the equilibrium film thickness is measured as a function of the equilibrium disjoining pressure:

$$\Pi = P_{appl} + P_{cap} - P_{hydr} = P_{appl} + 2\sigma_l/R_{cap} - g(\rho_w H_w - \rho_o H_o), \quad (6.22)$$

where P_{cap} is the capillary pressure in a capillary with a radius R_{cap}, P_{hydr} is the hydrostatic balance between the liquid heights of water H_w and oil H_o above the film level, ρ_w and ρ_o are the respective liquid densities, $g \approx 9.8$ m/s^2 is the gravitational acceleration. Thus, knowing the corrections terms P_{cap} and P_{hydr} (at any H_w and H_o), the disjoining pressure in the film is solely determined by P_{appl}, which can be readily controlled and measured by an appropriate pressure detector.

The interferometrically measured film thickness h_w is an 'equivalent thickness', which for an optically homogeneous emulsion film with uniform refractive index n_f (and refractive index n_s of the surrounding immiscible liquid) reads[16,32,76]

$$h_w = \frac{\lambda}{2\pi n_f}\left(\arcsin\sqrt{\frac{\Delta}{1+\frac{4N}{(1-N)^2}(1-\Delta)}}\right), \quad (6.23)$$

where λ is the wavelength of the illuminating monochromatic light beam, $N \equiv \left(\frac{n_f - n_s}{n_f + n_s}\right)^2$, and $\Delta \equiv I/I_{max}$, with I_{max} and I being the intensities of the interference maximum and of the reflected beam, respectively. Note that usually n_f is approximated to the refractive index n of the bulk solution from which the film is formed. Hence, for a foam film (surrounded by air with $n_s = 1$) or a O/W/O emulsion film (surrounded by oil with $n_s > 1$), n_f can be approximated to the refractive index of water ($n_f \approx 1.33$), while for a W/O/W emulsion film one assumes n_f to be that of the oil used.

In fact, the equivalent thickness h_w obtained from measurements may differ from the real physical thickness of the film h, because of the refractive index n_{ads} of the adsorption layers at the film's surfaces. This difference may become appreciable for film thicknesses below ca. 30 nm, whereas for thicker films it is frequently assumed that $h \approx h_w$. The quantities h_w and h can be related via application of a model of the structure of the film. The most commonly used one is the three-layer sandwich model,[1,16,77,78] which assumes two equal adsorption layers with thickness h_{ads} (and n_{ads}) and an aqueous film's core with thickness h_{core} (and n_{core}), hence,

$$h = h_{core} + 2h_{ads}, \quad (6.24)$$

where

$$h_{core} = h_w - 2h_{ads}\left(\frac{n_{ads}^2 - n_f^2}{n_f^2 - n_s^2}\right). \quad (6.25)$$

Obviously, proper analysis of the interferometric measurements requires values for the adsorption layers' parameters h_{ads} and n_{ads}, which are not always available. Moreover, the adsorption layer can be thought of displaying a two-slab structure with

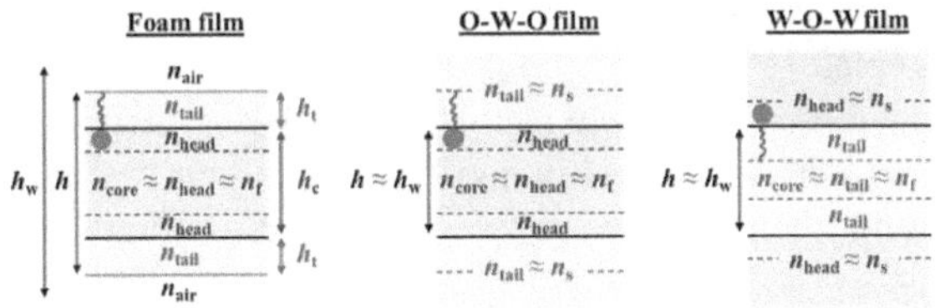

FIGURE 6.9 Schematic structure of foam and emulsion films. Denotations for refractive indexes and thicknesses: surrounding fluid – n_s (air – n_{air}); film core – n_{core}, h_{core} ($n_{core} \approx n_f$); surfactant hydrophilic heads – n_{head}, h_{head}; surfactant hydrophobic tails – n_{tail}, h_{tail}.

a hydrophilic part comprising surfactant heads (n_{head}, h_{head}) and a hydrophobic part comprising surfactant tails (n_{tail}, h_{tail}). In such a case, the three-layer film model is decomposed into a five-layer model,[16] as schematized in Figure 6.9. However, reasonable assumptions can be applied to ease the modeling of the film structure. For instance, the surfactant heads can be well assumed optically indistinguishable from water. Hence, for a foam film or a O/W/O emulsion film, with an aqueous core, one may use $n_{head} \approx n_{core} \approx n_f \approx 1.33$ and then the five-layer model reduces to a three-layer model, where the surfactant tails are taken into account by Equation (6.8). For emulsion films, the situation can be further eased by assuming that the surfactant tails are optically indistinguishable from the oil phase, and therefore, for such a single-layer film, the approximation $h \approx h_w$ is fairly valid.[72,74,77]

An original approach for obtaining and investigating a single emulsion film was proposed by Gabrieli et al.[79] In their experiments, the authors obtained a spherical cap-shaped aqueous emulsion film between a drop and a continuous phase of decane. Such spherical O/W/O films have a radius of around 1.5 mm (area of around 27 mm^2) and the film thickness is evaluated by an optical evanescent wave technique.

In analogy to the classical technique of vertical foam films in a frame drawn from a solution,[1,80,81] Kruglyakov et al.[82] and Horozov et al.[83] used small circular frames (diameters of 5–6 mm) to obtain either O/W/O or W/O/W films stabilized by solid particles. While the thickness of vertical foam films has been measured by reflectometric techniques based on light,[1,80] X-rays, or neutrons,[81] it seems that such measurements for the relevant emulsion films are challenging. However, in Ref. 82, the thickness of the investigated O/W/O films is measured by conductometry.

Gunning et al.[84] and Dagastine et al.[85] published nearly at the same time seemingly independent reports on their atomic force microscopy (AFM) results from experiments with emulsion films. In both studies, a similar experimental approach has been applied, namely, an aqueous SDS emulsion film is formed between two oil droplets: one being immobilized at a solid support and the other attached to tipless AFM cantilever. Force-distance profiles were measured as a function of either the approaching or the retracting velocity of the cantilever. Apart from the action of surface forces, hydrodynamic effects were accounted for and difficulties with the proper quantification of the obtained data were emphasized and discussed in Ref. 85. In Ref. 84, droplet coalescence was monitored as affected by changes in the solution ionic strength I. This AFM-based technique was later developed to study either O/W/O films (between two oil droplets) or W/O/W films (between two water droplets), as well as foam films (between two air bubbles) and wetting films (between an air bubble and a solid plate).[58]

The interaction forces and breakage of the O/W/O film between two isolated oil droplets can be analyzed experimentally by bringing the droplets into contact and then separating them in well-defined approach-retract cycles using optical tweezers.[30,86,87]

Bibette et al. used the *magnetic chain technique* (MCT) and small-angle neutron scattering (SANS) to access the surface forces and the thickness, respectively, of O/W/O films in real O/W emulsions.[7,88,89] MCT relies on the use of ferromagnetic Fe_2O_3 nanoparticles ($\approx$10 nm), which have to be stabilized by oleic acid against aggregation in the octane phase used as oil in the investigated emulsions.[7,88] By this technique, force-distance, *F-H*, profiles between nanoscopic monodisperse oil droplets aligned in chains are obtained. The strong Bragg diffraction of white light arising from the chains of ordered droplets allows for performing accurate optical measurements of the inter-droplet spacing, *H*, while the inter-droplet repulsive force, *F*, is measured as the balance to the applied magnetic field. Notably, the DLVO theory was well fitted to the measured *F-H* profiles for several SDS concentrations yielding values for the double-layer potential φ_0 as the only fitting parameter. Advantages of this methodology are the possibility of measurements of very weak forces (on the order of 10^{-13} N) as well as the fact that the measurements are ensemble averaging over a great number of inter-droplet contacts. The latter advantage holds also in the case of the SANS experiments with bulk samples of concentrated O/W emulsions reported in Ref. 89. In that SANS study, the authors deduced an average inter-droplet spacing of $H \approx 2.9$ nm, which is independent of the inter-droplet adhesion energy (estimated from contact angle measurements) in the studied SDS emulsions, which in fact gives evidence that the corresponding O/W/O films between emulsion droplets are of the NBF type.

Concerning W/O/W films, a method for estimation of the film thickness, based on the measurements of the film electrical capacitance, have been elaborated in the course of studies on BLMs[12,90,91] and then applied to oil films stabilized by polymers.[13] Notably, this technique is most sensitive to the thickness of the organic region of the film, which is in the order of 3–5 nm for lipids,[91] while information about the hydrophilic headgroups emerging from the organic region may be accessed by methods based on optical reflectivity.[19,78,91] The approach of application of electrical potential on W/O/W films[92] was further developed in relation to electrocoalscence of bitumen emulsions.

6.4 O/W/O EMULSION FILMS

Historically, the research on O/W/O films started with simple observations on the lifetime of a film formed between a freestanding oil droplet attached to a planar water-oil interface (Figure 6.6a).[3,26,61–63] As mentioned above, such experiments are nowadays well technically suited, allowing also for direct film thickness measurements.[66] Such studies together with other studies using similar experimental approaches are discussed in Section 6.4.1. Further, quantitate analysis was obtained by light interferometry measurements of the equivalent film thickness h_w

at constant capillary pressure P_C for films between two oil droplets,[5,67–70] as shown in Figure 6.6b. However, detailed information about the interaction surface forces in O/W/O films came from disjoining pressure isotherms $\Pi(h)$ measurements by the MCT,[7,23,88] LSFA,[16,72,93] and TFPB[16-19,23,74,93–96] methods.

6.4.1 O/W/O Emulsion Films between an Oil Droplet and a Planar Water-Oil Interface

The experimental investigation of a thin emulsion film formed at the collision of a freely rising single oil droplet with a water-oil interface offers unique opportunities for considering the so-called 'droplet history' regarding emulsion film stability and its drainage dynamics. Such experiments require a suitably designed experimental setup that allows for mimicking all fundamental steps of emulsion film formation under real-life conditions. The experiments include controlling the formation and release of a single oil droplet in aqueous phase, and subsequently monitoring its rising velocity, as explained in more details in Chapter 3. Such a rising oil drop collides with a water-oil interface, which can be either a planar interface, as shown in Figure 6.6a,[66] or in the form of a second droplet attached to a needle.[97] The term 'droplet history' encompasses all the aforementioned steps prior to emulsion film formation, primarily associated with the aging time of a droplet at an orifice before release and the degree of mobility of the droplet surface prior to collision with the water-oil interface, and subsequent formation of an emulsion film. In solutions containing surface-active species, the droplet surface may become partially or fully immobilized due to the formation of motion-induced nonuniform adsorption layer, leading to surface tension gradients and the Marangoni effect. The former determines the initial adsorption coverage at the droplet surface, while the latter can be controlled by managing the distance traveled by the droplet before reaching the water-oil interface. Moreover, by employing suitable protocols for droplet formation in appropriately designed equipment for single droplet generation, the magnitude of liquid flow (which can be quantified by the Reynolds number) can be easily controlled by adjusting the droplet diameter, providing an opportunity to study the impact of influencing hydrodynamic factors on the emulsion film stability.

The majority of above-mentioned experiments were performed in a qualitative manner, involving the assessment of the colliding droplets coalescence time only. In addition to the coalescence time, various parameters could be determined during the collision process, depending on the acquisition frequency of the video camera used. For instance, droplet shape deformation or variations in droplet velocity, as well as the evolution of the liquid/liquid interface shape during the act of coalescence, could be observed. The qualitative (limited) approach was associated with significant difficulties in the reproducibility of forming droplets and focusing them in the camera's field of view under dynamic conditions. Furthermore, acquiring emulsion film images of reasonable quality was challenging due to the relatively small differences between the refractive indexes of water and the oil used. A novel approach that allows for overcoming most of these obstacles was recently demonstrated by Gawel and Zawala[66]. They employed a method initially introduced by Delacotte et al.[98] to measure the

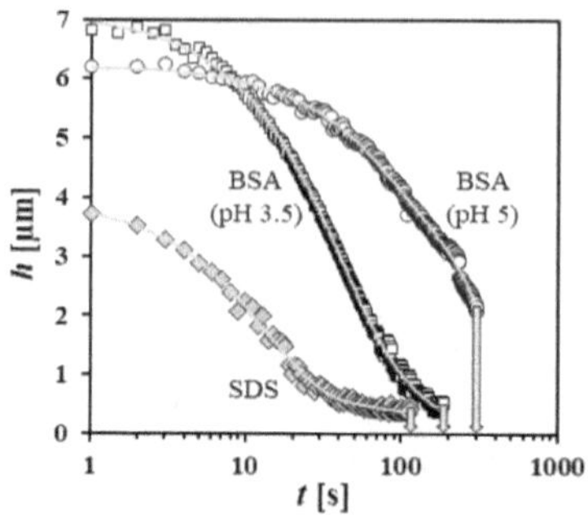

FIGURE 6.10 Film drainage, in terms of film thickness h vs. time, of O/W/O emulsion films formed between an oil drop and a planar water-oil interface; symbols are experimental data adapted from Gawel and Zawala[66] with permission under CC-BY 4.0 Attribution License, lines are guides for the eye, the arrows indicate film rupture. Aqueous solutions of 7.5 μM BSA or 5 mM SDS.

thickness of a liquid layer on a solid plate during its velocity-controlled withdrawal from a bath of aqueous surfactant solution. By introducing a suitably designed reflection probe immersed in the oil layer and coupled with a multiwavelength light source (200-900 nm), as well as by applying a simple approach for self-centering the colliding drop at the water-oil interface, it was possible to acquire spectra of waves reflected from both the emulsion film interfaces. Simple post-processing of the spectra allows for extraction of the time evolution of the average emulsion film thickness. A couple of such dependences, obtained for a 7.5 μM bovine serum albumin (BSA) solution at pH 3.5 or pH 5.0 (near isoelectric conditions) is presented in Figure 6.10; for comparison purposes, data for SDS emulsion film are shown too. Clear differences in drainage dynamics, initial thickness, and critical thickness of rupture, depending on the physicochemical conditions and the type of emulsifier, could be determined and compared with changes in the hydrodynamic boundary conditions assessed, based on the rise velocity variations for droplets of a wide range of diameters. Note that significant thickness inhomogeneities in thin films are typical for both types of foam films[24,59,60] and emulsion films[23,24,94] formed from protein solutions.

6.4.2 Planar Microscopic O/W/O Emulsion Films

6.4.2.1 DLVO Surface Forces

The necessary prerequisite for the action of effective electrostatic disjoining pressure Π_{el} in aqueous (foam and emulsion) films is the existence of extended diffuse part of the double layer at the film's surface.[1,15,32,36] The origin of the electric charge of bare water-fluid interface is a debatable theme, as discussed for instance in Refs[16,96,99]. However, a number of experimental studies with water-air[1,16] and water-oil[16,96,99] interfaces give evidence that the mechanism of charge generation at such interfaces (being either bare or covered with adsorbed nonionic amphiphiles) is adsorption of hydroxyl ions OH^-. There are at least three ways to effectively reduce or totally suppress Π_{el} in water-air (oil)/nonionic amphiphile TLF:

1. To screen the double-layer potential φ_0 by increase of I to/above a critical value I_{cr}, for instance, corresponding to a critical electrolyte concentration $c_{el,cr}$ [1,16,17,31,74,95] (e.g., Figure 6.11a); note that for a 1:1 indifferent electrolyte, e.g., NaCl, one can use the approximation $I \approx c_{el}$;
2. To obliterate the surface charge by setting pH to/below a critical value pH_{cr} [1,16,31,96,99] (e.g., Figure 6.11b);
3. To reduce the surface charge density (surface concentration of OH^-) by competitive adsorption of increasing amphiphile amounts.[16]

These mechanisms were experimentally and theoretically investigated for O/W/O emulsion films obtained from aqueous solutions of either LMW surfactants[1,16] or polymers[17,74,95,96] with nonionic nature. Such exemplary results are presented in Figure 6.11 for the case of PEO-PPO-PEO block copolymers[17,95,96] (further discussion on the steric forces Π_{st} in these films is provided in the next section).

For ionic amphiphiles, the overall charge of a water-fluid interface is strongly modulated by the electric charges of ionized functional groups in adsorbed molecules. Note that for proteins, effectively the pH-dependent net electric charge plays a role,[100] but only a few investigations with protein O/W/O emulsion films at different pH conditions were published.[25,94] However, at the background of the above-discussed intrinsic negative charge of bare water-fluid interfaces, the adsorption of cationic species seems an intriguing case. Indeed, a short communication by Kolarov et al.,[101] presenting results from experiments with foam films, gave evidence that

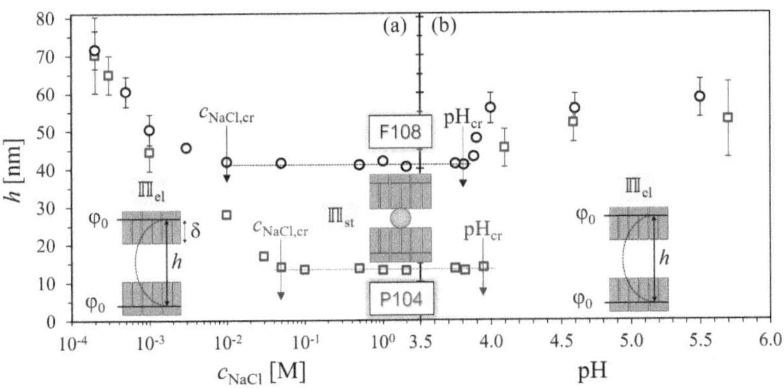

FIGURE 6.11 O/W/O emulsion films from aqueous solutions of 7 μM PEO-PPO-PEO block copolymers F108 (○) or P104 (□) measured in a *Scheludko cell* at constant capillary pressure of $P_C \approx 23$ Pa and $P_C \approx 15$ Pa, respectively. (a) Film thickness h vs. NaCl concentration, c_{NaCl}, at constant pH 5.5. Adapted from Exerowa et al.,[95] Copyright (2009), with permission from Elsevier. (b) h vs. pH at constant $I = 0.5$ mM (NaCl + HCl). Adapted from Gotchev et al.,[96] Copyright (2010), with permission from Elsevier. The vertical arrows indicate the parameters $c_{NaCl,cr}$ or pH_{cr}, which characterize transitions from electrostatic Π_{el} to steric Π_{st} stabilization of the films. The horizontal dashed lines are guides to the eye and indicate the average film thicknesses of sterically stabilized films for each copolymer from the plateau regions in both dependences $h(c_{NaCl})$ and h(pH). δ is PEO brush thickness; φ_0 is the electrical double-layer potential at the film's surfaces; further details about the sketches of the film structure are given in Figure 6.12.

the adsorption of the cationic surfactant $C_{16}TA^+Br^-$ leads to a charge reversal at the water-air interface, which is reflected by the thickness and the stability of the corresponding foam films. We are not aware of any studies on emulsion films that address this issue. However, charge reversal at the water-dodecane interface by the cationic surfactant $C_8TA^+Br^-$ was detected by flocculation transition experiments with real emulsions.[102] Concerning the corresponding water-oil interfaces, relevant studies were discussed in Chapter 3, including experiments with *vibrational sum frequency generation* spectroscopy (SFG). The SFG method is capable of measuring the magnitude of the interfacial electric field and thus detecting interfacial charge reversal,[100] as observed for water-air interfaces, but also for water-oil interfaces[103] (see also in Chapter 3 of this book and further below in Section 6.5.2). SFG can be applied for analyzing the charging state of water-fluid interfaces as affected by various factors, such as amphiphile load, electrolyte content, ion specific effects, pH, etc.,[100,104,105] and was fruitfully employed in combined SFG-foam film studies.[104,105] Although SFG was employed in investigations on water-oil interfaces, we are not aware of combined SFG-emulsion film studies, which is definitely a promising perspective.

Quantification of the action of Π_{el} in O/W/O emulsion films has been done by comparing the DLVO theory (Equations (6.7–6.9) with the double-layer potentials φ_0 as a fitting parameter) to experimental disjoining pressure isotherms for films from LMW surfactants,[7,16,72,73,88,93] polymers,[17,74,95] or proteins[7,23,94] measured with the above-considered methods (MCT, LSFA, and TFPB). Notably, the obtained φ_0 values depend on the concentration and ionic-nonionic nature of the amphiphile used as well as on the solvent conditions (I and/or pH). pH is an influencing factor that may rule out different surface charge controlling mechanisms – on the one hand, as discussed above, pH controls the intrinsic surface charge of bare water-oil interfaces,[96,99] and on the other hand, it controls the molecular net charge of proteins (and presumably other zwitterionic species).[94,100] It is noteworthy that the $\Pi(h)$ data for electrostatically stabilized protein (β-casein or BSA) emulsion films obtained with TFPB are in very good agreement with the *F-H* profiles for corresponding emulsions obtained with the MCT.[23]

Under the conditions of low surface charge or/and strong double-layer screening by relatively high $I < I_{cr}$, a TLF may thin down to an equilibrium CBF,[1,16,19,25,77,74,96] usually through the formation and expansion of black spots (see in Figures 6.8b and c); note that such a CBF remains stable only at sufficiently high amphiphile concentrations.[1,25] A CBF possesses a bulk aqueous core[1,80,81] and is stabilized by weak, but still operative, electrostatic forces. The $\Pi(h)$ isotherms for CBFs typically exhibit steep runs leading either to film rupture[1,16,19,23,72,93,94] or to the formation of a NBF.[1,16,74,95]

6.4.2.2 Non-DLVO Surface Forces

In the absence of Π_{el}, a TLF either ruptures during drainage or thins down to the formation of an equilibrium black film. Notably, the thickness and structure of such black films strongly depend on the molecular properties of the stabilizing amphiphilic species used. In the case of ionic or nonionic LMW surfactants, a NBF is readily formed at relatively low capillary pressure P_C in experiments with a *Scheludko cell* and its thickness remains independent of further pressure increase in a *porous plate cell*.[1,16,19] Such a NBF has a highly ordered bilayer structure[81] and is stabilized by short-range repulsive interactions, the nature of which is not yet fully understood.[1,19]

Formation of NBFs occurs also in foam and emulsion films in the presence of macromolecules, e.g., polymer emulsion NBFs,[17,18,95,96] polymer/surfactant foam NBFs,[18] as well as foam NBFs,[25,59,81,104] and emulsion NBFs[94] from protein solutions. However, there are some marked differences between films from LMW surfactants and macromolecules which manifest in the way of formation of the NBF and its structure.

Concerning proteins, it is well-known that they may form multilayer structures at soft interfaces.[81] For protein black films it was discussed that such multilayer interfacial structures may have impact on the structure of the film's interior; this issue is discussed in Refs 25 and 106. On the other hand, protein material may be spontaneously expelled from the film, and in such situations the film contains portions of NBF (protein bilayer) in coexistence with portions with larger thicknesses as observed for foam[59] and emulsion[94] films from β-lactoglobulin solutions. Such protein expulsion scenario is supported by experiments, where the film-forming β-lactoglobulin solution is exchanged by protein-free buffer (leaving only the protein adsorption layers at the solution surfaces) and then a foam NBF is readily obtained instead of the CBF formed from the original solution.[107] Furthermore, eventual stratification phenomena were discussed in Refs 25 and 81.

We now discuss some aspects of sterically stabilized O/W/O emulsion films from solutions of amphiphilic copolymers. Figure 6.12 presents schemes of the structures of adsorption layers from copolymers of the PEO-PPO-PEO block or AB_n graft type, as well as the structures of corresponding O/W/O emulsion films.

For the case of PEO-PPO-PEO copolymers, it was observed that in the absence of Π_{el}, foam films[1,31] and emulsion films[17,18,95,96] studied at low P_C in a *Scheludko cell* adopt thicknesses which are larger than the overall thickness of two polymer brush layers, i.e., $h > 2\delta$ (Figures 6.11 and 6.12b1). We should note here that such behavior was clearly observed for the high molecular weight (M_w) copolymer F108, but

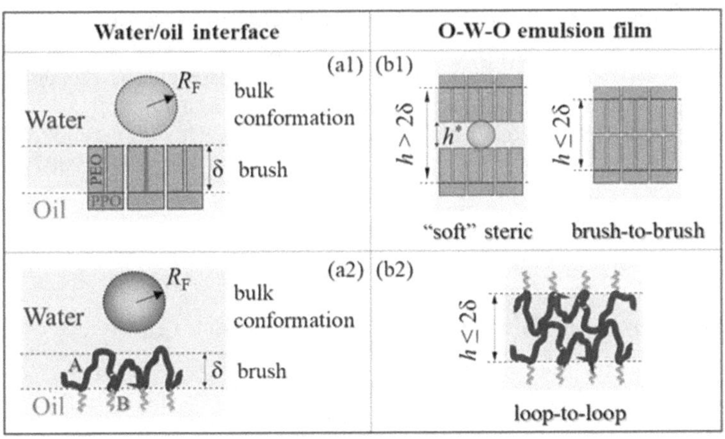

FIGURE 6.12 Schematic representations (not to scale) of amphiphilic copolymers at water-oil interfaces (a1,a2) and in O/W/O emulsion films (b1,b2); h is the film thickness, δ is the polymer brush thickness, $h^* = h - 2\delta$ is the inter-brush spacing, and R_F is the Flory radius. (a1,b1) PEO-PPO-PEO block copolymer. (a2,b2) AB_n graft copolymer, A is an inulin (polyfructose) chain and B is a grafted alkyl chain.

somewhat not clearly detected for films from copolymers with lower M_w such as P65, P104, and F68 (Figure 6.13 and Table 6.1).

Figure 6.13 presents disjoining pressure isotherms $\Pi_{st}(h)$ for sterically stabilized O/W/O emulsion films from four PEO-PPO-PEO block copolymers (Table 6.1). The data are divided into three regions:

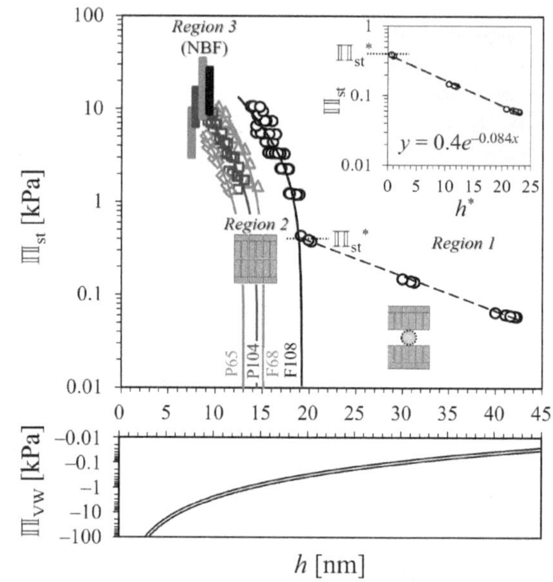

FIGURE 6.13 (Top graph) Steric disjoining pressure isotherms $\Pi_{st}(h)$ for O/W/O emulsion films from PEO-PPO-PEO copolymers, symbols are experimental data for 5 mM P65, 20 μM P104, 70 μM F68, and 7 μM F108, and the solid lines are best fits by Equation (6.10). Adapted from Gotchev et al.,[17] Copyright (2011); Exerowa et al.,[95] Copyright (2009); and Gotchev et al.,[96] Copyright (2010), with permission from Elsevier. The inset graph presents the data for F108 in *Region 1* in terms of Π_{st} versus the inter-brush spacing h^* (unpublished data). (Bottom graph) Theoretical $\Pi_{VW}(h)$ isotherm calculated via Equation (6.7) with $n = 3$ and $A = 5 \times 10^{-21}$ J. Adapted from Gotchev et al.,[17] Copyright (2011), with permission from Elsevier, and from Aveyard et al.,[72] Copyright (1997); Exerowa et al.,[74] Copyright (2007), and Binks et al.,[93] Copyright (1997) with permission from the American Chemical Society. For further details, see in the main text and in Table 6.1.

TABLE 6.1
Molecular Parameters for Some PEO-PPO-PEO Block Copolymers

Copolymer	M_W (g/mol)	N_{EO} (–)	N_{PO} (–)	R_F (nm)	δ^* (nm)
P65	3,400	2 × 18	30	~2	6.5
P104	5,900	2 × 31	54	~3	7.2
F68	8,400	2 × 75	30	~4	7.6
F108	14,600	2 × 128	54	~5	9.6

Denotations: M_W – molecular weight; N_{EO} – number of ethylene oxide units; N_{PO} – number of propylene oxide units; δ – brush thickness (* values from the best fits in Figure 6.13).

- *Region 1:* 'Soft' steric interactions in F108 films; the dashed line is an exponential regression up to Π_{st}^*, which denotes the end of *Region 1* and the onset of *Region 2*. It was discussed in the literature that such films are stabilized by specific 'soft' steric forces, whose action is dependent on the bulk copolymer concentration, and it can be assumed that at sufficiently high copolymer concentrations the inter-brush spacing h^* scales with the Flory radius R_F.[1,31] The latter fact suggests that such 'soft' steric interaction is fundamentally different from the brush-to-brush repulsion. The inset in Figure 6.13 presents some unpublished data, which reveal that in the studied F108 emulsion films such 'soft' steric repulsion is rather weak, and it is overcome at relatively weak compression at $\Pi_{st}^* \approx 0.4$ kPa.
- *Region 2:* Brush-to-brush steric repulsion operates at $h < 2\delta$ and may be described by the de Gennes scaling theory.[1,17,32,43,95,96] Figure 6.13 shows comparison between the experimental data and Equation (6.10) – the solid lines are best fits with the brush thickness δ as the only adjustable parameter – the obtained δ values are listed in Table 6.17. The data demonstrate different branches of brush-to-brush interaction in the $\Pi_{st}(h)$ isotherms for the different copolymers, depending on δ for each copolymer.
- *Region 3* NBF*:* The increase of the pressure applied on the film surfaces leads to compression of the film thickness and ultimately the formation of a NBF is observed (data presented as vertical ribbons in Figure 6.13). The thickness h_{NBF} of such a polymer bilayer remains virtually constant at higher pressures and its value is determined by the brush thickness – the smaller δ the smaller h_{NBF}.

Formation of emulsion NBF was found also for AB_n graft copolymers based on inulin (inulin chain A with n number of grafted alkyl chains B).[74,95] In such films, the steric repulsion arises from the physical contact between inulin loops and was called 'loop-to-loop' interaction (Figure 6.12b2) in analogy to the de Gennes' brush-to-brush concept.

Another type of amphiphilic AB_n graft copolymers are the so-called star-like polymers that consist of a core entity A, for instance diethylentriamine (DETA), to which a different number n of polymeric chains B are grafted.[18,108] Notably, it was found that variations in the number of grafted chains ($n = 4$, 6, or 9 as studied in Ref. 108) induce strong effects on the adsorption behavior of these polymers as well as on the thickness, stability, and other properties of TLFs. In addition, these molecular architecture effects are sensitive to the type of the film – foam film, emulsion film, or wetting film.[18,108]

6.5 W/O/W EMULSION FILMS

The interest to W/O/W emulsion films[12,90,91,109] has seemingly arisen in relation to biological membranes.[90,110] The early studies by Haydon and coworkers[12,90,91] involved films from hydrocarbon oils stabilized by lipids as a simplified model system to mimic biological membranes. Estimation of the film thickness was obtained via measurements of the film-specific electrical capacitance. In later studies, optical methods were involved in investigations of W/O/W films stabilized by lipids[78,91,109,111,112] and other types of LMW surfactants[14,56,113,114] as well as by amphiphilic polymers.[113,114]

Notably, the film drainage experimental data were theoretically analyzed in Ref. 56. In relation to petroleum emulsions, systematic studies of bitumen films gained significant attention only in the last couple of decades.[18,76,115-124] The delayed inception of W/O/W film studies can be attributed to several technical challenges that render this system more difficult to investigate.[108,114] One notable challenge arises from the inherent hydrophilicity of glass film holders, which makes them unsuitable for retaining the oil phase and achieving the required wettability to form a thin oil film in aqueous environment. This technical impediment was overcome in two ways: (1) by employing innately hydrophobic materials as oil film holders;[12,13,82,83,111,112] and (2) by pre-hydrophobization of the glass tube of a *Scheludko cell*[14] or the glass frit of a *porous plate cell*.[76,92,108,114–124] For instance, Taylor et al.[76,115] utilized the *porous plate cell* to investigate the surface forces and drainage dynamics of toluene- and heptane-diluted-in-bitumen films in water. In these studies, the hydrophobic modification of the fritted glass film holder was achieved by soaking it in bitumen solution, while Saint-Jalmes et al.[114] silanized the glass frit and applied it to dodecane films stabilized by a LMW surfactant or a polymer, as well as to a specific extract from nuclear waste treatment process.

6.5.1 Non-Bitumen Oil Thin Films

W/O/W films are readily stabilized by LMW surfactants,[14,56,113,114] including the specific case of lipids (in BLM).[12,19,78,90,91,109–112] It was concluded in Ref. 56 that the drainage and lifetime of benzene films, as determined by the concentration of the oil-soluble surfactant (lauryl alcohol) used, are not significantly influenced by the presence of a water-soluble surfactant (sodium octylsulfonate) in the dispersed phase. This experimental result is in full agreement with the theoretical scaling considerations presented in Section 6.2.2. The data from film thickness measurements (80–200 nm) revealed that the film thinning dynamics follows a linear dependence $h^{-2}(t)$ as its slope decreases with increase of the surfactant concentration, indicating decelerated film drainage due to immobilization of the film's surfaces.

Politova et al.[14] studied hexadecane or Isopar™ V (branched-chain alkanes) films stabilized by Span80 (sorbitane monooleate). Appreciable film lifetimes were observed only at surfactant concentrations above the critical micellar concentration and results for two characteristic film lifetime parameters τ_L and τ_S are presented in Figure 6.14. Increase of the surfactant concentration leads to an increase of the overall film lifetime τ_L due to increase of both the time of drainage and the lifetime τ_S of black spots emerging in the film. The results revealed an effect of the oil phase on the film stability – Isopar™ V produced more stable films as compared to hexadecane. The authors observed that black spots appear in the course of film drainage when the film thickness diminishes toward a critical value of about $h_{cr} \approx 45$ nm. The thickness of the black spots and the resulting overall black film was estimated as $h < 10$ nm. For comparison, the thickness of a decane black films stabilized by the LMW surfactant Arlacel83 (sorbitan sesquioleate, called also Span83) was found to be $h \approx 6$ nm,[113] which is in good agreement with the thickness of analogous decane films stabilized by lipids (BLM).[91] Such thickness range suggests the operation of short-range forces (in amphiphile bilayers[19]) arising at the physical contact between the alkyl chains

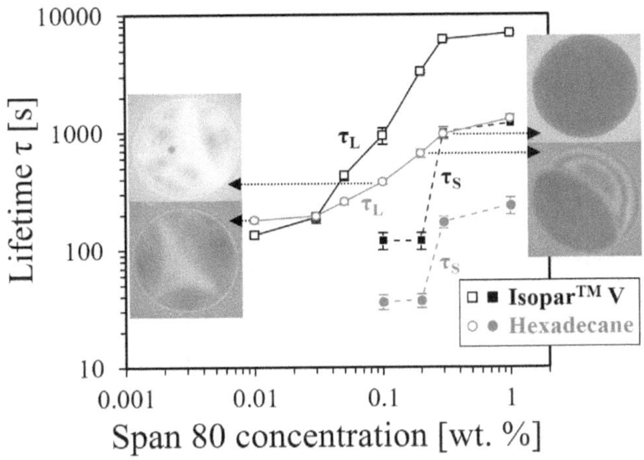

FIGURE 6.14 Characteristic lifetimes τ_L and τ_S for W/O/W films from hexadecane or Isopar™ V, stabilized by Span 80 (water phase contains 150 mM NaCl) at surfactant concentrations above the critical micellar concentration. τ_L is overall film lifetime and τ_S is lifetime of black spots; shown are also microphotographs of hexadecane films taken just before film rupture. Reproduced from Politova et al.,[14] Copyright (2017), with permission from Elsevier.

of the surfactant molecules adsorbed at the film's surfaces. This mechanism of film stabilization was discussed earlier on the basis of experimental results obtained by either electrical[12,90] or optical[19,56,113] methods mainly in BLM studies. Andrews et al.[12] employed the electrical capacitance method to measure $\Pi_{st}(h)$ isotherms for oil films stabilized by glyceryl mono-oleate and reported that the magnitude of the detected steric repulsion is sensitive to the nature of the oil used (various alkanes or CCl_4).

W/O/W films stabilized by oil-soluble polymeric surfactants have also been studied and the film thickness was measured by either electrical[13] or optical[113,114] methods. Aston et al.[113] reported $h \approx 15$ nm for decane or Isopar™ M black films stabilized by an ABA block copolymer (A is poly(12-hydroxystearic acid) and B is poly(ethyleneoxide)). The authors found that there was a relatively small proportion of oil present in the films (≈20 %) and the A-chains are presumably extended in the film, thus, they concluded that such films are sterically stabilized without discussing specificities of such steric forces. Such steric interaction in oil films can be classified as of brush-to-brush type, but in Ref. 113, the authors assumed that some penetration of the A-chains 'brushes' occurs. This scenario of interacting 'oily brushes' was also considered by Anklam et al.[13] for decane films stabilized by a comb copolymer (Abil EM-90 with a poly(dimethylsiloxane) backbone and grafted hydrophobic and hydrophilic chains). The authors calculated the thickness of the films from the measured film capacitance and proposed a theory for the interpretation of the film capacitance vs. applied electric field data in terms of disjoining pressure, and reported 'pseudo' disjoining pressure isotherms. In fact, an electrical compressive force was measured, revealing that the films studied are highly compressible (in the range of 32–62 nm). However, based on the thickness data, the authors discussed that the scenario of interacting 'oily brushes' may be alternatively considered as a result of stratification phenomena in analogy with O/W/O films from surfactant micellar solutions. Another explanation may lay on hydrodynamic phenomena as discussed by Saint-Jalmes et al.[114] The authors used TFPB to measure 'true' disjoining pressure isotherms for dodecane films stabilized by Abil EM-90, which revealed comparatively large thickness range (ca. 40–130 nm). They also observed coexistence of black portions (≈10 nm) and thicker portions in the overall films, which exhibited certain exchange dynamics (instabilities).

6.5.2 Bitumen Thin Films

As mentioned above, electrostatic interactions are not effectively involved in the stabilization mechanisms of W/O emulsions. However, here we just briefly comment on the interfacial charge of native water-bitumen interfaces, which might shed additional light on the state of the adsorption layers and their impact to the behavior of W/O emulsions. Using phase-sensitive SFG, Hu and Chou[103] observed that increase of the pH of water from 6 to 9 leads to increase of the SFG intensity for a native water-bitumen interface, meaning increase of the magnitude of the interfacial electric field. The authors discussed that the negative interfacial charge may be generated not only by OH^- adsorption (from the aqueous phase), but also by hydrolyzed acidic groups of endogenous substances (e.g., naphthenic acids) adsorbed from the bitumen phase. Furthermore, the native water-bitumen interface undergoes charge reversal upon adsorption of the cationic surfactant $C_{12}TA^+Cl^-$ (from the aqueous phase).

The initial studies of W/O/W films were followed by a series of systematic investigations focused on the understanding of the mechanisms of stabilization-destabilization of heavy crude oil (bitumen) emulsions.[20,116,123] Progress in water-in-diluted bitumen thin films were reviewed in Refs 18 and 124, so here we focus on some newer developments and more recent results. These studies, besides yielding important fundamental and technological results for the understanding of petroleum emulsions, have also prompted technical advancements in instrumentation and modifications to the classical thin liquid film technique using *porous plate cells.*

One notable advancement is the introduction by Panchev et al.[92] of a novel cell designed for applying an electrical field to W/O/W films. This modification enables routine measurements of the critical voltage required for film rupture, which is proposed as an indicator of film (emulsion) stability. Mostowfi et al.[123] used this new cell to investigate the mechanism of destabilization of W/O/W films in electric field, which is relevant to electro-enhanced coalescence treatment of W/O petroleum emulsions. It was observed that the application of electric field can induce electro-hydrodynamic instabilities in emulsion films. These instabilities are proposed as the principal mechanism leading to the rupture of relatively thick W/O/W films (thickness range of 50–100 nm) during the electro-induced coalescence of water droplets in petroleum emulsions.

Yang et al.[120] proposed a modification of the classical *porous plate cell* that allows for the addition of chemicals to already formed and equilibrated emulsion films (Figure 6.15). The motivation behind this improvement was to mimic industrial processes, where chemical aids are added to the emulsion. One such process, for example, involves the addition of demulsifiers to resolve W/O emulsions. In that work, the authors studied the effect of a biodegradable ethylcellulose polymeric

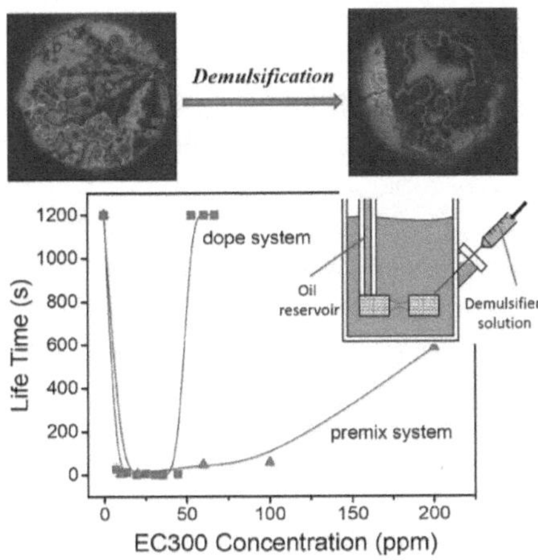

FIGURE 6.15 Schematic of the modified TLF cell with an option for dosing additives to already-formed stable film. Graph: Film lifetimes of 10 g/L asphaltene-in-(5:5 heptol) emulsion films as a function of EC300 concentration added by premixing or dosing method. Inset artwork: schematics of the modified *porous plate cell.* The microphotographs depict asphaltene-in-(5:5 heptol) stabilized films before (left) and after (right) dosing of 23 ppm demulsifier. Adapted from Yang et al.,[120] Copyright (2018), with permission from Elsevier.

demulsifier (EC300) with M_w = 182 Da on the stability of asphaltene-in-heptol (1:1 by volume blend of heptane and toluene). Two emulsion systems were compared: (1) a premixed system where a certain amount of EC300 was added to the asphaltene-in-heptol solution before loading the cell, and (2) an asphaltene-in-heptol emulsion film was formed in the cell and left to equilibrate, then EC300 was dosed into the film. The first procedure resulted in thinner films with fluid interfaces, most likely mixed asphaltene-demulsifier adsorption layers. In the second procedure, thicker and inhomogeneous stable films were formed, stabilized by asphaltenes forming 'rigid' interfaces. The addition of EC300 within the optimal concentration interval led to 'cracking' of the 'rigid' (or 'solid-like') asphaltene interfacial layer and destabilization of the films, rupturing within 20 seconds. These experiments revealed the overdosing effect of EC300 demulsifier. It was also found that the overdosing effect occurred at lower concentrations for the system by dosing the surfactant to already established films, as seen in Figure 6.15.

6.6 CONCLUDING REMARKS

The history of the TLF science is long[35] and our current understanding of both theoretical and experimental aspects of this central topic in the colloid chemistry is already on a high level as summarized, for instance, in Refs 1, 15–19, 25, 27, 32, 36, 37, 40–42, 125, and 126. It is obvious that studies on foam films predominate over studies on emulsion films. The reason for this situation should be related to experimental difficulties, as well as to complications with theoretical treatments, arising from the liquid nature of the film-surrounding fluid in emulsion systems as compared to liquid-gas systems (foam), as noted in some parts of this chapter. Furthermore, migration of emulsifier molecules across liquid/liquid interfaces (which for nonvolatile surfactants does not occur across liquid-gas interfaces) may have additional impact to emulsion film dynamics. Although aqueous (foam and O/W/O) films share many similarities, there are some clear differences observed.[25,32,77] For example, the drainage process in a microscopic TLF frequently exhibits the formation of a so-called 'dimple', which for a given aqueous solution formulation drains out slower in O/W/O films than in corresponding foam films.[25] The formation and stability of black films are also different for foam and O/W/O films in terms of surfactant concentration c – for a given appropriate c, the lifetime of foam films is longer than for corresponding O/W/O films.[77]

It becomes more and more clear how the complex interplay of capillarity, film drainage, disjoining pressure, transport phenomena, thermal fluctuations, and surface rheology control TLF dynamics and stability.[27,32,37,42,125] Concerning the impact of TLFs to the stability of foam or emulsion systems, two cases can be distinguished: film drainage related to bulk and interfacial effects within an anizodiametric space, and coupling of the drainage with the onset of surface forces (disjoining pressure). In the former case, the mobility of the interfaces (through the viscosity ratio of the contiguous bulk phases) and the mass transfer of surfactants (due to the flow-driven surface tension gradients) are the two basic elements to determine the TLF drainage performance. In the latter case, the prevailing factors are the specific interfacial structural properties and the onset of overlap between the interfacial regions within the film. In both cases, the effect of the droplet bulk phases is almost the same. In equilibrium TLFs the disjoining pressure components regulate film stability, and thus define the foam/emulsion stability, as related to both TLF-specific structural peculiarities and the respective impact of droplet bulk phases. Depending on the balance/coupling of the kinetic and thermodynamic stability factors, it appears that the mechanisms of coalescence (film rupture) in transient and in long-lasting foams/emulsions might be completely different.[27,125]

ACKNOWLEDGMENTS

Special thanks to S. Tcholakova for providing the original materials in Figure 6.14.

REFERENCES

1. (a) Exerowa, D.; Kruglyakov, P.M. *Foam and Foam Films: Theory, Experiment, Application.* Elsevier, 1998; Hardback ISBN: 9780444819222; eBook ISBN: 9780080531809. (b) Exerowa, D.; Gochev, G.; Platikanov, D.; Liggieri, L.; Miller, R. (Eds.) *Foam Films and Foams: Fundamentals and Applications.* CRC Press, 2018. Hardback ISBN: 9781466587724; eBook ISBN: 9781351117746.
2. Pickering, S.U. CXCVI.—Emulsions. *J. Chem. Soc. Trans.* **1907**, *91*, 2001–2021.
3. Rehbinder, P.; Wenström, E. Stabilisierende Wirkung von Adsorptionsschichten grenzflächenaktiver Stoffe auf disperse Systeme. II. Stabilität von Blasen und Tropfen an Trennungsflächen. (Eng. Stabilizing effect of adsorption layers of surface-active substances on disperse systems. II. Stability of bubbles and drops at interfaces) *Kolloid-Zeitschrift* **1930**, *53*, 145–158.

4. Dickinson, E. Emulsions. *Annu. Rep. Prog. Chem. Sect. C: Phys. Chem.* **1986**, 83, 31–58.
5. Izmailova, V.N.; Yampolskaya; G.P. Concentrated emulsions stabilized by macromolecules and the contributions of Hans Sonntag to this scientific field. *Colloids Surf. A* **1998**, *142*, 125–134.
6. Sjöblom, J. (Ed.). *Emulsions and Emulsion Stability*, 2nd ed.; CRC Press: Boca Raton, 2005.
7. Calderon, F.L.; Bibette, J.; Schmitt, V. (Ed.). *Emulsion Science, Basic Principles*, 2nd ed.; Springer Science + Business Media, LLC, 2007.
8. Tadros, T.F. (Ed.). *Emulsions: Formation, Stability, Industrial Applications*. de Gryuter: Berlin, 2016.
9. Salonen, A. Mixing bubbles and drops to make foamed emulsions. *Curr. Opin. Colloid Interface Sci.* **2020**, *50*, 101381.
10. Wasan, D.T.; Nikolov, A.D.; Huang, D.D.; Edwards, D.A. Foam stability: Effects of oil and film stratification. In *Surfactant Based Mobility Control*, Smith, D.H., Ed.; *ACS Symp. Ser.* **1988**, *373*, 136–162; Chapter 7.
11. Bergeron, V.; Radke, C.J. Disjoining pressure and stratification in asymmetric thin-liquid films. *Colloid Polym. Sci.* **1995**, *273*, 165–174.
12. Andrews, D.M.; Manev, E.D.; Haydon, D.A. Composition and energy relationships for some thin lipid films, and the chain conformation in monolayers at liquid-liquid interfaces. *Spec. Discuss. Faraday Soc.* **1970**, *1*, 46–56.
13. Anklam, M.R.; Saville, D.A.; Prud'homme, R.K. Disjoining pressure and film tension in comb–graft copolymer-stabilized oil films. *Langmuir* **1999**, *15*, 7299–7307.
14. Politova, N.; Tcholakova, S.; Denkov, N.D. Factors affecting the stability of water-oil-water emulsion films. *Colloids Surf. A* **2017**, *522*, 608–620.
15. Kralchevsky, P.A.; Danov, K.D.; Denkov, N.D. Chemical physics of colloid systems and interfaces. In *Handbook of Surface and Colloid Chemistry*, Second Expanded and Updated Edition, Birdi, K.S., Ed.; CRC Press: New York, 2002; Chapter 5.
16. Stubenrauch, C.; von Klitzing, R. Disjoining pressure in thin liquid foam and emulsion films—new concepts and perspectives. *J. Phys.: Condens. Matter* **2003**, *15*, R1197–R1232.
17. Gotchev, G.; Kolarov, T.; Khristov, K.; Exerowa, D. Electrostatic and steric interactions in oil-in-water emulsion films from Pluronic surfactants. *Adv. Colloid Interface Sci.* **2011**, *168*, 79–84.
18. Gochev, G. Thin liquid films stabilized by polymers and polymer/surfactant mixtures. *Curr. Opin. Colloid Interface Sci.* **2015**, *20*, 115–123.
19. Exerowa, E.; Kashchiev, D.; Platikanov D. Stability and permeability of amphiphile bilayers. *Adv. Colloid Interface Sci.* **1992**, *40*, 201–256.
20. Tchoukov, P.; Yang, F.; Xu, Z.; Dabros, T.; Czarnecki, J.; Sjöblom, J. Role of asphaltenes in stabilizing thin liquid emulsion films. *Langmuir* **2014**, *30*, 3024–3033.
21. Murray, B.S., Ettelaie, R., Sarkar, A., Mackie, A.R., Dickinson, E. The perfect hydrocolloid stabilizer: Imagination versus reality. *Food Hydrocolloids* **2021**, *117*, 106696.
22. Davis H.T. Factors determining emulsion type: Hydrophile—lipophile balance and beyond. *Colloids Surf. A* **1994**, *91*, 9–24.
23. Dimitrova, T.D.; Leal-Calderon, F.; Gurkov, T.D.; Campbell, B. Surface forces in model oil-in-water emulsions stabilized by proteins. *Adv. Colloid Interface Sci.* **2004**, *108–109*, 73–86.
24. Dimitrova, T.D.; Leal-Calderon, F. Rheological properties of highly concentrated protein-stabilized emulsions. *Adv. Colloid Interface Sci.* **2004**, *108–109*, 49–61.
25. Yampolskaya, G.; Platikanov, D. Proteins at fluid interfaces: Adsorption layers and thin liquid films. *Adv. Colloid Interface Sci.* **2006**, *128–130*, 159–183.

26. Jeffreys, G.F.; Davies, G.A. Coalescence of liquid droplets and liquid dispersion. In *Recent Advances in Liquid–Liquid Extraction*, Hanson, C., Ed.; Pergamon Press, 1971, ISBN 978-0-08-015682-8; Chapter 14, pp. 495–584.
27. Langevin, D. Coalescence in foams and emulsions: Similarities and differences. *Curr. Opin. Colloid Interface Sci.* **2019**, *44*, 23–31.
28. Rekvig, L.; Hafskjold, B.; Smit, B. Molecular simulations of surface forces and film rupture in oil/water/surfactant systems. *Langmuir* **2004**, *20*, 11583–11593.
29. Rekvig, L.; Frenkel, D. Molecular simulations of droplet coalescence in oil/water/surfactant systems. *J. Chem. Phys.* **2007**, *127*, 134701.
30. Aarøen, O.; Riccardi. E.; van Erp T.S.; Sletmoen, M. Thin film breakage in oil–in–water emulsions, a multidisciplinary study. *Colloids Surf. A* **2022**, *632*, 127808.
31. Sedev, R. Exerowa, D. DLVO and non-DLVO surface forces in foam films from amphiphilic block copolymers. *Adv. Colloid Interface Sci.* **1999**, *83*, 111–136.
32. Platikanov, D.; Exerowa, D. Symmetric thin liquid films with fluid interfaces. In Sjöblom, J. (Ed.). *Emulsions and Emulsion Stability*, 2nd ed.; CRC Press: Boca Raton, 2005. Chapter 3.
33. Ikeda, N.; Krustev, R.; Müller, H.J. Thermodynamic consideration on single oil in water emulsion film stabilized by cationic surfactant. *Adv. Colloid Interface Sci.* **2004**, *108–109*, 273–286.
34. Narsimhan, G. Stability of thin emulsion film between two oil phases with a viscoelastic liquid-liquid interface. *J. Colloid Interface Sci.* **2009**, *330*, 494–500.
35. Gochev, G.; Platikanov, D.; Miller, R. Chronicles of foam films. *Adv. Colloid Interface Sci.* **2016**, *233*, 115–125.
36. Derjaguin, B.V.; Churaev, N.V.; Muller, V.M. *Surface forces.* Springer: NY, 1987.
37. Mileva, E.; Radoev, B. Hydrodynamic interactions and stability of emulsion films. In *Emulsions: Structure Stability and Interactions*, Petsev D.N., Ed.; Elsevier, 2004; Chapter 6; Hardback ISBN: 9780120884995, eBook ISBN: 9780080472652.
38. Krotov, V.V.; Rusanov, A.I. *Physicochemical Hydrodynamics of Capillary Systems.* Imperial College Press, London, UK, 1999.
39. Scheludko, A. Thin liquid films. *Adv. Colloid Interface Sci.* **1967**, *1*, 391–464.
40. Boinovich, L. DLVO forces in thin liquid films beyond conventional DLVO theory. *Curr. Opin. Colloid Interface Sci.* **2010**, *15*, 297–302.
41. Danov, K. Effect of surfactants on drop stability and thin film drainage. In *Fluid Mechanics of Surfactant and Polymer Solutions*, Starov, V.M., Ivanov, I.B., Eds.; International Centre for Mechanical Sciences, Springer: Vienna, 2004, Volume 463, pp. 1–38.
42. Chatzigiannakis, E.; Jaensson, N.; Vermant, J. Thin liquid films: Where hydrodynamics, capillarity, surface stresses and intermolecular forces meet. *Curr. Opin. Colloid Interface Sci.* **2021**, *53*, 101441.
43. De Gennes, P.G. Polymers at an interface; a simplified view. *Adv. Colloid Interface Sci.* **1987**, *27*, 189–209.
44. Haber, S.; Hetsroni G.; Solan, A. On the low Reynolds number motion of two droplets. *Int. J. Multiphase Flow* **1973**, *1*, 57–71.
45. Mileva, E.; Radoev, B. Hydrodynamical interaction of two emulsion droplets at small separations. *Colloid Polym. Sci.* **1985**, *263*, 587–593.
46. Mileva, E.; Radoev, B. Effective zones of hydrodynamical interaction of fluid particles at small separations. *Colloid Polym. Sci.* **1986**, *264*, 823–828.
47. Sedov, L. *Similarity and Dimensional Methods in Mechanics.* CRC Press, 1993.
48. Radoev, B.; Dimitrov, D.; Ivanov, I. Hydrodynamics of thin liquid films effect of the surfactant on the rate of thinning. *Colloid Polym. Sci.* **1974**, *252*, 50–55.
49. Ozan, S.C.; Jakobsen, H.A. Effect of surface viscoelasticity on the film drainage and the interfacial mobility. *Int. J. Multiphase Flow* **2020**, *130*, 103377.

50. Traykov, T.T.; Ivanov, I.B. Hydrodynamics of thin liquid films. Rate of thinning of emulsion films from pure liquids. *Int. J. Multiphase Flow* **1976**, *2*, 397–410.
51. Levich, V. *Physicochemical Hydrodynamics*, Prentice-Hall, Englewood Cliffs, N.J., 1962; ISBN-13: 978-0136744405.
52. Lee, J.; Hodgson, T. Film flow and coalescence – I. Basic relations, film shape and criteria for interface mobility. *Chem. Eng. Sci.* **1968**, *23*, 1375–1397.
53. Mileva, E.; Radoev, B. The influence of surfactants on the hydrodynamical interaction in emulsion systems. *Colloid Polym. Sci.* **1986**, *264*, 965–971.
54. Mileva, E.; Radoev, B. Mass transfer of surfactants and hydrodynamic interaction at small separations in emulsion systems 1. Scaling concepts. *Colloids Surf. A* **1993**, *74*, 259–266.
55. Mileva, E.; Nikolov, L. Mass transfer of surfactants and hydrodynamic interaction at small separations in emulsion systems 2. Potential theory. *Colloids Surf. A* **1993**, *74*, 267–273.
56. Traykov, T.T.; Manev, E.D.; Ivanov, I.B. Hydrodynamics of thin liquid films. Experimental investigation of the effect of surfactants on the drainage of emulsion films. *Int. J. Multiphase Flow* **1977**, *3*, 485–494.
57. Mileva, E.; Tchoukov, P. Surfactant nanostructures in foam films. In *Colloid Stability. The role of surface forces – Part 1.* Tadros, Th., Ed.; Wiley-VCH, 2007; Chapter 8.
58. Xie L.; Shi, C.; Zeng, H. Surface forces and interaction mechanisms of emulsion drops and gas bubbles in complex fluids. *Langmuir* **2017**, *33*, 3911–3925.
59. Gochev, G.; Retzlaff, I.; Exerowa, D.; Miller, R. Electrostatic stabilization of foam films from β-lactoglobulin solutions. *Colloids Surf. A* **2014**, *460*, 272–279.
60. Chatzigiannakis, E.; Yang, J.; Sagis, L.; Nikiforidis, C.V. Thin *Liquid Films Stabilized by Plant Proteins: Implications for Foam Stability.* SSRN Preprint DOI: 10.2139/ssrn.4928284.
61. Cockbain, E.G.; McRoberts, T.S. The stability of elementary emulsion drops and emulsions. *J. Colloid Sci.* **1953**, *18*, 440–451.
62. Biswas, B.; Haydon, D.A. The coalescence of droplets stabilised by viscoelastic adsorbed films. *Kolloid-Z. Z. Polym.* **1962**, *185*, 31–38.
63. Velev, O.D.; Gurkov, T.D.; Chakarova, Sv.K.; Dimitrova, B.I.; Ivanova, I.B.; Borwankar, R.P. Experimental investigations on model emulsion systems stabilized with non-ionic surfactant blends. *Colloids Surf. A* **1994**, *83*, 43–45.
64. Basheva, E.S.; Gurkov, T.D.; Ivanov, I.B.; Bantchev, G.B.; Campbell, B.; Borwankar, R.P. Size dependence of the stability of emulsion drops pressed against a large interface. *Langmuir* **1999**, *15*, 6764–6769.
65. Politova, N.I.; Tcholakova, S.; Tsibranska, S.; Denkov, N.D.; Muelheims, K. Coalescence stability of water-in-oil drops: Effects of drop size and surfactant concentration. *Colloids Surf. A* **2017**, *531*, 32–39.
66. Gawel, D.; Zawala, J. Stability of liquid films formed by a single bubble and droplet at liquid/gas and liquid/liquid interfaces in bovine serum albumin solutions. *ACS Omega* **2021**, *6*, 18289–18299.
67. Van den Tempel M. Distance between emulsified oil globules upon coalescence. *J. Colloid Sci.* **1958**, *13*, 125–133.
68. (a) Sonntag, H.; Netzel, J.; Klare, Jr. H. Untersuchungen über die Stabilität von Emulsionen in Gegenwart grenzflächenaktiver Stoffe (Eng. Studies on the stability of emulsions in the presence of surface-active substances). *Kolloid-Z. Z. Polym.* **1966**, *211*, 121. (b) Sonntag, H.; Netzel, J.; Unterberger, B. Interaction energy of emulsion droplets and the influence of adsorbed layers on it. *Spec. Disc. Faraday Soc.* **1970**, *1*, 57–63.

69. Sonntag, H.; Unterberger, B.; Zimontkowski, S. Experimentelle Untersuchung der Sterischen Stabilisierung von Emulsionen durch Polyvinylalkohol. (Eng. Experimental investigation of the steric stabilization of emulsions by polyvinyl alcohol) *Colloid Polym. Sci.* **1979**, *257*, 286–291.
70. (a) Sonntag, H.; Klare Jr, H. Die Brechung von Wasser-in-Erdöl-Emulsionen mit dem Demulgator Dispersal BO im Zusammenhange mit der Veränderung der Wechselwirkungskräfte zwischen den Wassertröpfchen. (Eng. Breakage of water-in-crude oil emulsions by the demulsifier Dispersal BO with respect to changes in the interaction forces between water droplets.). *Kolloid-Z. Z. Polymere* **1964**, *195*, 35–39. (b) Sonntag, H.; Klare, H. Beeinflussung der Koaleszenzstabilität von W/O-Emulsionen durch wasserlösliche Tenside, äußere elektrische Felder und Temperaturerhöhung. (Eng. Influence of water-soluble surfactants, external electric fields and temperature increase on the coalescence stability of W/O emulsions). *Tenside Surfactants Detergents* **1967**, *4*, 104–108.
71. Won J.Y.; Krägel, J.; Makievski, A.V.; Javadi, A.; Gochev, G.; Loglio, G.; Pandolfini, P.; Leser, M.E.; Gehin-Delval, C.; Miller, R. Drop and bubble micro manipulator (DBMM)—A unique tool for mimicking processes in foams and emulsions. *Colloids Surf. A* **2014**, *441*, 807–814.
72. Aveyard, R.; Binks, B.P.; Cho, W.-G.; Fisher, L.R.; Fletcher, P.D.I.; Klinkhammer F. Investigation of the force-distance relationship for a small liquid drop approaching a liquid-liquid interface. *Langmuir* **1996,** *12,* 6561–6569.
73. Cho, W.-G.; Fletcher, P.D.I. Investigation of the force–distance relationship for a small liquid drop approaching an adhesive liquid/liquid interface. *Langmur* **1997**, *93*, 1389–1395.
74. Exerowa, D.; Gotchev, G.; Kolarov, T.; Khristov, K.; Levecke, B.; Tadros, T. Interaction forces in thin liquid films stabilized by hydrophobically modified inulin polymeric surfactant. 2. Emulsion films. *Langmuir* **2007,** *23,* 1711–1715.
75. Wierenga, P.A.; Basheva, E.S.; Denkov, N.D. Modified capillary cell for foam film studies allowing exchange of the film-forming liquid. *Langmuir* **2009**, *25*, 6035–6039.
76. Khristov, K., Taylor, S.D., Czarnecki, J., & Masliyah, J. Thin liquid film technique — application to water – oil – water bitumen emulsion films. *Colloids Surf. A* **2000**, *174*, 183–196.
77. Müller, H.J.; Balinov, B.B. Exerowa, D.R. Black foam and emulsion films from non-ionic surfactant solutions: Rupture by hole formation. *Colloid Polymer Sci.* **1988**, *266*, 921–925.
78. Möglich, P.M., Kretzschmar, G. Optical determination of the thickness of single planar thin lipid films using the symmetrical three-layer model. *Colloid Polymer Sci.* **1988**, *266*, 739–744.
79. Gabrieli, R.; Loglio, G.; Pandolfini, P.; Fabbri, A.; Simoncini, M.; Kovalchuk, V.I.; Noskov, B.A.; Makievski, A.V.; Krägel, J. ; Miller, R.; Ravera, F.; Liggieri, L. Spherical cap-shaped emulsion films: Thickness evaluation at the nanoscale level by the optical evanescent wave effect. *Colloids Surf. A* **2012**, *413*, 101–107.
80. Mysels, K.J.; Shinoda, C.; Frankel, S. *Soap films: Studies on their thinning and a bibliography.* Pergamon, 1959.
81. Gochev, G.G.; Campbell, R.A.; Schneck, E.; Zawala, J.; Warszynski, P. Exploring proteins at soft interfaces and in thin liquid films - From classical methods to advanced applications of reflectometry. *Adv. Colloid Interface Sci.* **2024**, *329*, 103187.
82. Nushtaeva, A.V.; Kruglyakov, P.M. Investigation of model emulsion films stabilized by solid particles: Thickness of films, their stability, and interfacial tension. *Colloid J.* **2004**, *66*, 456–465.

83. Horozov, T.S.; Aveyard, R.; Clint, J.H.; Neumann, B. Particle zips: Vertical emulsion films with particle monolayers at their surfaces. *Langmuir* **2005**, *21*, 2330–2341.
84. Gunning, A.P.; Mackie, A.R.; Wilde, P.J.; Morris, V.J. Atomic force microscopy of emulsion droplets: Probing droplet-droplet interactions. *Langmuir* **2004**, *20*, 116–122.
85. Dagastine, R.R.; Stevens, G.W.; Chan, D.Y.; Grieser, F. Forces between two oil drops in aqueous solution measured by AFM. *J. Colloid Interface Sci.* **2004**, *273*, 339–342.
86. Bauer, W.-A.C.; Kotar, J.; Cicuta, P.; Woodward, R.T.; Weaver, J.V.; Huck, W.T. Microfluidic production of monodisperse functional o/w droplets and study of their reversible pH dependent aggregation behavior. *Soft Matter* **2011**, *7*, 4214–4220.
87. Aarøen, O.; Riccardi, E.; Sletmoen, M. Exploring the effects of approach velocity on depletion force and coalescence in oil-in-water emulsions. *RSC Adv.* **2021**, *11*, 8730–8740.
88. Calderon, F.L.; Stora, T.; Mondain Monval, O.; Poulin, P.; Bibette, J. Direct measurement of colloidal forces. *Phys. Rev. Lett.* **1994**, *72*, 2959–2962.
89. Poulin, P.; Nallet, F.; Cabane, B.; Bibette, J. Evidence for Newton black films between adhesive emulsion droplets. *Phys. Rev. Lett.* **1996**, *77*, 3248–3251.
90. (a) Hanai, T.; Haydon, D.A.; Taylor, J. An investigation by electrical methods of lecithin-in-hydrocarbon films in aqueous solutions. *Proc. R. Soc. Lond. A* **1964**, *281*, 377–391. (b) Hanai, T.; Haydon, D.A.; Taylor, J. Polar group orientation and the electrical properties of lecithin bimolecular leaflets. *J. Theor. Biol.* **1965**, *9*, 278–296.
91. Dilger, J.P.; Fisher, L.R.; Haydon, D.A. A critical comparison of electrical and optical methods for bilayer thickness determination. *Chem. Phys. Lipids* **1982**, *30*, 159–176.
92. Panchev, N.; Khristov, K.; Czarnecki, J.; Exerowa, D.; Bhattacharjee, S.; Masliyah, J. A new method for water-in-oil emulsion film studies. *Colloids Surf. A* **2008**, *315*, 74–78.
93. Binks, B.P.; Cho, W.-G.; Fletcher, P.D.I. Disjoining pressure isotherms for oil-water-oil emulsion films. *Langmuir* **1997**, *13*, 7180–7185.
94. Basheva, E.S.; Gurkov, T.D.; Christov, N.C.; Campbell, B. Interactions in oil/water/oil films stabilized by β-lactoglobulin; role of the surface charge. *Colloids Surf. A* **2006**, *282–283*, 99–108.
95. Exerowa, D.; Gotchev, G.; Kolarov, T.; Khristov, K.; Levecke, B.; Tadros, T. Comparison of oil-in-water emulsion films produced using ABA or AB_n copolymers. *Colloids Surf. A* **2009**, *335*, 50–54.
96. Gotchev, G.; Kolarov, T.; Khristov, K.; Exerowa, D. On the origin of electrostatic and steric repulsion in oil-in-water emulsion films from PEO-PPO-PEO triblock copolymers. *Colloids Surf. A* **2010**, *354*, 56–60.
97. Villwock, J.; Gebauer, F.; Kamp, J.; Bart, H.-J.; Kraume, M. Systematic analysis of single droplet coalescence. *Chem. Eng. Technol.* **2014**, *37*, 1103–1111.
98. Delacotte, J.; Montel, L.; Restagno, F.; Scheid, B.; Dollet, B.; Stone, H.A.; Langevin, D., Rio. E. Plate coating: Influence of concentrated surfactants on the film thickness. *Langmuir* **2012**, *28*, 3821–3830.
99. Marinova K.G.; Alargova R.G.; Denkov, N.D.; Velev, O.D.; Petsev, D.N.; Ivanov, I.B.; Borwankar, R.P. Charging of oil–water interfaces due to spontaneous adsorption of hydroxyl ions. *Langmuir* **1996**, *12*, 2045–2051.
100. Guckeisen, T.; Hosseinpour, S.; Peukert, W. Isoelectric points of proteins at the air/liquid interface and in solution. *Langmuir* **2019**, *35*, 5004–5012.
101. Kolarov, T.; Yankov, R.; Esipova, N.E.; Exerowa, D.; Zorin Z.M. Charge reversal at the air/water interface as inferred from the thickness of foam films. *Colloid Polym. Sci.* **1993**, *271*, 519–520.
102. Aveyard, R.; Binks, B.P.; Esquena, J.; Fletcher, P.D.I.; Bault, P.; Villa, P. Flocculation transitions of weakly charged oil-in-water emulsions stabilized by different surfactants. *Langmuir* **2002**, *18*, 3487–3494.

103. Hu, D.; Chou, K.C. Surface charge at the Bitumen/water interface investigated by phase-sensitive sum frequency generation vibrational spectroscopy: Effects of pH, Ions, and surfactants. *Energy Fuels* **2015**, *29*, 7885–7888.
104. Braunschweig, B.; Schulze-Zachau, F.; Nagel, E.; Engelhardt, K.; Stoyanov, S.; Gochev, G.; Khristov, K.; Mileva, E.; Exerowa, D.; Miller, R.; Peukert, W. Specific effects of Ca2+ ions and molecular structure of β-lactoglobulin interfacial layers that drive macroscopic foam stability. *Soft Matter* **2016**, *12*, 5995–6004.
105. García Rey, N.; Weißenborn, E.; Schulze-Zachau, F.; Gochev, G.; Braunschweig, B. Quantifying double-layer potentials at liquid-gas interfaces from vibrational sum-frequency generation. *J. Phys. Chem. C* **2019**, *123*, 1279–1286.
106. Tcholakova, S.; Denkov, N.D.; Ivanov, I.B.; Campbell, B. Coalescence stability of emulsions containing globular milk proteins. *Adv. Colloid Interface Sci.* **2006**, *123–126*, 259–293.
107. Delahaije, R.J.; Wierenga, P.A.; van Nieuwenhuijzen, N.H.; Giuseppin, M.L.; Gruppen, H. Protein concentration and protein-exposed hydrophobicity as dominant parameters determining the flocculation of protein-stabilized oil-in-water emulsions. *Langmuir* **2013**, *29*, 11567–11574.
108. Khristov, K.; Petkova, H.; Alexandrova, L.; Nedyalkov, M.; Platikanov, D.; Exerowa, D.; Beetge, J. Foam, emulsion and wetting films stabilized by polyoxyalkylated diethylenetriamine (DETA) polymeric surfactants. *Adv. Colloid Interface Sci.* **2011**, *168*, 105–113.
109. (a) Ti Tien, H.; Dawidowicz, A.L. Black lipid films in aqueous media: A new type of interfacial phenomenon: Experimental techniques and thickness measurements. *J. Colloid Interface Sci.* **1966**, *22*, 438–456. (b) Ti Tien, H.; Diana, A.L. Bimolecular lipid membranes: A review and a summary of some recent studies. *Chem. Phys. Lipids* **1968**, *2*, 55–101.
110. (a) Mueller, A.L.; P., Rudin, D.; Ti Tien, H.; Wescott, W.C. Reconstitution of cell membrane structure *in vitro* and its transformation into an excitable system. *Nature* **1962**, *194*, 979–980. (b) Ti Tien, H.; Ottova, A.L. The lipid bilayer concept and its experimental realization: From soap bubbles, kitchen sink, to bilayer lipid membranes. *J. Membrane Sci.* **2001**, *189*, 83–117.
111. Huang, C.; Wheeldon, L.; Thompson, T.E. The properties of lipid bilayer membranes separating two aqueous phases: Formation of a membrane of simple composition. *J. Mol. Biol.* **1964**, *8*, 148–160.
112. (a) Cherry, R.J.; Chapman, D. Optical properties of black lecithin films. *J. Mol. Biol.* **1969**, *40*, 19–32. (b) Cherry, R.J.; Chapman, D. Optical determination of the thickness of thin lipid films. *J. Theor. Biol.* **1969**, *24*, 137–146.
113. Aston, MS.; Herrington, T.M.; Tadros, Th.F. Stability of thin oil liquid films using mixed surfactants. *Colloids Surf.* **1989**, *40*, 49–61.
114. Saint-Jalmes, A.; Zemb, T.; Langevin, D. Water/oil/water thin films: construction and applications. *Prog. Colloid Polymer Sci.* **2001**, *118*, 1–4.
115. Taylor, S.D.; Czarnecki, J.; Masliyah, J. Disjoining pressure isotherms of water-in-Bitumen emulsion films. *J. Colloid Interface Sci.* **2002**, *252*, 149–160.
116. Czarnecki, J.; Tchoukov, P.; Dabros, T.; Xu, Z. Role of asphaltenes in stabilisation of water in crude oil emulsions. *Can. J. Chem. Eng.* **2013**, *91*, 1365–137.
117. Tchoukov, P.; Czarnecki, J.; Dabros, T. Study of water-in-oil thin liquid films: Implications for the stability of petroleum emulsions. *Colloids Surf. A* **2010**, *372*, 15–21.
118. Yang, F.; Tchoukov, P.; Pensini, E.; Dabros, T.; Czarnecki, J.; Masliyah, J.; Xu, Z. Asphaltene subfractions responsible for stabilizing water-in-crude oil emulsions. Part 1: Interfacial behaviors. *Energy Fuels* **2014**, *28*, 6897–6904.

119. Bi, J.; Yang, F.; Harbottle, D.; Pensini, E.; Tchoukov, P.; Simon, S.; Sjöblom, J.; Dabros, T.; Czarnecki, J.; Liu, Q.; Xu, Z. Interfacial layer properties of a polyaromatic compound and its role in stabilizing water-in-oil emulsions. *Langmuir* **2015**, *31*, 10382–10391.
120. Yang, F.; Tchoukov, P.; Qiao, P.; Ma, X.; Pensini, E.; Dabros, T.; Czarnecki, J.; Xu, Z. Studying demulsification mechanisms of water-in-crude oil emulsions using a modified thin liquid film technique. *Colloids Surf. A* **2018**, *540*, 215–223.
121. Hristova, E.; Tchoukov, P.; Stoyanov, S.R. Coalescence inhibition and agglomeration initiation near the critical dilution of asphaltene precipitation. *Colloids Surf. A* **2021**, *629*, 127400.
122. Hristova, E.; Tchoukov, P.; Stoyanov, S.R.; McFarlane, R. Asphaltene precipitation onsets in relation to the critical dilution of Athabasca Bitumen in paraffinic solvents. *Energy Fuels* **2022**, *36*, 1832–1841.
123. Mostowfi, F.; Tchoukov, P.; Panchev, N.; Dabros, T.; Czarnecki, J. Electrohydrodynamic instabilities in free emulsion films. *Colloids Interfaces* **2021**, *5*, 36.
124. Czarnecki, J.; Khristov, K.; Masliyah, J.; Panchev, N.; Taylor, S.D.; Tchoukov, P. Application of Scheludko–Exerowa thin liquid film technique to studies of petroleum W/O emulsions. *Colloids Surf. A* **2017**, *519*, 2–10.
125. Langevin, D. Recent advances on emulsion and foam stability. *Langmuir* **2023**, *39*, 3821–3828.
126. Zawala, J.; Miguet, J.; Rastogi, P.; Atasi, O.; Borkowski, M.; Scheid, B.; Fuller, G.G. Coalescence of surface bubbles: The crucial role of motion-induced dynamic adsorption layer. *Adv. Colloid Interface Sci.* **2023**, *317*, 102916.

Section II

Emulsion preparation and characterization

7 Experimental Methods for Emulsion Production

Eduardo Guzmán

7.1 INTRODUCTION

Emulsions are systems that are of great interest for many industries, including the food industry, pharmaceuticals, cosmetics and personal care, paints and coatings, oil and gas industry, agriculture, and textiles.[1–5] The preparation of emulsions typically involves the incorporation of an emulsifier into the continuous phase, followed by the gradual addition of the dispersed phase under moderate stirring.[6,7] Numerous factors contribute to the instability of emulsion systems, including phenomena such as flocculation, phase separation, coalescence, and Ostwald ripening. These mechanisms have a significant impact on the applicability of emulsion-based products. Consequently, the production of stable emulsions requires the use of highly effective emulsifiers. However, this is not enough for a suitable production of emulsions with long-term stability, and the introduction of efficient methodologies to ensure the correct mixtures of the phases and the emulsifier is essential.[8]

A wide range of emulsification techniques are currently available, providing tools for the preparation of emulsions under different conditions. These methods include high-pressure homogenization, rotor–stator homogenization, microchannel emulsification, membrane emulsification, microfluidics, and ultrasound emulsification, among others.[7,8]

The broad interest in emulsions has made the design of scaling-up strategies a key challenge to produce emulsions. However, this is not always straightforward and requires in most cases the use of methodologies that need an elevated energy consumption ('high-energy emulsification' methods) such as rotor–stator system and high-pressure homogenization.[2] However, different methodologies present interest in improving and optimizing the characteristics of the products. These methodologies include different low-shear rate methodologies, including membrane, microchannel, and microfluidic emulsification. In particular, the latter family of methodologies is of particular interest because it presents remarkable capability in generating highly monodisperse emulsions and fabricating individual droplets with precision.[9] Nevertheless, to date, these emulsification systems have primarily been investigated on a laboratory scale, necessitating further exploration for their viability in larger-scale applications. It should be noted that each emulsification method presents specific advantages and disadvantages.[10,11]

This chapter thoroughly explores the diverse landscape of emulsification methods, meticulously analyzing their strengths and limitations in shaping essential emulsion

DOI: 10.1201/9781032636184-9

properties. By delving into the intricate interplay of process variables and their influence on emulsion properties, it reveals recent breakthroughs and illuminates new avenues for future progress. The insights gained from this comprehensive overview provide an invaluable compass for researchers, scientists, and industry practitioners navigating the field of emulsification. As a compass for navigating the complex terrain of emulsification, this chapter equips professionals with the knowledge to make informed decisions when selecting techniques for specific applications. It also ignites the spirit of innovation, paving the way for continued progress and innovation in this dynamic field.

7.2 CLASSIFICATION OF THE EMULSIFICATION METHODS

Preparing emulsions involves different techniques due to variations in free energies within the systems. Microemulsions, due to their energetically favorable nature, can be prepared spontaneously by directly mixing the different components. On the other hand, nanoemulsions, due to their thermodynamic instability, require an external energy source to exceed the positive value of the Gibbs free energy.[12]

The most common division of the emulsification techniques depends on the energy required for the process. Thus, it is possible to define high-energy and low-energy emulsification methods. The selection of the appropriate emulsification device depends on several factors, including the volume of the mixture, the characteristics of the initial ingredients, the intended physicochemical properties of the resulting emulsion, and the overall cost associated with the emulsification process.[13–16]

Low-energy emulsification techniques are gaining significance in the preparation of emulsions. In general, low-energy emulsification processes operate under laminar and low-energy conditions,[17] offering economic appeal and the potential to guarantee the protection of sensitive molecules.[18] In fact, methodologies using low-energy emulsification leverage the intrinsic chemical energy of the system, utilizing the chemical potential of the components involved. Therefore, the performance of this type of techniques is mainly governed by the physicochemical properties of emulsifiers which makes necessary a careful selection of the latter.[19] Low-energy methods generally facilitate the production of small droplet sizes comparable to those achieved by high-energy approaches.[20] This phenomenon results from the spontaneous curvature shifting of the surfactant, commonly referred to as self-emulsification (SE).[21]

On the other hand, high-energy emulsification processes allow controlling the size of the emulsion droplets, facilitating the production of fine emulsions.[22] This family of techniques allows direct control of process parameters such as mechanical energy input and time within the equipment. In addition, factors such as the amount of surfactant, component properties, and phase viscosity have a significant impact on high-energy emulsification.[23,24] Figure 7.1 displays a classification of different emulsification methodologies as a function of the energy requirements.

High-energy emulsification involves simultaneous processes during emulsion formation. The application of high shear stress provides significantly more energy than the Laplace pressure gradient. This causes droplet deformation and a subsequent

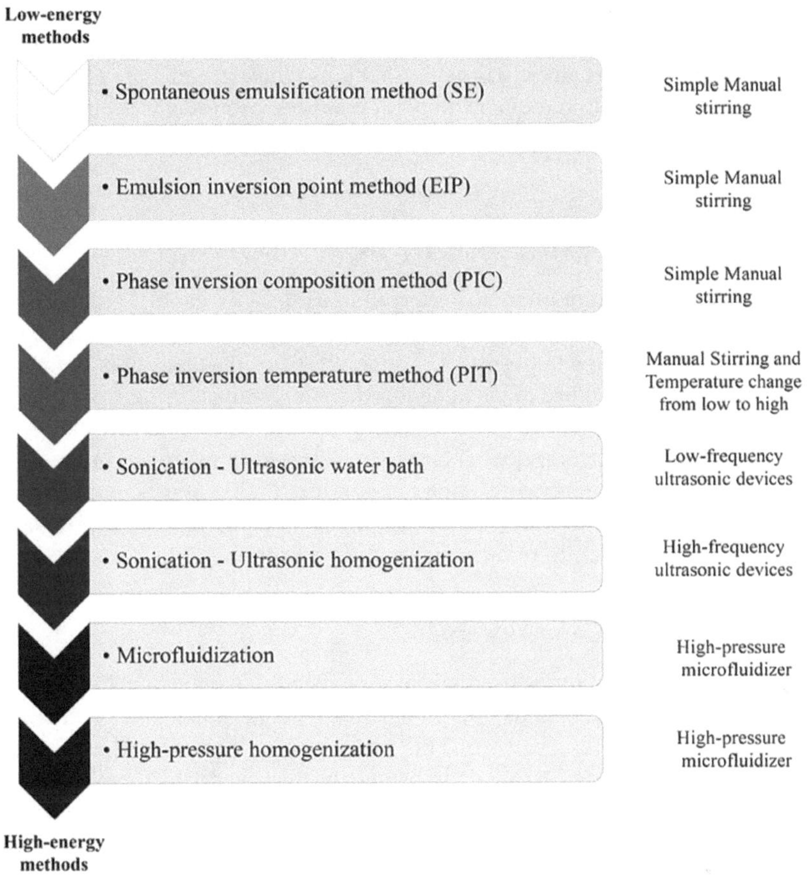

FIGURE 7.1 Classification of emulsification methodologies as a function of the required energy. Reprinted from Lucia and Guzmán,[1] Copyright (2021), with permission from Elsevier.

increase in specific surface area until rupture occurs. Emulsifiers can stabilize this new interface. Whether or not droplet collision leads to re-coalescence depends on the protective effect of the emulsifier and its adsorption on the droplet surface.[25,26] Despite extensive research, the full understanding of the mechanisms involved in droplet breakup remains incomplete. The homogenization process involves simultaneous cavitation in addition to laminar and turbulent flow conditions. However, there is no consensus on how the flow conditions vary with different geometries and emulsifier systems. In most cases, except smaller laboratory setups, turbulent shear forces within the gap and/or at the outlet typically play the most significant role in droplet disruption,[25,27] although cavitation and laminar shear may also exert an influence.[25] In laminar flow, viscous forces dominate the flow, with shear forces being the sole cause of droplet deformation and disruption.[28] Turbulent flows are characterized by vortices and unpredictable velocity fluctuations. Inertial forces typically dominate, but viscous forces can also be involved in the homogenization process.[28] Cavitation involves the formation of gas- or vapor-filled cavities due to a decrease in static pressure. Their subsequent collapse upon pressure increase generates powerful shock waves responsible for droplet disruption.[29] Various factors contribute to the decrease in static pressure, including reduced ambient pressure, oscillatory effects, or specific flow conditions that convert pressure energy into kinetic energy, thereby inducing hydrodynamic cavitation.[30]

7.3 LOW-ENERGY EMULSIFICATION METHODS

Low-energy emulsification methods, referred to as the low-energy methods, rely on the release of chemical energy during emulsification, which is considered the driving force behind the phenomena observed. The key factor contribution is the alteration in the spontaneous curvature of surfactant molecules, transitioning from negative to positive (oil-in-water) or positive to negative (water-in-oil).[31] The low-energy methods encompass various techniques such as spontaneous emulsification, phase inversion composition (PIC), phase inversion temperature (PIT), microemulsion dilution, and D-phase emulsification (DPE). Notably, these methods involve stirring at a low rate (1,600 rpm), resulting in reduced energy consumption.

7.3.1 Spontaneous Emulsification

Spontaneous emulsification is the natural occurrence of emulsion formation typically observed when two liquids that do not readily mix, namely, oil and water, are combined at a certain temperature. This process does not require any additional equipment.[32] Its effectiveness depends primarily on several factors, including the concentration and molecular structure of the surfactant, the interfacial tension, the phase transition region, and the interfacial and bulk viscosity.[33] Therefore, the understanding of spontaneous emulsification phenomena requires combining considerations related to phase behavior, numerical calculations, and an examination of the droplet formation process.[34] Figure 7.2 shows a sketch of the main mechanistical aspect involved in spontaneous emulsification.

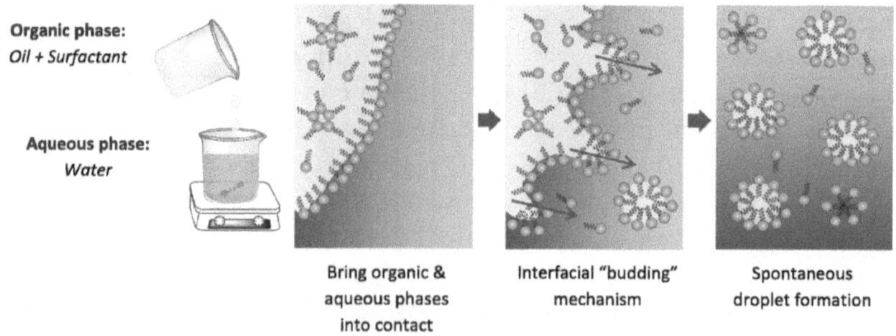

FIGURE 7.2 Sketch of the most accepted mechanism for spontaneous emulsification. Adapted from McClemens and Rao,[40] Copyright (2011), with permission from Taylor and Francis.

The spontaneous emulsification is based on the migration of water-miscible elements, such as solvent, surfactant, and cosurfactant, from an organic phase to an aqueous phase. The process is initiated by introducing an organic phase consisting of oil and surfactant into an aqueous phase containing water and cosurfactant. The rapid movement of water-miscible components into the aqueous phase induces significant turbulence at the interface of the two phases and an increase in the oil–water interfacial area.[35–37] Consequently, this turbulence leads to the spontaneous formation of fine oil droplets when the bicontinuous microemulsion phase breaks up through a budding process, resulting in the phenomenon of spontaneous emulsification. The use of solvents can facilitate this process, whether surfactants are present or not. This phenomenon is known as the Ouzo effect.[38] The order of the mixing components has not been shown to have a critical effect on the process, as nanoemulsions form spontaneously.[39]

7.3.2 Emulsion-Phase Inversion (EPI) Methods

The principle of EPI contrasts with SE. In fact, the EPI method relies on the introduction of an aqueous phase into a continuously stirred organic phase, typically consisting of oil and surfactant. This experimental setup represents a reversal of the setup seen in self-emulsification, where an aqueous phase is added to a stirred organic phase. As the EPI approach involves the mixing of water with oil, it is alternatively referred to as catastrophic phase inversion (CPI).[40,41] The phase transitions occur due to changes in the spontaneous curvature of the surfactant and can be induced at a constant temperature by adjusting the chemical composition of the system using the PIC method. Alternatively, the PIT method can be used to maintain a constant chemical composition and vary the temperature. While these technologies offer unique advantages, they also have certain drawbacks, including complexity, the need for precision in execution, and reliance on artificial surfactants. Recognition of these limitations is essential in the quest to refine and advance emulsion-phase inversion methods.[40]

7.3.2.1 Phase Inversion Temperature Method

Temperature plays a crucial role in the production of emulsions by the PIT method.[42,43] This approach involves changing the temperature of the system while keeping the composition constant. This method allows emulsions to be produced by gradually heating a mixture of surfactant, oil, and water at room temperature until the phase inversion temperature is reached. Therefore, the affinity of the surfactant to the water and oil phase is modified.[44] As the temperature rises, the mixture exhibits a positive curvature, and the surfactant becomes hydrophobic (due to dehydration) and soluble in oil.[12] Once the mixture reaches the PIT, it is rapidly cooled by transfer to an ice bath, resulting in the formation of emulsions. In particular, oil-in-water emulsions are obtained when the PIT is approximately 20–65°C higher than the storage temperature, whereas water-in-oil emulsions are formed at a temperature range of 10–40°C lower than the storage temperature. The use of nonionic surfactants, whose molecular geometry changes with temperature, is effective in stabilizing emulsions prepared by the PIT method.[45]

It is worth noting that as the curvature approaches zero at intermediate temperatures, termed the hydrophilic–lipophilic balance (HLB) temperature that indicates the equilibrium of the hydrophilic and lipophilic properties of the surfactant), the result is the formation of highly unstable emulsions.[33] To ensure the production of stable emulsions, a rapid alteration in temperature is viable; a shift in the HLB temperature by 25–30°C, either upward or downward, effectively prevents droplet coalescence. However, a significant limitation of the PIT method is its heightened vulnerability to droplet coalescence with an increase in the temperature of the dispersed phase (droplets). In this context, a comprehensive understanding of the surfactant phase behavior, hydrophilic–lipophilic balance, solvent solubility, viscosity, homogenization speed, shear force, and temperature emerges as crucial. The sensitivity of the system stability near the PIT can be partially solved by the addition of cosurfactants.[12]

7.3.2.2 Phase Inversion Composition Method

The PIC method operates on the principle of inducing a phase transition in the emulsion mixture, specifically between oil-in-water and water-in-oil (w/o) configurations, or vice versa. This transition occurs in response to a deliberate change in the composition of the emulsion mixture, all while maintaining a constant temperature. Essentially, the PIC method is based on manipulating the composition of the emulsion to induce a phase transition, thus facilitating the creation of either oil-in-water or water-in-oil emulsions.[37]

The preparation of emulsions according to the PIC method relies on the gradual introduction of either water or oil into an oil–surfactant or water–surfactant mixture.[46] The PIC method offers a practical advantage over the PIT method for large-scale production of nanoemulsions, as it involves the gradual addition of components to the emulsion, eliminating the need for abrupt temperature changes. However, the efficiency of the PIC method as an emulsification method is one-third of that of the PIT method.[37] As water or oil is continuously added, the volume fraction of water in the system changes, resulting in a transient composition. Essentially, the surfactant chains undergo changes in their hydration leading to the shift of the spontaneous curvature of the droplets to zero.[12] This transition triggers the initiation of a bicontinuous phase as phase inversion takes place. The formation of droplets within this bicontinuous structure is likely to be due to the entrapment of one phase into the other, which occurs as one of the two liquids is added to the mixture.[47]

In summary, the PIC method, with its gradual and controlled approach to compositional modification, proves advantageous for the large-scale production of emulsions. The intricate interplay of surfactant properties and compositional changes during phase inversion contributes to the formation of stable and well-defined emulsions, providing insights into the scalable and efficient production of emulsions for various applications.

7.3.3 Self-Emulsification Method

The self-emulsification method involves the preparation of emulsions through a dilution process at a constant temperature. Starting from a microemulsion characterized by an oil-in-water composition, the system is subjected to rapid dilution with a significant volume of water. This dilution leads to a reduction in the surfactant concentration, a key factor in maintaining thermodynamic stability. A notable advantage of this method is its scalability, making it suitable for large-scale production. In addition, it has low energy consumption, mainly due to the absence of temperature changes and the lack of need for high stirring rates. The appeal of the self-emulsification method lies in its simplicity and efficiency, allowing emulsions to be produced on a larger scale without the need for complex equipment or high-energy input, which contributes to its practicality and adaptability in various industrial settings.[48,49]

7.3.4 D-Phase Emulsification

DPE uses a system of surfactant, water, and oil, similar to other methods.[50] However, an alkyl polyol is added as an additional component to facilitate emulsion formation. Unlike other low-energy methods, the DPE method requires a low concentration of surfactant, does not require strict adherence to hydrophilic–lipophilic scales or precise surfactant mixtures, eliminates the need for solvents, and has been reported to have lower energy consumption than the phase inversion composition method.[51–53]

7.4 HIGH-ENERGY EMULSIFICATION METHODS

The production of emulsions by high-energy methods involves the use of mechanical devices such as microfluidizers, high-pressure homogenizers (HPH), and ultrasonic devices. These mechanical devices exert large disruptive forces, such as compression, collision, and cavitation, to produce droplets.[54] Emulsifying methodologies are generally classified based on their effectiveness in disrupting droplets, evaluated in terms of the mechanical energy input per unit volume within the zone where droplets are disrupted.[55] Table 7.1 presents a summary of common high-energy emulsification methodologies.

Droplet size in emulsions is influenced by several factors, including equipment, production conditions, sample composition and characterization, and time and temperature.[40,56] High-energy techniques, which require sophisticated machinery and consume considerable energy, tend to be expensive. Although effective for particle size reduction, these methods may not be suitable for temperature-sensitive substances.[47] This limitation is due to the inherently elevated temperatures and pressures associated with high-energy technologies, which have the potential to adversely affect thermolabile or sensitive substances.[12] Table 7.2 reports the operational conditions of some common high-energy emulsification devices.

7.4.1 Rotor–Stator Emulsification

Rotor–stator homogenizers are devices consisting of a tool with rotating blades that rotate around its central axis, called the internal rotor. The internal rotor rotates at high speed within a fixed outer casing known as the stator. The homogenization process involves mechanical tearing and the application of shear forces to the

TABLE 7.1
Summary of Common High-Energy Emulsification Devices, Including Their Operation Mechanisms and the Characteristics of the Produced Emulsions

Emulsification Device	Emulsification Mechanism	Applications
High-shear mixers	The head of the mixer generates rotational, longitudinal, and radial velocity gradients in the system, which breaks the larger droplets into smaller ones, disrupts the oil–water interface, and intermingles the liquids	Coarse emulsion preparation and dissolution of powdered ingredients
Ultrasound homogenizers	The physical effects of ultrasound induced by acoustic cavitation can boost the disruption of oil droplets, facilitating the formation of stable oil-in-water emulsions with small droplets size	Emulsions and nanoemulsions preparation
Microfluidizers	The fluids are brought together from two inlets at a high velocity (using pumping devices) that disrupt the oil droplets and produce fine emulsions after a single pass	Production of finely dispersed emulsions
Membrane emulsification	One immiscible liquid is forced (by pressure) to pass through a microporous membrane into the continuous phase. An emulsifier is usually dissolved in the continuous phase. The size of the droplets could be controlled by the size of the membrane pores	Mainly used at the lab scale to produce emulsions with high dispersed phase fraction and specific dimensions and structures

Source: Adapted from Taha et al.,[8] Copyright (2020), with permission from Elsevier.

TABLE 7.2
Operational Conditions of Some Common High-Energy Emulsification Devices

Emulsification Device	Throughput	Average Droplet Diameter (μm)	Final Product Viscosity	Energy Efficiency	Flow Regime of the Fluid
High-shear mixers (rotor–stator homogenizers)	Batch or continuous	15–25	High	Low	Turbulent-viscous (TV), laminar-viscous (LV)
High-pressure homogenizers	Continuous	0.1–1	Low to medium	High	TV, LV
Ultrasound homogenizers	Batch or continuous	0.1–2	Low	High	Cavitational (CI)
Microfluidizers	Continuous	0.1–1	Low to medium	High	TV
Membrane emulsification	Batch or continuous	0.1–2	Low to medium	Very high	Injection

Source: Adapted from Taha et al.,[8] Copyright (2020), with permission from Elsevier.

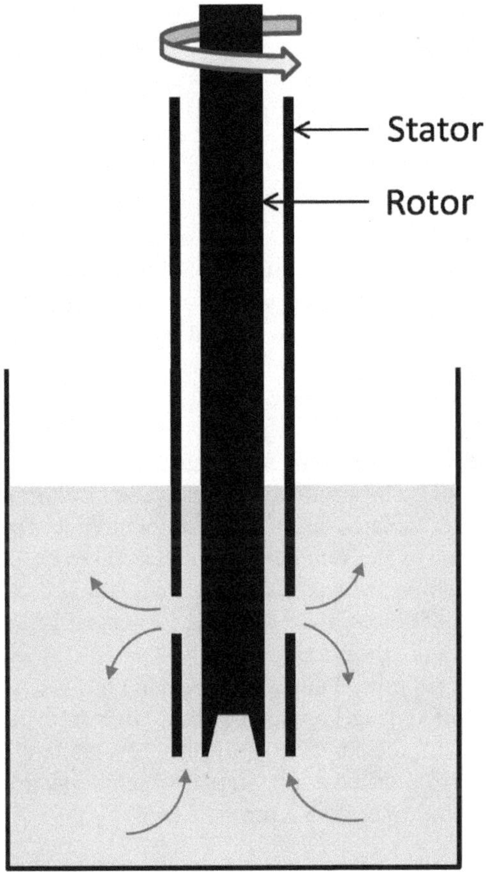

FIGURE 7.3 Sketch of a rotor–stator homogenizer and its operation. Reprinted from Albert et al.,[59] Copyright (2019), with permission from Elsevier.

samples.[57,58] The operating mechanism of this type of mixers is based on the rotational movement of its blades, which creates a depression. This causes the liquid sample to be attracted to one end of the mixture and forcefully ejected through the orifices of the stator.[58,59] The difference in the rotational speeds of the rotor and stator induces substantial hydraulic cutting. This facilitates a fast homogenization, resulting in the formation of small droplets within the emulsions.[60] Figure 7.3 displays a sketch showing the common aspect of a rotor–stator homogenized and its operation.

The choice of the rotational velocity and the duration of the homogenization process play a key role in the control of the dimension of the obtained droplet.[61] Therefore, the control of the shear applied during the emulsification process allows tuning the size of the droplets produced. This is possible because the shear has a direct impact on the interfacial area, stress, and, consequently, stability. In fact, the emulsification process produces smaller droplets with increased interfacial area and reduced tension as the applied shear increases. This in turn increases the stability of the emulsion. However, excessive shear resulting from an excess of applied energy can induce droplet coalescence, which is detrimental to the stability of the obtained emulsions.[58,60] The minimum droplet size achievable is closely related to the design of the mixing head, the geometry of the vessel, the number of passes through the mixing zone, and the type of stabilization.[62] Therefore, rotation speed and homogenization time are key factors in controlling emulsion droplet size.[59]

Rotor–stator mixers offer distinct advantages over alternative methods for the preparation of emulsions. These include ease of installation, ease of operation and cleaning, low capital cost, relatively high yield, and rapid processing speed, with an emulsion typically being obtained within a few minutes.[63] Notable advantages include minimal liquid requirements, a wide droplet size distribution, and the high shear rate generated between the rotor and stator. In addition, these homogenizers are capable of handling large volumes of emulsion.[64,65]

The versatility of rotor–stator homogenizers extends to the emulsification of different types of liquids, making them widely used in the formulation of cosmetic emulsions and other industrial products.[58,60] However, it is important to recognize certain disadvantages associated with rotor–stator homogenizers. These are first related to the limitation in the number of samples that can be limited at a time. However, advanced high-performance devices, that can operate in continuous, are currently available. Other potential drawbacks of rotor–stator homogenizers include (i) potential heterogeneity in the obtained samples, particularly when operating close to the volume limit of the probe. This problem can be mitigated by repositioning the probe within the sample during homogenization; (ii) risk of temperature increase, mainly due to frictional forces, which can lead to destabilization for temperature-sensitive systems, and (iii) limited energy input, which only allows the production of droplets with dimensions larger than 1 μm.[59]

7.4.2 High-Pressure Homogenization

High-pressure homogenization is probably the most widely used continuous emulsification method. It is used in a wide range of industries, from pharmaceuticals to food and beyond, which rely on this method to produce stable and fine-tuned emulsions suitable for a wide range of applications. This widely used technique involves subjecting a substance to elevated pressure to produce a finely dispersed emulsion. One of the key advantages of high-pressure homogenization is its continuous nature, which ensures a continuous and uniform emulsification process. In addition, high-pressure homogenization offers a level of control and precision that meets the stringent requirements of modern industrial processes. Its adaptability allows for customization, making it suitable for a wide range of products with varying viscosities and compositions. However, high-pressure homogenizers are mostly used to prepare emulsions characterized by low viscosity and moderate amount of disperse phase.[66]

High-pressure homogenization typically involves the use of a high-pressure pump coupled with a homogenizing nozzle. To improve the process, a pre-emulsification step is recommended to obtain a primary coarse emulsion. Typically, the initial pre-emulsification phase is conducted using either a rotor–stator mechanism or a vortex mixer. The fine emulsion is then obtained at the outlet of the homogenizer as a result of the increase of the pressure (in the range of 3–500 MPa),[67] which helps with the injection of the pre-emulsified mixture in the homogenizing nozzle. This makes it possible to disrupt the droplets, inducing the emulsification. This strategy of combining pre-emulsification with high-pressure homogenization helps to produce finely dispersed emulsion droplets in the final stage.[59] In addition to the effect of pressure, manipulation of the intensity and duration of homogenization plays a significant role in influencing droplet size distribution, morphology, and overall droplet size.[68] There is also the option of subjecting the emulsion to multiple passes through the homogenizer, effectively reducing the droplet size to the nanometer range.[69] Therefore, the size of emulsion droplets can be modulated throughout the emulsification process by adjusting both the applied pressure level and the number of homogenization cycles

Commonly, droplets produced by high-pressure homogenization tend to be smaller than those produced by rotor–stator homogenization. This difference in droplet size is due to the different mechanisms employed by each method. In high-pressure homogenization, the emulsion is subjected to increased pressure, resulting in a strong disruption of the interface which in turn leads to the production of finer droplets. On the other hand, rotor–stator homogenization relies on the mechanical action of a rotating element.[69,70]

The main advantages of using high-pressure homogenization include (i) the ability to process large sample volumes in a continuous and reproducible manner; (ii) the ability to achieve exceptionally small droplets, potentially reaching dimensions in the hundreds of nanometers, and (iii) the flexibility to adjust droplet size by varying either the pressure level or the number of homogenization cycles.[71] However, this method also presents several drawbacks: (i) the associated high energy consumption results in increased operating costs; (ii) a minimum volume requirement, typically in the tens of milliliters (larger than that of a rotor–stator homogenizer), which contributes to increased costs for emulsions with expensive components; (iii) difficult cleaning procedures that can lead to cross-contamination; (iv) a temperature rise requiring a cooling system to prevent emulsion destabilization, similar to rotor–stator homogenization, and (v) the generation of a broad droplet size distribution.[64,71]

7.4.3 Ultrasonic Emulsification

Various ultrasonic devices are currently used to produce emulsions, with the most widely used being the ultrasonic probe. In this device, a titanium probe vibrates through a transducer containing a piezoelectric crystal, converting electrical energy into high-frequency mechanical motion. The ultrasonic probe then transfers this energy to the surrounding sample, primarily promoting emulsification through mechanisms such as cavitation and ultrasonic forces.[8,72] Figure 7.4 represents a schematic view of the ultrasonic emulsification process.

Only high-power ultrasound, operating in the frequency range of 16–100 kHz (and to a lesser extent between 100 kHz and 1 MHz), shows the ability to interact with matter, making it suitable for emulsification.[72] The widespread adoption

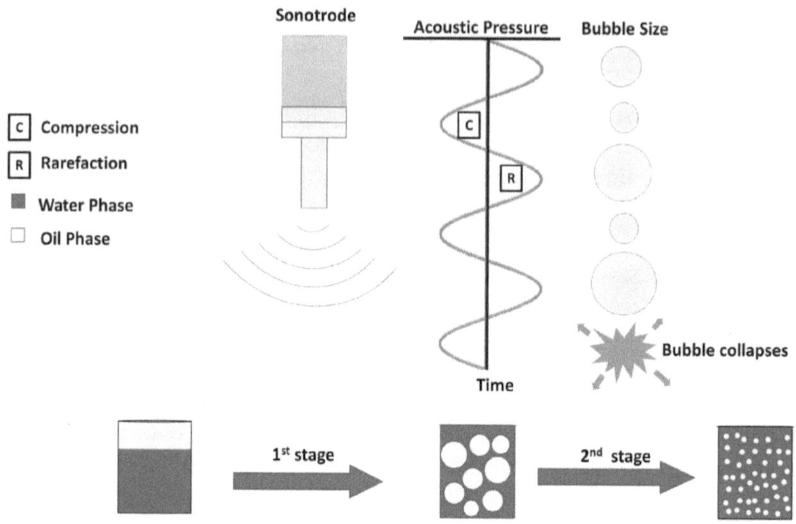

FIGURE 7.4 Sketch of the general mechanism of ultrasonic emulsification. Reprinted from Taha et al.,[8] Copyright (2020), with permission from Elsevier.

of high-intensity ultrasound (HIU) emulsification can be attributed to its distinct advantages. Indeed, these devices have proven to be highly versatile and offer several advantages over conventional emulsification methods. One notable advantage is their remarkable energy efficiency, enabling more resource-efficient processes. In addition, HIU emulsification contributes to the production of emulsions with exceptional stability. The ease of operation and cleaning associated with these units further enhances their appeal, simplifying the overall emulsification process and maintenance procedures.[73,74] The droplet size in emulsification is primarily affected by the ultrasound frequency, amplitude, and emulsification duration. Similar to the high-pressure homogenizer, a pre-emulsification step can contribute to the formation of smaller emulsion droplets. Ultrasonic homogenization is applicable across a broad spectrum of volumes, spanning from hundreds of microliters to several hundred milliliters.[72,75]

It should be noted that although the use of ultrasound offers many opportunities for the preparation of emulsions, it also has several drawbacks. These include the challenges associated with the industrial scale-up of this technique, the achievement of a broad droplet size distribution, and the significant temperature increase during the emulsification process, which poses potential problems, particularly for emulsion stability. However, the latter drawback can be overcome by using ultrasound in a pulsed mode or by incorporating a cooling system.[64,72]

7.4.4 Emulsification Using Microfluidic Devices

The use of microfluidic devices enables the production of very homogeneous emulsions, often characterized by a coefficient of variation (CV) of less than 5%. This type of methodologies has been widely adopted to study the stabilization mechanisms of emulsions, but also in the preparation of materials with unique structures derived from emulsions.[76–79] In general, microfluidic devices for the preparation of emulsions consist of an air-driven intensifier pump with pressures up to 2,000 psi and an interaction chamber.[80]

The interaction chamber consists of micrometer-sized channels with specific geometries through which fluids circulate. These channels facilitate the formation of droplets of one liquid into another, allowing the formation of emulsions. Under laminar flow conditions, droplets undergo deformation and breakup due to simple shear or stretching flow. Droplet breakup can occur by processes such as extension, tip streaming, or trailing. Various microfluidic devices have demonstrated the ability to produce emulsions, including T-junction devices where the dispersed phase is directed through a small orifice into the perpendicular continuous phase. Flow-focusing devices work by concentrating the flowing dispersed phase between two perpendicular streams of continuous phase, inducing the formation of a jet and subsequent droplets. In addition, terrace (or plateau) devices have a restricted microchannel with a step in which the dispersed phase circulates, surrounded by the continuous phase. Upon reaching the step, a reduction in Laplace pressure induces droplet formation.[81–83] The control of the droplet size can be achieved by altering the flow rate[84] or modifying the microchannel geometry. A rise in the flow rate leads to an increase in droplet size.[85] Notably, this emulsification technique enables the production of multiple emulsions while maintaining precise control over the number of encapsulated inner droplets and their encapsulation rate.[83]

Concerning microfluidic emulsification methods, their attractiveness is based on several advantages: (i) the absence of extensive mechanical shear, which minimizes disruptive effects ;[86,87] (ii) the achievement of excellent control of droplet size, with better monodispersity compared to membrane emulsification and consequently homogenization techniques;[64,86–88] (iii) low energy consumption, leading to cost-effective operation; (iv) low liquid volume requirement; and (v) absence of heat generation during the emulsification process, eliminating the risk of destabilization for emulsions. However, the use of microfluidic devices for emulsification comes with certain limitations: (i) a low preparation flux resulting in a low-throughput production, posing challenges for potential industrialization;[85,89] (ii) the potential risk of interaction between the droplets and the microchannel;[83,90] and (iii) the restriction to liquids with low viscosities capable of flowing through the microchannel.[84]

7.4.5 Membrane Emulsification

Membrane emulsification is a technique used mainly to produce emulsions on a laboratory scale. It relies on the application of pressure to force the dispersed phase through the pores of a microporous glass or ceramic membrane into the continuous phase. Typically, the emulsifier is dissolved in the continuous phase and adsorbed at the interface to stabilize the droplets and prevent aggregation. This methodology is commonly called direct membrane emulsification (DME). As an alternative, emulsification can also be obtained by passing through the membrane a pre-emulsion. In this latter case, the process is known as premix membrane emulsification (PME).[91]

Methods based on the DME principle, in which low shear forces are applied to detach droplets from the membrane surface, have led to the development of several techniques. These include stirred cell membrane emulsification (SCME),[64,88] rotational membrane emulsification (RME),[86,92] vibrational membrane emulsification (VME),[93] and crossflow membrane emulsification (XME).[86] In SCME, additional mechanical agitation occurs in the receptor chamber.[64] In RME and VME, the dispersed phase is forced into the continuous phase by a rotating or vibrating membrane.[86] In XME, the dispersed phase is forced through a membrane tube within a flowing continuous phase. In all these methods, agitation plays a crucial role in detaching droplets from the membrane, resulting in reduced droplet size.[86]

The properties of the resulting emulsions obtained by membrane emulsification can be influenced by factors such as the type and concentration of the emulsifier and the properties of the membrane. A major advantage of membrane emulsification is the ability to control droplet size by selecting the pore size of the membrane. In fact, it allows the formation of emulsions characterized by small, size-controlled, and uniform droplets with minimal polydispersity.[86] Moreover, the energy consumption of this methodology is low, resulting in cost-effective operation, and it does not generate heat during the emulsification process, thereby reducing the risk of destabilization for emulsions.[64] However, it is important to note a notable drawback associated with this method, which is the relatively low production rate. Despite this limitation, the simplicity and precision of membrane emulsification make it a valuable tool for laboratory-scale emulsion production, offering a degree of control over droplet size that is crucial for certain applications.[94]

Table 7.3 provides an overview of the key advantages associated with each high-energy emulsification technique. It should be stressed that the overcoming of scale-up challenges would render low-shear processes such as membrane and microfluidic emulsification particularly suitable for formulating pharmaceutical emulsions. These processes offer the advantage of producing droplets with low polydispersity, eliminating the risk of temperature increase—critical considerations for temperature-sensitive systems. Additionally, they facilitate droplet formation with low energy consumption, a pivotal criterion for future industrialization efforts.

7.5 CONCLUDING REMARKS

The journey through emulsion preparation methodologies is not static but dynamic, characterized by continuous exploration, adaptation, and innovation. This chapter has provided a comprehensive overview of emulsion preparation methodologies, providing a roadmap for researchers and industry professionals in the field of emulsion science. The complex interplay of factors such as dispersed and continuous phase properties, process scalability, and desired emulsion properties requires a nuanced understanding.

Embracing the diversity of available methodologies is a fundamental premise for the advancement of emulsion science. The range of methodologies explored in this chapter serves as a toolkit from which scientists can draw inspiration for tailored approaches. By recognizing the strengths and limitations of each method, researchers can make informed decisions and tailor their strategies to meet the specific needs

TABLE 7.3
Advantages and Disadvantages of the Different Emulsification Methodologies

Emulsification methodologies	Easy set-up	Quick process	Risk of temperature rise	Possibility to obtain submicrometric droplets	Low polydispersity	Low energy consumption	Current industrialization possibilities
Rotor–stator homogenizers	+	+	+	–	–	+/–	+
High-pressure homogenizers	–	+	+	+	–	–	+
Ultrasound homogenizers	+	+	+	+	–	+/–	–
Microfluidizers	–	–	–	+	+	+	–
Membrane emulsification	–	–	–	+	+	+	–

The symbols indicate: + = yes; – = no, and +/– = intermediate.
Source: Adapted from Albert et al.,[59] Copyright (2019), with permission from Elsevier.

of their unique emulsion systems, driving advancements in diverse fields from food and pharmaceuticals to cosmetics and beyond.

The future of emulsion science is promising, driven by the rapid pace of technological advancement and the promotion of interdisciplinary collaborations. As cutting-edge innovations emerge from the convergence of different scientific disciplines, new methods of emulsion preparation are likely to emerge. The continuous refinement of emulsion preparation methods not only enhances product quality and efficacy but also opens the door to new applications that meet today's challenges and evolving consumer demands.

ACKNOWLEDGMENTS

This work was funded by grants PID2023-147156NB-I00 funded by MCIN/AEI/10.13039/501100011033 (Spain) and PR12/24-31566 (Ayudas para la Financiación de Proyectos de Investigación UCM 2023), and the European Innovative Training Network-Marie Sklodowska-Curie Action NanoPaInt (grant agreement 955612), funded by the E.U.

REFERENCES

1. Lucia, A.; Guzmán, E. Emulsions containing essential oils, their components or volatile semiochemicals as promising tools for insect pest and pathogen management. *Adv. Colloid Interface Scie.* **2021**, *287*, 102330.
2. Guzmán, E.; Ortega, F.; Rubio, R.G. Pickering emulsions: A novel tool for cosmetic formulators. *Cosmetics* **2022**, *9*, 68.
3. Faisal, W.; Almomani, F. A critical review of the development and demulsification processes applied for oil recovery from oil in water emulsions. *Chemosphere* **2022**, *291*, 133099.
4. Zhou, Y.; Yin, D.; Chen, W.; Liu, B.; Zhang, X. A comprehensive review of emulsion and its field application for enhanced oil recovery. *Energy Sci. Eng.* **2019**, *7*, 1046–1058.
5. Tan, C.; McClements, D.J. Application of advanced emulsion technology in the food industry: A review and critical evaluation. *Foods* **2021**, *10*, 812.
6. Esquena, J.; Sankar; Solans, C. Highly concentrated w/o emulsions prepared by the PIT method as templates for solid foams. *Langmuir* **2003**, *19*, 2983–2988.
7. Zhu, Q.; Pan, Y.; Jia, X.; Li, J.; Zhang, M.; Yin, L. Review on the stability mechanism and application of water-in-oil emulsions encapsulating various additives. *Compr. Rev. Food Sci. Food Saf.* **2019**, *18*, 1660–1675.
8. Taha, A.; Ahmed, E.; Ismaiel, A.; Ashokkumar, M.; Xu, X.; Pan, S.; Hu, H. Ultrasonic emulsification: An overview on the preparation of different emulsifiers-stabilized emulsions. *Trends Food Sci. Technol.* **2020**, *105*, 363–377.
9. Steegmans, M.L.J.; Schroën, K.G.P.H.; Boom, R.M. Characterization of emulsification at flat microchannel y junctions. *Langmuir* **2009**, *25*, 3396–3401.
10. Modarres-Gheisari, S.M.M.; Gavagsaz-Ghoachani, R.; Malaki, M.; Safarpour, P.; Zandi, M. Ultrasonic nano-emulsification – A review. *Ultrason Sonochem* **2019**, *52*, 88–105.
11. Perazzo, A.; Preziosi, V.; Guido, S. Phase inversion emulsification: Current understanding and applications. *Adv. Colloid Interface Sci.* **2015**, *222*, 581–599.

12. Soni, M.; Yadav, A.; Maurya, A.; Das, S.; Dubey, N.K.; Dwivedy, A.K. Advances in designing essential oil nanoformulations: An integrative approach to mathematical modeling with potential application in food preservation. *Foods* **2023**, *12*, 4017.
13. McClements, D.J. *Food Emulsions*, CRC Press: Boca Raton, FL, USA, 2015.
14. Fernandez-Avila, C.; Trujillo, A.J. Ultra-high pressure homogenization improves oxidative stability and interfacial properties of soy protein isolate-stabilized emulsions. *Food Chem.* **2016**, *209*, 104–113.
15. Li, W.; Leong, T.S.H.; Ashokkumar, M.; Martin, G.J.O. A study of the effectiveness and energy efficiency of ultrasonic emulsification. *Phys. Chem. Chem. Phys.* **2018**, *20*, 86–96.
16. Taha, A.; Hu, T.; Zhang, Z.; Bakry, A.M.; Khalifa, I.; Pan, S.; Hu, H. Effect of different oils and ultrasound emulsification conditions on the physicochemical properties of emulsions stabilized by soy protein isolate. *Ultrasonics Sonochem.* **2018**, *49*, 283–293.
17. Fradette, L.; Brocart, B.; Tanguy, P.A. Comparison of mixing technologies for the production of concentrated emulsions. *Chem. Eng. Res. Design* **2007**, *85*, 1553–1560.
18. Anton, N.; Benoit, J.-P.; Saulnier, P. Design and production of nanoparticles formulated from nano-emulsion templates—A review. *J. Control. Rel.* **2008**, *128*, 185–199.
19. Seekkuarachchi, I.N.; Tanaka, K.; Kumazawa, H. Formation and characterization of submicrometer oil-in-water (O/W) emulsions, using high-energy emulsification. *Ind. Eng. Chem. Res.* **2006**, *45*, 372–390.
20. Safaya, M.; Rotliwala, Y.C. Nanoemulsions: A review on low energy formulation methods, characterization, applications and optimization technique. *Mater. Today Proc.* **2020**, *27*, 454–459.
21. Singh, I.R.; Pulikkal, A.K. Preparation, stability and biological activity of essential oil-based nano emulsions: A comprehensive review. *OpenNano* **2022**, *8*, 100066.
22. Jafari, S.M.; Assadpoor, E.; He, Y.; Bhandari, B. Re-coalescence of emulsion droplets during high-energy emulsification. *Food Hydrocoll* **2008**, *22*, 1191–1202.
23. McClements, D.J.; Rao, J. Food-grade nanoemulsions: Formulation, fabrication, properties, performance, biological fate, and potential toxicity. *Crit. Rev. Food Sci. Nutrit.* **2011**, *51*, 285–330.
24. Qian, C.; McClements, D.J. Formation of nanoemulsions stabilized by model food-grade emulsifiers using high-pressure homogenization: Factors affecting particle size. *Food Hydrocoll* **2011**, *25*, 1000–1008.
25. Floury, J.; Desrumaux, A.; Lardières, J. Effect of high-pressure homogenization on droplet size distributions and rheological properties of model oil-in-water emulsions. *Innov. Food Sci. Emerg. Technol.* **2000**, *1*, 127–134.
26. Windhab, E.J.; Dressler, M.; Feigl, K.; Fischer, P.; Megias-Alguacil, D. Emulsion processing—from single-drop deformation to design of complex processes and products. *Chem. Eng. Sci.* **2005**, *60*, 2101–2113.
27. Håkansson, A.; Trägårdh, C.; Bergenståhl, B. Studying the effects of adsorption, recoalescence and fragmentation in a high pressure homogenizer using a dynamic simulation model. *Food Hydrocoll* **2009**, *23*, 1177–1183.
28. Stang, M.; Schuchmann, H.; Schubert, H. Emulsification in high-pressure homogenizers. *Eng. Life Sci.* **2001**, *1*, 151.
29. Santana, R.C.; Perrechil, F.A.; Cunha, R.L. High- and low-energy emulsifications for food applications: A focus on process parameters. *Food Eng. Rev.* **2013**, *5*, 107–122.
30. Freudig, B.; Tesch, S.; Schubert, H. Production of emulsions in high-pressure homogenizers – Part II: Influence of cavitation on droplet breakup. *Eng. Life Sci.* **2003**, *3*, 266–270.
31. Jadhav, C.; Kate, V.; Payghan, S.A. Investigation of effect of non-ionic surfactant on preparation of griseofulvin non-aqueous nanoemulsion. *J. Nanostruct. Chem.* **2015**, *5*, 107–113.

32. Sokolov, Yu. V. Nanoemulsion formation by low-energy methods: A review. *News Pharm.* **2014**, *3*, 16–19.
33. Setya, S.; Talegaonkar, S. Nanoemulsions: Formulation methods and stability aspects. *World J. Pharm. Pharm. Sci.* **2014**, *3*, 2214–2228.
34. Solans, C.; Morales, D.; Homs, M. Spontaneous emulsification. *Curr. Opin. Colloid Interface Sci.* **2016**, *22*, 88–93.
35. Saberi, A.H.; Fang, Y.; McClements, D.J. Thermal reversibility of vitamin e-enriched emulsion-based delivery systems produced using spontaneous emulsification. *Food Chem.* **2015**, *185*, 254–260.
36. Lefebvre, G.; Riou, J.; Bastiat, G.; Roger, E.; Frombach, K.; Gimel, J.-C.; Saulnier, P.; Calvignac, B. Spontaneous nano-emulsification: Process optimization and modeling for the prediction of the nanoemulsion's size and polydispersity. *Int. J. Pharm.* **2017**, *534*, 220–228.
37. Jintapattanakit, A. Preparation of nanoemulsions by phase inversion temperature (PIT). *Pharm. Sci. Asia* **2018**, *42*, 1–12.
38. Komaiko, J.S.; McClements, D.J. Formation of food-grade nanoemulsions using low-energy preparation methods: A review of available methods. *Compr. Rev. Food Sci. Food Saf.* **2016**, *15*, 331–352.
39. Shafiq-un-Nabi, S.; Shakeel, F.; Talegaonkar, S.; Ali, J.; Baboota, S.; Ahuja, A.; Khar, R.K.; Ali, M. Formulation development and optimization using nanoemulsion technique: A technical note. *AAPS PharmSciTech* **2007**, *8*, E12–E17.
40. McClements, D.J.; Rao, J. Food-grade nanoemulsions: Formulation, fabrication, properties, performance, biological fate, and potential toxicity. *Crit. Rev. Food Sci. Nutrit.* **2011**, *51*, 285–330.
41. Li, Y.; Zhang, Z.; Yuan, Q.; Liang, H.; Vriesekoop, F. Process optimization and stability of d-limonene nanoemulsions prepared by catastrophic phase inversion method. *J. Food Eng.* **2013**, *119*, 419–424.
42. Shinoda, K.; Saito, H. The stability of o/w type emulsions as functions of temperature and the HLB of emulsifiers: The emulsification by PIT-method. *J. Colloid Interface Sci.* **1969**, *30*, 258–263.
43. Saito, H.; Shinoda, K. The stability of w/o type emulsions as a function of temperature and of the hydrophilic chain length of the emulsifier. *J. Colloid Interface Sci.* **1970**, *32*, 647–651.
44. Yukuyama, M.N.; Ghisleni, D.D.M.; Pinto, T.J.A.; Bou-Chacra, N.A. Nanoemulsion: Process selection and application in cosmetics – a review. *Int. J. Cosmetic Sci.* **2016**, *38*, 13–24.
45. Rao, J.; McClements, D.J. Stabilization of phase inversion temperature nanoemulsions by surfactant displacement. *J. Agric. Food Chem.* **2010**, *58*, 7059–7066.
46. Solans, C.; Solé, I. Nano-Emulsions: Formation by low-energy methods. *Curr. Opin. Colloid Interface Sci.* **2012**, *17*, 246–254.
47. Azmi, N.A.N.; Elgharbawy, A.A.M.; Motlagh, S.R.; Samsudin, N.; Salleh, H. Mohd. Nanoemulsions: Factory for food, pharmaceutical and cosmetics. *Processes* **2019**, *7*, 617.
48. Wang, L.; Li, X.; Zhang, G.; Dong, J.; Eastoe, J. Oil-in-water nanoemulsions for pesticide formulations. *J. Colloid Interface Sci.* **2007**, *314*, 230–235.
49. Feng, J.; Zhang, Q.; Liu, Q.; Zhu, Z.; McClements, D.J.; Jafari, S.M. Application of nanoemulsions in formulation of pesticides. In *Nanoemulsions: Formulation, Applications, and Characterization*, McClements, D.J., Jafari, S.M., Eds.; Academic Press: Cambridge, MA, USA, 2018; pp. 379–413.
50. Sagitani, H.; Hattori, T.; Nabeta, K.; Nagai, M. Formation of o/w emulsion having and uniform droplets by the surfactant (D) phase emulsification method. *Nippon Kagaku Kaishi* **1983**, *1983*, 1399–1404.

51. Yukuyama, M.N.; Oseliero, P.L.F.; Kato, E.T.M.; Lobënberg, R.; de Oliveira, C.L.P.; de Araujo, G.L.B.; Bou-Chacra, N.A. High internal vegetable oil nanoemulsion: D-phase emulsification as a unique low energy process. *Colloids Surf. A* **2018**, *554*, 296–305.
52. Sagitani, H.; Nabeta, K.; Nagai, M. A new preparing method for fine o/w emulsions by D phase emulsification and their application to cosmetic industry. *J. Japan Oil Chem. Soc.* **1991**, *40*, 988–994.
53. Kawanami, T.; Togashi, K.; Fumoto, K.; Hirano, S.; Zhang, P.; Shirai, K.; Hirasawa, S. Thermophysical properties and thermal characteristics of phase change emulsion for thermal energy storage media. *Energy* **2016**, *117*, 562–568.
54. Anton, N.; Benoit, J.-P.; Saulnier, P. Design and production of nanoparticles formulated from nano-emulsion templates—A review. *J. Control. Releas.* **2008**, *128*, 185–199.
55. Karbstein, H.; Schubert, H. Developments in the continuous mechanical production of oil-in-water macro-emulsions. *Chem. Eng. Process.: Process Intensif.* **1995**, *34*, 205–211.
56. Liu; Huang; Chen; Lin; Wang. Food-grade nanoemulsions: Preparation, stability and application in encapsulation of bioactive compounds. *Molecules* **2019**, *24*, 4242.
57. Maa, Y.-F.; Hsu, C. Liquid-liquid emulsification by rotor/stator homogenization. *J. Control. Release* **1996**, *38*, 219–228.
58. Coutinho, V.P.; Santana Jr., D.B.; de Oliveira Santos, N.A.; Sodre da Silva, C.A.; de Castro Santana, R. Efeito Da Velocidade Do Processo de Homogeneizaçao Nas Propriedades de Emulsoes Cosméticas. *J. Eng. Exact Sci.* **2018**, *4*, 0240–0245.
59. Albert, C.; Beladjine, M.; Tsapis, N.; Fattal, E.; Agnely, F.; Huang, N. Pickering emulsions: Preparation processes, key parameters governing their properties and potential for pharmaceutical applications. *J. Control. Release* **2019**, *309*, 302–332.
60. Banaszek, C. Mixing fine emulsions. *NutraCos Cosmetics* **2011**, *1*, 15–17.
61. Maa, Y.-F.; Hsu, C. Liquid-liquid emulsification by rotor/stator homogenization. *J. Control. Release* **1996**, *38*, 219–228.
62. Ouzineb, K.; Lord, C.; Lesauze, N.; Graillat, C.; Tanguy, P.A.; McKenna, T. Homogenisation devices for the production of miniemulsions. *Chem. Eng. Sci.* **2006**, *61*, 2994–3000.
63. Urban, K.; Wagner, G.; Schaffner, D.; Röglin, D.; Ulrich, J. Rotor-stator and disc systems for emulsification processes. *Chem. Eng. Technol.* **2006**, *29*, 24–31.
64. Thompson, K.L.; Armes, S.P.; York, D.W. Preparation of pickering emulsions and colloidosomes with relatively narrow size distributions by stirred cell membrane emulsification. *Langmuir* **2011**, *27*, 2357–2363.
65. Gazolu-Rusanova, D.; Lesov, I.; Tcholakova, S.; Denkov, N.; Ahtchi, B. Food grade nanoemulsions preparation by rotor-stator homogenization. *Food Hydrocoll* **2020**, *102*, 105579.
66. Stang, M.; Schuchmann, H.; Schubert, H. Emulsification in high-pressure homogenizers. *Eng. Life Sci.* **2001**, *1*, 151.
67. Salehi, F. Physico-chemical and rheological properties of fruit and vegetable juices as affected by high pressure homogenization: A review. *Int. J. Food Propert.* **2020**, *23*, 1136–1149.
68. Ren, Z.; Chen, Z.; Zhang, Y.; Lin, X.; Li, B. Characteristics and rheological behavior of pickering emulsions stabilized by tea water-insoluble protein nanoparticles via high-pressure homogenization. *Int. J. Biol. Macromol.* **2020**, *151*, 247–256.
69. Gupta, R.; Rousseau, D. Surface-active solid lipid nanoparticles as pickering stabilizers for oil-in-water emulsions. *Food Funct.* **2012**, *3*, 302.
70. Cossu, A.; Wang, M.S.; Chaudhari, A.; Nitin, N. Antifungal activity against candida albicans of starch pickering emulsion with thymol or amphotericin B in suspension and calcium alginate films. *Int. J. Pharm.* **2015**, *493*, 233–242.

71. Köhler, K.; Santana, A.S.; Braisch, B.; Preis, R.; Schuchmann, H.P. High pressure emulsification with nano-particles as stabilizing agents. *Chem. Eng. Sci.* **2010**, *65*, 2957–2964.
72. Canselier, J.P.; Delmas, H.; Wilhelm, A.M.; Abismaïl, B. Ultrasound emulsification—An Overview. *J. Disper. Sci. Technol.* **2002**, *23*, 333–349.
73. Chemat, F.; Zill-e-Huma; Khan, M.K. Applications of ultrasound in food technology: Processing, preservation and extraction. *Ultrasonics Sonochem.* **2011**, *18*, 813–835.
74. Li, W.; Leong, T.S.H.; Ashokkumar, M.; Martin, G.J.O. A study of the effectiveness and energy efficiency of ultrasonic emulsification. *Phys. Chem. Chem. Phys.* **2018**, *20*, 86–96.
75. Kaltsa, O.; Gatsi, I.; Yanniotis, S.; Mandala, I. Influence of ultrasonication parameters on physical characteristics of olive oil model emulsions containing xanthan. *Food Bioprocess Technol.* **2014**, *7*, 2038–2049.
76. Nie, Z.; Park, J. Il; Li, W.; Bon, S.A.F.; Kumacheva, E. An "inside-out" microfluidic approach to monodisperse emulsions stabilized by solid particles. *J. Am. Chem. Soc.* **2008**, *130*, 16508–16509.
77. Priest, C.; Reid, M.D.; Whitby, C.P. Formation and stability of nanoparticle-stabilised oil-in-water emulsions in a microfluidic chip. *J. Colloid Interface Sci.* **2011**, *363*, 301–306.
78. Hassan, N.; Stocco, A.; Abou-Hassan, A. Droplet liquid/liquid interfaces generated in a microfluidic device for assembling janus inorganic nanohybrids. *J. Phys. Chem. C* **2015**, *119*, 10758–10765.
79. Subramaniam, A.B.; Abkarian, M.; Stone, H.A. Controlled assembly of jammed colloidal shells on fluid droplets. *Nat. Mat.* **2005**, *4*, 553–556.
80. Modarres-Gheisari, S.M.M.; Gavagsaz-Ghoachani, R.; Malaki, M.; Safarpour, P.; Zandi, M. Ultrasonic nano-emulsification – A review. *Ultrasonics Sonochem.* **2019**, *52*, 88–105.
81. Vladisavljević, G.T.; Kobayashi, I.; Nakajima, M. Production of uniform droplets using membrane, microchannel and microfluidic emulsification devices. *Microfluid. Nanofluid.* **2012**, *13*, 151–178.
82. Seemann, R.; Brinkmann, M.; Pfohl, T.; Herminghaus, S. Droplet based microfluidics. *Rep. Progr. Phys.* **2012**, *75*, 016601.
83. Engl, W.; Backov, R.; Panizza, P. Controlled production of emulsions and particles by milli- and microfluidic techniques. *Curr. Opin. Colloid Interface Sci.* **2008**, *13*, 206–216.
84. Nie, Z.; Park, J. Il; Li, W.; Bon, S.A.F.; Kumacheva, E. An "inside-out" microfluidic approach to monodisperse emulsions stabilized by solid particles. *J. Am. Chem. Soc.* **2008**, *130*, 16508–16509.
85. Holtze, C. Large-scale droplet production in microfluidic devices—an industrial perspective. *J. Phys. D: Appl. Phys.* **2013**, *46*, 114008.
86. Yuan, Q.; Cayre, O.J.; Manga, M.; Williams, R.A.; Biggs, S. Preparation of particle-stabilized emulsions using membrane emulsification. *Soft Matter* **2010**, *6*, 1580–1588.
87. Xu, Q.Y.; Nakajima, M.; Binks, B.P. Preparation of particle-stabilized oil-in-water emulsions with the microchannel emulsification method. *Colloids Surf. A* **2005**, *262*, 94–100
88. Sun, G.; Qi, F.; Wu, J.; Ma, G.; Ngai, T. Preparation of uniform particle-stabilized emulsions using SPG membrane emulsification. *Langmuir* **2014**, *30*, 7052–7056.
89. Schroën, K.; Bliznyuk, O.; Muijlwijk, K.; Sahin, S.; Berton-Carabin, C.C. Microfluidic emulsification devices: From micrometer insights to large-scale food emulsion production. *Curr. Opin. Colloid Interface Sci.* **2015**, *3*, 33–40.

90. Chu, L.; Utada, A.S.; Shah, R.K.; Kim, J.; Weitz, D.A. Controllable monodisperse multiple emulsions. *Ang. Chem. Int. Ed.* **2007**, *46*, 8970–8974.
91. Joscelyne, S.M.; Trägårdh, G. Membrane emulsification — a literature review. *J. Membrane Sci.* **2000**, *169*, 107–117.
92. Manga, M.S.; Cayre, O.J.; Williams, R.A.; Biggs, S.; York, D.W. Production of solid-stabilised emulsions through rotational membrane emulsification: Influence of particle adsorption kinetics. *Soft Matter* **2012**, *8*, 1532–1538.
93. Kelder, J.D.H.; Janssen, J.J.M.; Boom, R.M. Membrane emulsification with vibrating membranes: A numerical study. *J. Membrane Sci.* **2007**, *304*, 50–59.
94. van der Graaf, S.; Schroen, C.G.P.H.; Boom, R. Preparation of double emulsions by membrane emulsification? A review. *J. Membrane Sci.* **2005**, *251*, 7–15.

8 Microemulsions, Emulsions, and Nanoemulsions Formulation

Jesús F. Ontiveros and Ronald Marquez

8.1 MICROEMULSIONS AND THE HYDROPHILIC–LIPOPHILIC DEVIATION (HLD) CONCEPT

8.1.1 MICROEMULSIONS DEFINITION

The term 'microemulsion' was introduced by Schulman et al.[1] to describe any thermodynamically stable, isotropic liquid system within a SOW system.[2] Oil and water domains within the microemulsion are significantly smaller than the wavelength of visible light, which explains its transparent appearance.[3–5] The assessment of the middle-phase structure is most directly accomplished through the measurement of self-diffusion coefficients of water (D_w), oil (D_o), and surfactant (D_s).[6] In swollen micelle solutions (Type I), D_s approximates D_{oil}, markedly lower than D_w, whereas in swollen inverse micelle (Type II) solutions, D_w is comparable to D_s, significantly lower than D_{oil}. Bicontinuous middle phases exhibit diffusion coefficients for water and oil similar to those of their pure components, indicating no correlation among the three components.[7]

Bicontinuous microemulsions are characterized by the formation of structures where oil and water are interdispersed within a continuous matrix. The naming conventions introduced by Talmon and Zemb – such as 'interdispersed domains tessellation' and 'disordered open connected cylinders' – indicate the presence of interconnected networks that extend beyond simple spherical domains to encompass a variety of geometric forms.[8,9]

Bicontinuous microemulsions exhibit regions of varying curvature, including saddle-shaped interfaces with zero mean curvature (Figure 8.1).[10,11] Additionally, they can exhibit both stable thermotropic and potentially unstable lyotropic behaviors, influenced by the specific arrangement and interaction of their constituent molecules. Scattering techniques (light, X-ray, and neutron) have revealed the presence of disordered but interconnected domains, supporting the concept of a tessellated network.[7,12,13]

 DOI: 10.1201/9781032636184-10

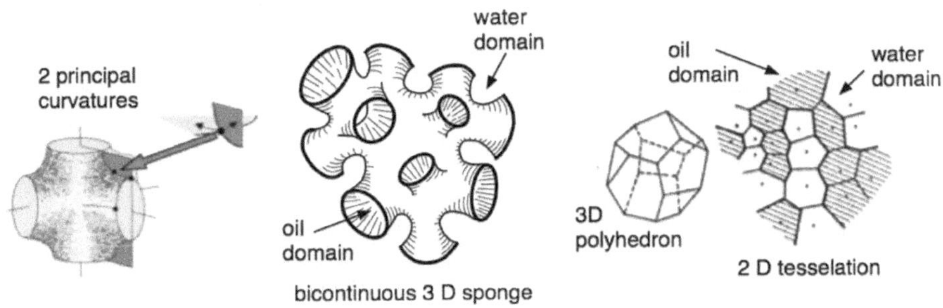

FIGURE 8.1 Bicontinuous shapes percolated-like and more or less merged micelles and inverse micelles, sometimes called mixture of worm-like elongated structures as bicontinuous sponge shape or randomly interdispersed tessellations shown in 2D and 3D space. Reproduced from Salager et al.[14] under a Creative Commons Attribution [CC-BY] License.

8.1.2 Formulation of Surfactant–Oil–Water (SOW) Systems and the HLD_N Equation

Research on the formulation of SOW systems dates back to the mid-20th century. Griffin's hydrophilic–lipophilic balance (HLB) was an early method to quantify the surfactant balance between hydrophilic and lipophilic properties with a single value.[15,16] However, its reliance on arbitrary assumptions limits its utility, as it fails to account for variables such as temperature and salinity. Around the same time, Winsor introduced basic physicochemical formulation concepts, correlating the SOW system phase behavior with surfactant molecules interactions at the oil–water interface.[17,18] Winsor classified SOW systems into three equilibrium types based on the affinity ratio I of surfactant I molecules for oil (O) and water (W), $R = A_{co}/A_{cw}$, leading to distinct micellar structures or a three-phase behavior in Winsor III systems, characterized by equal surfactant affinities for oil and water.

The original R relationship, while foundational, proves challenging in practice due to the complexity of determining the interaction parameters A_{co} and A_{cw}.[3,19,20] Davies[21] introduced a correlation between the HLB value and surfactant partitioning between water and oil phases, based on the free energy transfer of a surfactant molecule from water to oil. Shinoda expanded on this correlation to explain the relationship between phase inversion temperature (PIT) and ethylene oxide number (EON) of nonionic surfactants.[22]

The free energy transfer of a surfactant molecule from water to oil,

$$\Delta G_{W \to O} = RT \ln \frac{C_W}{C_O}, \tag{8.1}$$

where C_W and C_O denote the activities of the surfactant, typically assumed as their concentrations in water and oil, respectively. Shinoda later applied this correlation to elucidate the linear association between the change in PIT and the EON of nonionic surfactants. By the late 1970s, the surfactant affinity difference (*SAD*) concept was introduced, defined as the difference between the negative standard chemical potential of the surfactant in the oil phase (μ_O^*) and in the water phase (μ_W^*),[23,24] formulated as follows:

$$SAD = -\mu_O^* - \left(-\mu_W^*\right) = \mu_W^* - \mu_O^*. \quad (8.2)$$

Assuming an activity coefficient of unity, indicative of low surfactant concentration in the excess phases of Winsor type III systems, the equilibrium between water and oil phases can be represented through the surfactant's chemical potentials:

$$\mu_W = \mu_O = \mu_W^* + RT\,\ln\frac{C_W}{C_{W(\mathrm{ref})}} = \mu_O^* + RT\,\ln\frac{C_O}{C_{O(\mathrm{ref})}}, \quad (8.3)$$

where the asterisk represents standard chemical potentials, and the subscript ref indicates reference concentrations. Defining the partition coefficient as $K = C_W/C_O$ leads to

$$RT\ln K + \text{constant} = \mu_O^* - \mu_W^* = \Delta G_{W\to O} = -SAD. \quad (8.4)$$

The constant K represents the partition coefficient between two reference states and is significant across formulation variables affecting standard chemical potentials.

The *SAD* was defined in the early 1980s as the difference in standard chemical potential of the surfactant in oil versus water phases.[23,24] This concept simplifies the calculation of equilibrium between water and oil phases based on surfactant chemical potentials. The *SAD* facilitates the understanding of formulation at equilibrium by subtracting chemical potential terms rather than calculating their ratio.[23,25]

The *HLD* concept, introduced later, serves as a generalized, dimensionless criterion for the formulation, incorporating variables such as temperature, surfactant characteristics, and salinity.[26,27] The *HLD* usefulness is evident in its application to both ionic and nonionic surfactants, despite the challenges in calculating partition coefficients for surfactants with a range of molecular weights:

$$HLD = \frac{SAD}{RT} = -\frac{\Delta G_{W\to O}}{RT} = \frac{1}{RT}\left(\mu_W^* - \mu_O^*\right) = \ln K + \text{constant}. \quad (8.5)$$

This equation comprises the partition coefficient $K = C_W/C_O$ and includes a constant term for chosen reference parameters. The ln K term, associated with formulation variables, demonstrates a linear relationship with the *HLD* expression by assessing the partition coefficient across various systems, altering oil types, surfactant properties, alcohol content, brine salinity and temperature. This method is effective for ionic surfactants but differs for nonionic ones, which may exhibit a range of molecular weights and partitioning, making difficult the calculation of K coefficients.

The concept of normalized hydrophilic–lipophilic deviation (HLD_N) simplifies the hydrophilic–lipophilic deviation calculation by normalizing the equation through the coefficient of the alkane carbon number (*ACN*) variable. This adjustment allows the surfactant parameter for both anionic and nonionic surfactants to be standardized and expressed as a preferred alkane carbon number (*PACN*) or surfactant contribution parameter (*SCP*). The HLD_N equation is instrumental in determining the optimum formulation for simple SOW mixtures and in understanding complex systems in applications such as enhanced oil recovery, detergency, dewatering, or emulsification, where surfactant mixtures are frequently used. The HLD_N equation is given as follows:[19,25,28]

$$HLD_N = k_S \ln(S) - EACN - \varnothing(A) + PACN - k_T(T-25), \text{ for anionic surfactants,} \tag{8.6}$$

and

$$HLD_N = k_S S - EACN - \varnothing'(A) + PACN - k_T(T-25), \text{ for nonionic surfactants,} \tag{8.7}$$

where *EACN* represents the alkane carbon number when the oil is an *n*-alkane or the equivalent alkane carbon number (*EACN*) for other oils. *S* denotes the salinity, expressed as wt/vol% of NaCl; $\varnothing(A)$ and $\varnothing'(A)$ are functions depending on the alcohol type and concentration; *T* is the temperature (°C). The constants k_S and k_T have numerical values specified in Table 8.1. The HLD_N approach facilitates the use of a common surfactant parameter reference for different types of surfactants, enhancing the equation's applicability across a wide range of domains.

Formulation scans, a method to systematically modify the physicochemical interactions within SOW systems, involve adjusting variables involved in the HLD_N equation. While this method has broad application in identifying the optimum formulation, it demands considerable time unless executed adequately. Unidimensional scans, altering a single variable sequentially, offer an efficient strategy to determine the optimum formulation with minimum interfacial tension,[28–30] as schematized in Figure 8.2.[31]

Figure 8.3 presents a graphical representation of the phase behavior in a SOW system, showing changes in the Winsor III middle phase as a function of formulation in a scan. The optimum formulation corresponds to the point where a minimum interfacial tension, maximum solubilization, or equal volumes in the excess phases in a Winsor III triphasic system are attained.

Minor adjustments in the formulation can shift the phase behavior at a specific composition point within a ternary diagram, potentially altering the optimum formulation characterized by the minimum interfacial tension or a three-phase behavior with equal excess volumes. These changes indicate the nature of surfactant partitioning across phases, influenced by variables such as salinity, and highlight the critical role of bicontinuous structures in achieving the optimum formulation. Each variable within the HLD_N equation plays a crucial role in determining the system's overall phase behavior.[25,33] Table 8.1 presents a summary of the values of the main

TABLE 8.1
Parameters of the HLD_N Equation: *SCP*, k_S, k_T, and *EACN*

Conventional Surfactants	*SCP* (*ACN* units±2)
Tetradecyl sulfate, sodium salt Na	−22
Tetradecanoic acid, sodium salt	−18
Tetradecyl benzene sulfonate, sodium salt	2
Nonyl benzene sulfonate, sodium salt	−10
Tetradecyl trimethyl ammonium chloride	−14
Octadecyl trimethyl ammonium chloride	−4
Nonyl *ortho*-xylene sulfonate	−4
Pentadecyl *ortho*-xylene sulfonate, sodium salt	10
Dodecyl 14PO 2EO sulfate, sodium salt	−1
Dodecyl 10PO 2EO sulfate, sodium salt	−7.5
Dodecyl 6PO 2EO sulfate, sodium salt	−15
C_6E_4, high purity	−1.5
C_8E_4, high purity	3.5
$C_{10}E_4$, high purity	8.2
$C_{12}E_4$, high purity	12
$C_{12}E_5$, high purity	5.5
$C_{12}E_6$, high purity	−0.5
Conventional Surfactants	*SCP* (*ACN* units ± 2)
nC_8-β-Glucoside	−15.9
nC_{10}-β-Glucoside	−6.6
nC_{12}-β-Glucoside	−0.8
nC_{10}-β-Maltoside	−28.9
nC_{12}-β-Maltoside	−26.6
nC_{14}-β-Maltoside	−16.8
nC_{16}-β-Maltoside	−14.8

Surfactant	k_S ± 10%	k_T (°C^{-1}) ± 10%
Alkyl benzene sulfonate, sodium salt	6.2	-0.06
n-Alkyl sulfate/carboxylate, sodium salt	10	−0.1
n-Alkyl trimethyl ammonium chloride	5.3	−0.1

(*Continued*)

TABLE 8.1 (CONTINUED)
Parameters of the HLD_N Equation: SCP, k_S, k_T, and $EACN$

n-Alkyl ammonium chloride	5.9	–0.12	
$C_{12}E_4$, high purity	0.9	0.33	
$C_{12}E_6$, high purity	0.9	0.36	
$C_{12}E_8$, high purity	0.9	0.35	
C_6E_4, high purity	—	0.18	
$C_{10}E_4$, high purity	0.53	0.4	
Oil	*EACN* ± 10%	Oil	*EACN* ± 10%
1,4-Cyclohexadiene	–4.0	1-Chlorodecane	3.5
2-Octanone	–3.4	1-Dodecane	2
Phenyl-1-butyne	–3.3	Isopropylcyclohexane	5.5
1,3-Cyclohexadiene	–3.1	1-Decene	5.5
2,5-Norbornadiene	–3.2	Propylcyclohexane	5.9
1-Octyne	–1.8	Cyclodecane	5.6
p-Xylene	–2.0	*cis*-Decalin	5.3
1-Octyne	–2	Ethyl myristate	5
p-Xylene	–2	Decyl butyrate	5.0
2-Decanone	–2.1	Butylcyclohexane	7.4
Octanenitrile	–1.7	Dodecylbenzene	7.8
2-Undecanone	–1.3	Hexyl octanoate	6.2
Cyclohexene	–1.0	Dihexyl ether	6.2
p-Cymene	–0.5	Myristyl propionate	6.8
Decanenitrile	–0.6	Ethyl palmitate	6.8
2-Dodecanone	–0.6	Isopropyl myristate	7.3
Dodecanenitrile	0.3	Butyl dodecanoate	7.2
Terpinolene	0.3	Octyl octanoate	8.1
1-Decyne	0.1	Dioctyl ether	10.3
Benzene	0 or less	Hexyl dodecanoate	8.3
Butylbenzene	0.4	Ethyl oleate	7
Limonene	1.6	Bis(2-ethylhexyl) adipate	8.7
cis-Cyclooctene	1.5	Miglyol 840–diglyc. C11	9
γ-Terpinene	1.7	Myrcane	10.6
Ethyl decanoate	1.2.1	Glycerol trioctanato	12.3
β-Pinene	2.2	Tricaprilin	12.2
Diisopropylether	2.2	Squalene	13.8
Dibutylether	2.9	Tricaprin	13.4
Cyclohexane	2.5	Miglyol 812 triglyc. C11	14
Methylcyclohexane	3.5	Glycerol tridecanoate	14
α-Terpinene	1.3	Decylcyclohexane	14.4
α-Pinene	3.5	Dodecylcyclohexane	17.5
p-Menth-2-ene	3.4	Trimyristin	18.5
1-Octene	3.9	Soya oil triglyceride C18	18
Pinane	4.1	Triolein	21
Cyclooctane	4.1	Tripalmitin	21.2

(*Continued*)

TABLE 8.1 (CONTINUED)
Parameters of the HLD_N Equation: SCP, k_S, k_T, and $EACN$

Dipentylether	4.2	Trilaurin	16
Ethylcyclohexane	4.5	Tristearin	24
1,4-Dimethylcyclohexane	4.6	Squalane (branched C30)	24
Dibutyl ether	3.4	Squalane	24.5

Source: Data adapted with permission from Refs. 28 and 34–37. Copyright from the corresponding editorial.

parameters of the HLD_N equation according to the literature.[19,25,28,34,35–37] It is important to point out that the sign of these parameters indicates the tendency of the variable, i.e., the negative sign of the *SCP* of tetradecyl sodium sulfate shows that it is a hydrophilic surfactant compared to the positive sign of $C_{12}E_4$. Opposite signs in the k_T (temperature parameter) for some ionic and nonionic surfactants indicate the contrary phenomena involved when heating these molecules; for most ionic surfactants, heating enhances the hydrophilicity (negative sign) but for nonionic polyethoxylated, it increases its lipophilicity by the dehydration of ethylene groups (positive sign).

The hydrophobicity of oils is quantified as the *ACN* for alkanes. The bases for the *EACN* concept were introduced by Cayias et al.[38,39] when studying the oils that allows getting a minimal interfacial tension with other families of hydrocarbons different from n-alkanes. The *EACN* is defined as the number of carbon atoms of the linear alkane that has its optimum formulation in the same formulation conditions as the oil under study. The *EACN* of oils can be determined following different methods, but no matter the method, it always needs to compare the oil under study with a calibration curve of n-alkanes made with the same surfactant and using the same formulation variable. Typical methods are salinity scan using an anionic surfactant or temperature scan with polyethoxylated nonionic surfactants, which are more practical than interfacial tension measurements. Queste et al. [40] suggested to determine the *EACN* by comparing the T^* of a fish diagram (see the left panel in Figure 8.4) of the oil (temperature vs. concentration) using pure well-defined C_iE_j. Temperature T^* represents the critical value in which WI, WII, WIII, and WIV behaviors meet.

8.1.3 Partition of Surfactants and Oils Segregation in the Formulation of Microemulsions

Most surfactants and oils are not pure substances but mixtures. Indeed, when using commercial technical-grade surfactants their concentration can be an additional formulation variable because the partition of different oligomers between oil and water modifies the composition at the interface.[41–43] The differences between oligomers in technical-grade surfactants are most in their *EON* (for polyethoxylated nonionic surfactants) or in the alkyl chain. Figure 8.4a shows the phase behavior of

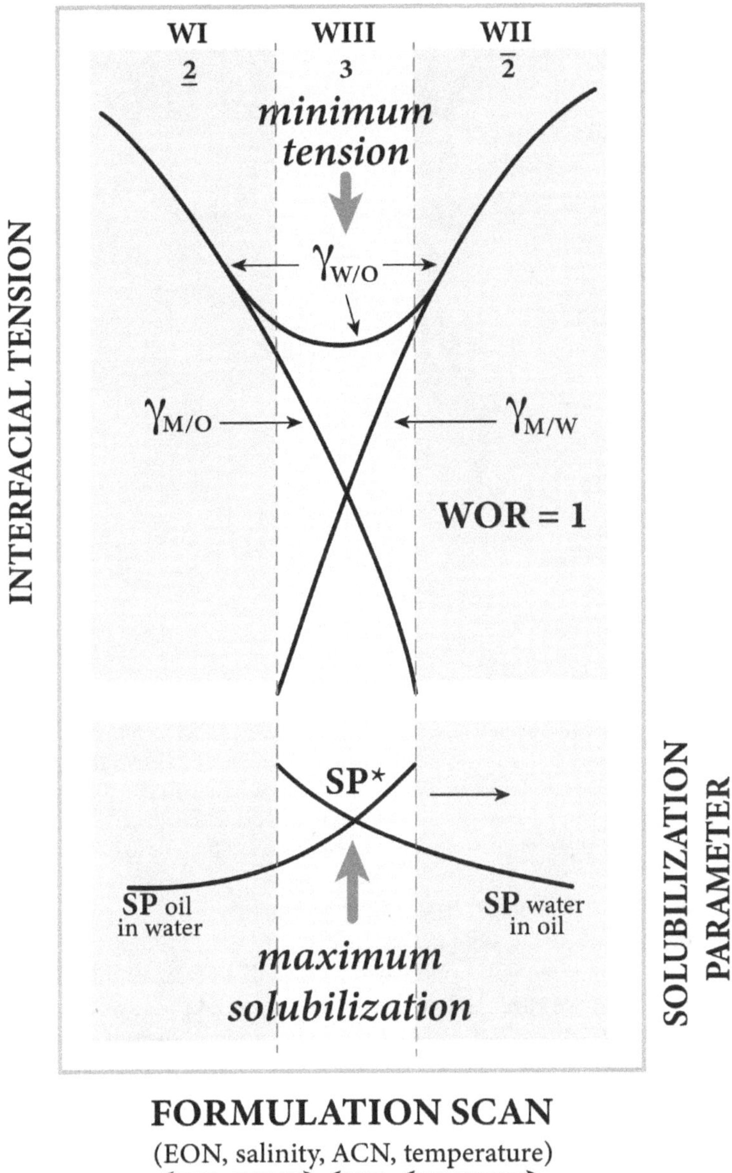

FIGURE 8.2 Properties of equilibrated systems at optimum formulation: minimum interfacial tension, which corresponds to the maximum solubilization WOR = 1, according to Chun Hu relation.[32] WI is characterized by swollen micelles formation in the aqueous phase, as depicted in the left test tube. WIII represents the system at minimum interfacial tension, indicating maximum solubilization, and presented by the three-phase behavior test tube with maximum solubilization (also, equal volumes of oil and water phases). WII corresponds to swollen micelles formation in the oil phase. Reproduced from Marquez et al.[31] under a Creative Commons Attribution [CC-BY 4.0] License.

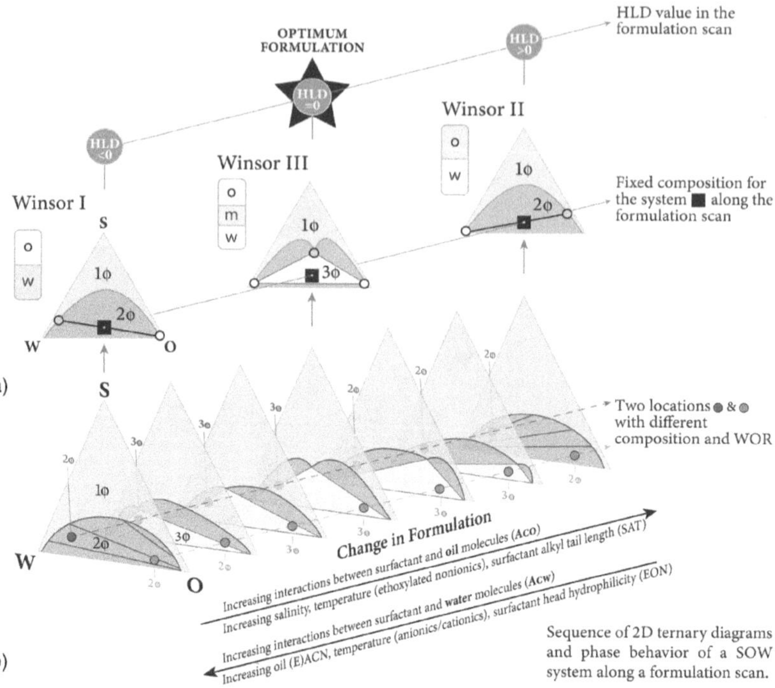

(a)

(b)

FIGURE 8.3 (a) Variation in the Winsor ternary diagram when a formulation change occurs (salinity in this instance, but also occurring with temperature, *EON*, alkyl chain number [*ACN*], etc.). (b) The morphological changes and the variation in the Winsor III triphasic zone in a SOW system upon changes in formulation, surfactant concentration, and *WOR*. Reproduced from Marquez et al.[31] under a Creative Commons Attribution [CC-BY 4.0] License.

the $C_{12}E_6$–isooctane–water system prepared with monodistributed $C_{12}E_6$ (black) and mixing equimolar quantities of monodistributed $C_{12}E_4$, $C_{12}E_6$, and $C_{12}E_8$ surfactants (blue). Different behaviors are indicated only for the pure $C_{12}E_6$ system and the phase boundaries for the WIII and WIV zones show the typical form of a horizontal fish. When mixing different surfactants, the shape of the fish diagram changes even if the 'average' *EON* is the same in both cases (*EON* = 6). At low concentrations, the partitioning of the most lipophilic species to the oil phase produces the interface that seems more hydrophilic and the temperature needed to get a WIII behavior increases. When concentration increases, the interface composition tends toward the average mixture and the partitioning becomes less important as is shown by the small differences observed in the fish tail. According to Graciaa et al.,[44] the partition coefficient depends on several factors such as the nature of the oil (aromatic oils enhance partition), the concentration (as shown in the fish diagram), the water–oil ratio, and the shape of the oligomer distribution in the mixture.

These results show that linear molar mixing rules can be used with precaution, and the linear fit is the exception and not the rule when mixing two nonionic or two

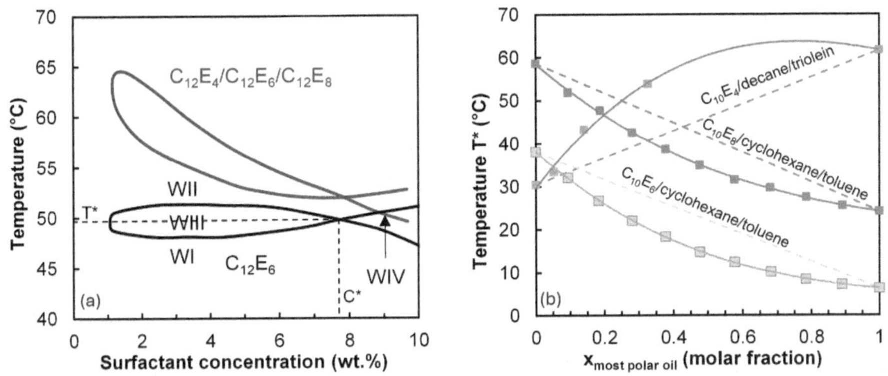

FIGURE 8.4 (a) Fish diagram (temperature vs. surfactant concentration) for the $C_{12}E_6$–iso-octane–water system at $\varnothing_{water}$= 0.5. Black line represents pure $C_{12}E_6$ surfactant and blue line represents an equimolar mixture of $C_{12}E_4$, $C_{12}E_6$, and $C_{12}E_8$; black dotted lines indicate the temperature and surfactant concentration at the fish tail T^* and C^*. Adapted from Ben Ghoulam et al.,[42] Copyright (2004), with permission from American Chemical Society. (b) Fish tail temperature for $C_{10}E_4$–decane–triolein–water (green), $C_{10}E_8$–cyclohexane–toluene–water (red), and $C_{10}E_6$–cyclohexane–toluene–water (yellow) systems at $\varnothing_{water}$= 0.5. Adapted from Engelskirchen et al.,[46] Copyright (2007), with permission from Elsevier, and from Burauer et al.[47] (unlicensed content).

anionic surfactants. Indeed, more complex cases arise when mixing surfactants with very different chemical natures like ionic and nonionic surfactants or anionic and cationic ones.[45] These cases are beyond the scope of this chapter but the reader can refer to Ref. 45.

Analogous to surfactants, oils mixtures are ubiquitous and triglycerides, essential oils, or crude oils are good examples of that. When oils of a similar chemical nature (i.e., alkanes) are mixed, a linear mixing rule can be applied to determine the final EACN:

$$EACN_{\text{mixture}} = \sum_{i=1}^{i} x_i EACN_i. \tag{8.8}$$

However, when the chemical structure of the mixed oils is very different, there is a segregation phenomenon at the interface, as pointed out by Graciaa et al.[48] when hexadecane and ethyl oleate are used as oil phase in an *EON* scan with technical-grade polyethoxylated octyl phenols. The composition of the oil layer close to the interface differs from that of the bulk oil; the more polar oil species (ethyl oleate) penetrate the interfacial layer. Consequently, the adsorbed surfactant interacts with an oil mixture whose composition differs from the overall oil phase modifying the 'optimum formulation' and the solubilization.

Figure 8.4b shows the fish tail temperature T^* for several systems mixing two pure well-defined oils as a function of the molar fraction of the most polar oil. In the three cases, the experimental value of T^* is different from the predicted value by a linear mixing rule (dotted lines) and seems closer to the value of the most polar oil. If the *EACN* of the polar oil is the highest, e.g., triolein (*EACN* = 21.2)/*n*-decane (*ACN* = 10), the T^* of the mixture (and its *EACN*) is higher than predicted T^*. However, in cyclohexane (*EACN* = 2.4)–toluene mixture, the *EACN* of the toluene is lower, and thus, T^* of the mixture is lower than the predicted T^* for both $C_{10}E_6$ and $C_{10}E_8$.[28] The nonlinearity of binary oil mixtures has also been reported with other formulation variables like salinity[49–51] and some models have been proposed to quantify this behavior.[49,52] In any case, the determination of *EACN* of oil mixtures must be done in conditions (temperature, nature, and concentration of surfactant) as close as possible to the final application to have similar segregation of the oil in both cases and assure the precision and validity of the calculated *EACN*.

8.1.4 SOW Phase Behavior at Equilibrium and Emulsions

In 1982, Salager et al.[53] demonstrated a direct relationship between the phase behavior of SOW systems at equilibrium and the corresponding emulsion properties. By performing scans of the various formulation variables (oil type, alcohol, salinity, and the proportion between two surfactants) at a water–oil ratio close to unity, they found that close to the optimum formulation (WIII systems), electrical conductivity showed a sharp change, indicating inversion of the continuous phase. Indeed, not only the morphology of the system changes when the HLD_N evolves, but also all the properties of emulsions, as will be described in the next section.

8.2 EMULSIONS AND FORMULATION: COMPOSITION MAPS

8.2.1 Generalities

Emulsions are kinetically stable dispersions of two immiscible liquids. When two immiscible liquids are dispersed to form an emulsion, the interfacial surface area increases dramatically and some energy must be added to the system to do this, with some of the work dissipated as heat. Emulsions are nonequilibrium thermodynamic systems (i.e., $\Delta G_{\text{formation}} > 0$), necessitating energy input for their formation, which can be defined as follows:

$$\Delta G_{\text{formation}} = \gamma \Delta A - T\Delta S, \tag{8.9}$$

where ΔA represents the change in surface area, with γ being the interfacial tension, and ΔS the change in entropy. Spherical droplets are the outcome of the system's natural attempt to reduce its energy upon production by decreasing the interfacial surface. When two pure immiscible liquids are mixed, the droplets often have such a strong interfacial tension that the system rapidly returns to two phases, the lowest energy configuration. Adding an emulsifier allows decreasing the interfacial tension and easily obtaining droplets, additionally providing a temporary barrier (electric or steric) against coalescence. Common emulsifiers are surfactants, some polymers, and even particles in so-called Pickering emulsions.[54]

The dispersed phase is known as the internal phase and the other phase is the continuous phase. The most common types of simple emulsions are water-in-oil (W/O) and oil-in-water (O/W) emulsions, but water-in-water (W/W)[55] and oil-in-oil (O/O)[56] emulsions also exist. When there is more than one interface between the internal phase and the continuous phase, the emulsion is multiple of order *n* (number of interfaces). An oil-in-water-in oil emulsion (O/W/O) is a multiple emulsion of order 2.

There are many variables to consider when formulating systems to obtain emulsions with specific desired properties. In 1910, Ostwald[57] concluded that the volume fraction of phases imposed a limit on the type of emulsion that could be formed, and based on the hypothesis of spherical and monodisperse droplets, the maximum fraction of dispersed phase would be 0.74. In 1913, Bancroft[58] studied the influence of surfactant affinity on the type of emulsion obtained during agitation and suggested that a hydrophilic colloid makes the aqueous phase like the continuous phase. These two 'rules' are well illustrated in the composition formulation maps.

The morphology and proportions between the continuous and dispersed phase are generally dictated by the application or the specifications of the final 'product'. The formulator scientist generally chooses the surfactant, the concentration, and the preparation protocol (order of incorporation of the products, type of agitation, agitation time, among others). There are different emulsifying machines using high-energy methods like high-speed stirrers, colloids mills, ultrasonic devices, microfluidizers, membrane emulsification, among others. All variables having an influence on the emulsion properties can be organized into formulation, composition, and protocol variables. This chapter will only focus on the formulation and composition variables when using surfactants as emulsifiers.

As explained in the previous section, the HLD_N allows taking into account all the formulation variables. When these variables change, the behavior of the surfactant also does at the interface. Properties like interfacial tension and the interfacial dilational elasticity[59–61] (also known as compression elasticity[3,62]) are both function of the HLD_N and, indeed, they have a minimum at $HLD_N = 0$.[62]

8.2.2 Emulsion Properties and Formulation

8.2.2.1 Morphology

The kind of emulsion is the common term to refer to the continuous or external phase of an emulsion. The morphology of an emulsion can be detected simply by placing a drop of a fluid emulsion in a small volume of water. If the continuous phase of the emulsion is water, the oil droplets will disperse in the water and if it is not, the emulsion will not disperse and according to its density will remain on the surface or the bottom of the vial. Using a water-soluble dye and an oil-soluble dye is another typical test: if the water-soluble dye dissolves into the emulsion, then water is the continuous phase. Similarly, if oil-soluble dye dissolves, then oil is the continuous phase.

Conductivity is a simple technique that allows online monitoring of the nature of the external phase. Distilled water conductivity at 25°C is near mS/cm and conductivity for most hydrocarbon oils is zero. Adding small amounts of salt allows maximizing this difference and easily distinguishing the morphology of an emulsion.

Bruggeman's[63] approximation establish that the conductivity κ_m of a dispersion, which consists of a continuous phase of conductivity κ_c, a dispersed phase of zero conductivity, and volume dispersed fraction '$\varnothing_d$. can be simplified to

$$\kappa_m = \kappa_c \left(1 - \varnothing_d\right)^{3/2}. \tag{8.10}$$

The w/O/W emulsions can be easily distinguished from O/W emulsion by conductivity if both aqueous phases w and W have similar conductivity. The volume continuous fraction $\left(1 - \varnothing_d\right)$ is different to the total volume continuous fraction because some of the water is not available as continuous phase, but as droplets in the oil phase, diminishing the conductivity of the dispersion.

Figure 8.5 shows the conductivity profiles when changing a formulation variable: the temperature (left) or the salinity (right) of the 3 wt% $C_{10}E_4$–*n*-octane–brine system at $f_w = 0.5$. Black dots represent the conductivity of samples agitated after 24 h of pre-equilibration. High conductivities indicate O/W emulsion and when conductivity is near to zero, the emulsion is W/O; in both cases, an increase of temperature or salinity produces a drastic change in the emulsion morphology. This change is known as emulsion inversion.

The increase of temperature diminishes the interaction between the ethylene oxide groups and the water and the HLD_N increase. The increase of NaCl concentration diminishes the water activity, increasing the affinity of the surfactant to the oily phase. In both cases, the drastic change of conductivity occurs in the vicinity of the 'optimum formulation' and the inversion happens in the Winsor III domain. This match does not happen in all water–oil ratios nor in all SOW systems, but for most well-defined pure surfactants with a water fraction near to 0.5, the behavior observed in Figure 8.5 is classical. The intermediate values of conductivity when the morphology of the emulsion change are related to the bicontinuous structure of the Winsor III observed at equilibrium.[65] Blue and red lines in Figure 8.5 indicate a continuous change of the formulation variable: the blue represents the increase of the

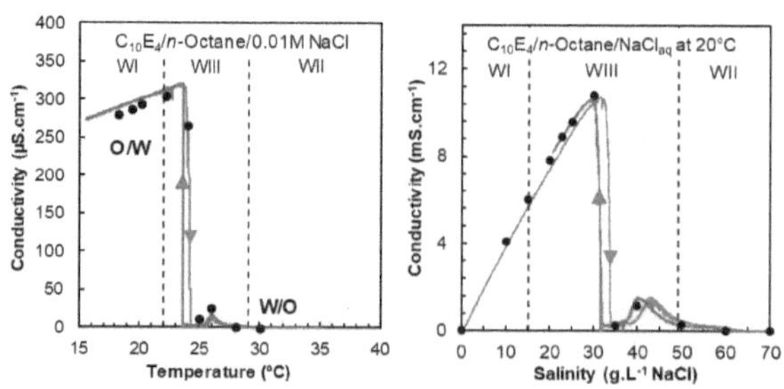

FIGURE 8.5 *Left panel:* Conductivity–temperature profile for the 3 wt% $C_{10}E_4$–*n*-octane–0.01 M NaCl system at $f_w = 0.5$. *Right panel:* Conductivity–salinity profile for the same system reported in the left panel. The equilibrium phase behavior (WI, WII, or WIII) zones at different temperatures or salinities are indicated in both figures (black dotted vertical lines). Adapted from Lemahieu et al.,[64] Copyright (2021), with permission from Elsevier.

temperature or salinity and the red one the decrease. Small hysteresis is observed in both cases, but the data obtained from equilibrated samples (black dots) and from dynamic protocol (lines) match well.

8.2.2.2 Stability

Emulsions are not thermodynamically stable, and what is called stability is the rate at which it changes its properties over time. Different phenomena can occur in an emulsion. Depending on the difference in density between the dispersed phase and the continuous phase, creaming (droplets going to the top of the container) or sedimentation (the opposite) may occur. According to the Hadamard–Rybczynski[66,67] equation, the terminal velocity v_t of a single dispersed fluid droplet moving in another fluid can be calculated as follows for low Reynolds numbers:

$$v_t = \frac{2(\rho_c - \rho_d)r^2}{3\mu_c}\left(\frac{\mu_c + \mu_d}{2\mu_c + 3\mu_d}\right). \tag{8.11}$$

The equation shows that the droplet size is an important factor and that if you compare a 0.1 μm droplet and a 10 μm droplet, the latter would have a speed 10,000 times more important. Emulsions with a smaller droplet size have smaller velocities and are less susceptible to settling. A high viscosity of the continuous phase can also reduce the velocity and adding a thickening agent is a common strategy in the formulation of commercial emulsions. Other factors must be considered in this process; the fact that there is not a single droplet but a distribution of droplets, the interactions when the dispersed volume fraction increases and the presence of surfactant at the droplet interface are some of the most important factors. In addition, droplets can aggregate leading to the formation of larger particles known as flocs. Flocculation can accelerate creaming or sedimentation by changing the size of aggregates from the original radius to the final bigger cluster.

There are two irreversible mechanisms in the destabilization of an emulsion: Ostwald ripening and coalescence. Ostwald ripening refers to the gradual change of size of droplets by the mass transfer through the continuous phase, occurring from the small droplets to the bigger ones (from high Laplace pressure to lower pressure). The Lifshitz, Slyozov, and Wagner theory predicts that the average particle size increases with the cube root of time:

$$\frac{dr^3}{dt} = \frac{8\gamma D_d V_M c_\infty}{9RT}, \tag{8.12}$$

where V_M, D_d, and c_∞ are the molar volume (mol/m^3), diffusivity (m^2/s), and the solubility of dispersed phase in the solvent expressed as volume fraction. The diffusion of the dispersed phase minimizes the total interfacial area and the total energy of the system; this equation predicts that size distribution becomes narrow, and it tends to an asymptotic behavior.

Coalescence is the process by which droplets come into contact and combine to produce a bigger droplet. This causes the overall number of droplets in the system to decrease and eventually the separation of the dispersed phase; furthermore, contrary to Ostwald ripening, coalescence increases polydispersity.[68] At the droplet surfaces, repulsive forces or stabilizing forces from emulsifying agents are subdued by attractive forces, such as van der Waals forces, which propel the process. As the droplets approach each other, they begin to deform and the fluid between thins. Kabalnov and Wennerström[69] proposed that coalescence begins with the creation of a thermally activated hole that acts like a passageway between two droplets. This 'channel' between the droplets is strongly curved and the activation energy of the coalescence depends on the value of the spontaneous curvature of the surfactant monolayer. There are other factors that affect activation energy like the interfacial dilational elasticity ε; a higher dilational elasticity enhances the film stability. Thus, there are many possible scenarios and phenomena for coalescence and predicting emulsion stability is extremely complex and challenging.[3,70]

Quantifying the stability of an emulsion should involve granulometric analysis of the emulsion as a function of time, which is impractical due to the possible segregation of droplets in the container as a function of size.[70] The monitoring of separated phases is more practical and the coalesced volume is a good parameter to follow the stability. In other applications, the volume of clarified phase after creaming or sedimentation is also important. When emulsions are very stable, accelerated test can be done and the emulsion is studied after centrifugation, even if the correlation between these results and the long-term stability is not always evident. In some cases, the stability is also monitored at higher temperature than storage temperature to accelerate the destabilization phenomena.

The evolution of the emulsion behavior in terms of phase separation can be followed visually or by other methods like static multiple light scattering (SMLS). Figure 8.6 represents the transmission and backscattering profiles for three 1 wt% dodecyl diglyceryl ether ($C_{12}Gly_2$) –oil–water systems at 25°C and the same volumes

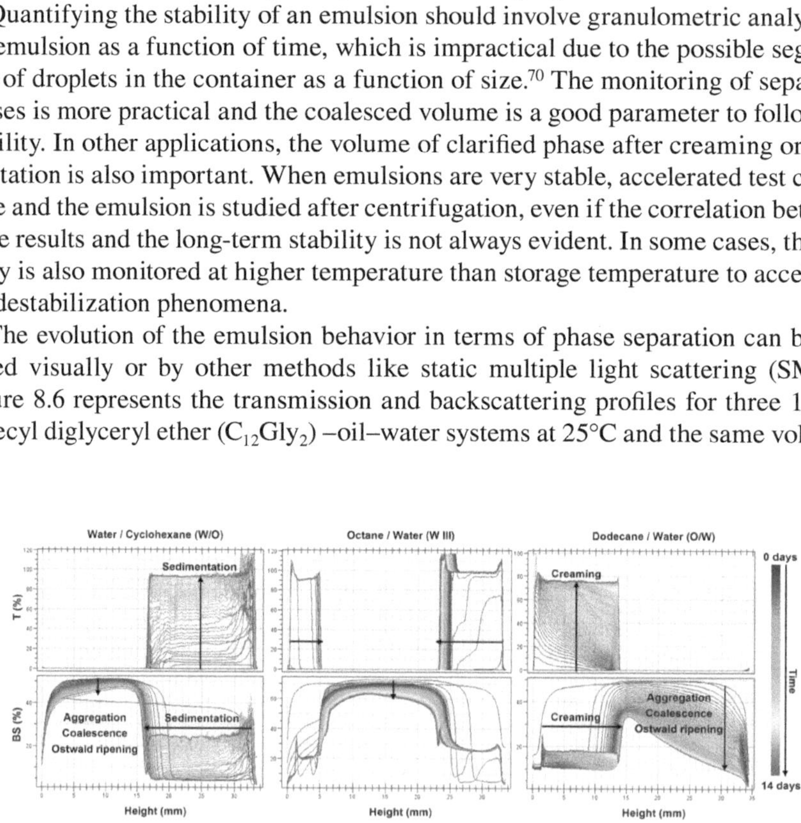

FIGURE 8.6 Transmission (*T*) and backscattering (*BS*) light signals over time for water–cyclohexane (left), octane–water (middle), and dodecane–water (right) emulsions (*WOR* = 50–50, 1% $C_{12}Gly_2$). Creaming and sedimentation fronts are represented by horizontal arrows on *BS* signals and vertical arrows on *T* evolution. Coalescence and/or aggregation and/or Ostwald ripening are visible as *BS* decreases over time. Reprinted from Delforce et al.,[36] Copyright (2023), with permission from Elsevier.

of water and oil. The signals are presented as function of the height of the tube in which emulsion was prepared and then stored for 14 days.

The influence of the nature of the oil on the morphology of the emulsion is clearly established in this system. For cyclohexane (*EACN* = 2.1), octane (*ACN* = 8), and dodecane (*ACN* = 12), the phase behavior at equilibrium is WII, WIII, and WI, respectively. After emulsification, the morphology of emulsions is W/O for cyclohexane and O/W with dodecane. Oil–oil interactions depend on the square of the number of carbons in the alkyl chain[71] and when the ACN increases, the HLD_N decreases because the affinity of the surfactant for the oily phase also does. In cyclohexane, the increase of the transmission signal in the top of the vial with time indicates that there is a sedimentation of water droplets, and the oily continuous phase is more and more clear in this zone. For dodecane, the transmission profile presents the same behavior but at the bottom of the vial, indicating a creaming phenomenon. In both cases, there is no macroscopic separation of the dispersed phase. The octane case is particular because profiles show clearly two quick fronts in the droplet's movements, at the top and in the bottom to finally stabilize in a three-phase zone: water in the bottom, the microemulsion in the middle, and the oil in the top.

Analyzing other system, with an anionic surfactant (sodium dodecyl benzene sulfonate [SDBS]) results are similar.[72] The left panel of Figure 8.7 shows that the time required to separate 60% of the dispersed phase in emulsions presents a minimum in the $HLD_N = 0$ condition, analogously to the system of Figure 8.6. The right panel of Figure 8.7 shows the interfacial tension and the interfacial elastic modulus as a function of salinity obtained from samples at equilibrium.

Why the emulsions formulated from systems in the vicinity of $HLD_N = 0$ are so unstable? At 'optimum formulation', the partition coefficient of the surfactant is close

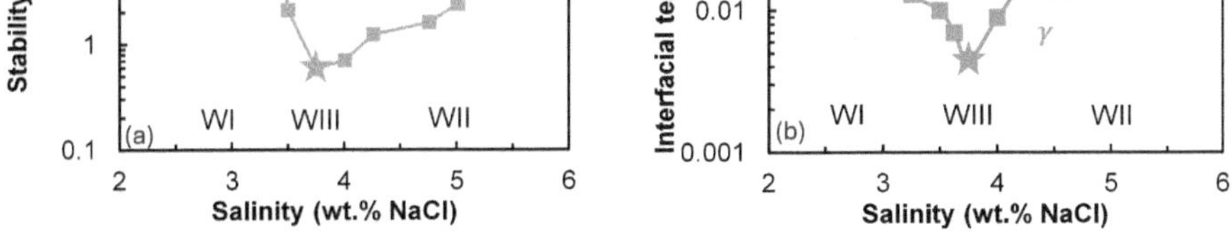

FIGURE 8.7 *Left panel:* Stability as a function of salinity. Stability is quantified as the time to observe the coalescence of 60% of the dispersed phase. *Right panel:* Interfacial tension (γ) and interfacial effective elastic modulus at 0.1 Hz (ε) as a function of salinity. Sodium dodecyl benzene sulfonate (0.02 wt%/*sec*-butanol (1 wt%)/cyclohexane/brine at T = 30°C. *WOR* = 1. The lines are guides to the eye. Adapted from Marquez et al.,[62] Copyright (2018), with permission from American Chemical Society.

to the unity and there is surfactant available in both water and oil, diminishing tension gradients in the thin film and in consequence the Marangoni forces. The minimum value of the effective elastic modulus at this condition also indicates that the interface has a low rigidity enhancing the instabilities at interface and coalescence.[62] To formulate stable emulsions, the condition of $HLD_N = 0$ must be avoided; however, this condition is important when broken emulsion is the objective of formulator like water in crude oil emulsions in the petroleum industry.[73]

8.2.2.3 Droplet Size

In the IUPAC definition of emulsion,[74] there is no precise range of sizes and terms like macroemulsions, miniemulsions, nanoemulsions, and the confusion with microemulsions is still present in literature. Usually, when the droplet size is higher than 0.5 μm, these systems are called macroemulsions or emulsions to simplify.

After agitation, emulsion is often a system in which different droplet sizes coexist. Several models have been proposed to predict droplet size distribution. These models are based on Kolmogorov–Hinze theory,[75,76] which takes into account the smallest eddies in turbulent flow to determine droplet size. Larger droplets than a critical length break up due to the effects of the turbulent inertial stress, while smaller ones remain unbroken. The balance between internal cohesive and external turbulent stresses determines the maximum stable droplet diameter, so the droplet size depends on the Reynolds and Weber numbers. Other mechanisms proposed are the droplets formation by fragmentation due to Rayleigh instabilities and surface tension. However, there is no 'unified picture of the drop size and its distribution'.[77]

In order to characterize the size distribution, several diameters exist. $D_{(V,0.5)}$ is the median diameter in volume (50% of the dispersed size is under this value). The Sauter diameter $D_{[3,2]}$ is the surface area mean, it represents the diameter of a sphere that has the same volume–surface ratio that all the droplets population have and it is sensitive to smaller sizes. The De Brouckere diameter $D_{[4,3]}$ is the volume mean, it is sensitive to larger sizes. For very monodispersed emulsions, these three diameters are almost equivalents, but for most emulsions, these diameters can present very different values according to their definitions:

$$D_{[3,2]} = \frac{\sum_i N_i D_i^3}{\sum_i N_i D_i^2} \tag{8.13}$$

and

$$D_{[4,3]} = \frac{\sum_i N_i D_i^4}{\sum_i N_i D_i^3}, \tag{8.14}$$

where N_i and Di are the number and diameter in a particulate size fraction i. Typical techniques to determine the size distribution are laser diffraction, dynamic light scattering (DLS), electrophoretic light scattering (ELS), acoustic spectroscopy, and optical microscopy (when size is higher than 1 μm).

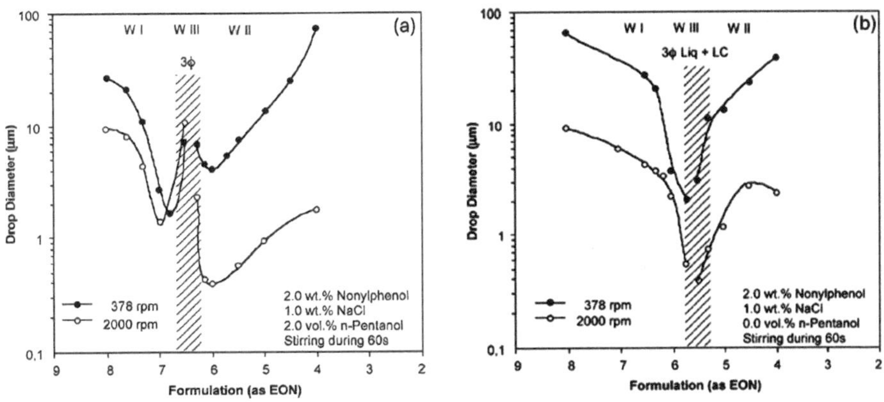

FIGURE 8.8 Drop diameter ($d_{V(50)}$) vs. *EON* for the 2 wt% nonylphenol/kerosene/water (1 wt% NaCl) agitated for 60 s. (a) 2% *n*-pentanol. (b) 0% *n*-pentanol. Reprinted from Tolosa et al.,[78] Copyright (2006), with permission from American Chemical Society.

Most physicochemical variables described by the HLD_N have an influence in the final droplet size of an emulsion. Figure 8.8 shows the influence of formulation variables in the drop average diameter $D_{(V,0.5)}$. The formulation variable is the *EON* in the ethoxylated nonylphenol surfactant (2 wt%)–kerosene–brine with and without *n*-pentanol (2 vol%). Diminishing the *EON* increases the lipophilicity of the system and the transition of the equilibrated systems is WI→WIII→WII. When these systems are agitated, the diameter of these emulsions are measured and represented in the figure. The decrease in the interfacial tension allows obtaining smaller droplets when the emulsion is near the $HLD_N = 0$; however, the minimum interfacial modulus ε enhances the coalescence and the size quickly increases (in both sides). In Figure 8.8a (the system with *n*-pentanol), there is a minimum droplet size in the vicinity of the optimum formulation ($HLD_N = 0$) in both sides of the diagram, for O/W and W/O emulsions. In these points, there is the 'best compromise between low tension and not too unstable emulsions'.[70]

In Figure 8.8b, the system without *n*-pentanol, there is a minimum droplet size in the WIII zone (in which excess water and oil, microemulsion, and lamellar liquid crystals coexist), for both O/W and W/O emulsions. In this case, the formation of lamellar liquid crystal layer around the droplets could prevent the coalescence and the absence of alcohol also probably increases the interfacial tension compared to the system with alcohol.[78]

8.2.2.4 Viscosity

Rheology and textural characterization of emulsions are closely related, and this is a key parameter in the cosmetic and food domains.[79] The consumer's experience with final product depends on the texture of the emulsion and parameters like mouthfeel, skin feel, spreading, and creaminess among others are very important. Furthermore, unit operations like pumping, gravity drainage, extrusion, spraying, and pouring of emulsions also depend on their rheology.[80] Viscosity of emulsions is often higher than viscosity of continuous and dispersed phases separately: the presence of droplets in the system increases the resistance to deformation. Additionally, its viscosity depends on the applied shear rate (non-Newtonian fluids) by the presence of deformable droplets and, as droplets size evolves with time, time-variable fluids.[70]

Viscosity of very diluted emulsions ($\varnothing_c < 0.02$) follow the Einstein equation[81] of viscosity defined as follows:

$$\eta_{(\varnothing)} = \eta_c \left(1 + 2.5\varnothing_d\right), \tag{8.15}$$

where $\eta_{(\varnothing)}$ is the viscosity of the suspension, η_c is the viscosity of the continuous phase, and $\varnothing_d$ is the dispersed volume fraction. Even if this equation assumes rigid spheres $(\eta_d \gg \eta_c)$ with extremely low concentration of dispersed phase, it shows that increasing the quantity of dispersed phase also increases the viscosity and that the viscosity of emulsion is related directly to the viscosity of the continuous phase. Several models have been proposed to describe the viscosity of more concentrated emulsions and its increase is exponential with $\varnothing_d$.[80] For example, the consistency of mayonnaise, an O/W emulsion with typically 70–75% of oil, is clearly explained by the high internal proportion of oil ($\varnothing_d$) compared to the external aqueous phase.

Other physical variables in the emulsion's rheology are the droplet size and distribution. The influence of the droplet size is related to the $\varnothing_d$; at lower $\varnothing_d$, the influence of the droplet size seems to be neglectable.[82] When $\varnothing_d$ increases, the influence of the droplet size is important and the viscosity is higher for emulsions with smaller size, which present a higher shear thinning behavior.[82,83] The increase in viscosity is due to the higher interactions between droplets (emulsions with smaller droplets size presents have higher tendency to flocculation)[83] and the increase in the interfacial area. The distribution is also important and polydispersed systems are less viscous than monodispersed ones with the same $\varnothing_d$ and droplet size.[80] Mixing monodispersed emulsions results in a bimodal system with a minimum in the viscosity at lower shear stress and some specific proportion between the two emulsions.[80,83]

At fixed dispersed phase fraction, the nature of the continuous phase and the droplet size are function of the formulation variables. When formulation changes, these variables change, and the viscosity of the emulsion also does. Figure 8.9 shows how viscosity evolves, at constant shear rate, when the nature of the oil (left panel) and temperature (right panel) change. For the SDS–oil–pentanol–brine emulsions, the viscosity is obtained from pre-equilibrated systems (samples were emulsified at least 24 h after preparation). For the Brij 30 (technical-grade $C_{12}E_4$)–isopropyl myristate (IPM)–brine system, the data is obtained from the same emulsion sample which temperature changes dynamically at 1° C/min.

In the two cases, no matter the system (ionic or nonionic), there is a minimum of viscosity in the three phases zone. The coincidence of the minimum viscosity in the zone of minimum interfacial tension and minimum stability indicates that the droplets of the dispersed phase are larger (by the higher coalescence rate) and more

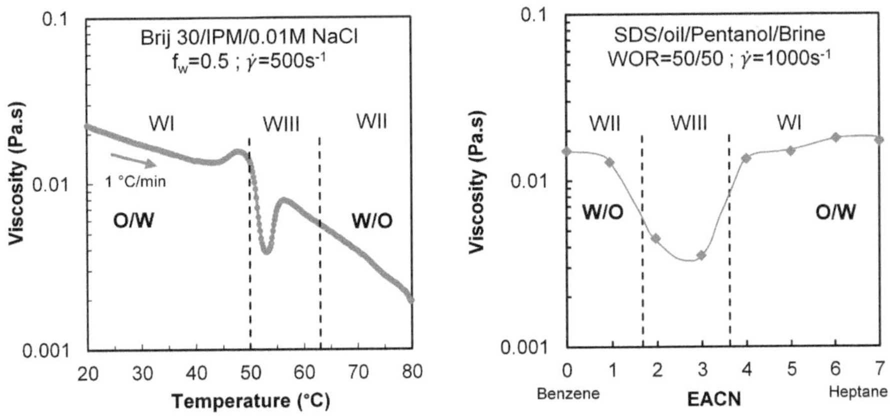

FIGURE 8.9 *Left panel:* Viscosity of emulsions vs. *EACN* of the oil phase (benzene–heptane mixtures) of the SDS (0.009 M)–oil/pentanol (6 vol%)–NaCl (1.7 wt%) system. Adapted from Salager et al., [84] with permission from Taylor & Francis Ltd. (http://www.tandfonline.com). *Right panel:* Viscosity–temperature profile for Brij30 (9 wt%)–isopropyl myristate–0.01 M $NaCl_{(aq)}$ emulsion. Adapted from Pierlot et al.,[85] Copyright (2016), with permission from American Chemical Society.

susceptible to shear deformation.[84] In contrast to the viscosity vs. *EACN* profile in which the values of viscosity in the W/O and O/W zones are relatively constants, in the profile for the Brij 30 system the viscosity of the emulsion decreases because the increase of temperature diminishes the viscosity of the liquids. The value of the viscosity is closely related to the preparation protocol and different values can be obtained for the same system if these conditions change. However, the shape of the profile remains and the value at which the minimum of the viscosity is obtained also does.

8.2.3 Formulation–Composition Maps

As described in the previous section, all properties of emulsions change when the formulation variables do. Most of the described systems are in the vicinity of equal amounts of dispersed and continuous phases to maximize the formulation effect; however, as established by Ostwald,[57] the volume fraction of dispersed phase imposes a limit on the type of emulsion that can be formed. In fact, the water–oil ratio (WOR) or the water weight fraction (f_w) is a very important variable in the emulsion's properties. As a matter of fact, most commercial emulsions in cosmetics or food domains are far away $WOR = 1$ or $f_w = 0.5$. Formulation–composition maps are graphical representations of emulsion properties as a function of formulation variable (HLD_N) and the water fraction in the system at fixed surfactant concentration. These maps were introduced by Salager et al.[86,87], they have been reported for several systems and this representation is very useful in the design emulsions with final desired properties.

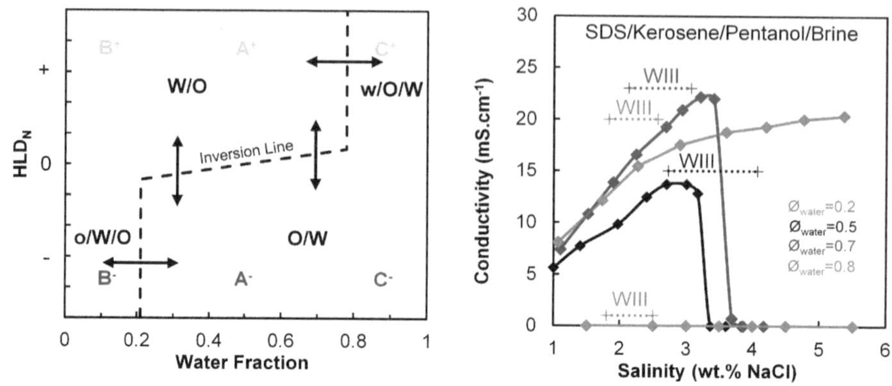

FIGURE 8.10 *Left panel:* Conductivity of emulsions vs. salinity of the SDS (0.02 M)–kerosene/pentanol (4.7 wt.%)–brine system at different water volume fractions. The dotted lines indicate the salinity range in which WIII systems are observed at equilibrium. Adapted from Salager et al.,[86] with permission from Taylor & Francis Ltd., (http://www.tandfonline.com). *Right panel:* Scheme indicating different morphologies observed in a formulation–composition map. Vertical arrow indicates a transitional inversion and horizontal arrow indicates catastrophic inversion.

The left panel of Figure 8.10 shows the conductivity profiles at different volume fractions of water and also the salinity interval in which the WIII behavior is obtained for the SDS (0.02 M)–kerosene–*n*-pentanol–brine system. When $\varnothing_{\text{water}}$ is 0.5, the inversion matches well with the WIII zone. When the oil quantity is very low, the continuous phase is always aqueous, no matter the salinity (even if at equilibrium, there is a WIII zone); on the contrary, when the quantity of the oil is very high ($\varnothing_{\text{water}} = 0.2$), the morphology of the emulsion is W/O. Scheme in the right panel of Figure 8.10 represents a simplification of the different morphologies that can be observed in the formulation–composition map.

When the HLD_N is positive, the affinity of the surfactant is higher for the oily phase, but depending on the f_w, the morphology of the emulsion can evolve from W/O (Zones A^+ and B^+) to O/W or W/O/W (Zone C^+). When the HLD_N is negative, the affinity is higher for the aqueous phase and the morphology can evolve from O/W/O or W/O (Zone B^-) to O/W (Zones A^- and C^-). Zones B^- and C^+ are called abnormal zones by the divergence in the morphology according to the formulation variables and the water–oil ratio. The limits are the inversion lines. When the inversion is related to the change of the formulation variables (vertical arrow), the inversion is transitional. When the inversion is related to the change of the water–oil ratio horizontal arrow), the inversion is catastrophic. Localization of limits of inversion are arbitrary in this schema, especially those related to the catastrophic inversion; indeed, the shape of the frontiers can be very different depending on the nature of studied SOW system.

Figure 8.11 represents the formulation–composition map for the $C_{12}Gly_2$(1 wt%)–oil–water system, including iso-granulometry data. The formulation variable is the

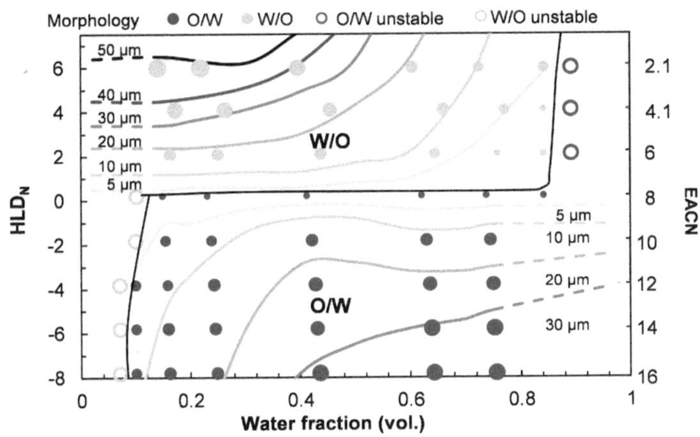

FIGURE 8.11 Iso-granulometry curves of emulsions containing 1 wt% $C_{12}Gly_2$ and prepared by varying the nature of the oil (EACN) and the water fraction (vol%) at 25°C. Droplet $D_{[4,3]}$ are measured 5 min after emulsification. O/W and W/O emulsions are represented by blue dots and yellow dots, respectively, and symbol size is varied with $D_{[4,3]}$ values. The standard inversion line separating O/W and W/O areas is represented as the continuous thick black line. Adapted from Delforce et al.,[36] Copyright (2023), with permission from Elsevier.

hydrophobicity of the oil (*EACN*) and when $\varnothing_{water}$ is in the vicinity of 0.5, the emulsion evolves from O/W for hydrophobic oils to W/O when shorter hydrocarbons are used (optimum formulation is obtained at *EACN* = 8.1). When *EACN* of oil is lower than 8 and $\varnothing_{water}<0.08$, the continuous phase is oil but the emulsions are unstable. When EACN of oil is higher than 8 and $\varnothing_{water}>0.87$, the continuous phase is aqueous, but the emulsions are also unstable. These conditions were used to place the catastrophic inversion lines. It is quite important to note that increasing the dispersed phase fraction induces a decrease in the droplet size in both O/W and W/O emulsions.

Figure 8.12 shows the formulation–composition map for the Brij30 (9 wt%)–isopropyl myristate–0.01 M $NaCl_{(aq)}$ system, including iso-viscosity values (determined at shear rate of 500 s^{-1}). In this case, the surfactant is a technical-grade alkyl-polyethoxylated alcohol and the effect of the water fraction is quite important in the value of the temperature at which the transitional inversion is located. This temperature is the so-called PIT.

When the water fraction decreases, the PIT increases because the partition in the oil and water bulk phases of different monomers present in Brij 30 also changes and the composition at the interface seems more hydrophilic.[41,44] Indeed, most hydrophobic monomers in Brij 30 are dissolved in the oil phase. At a fixed water fraction, the viscosity trend described in previous section is also observed here with a minimum viscosity in the vicinity of the phase inversion. When the dispersed fraction increases, in both sides of the map, the viscosity of the emulsion also does until a critical point in which viscosity decreases. At these points, the catastrophic inversion lines were placed.

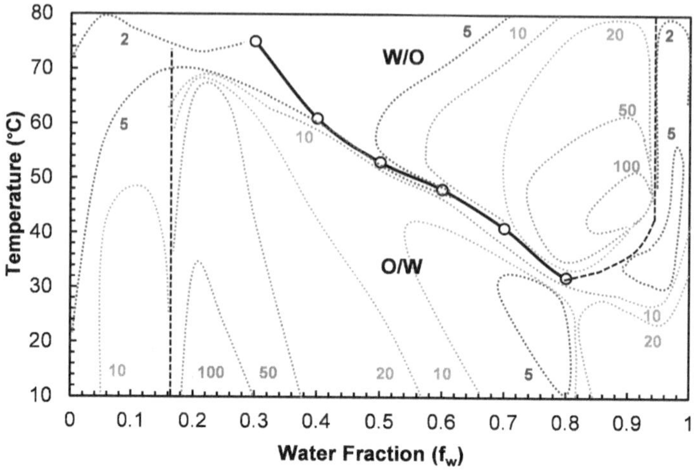

FIGURE 8.12 Iso-viscosity curves (indicated in mPa·s) of Brij30–IPM–10^{-2} M $NaCl_{(aq)}$ system in the temperature vs. water weight fraction (f_w). Viscosity was measured at shear rate = 500 s^{-1} from 10°C to 80°C at 1°C/min. PIT (O). Adapted from Pierlot et al.,[85] Copyright (2016), with permission from American Chemical Society.

8.2.4 Emulsion-Phase Inversion and Application of the Formulation–Composition Map

According to the protocol followed to achieve the inversion, there are two kinds of inversion: standard and dynamic. When each point in the formulation–composition map is formulated separately, equilibrated, and then emulsified, the inversion is standard. This is the case of the formulation–composition map of Figure 8.11. When the variable (formulation or water fraction) changes continuously upon stirring, the inversion is dynamic, as shown in Figure 8.12; in this case, the rate at which the variable changes must be indicated (1 °C/min). In both systems, the catastrophic inversion was detected indirectly by the drastic change in the stability (Figure 8.11) or in the viscosity (Figure 8.12). However, the dynamic inversion boundaries can present a delayed response, especially for a catastrophic inversion.[88] In this case, there is a hysteresis zone in which the morphology of the emulsion can be either W/O or O/W depending on the trajectory of the inversion (the initial point and the direction) and the protocol (stirring, flowrate).

The left panel in Figure 8.13 shows a scheme for the hysteresis zone. It is important to note that this zone becomes ‘smaller’ when the HLD_N is near zero and therefore the difference between the dynamic and the standard value depends on the initial HLD_N condition. If the initial point is P and oil with a lipophilic surfactant is continuously added to the system, the inversion happens at a water fraction that can be very different to the standard value (dotted line). When the flowrate increases, the inversion is ‘delayed’. To minimize the difference with the standard inversion, a small flowrate must be chosen. Multiple emulsion morphology (W/O/W in the case of PR trajectory) is associated with this inversion before it becomes a W/O emulsion

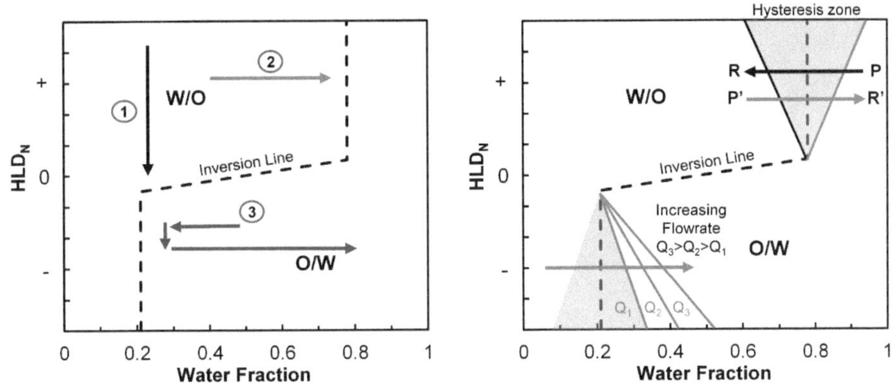

FIGURE 8.13 *Left panel:* Formulation–composition map schema showing the hysteresis zone in the catastrophic inversion and the influence of flowrate in the inversion frontier. *Right panel:* Different formulation paths without phase inversion. Adapted from Salager et al.,[70] with permission from Taylor & Francis Ltd. (http://www.tandfonline.com) and adapted from Zambrano et al.,[89] Copyright (2003), with permission from American Chemical Society.

(point *R* in the figure).[89] The influence of stirring is more complex and the delay passes through a maximum with an intermediate stirring energy.[90]

The catastrophic inversion is utilized in numerous industrial processes, particularly when working with highly viscous oil phases such as alkyd or epoxy resins to formulate O/W emulsions in paint manufacturing. Given the extreme viscosity of the resins (100–1,000 Pa·s at room temperature), conventional stirring techniques may not be effective in generating small droplets. The inversion technique entails the gradual addition of water to the viscous oil while employing slow mixing. In most instances, the surfactant is hydrophilic (HLD_N <0) and is initially dispersed in the oil phase. Adding water diminishes the concentration of the surfactant continuously and the formulation–composition map has only schematic purposes because the real inversion need a 3D diagram (several maps at different surfactant concentrations).[88,91] In other cases, the inversion process includes a formulation and a water–oil ratio change at the same time, and the inversion can be classified as combined inversion.

The formulation–composition map is used to study not only the phase inversion, but also other trajectories that have been applied to modulate the emulsion properties. The right panel of Figure 8.13 shows three practical examples. Process 1 corresponds to the breaking of a diluted water in crude oil (W/O) emulsion by adding a hydrophilic surfactant until a condition is near to the $HLD_N = 0$, diminishing the stability of the emulsion and allowing separation of the water. Process 2 is a W/O emulsion in which the dispersed phase increase by adding water in order to increase the viscosity of the system and its stability. In process 3, a diluted O/W stable emulsion must be formulated; at this condition, the agitation is quite inefficient to get small droplets and high-energy emulsifying methods are usually required. Process 3 has two steps; in the first one, oil is added to an O/W emulsion, increasing its viscosity and the efficacity of the stirring in order to get small droplets when approaching the inversion catastrophic locus. At this point, the HLD condition is slightly modified to move away from $HLD_N = 0$ and finally the emulsion is diluted until the required formulation avoiding creaming phenomena.[70]

8.3 NANOEMULSIONS

8.3.1 Nanoemulsions Definition

Nanoemulsions are dispersions comprising at least two liquid phases, typically characterized by a droplet size of less than 0.5 μm, usually ranging from 20 nm to 500 nm.[92–94] Over time and across different research groups, these systems have been referred to by various terms: ultrafine emulsions, miniemulsions, and submicrometric emulsions.[94–100] While these descriptors are accurate, they have fallen out of frequent use. Consequently, the term nanoemulsion has emerged as the most accepted and commonly utilized for dispersions with small droplet sizes (see Figure 8.14).[94]

8.3.2 Physicochemical Aspects of Nanoemulsions Formation and Stability

Nanoemulsions exhibit the typical properties of emulsions formulated in surfactant–oil–water systems, and thus follow the trends when using the HLD_N equation and the understanding of emulsion properties through formulation–composition maps (HLD_N vs. WOR).[36] However, there are some particular aspects in nanoemulsion physicochemical behavior, mainly related to the surfactant concentration and stabilization mechanisms, worth mentioning.[36,92]

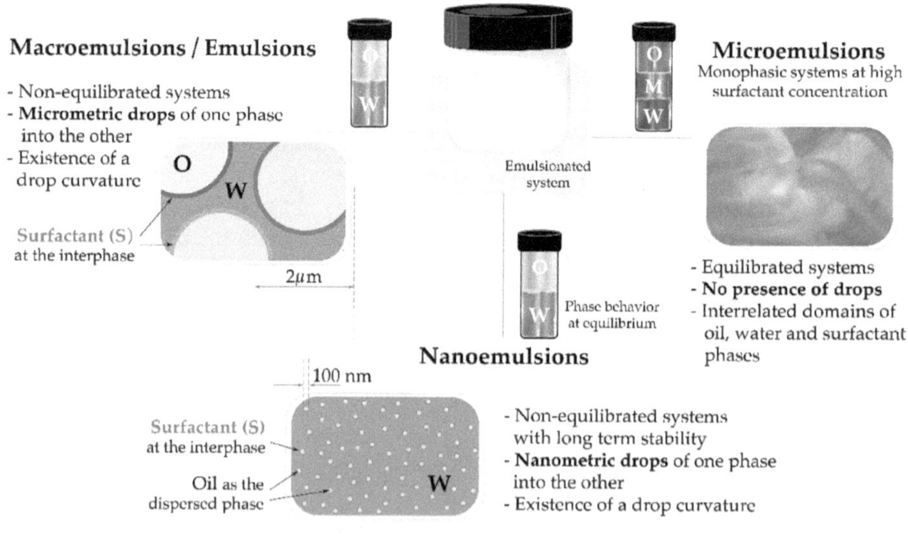

FIGURE 8.14 Differences between micro-, nano-, and macroemulsions.

Nanoemulsions exhibit a significantly larger interfacial area than macroemulsions due to their smaller droplet size; therefore, a higher quantity of surfactant is required for their stabilization. The surfactant quantity can be around 0.5–1.0 wt% when high shear energy methods are applied, and up to 2–4 wt% when low-energy emulsification are used.[92,94,100] However, the surfactant concentration needed for nanoemulsions remains lower than that required for the total co-solubilization of oil and water in bicontinuous microemulsions equilibrated systems, which usually is higher than 10 wt%.[94,96,101,102]

In nanoemulsions, the Laplace pressure within the droplets is considerably high, which makes the process of shearing energy demanding, to induce either elongation or deformation. Also, given their small droplet size and consequently large interfacial area, the production of these dispersed systems necessitates a large energy expenditure to generate such an area. This makes challenging to break them into smaller droplets, and therefore one of the methods to obtain them with decreased energy is to reduce significantly the water–oil interfacial tension.[37,92,103–105] It is noteworthy that when mechanical energy is applied to the system to produce droplet shearing, only a minor fraction, approximately 1% or less, is utilized for droplet formation. The remainder of the energy, provided by the emulsifying apparatus (be it a mixer, turbine, etc.), is largely dissipated as heat, representing a substantial efficiency loss in the process.[106,107]

The submicrometric droplet size of nanoemulsions confers a translucent, bluish appearance, and, depending on the droplet size distribution and the internal phase content, they can be near transparent.[97] Destabilization predominantly occurs through Ostwald ripening, generating an increase in droplet size, and subsequent rise in turbidity, usually to a whitish color as time progresses.[100] The relative stability of nanoemulsions can be from days to several months; however, the large interfacial area and the surfactant concentration employed significantly enhance their stability when compared to macroemulsions.[94,96,99,104] Due to their prolonged stability, nanoemulsions have emerged as highly attractive systems for pharmaceutical, cosmetic,[100,108,109] and edible applications.[106,110] The formation and stability of nanoemulsions are influenced by several factors: the physicochemical properties, composition of the surfactant–oil–water system, and the rheo-mechanical characteristics of the emulsification process.

In nanoemulsions, the specific surface area per unit volume of the dispersed phase inversely correlates with droplet size. The formation of droplets is related to the ΔA term or generation of interfacial area in a liquid–liquid (immiscible) system. As a result, energy must be supplied to the system,[107] and for this purpose, processes with high-energy input have been traditionally used in industrial settings.[107,109] The formation of nanoemulsions via these methods is relatively straightforward, as increased energy input generally leads to smaller droplet size. However, the level of energy required to achieve nanometer-scale droplets is significantly high, rendering the process cost-inefficient.[37,107,111] This has enabled low-energy emulsification methods to be applied. These methods are based on the physicochemical behavior of the system, and can be explained in formulation vs. composition maps, fish diagrams, or pseudoternary diagrams.[36,96,102,111,112]

8.3.3 Low-Energy Emulsification Methods

Low-energy emulsification methods, with a foundation on the physicochemical formulation, decrease in interfacial tension, formation of nano- microstructures, such as liquid crystals, spontaneous emulsification, or even droplets freezing, tend to be more energy-efficient than the high-energy ones.[113] These methods require only simple stirring and a change in temperature or composition,[105] typically producing smaller droplet sizes than high-energy methods.[92,94,114–117] These methods may be the most attractive when a decrease in the carbon footprint in emulsification processes is the goal. While there are works reviewing high-energy methods [107,118] and low-energy emulsification processes,[94,100,117,119] this section aims to provide an up-to-date, critical, and comprehensive overview of the generation of nanoemulsions via low-energy emulsification processes.

The production of nanoemulsions through low-energy methods relies on attaining low interfacial tension or promoting a near-zero curvature at the interface.[92,94] The first of these cases occur when the formulation is near the optimum ($HLD_N = 0$), wherein the interfacial tension is ultralow,[78,102] which can be schematically described in a formulation vs. composition diagram. The second case occurs when the curvature nears zero, e.g., when a liquid crystalline phase is attained through dilution, which can be represented in a pseudo-ternary diagram.[105,120–122] These methods involve manipulating formulation (e.g., temperature) or composition to assemble nanoscale oil droplets (see Table 8.2). This can involve techniques such as PIT, wherein a temperature change induces a phase transition, or dilution methods, which involve the gradual

TABLE 8.2
Summary of Low-Energy Nanoemulsion Formation Methods

Method	Mechanism	Application	Recent Advances	References
Phase inversion temperature (PIT)	Triggered by temperature changes altering surfactant curvature	Wide range of industries, including pharmaceuticals and cosmetics	Focus on small droplet size and low polydispersity; scale-up for industrial applications	120,124–127,127–129
Dilution methods: phase inversion composition (PIC)	Triggered by composition changes altering surfactant curvature	Pharmaceuticals, cosmetics, and broader industrial applications	Investigation into structural transformations; potential for large-scale production	94,102,105,109,130–132
Spontaneous emulsification	Diffusion of surfactant/solvent molecules without change in surfactant curvature	Pharmaceuticals, especially as carriers for lipophilic drugs	Optimization for low energy costs and effective drug delivery	103,133–138

addition of one phase into another. These approaches can induce structural instabilities (e.g., interfacial gradients or Laplace pressure gradients) which effectively result in a $\Delta G < 0$, indicative of spontaneous emulsification.[122,123] In Table 8.2 a literature review of the main works on nanoemulsions with low-energy methods is reported.

8.3.4 Phase Inversion Temperature Method

The PIT, introduced by Shinoda 50 years ago,[129,139] and used in the pioneering work by Förster et al.[126,140] to obtain nanoemulsions, involves transitional inversion due to a change in HLD_N (temperature) at constant surfactant concentration and water–oil ratio.[124,127,141,142] The understanding of the PIT mechanism has evolved over the years,[143] and it is currently accepted that the main structures responsible for nanodroplet formation are those with zero spontaneous curvature.[94,105,117] Morales et al.[132] showed that it is of major importance that the oil and surfactant coexist in the same phase at the PIT, and there could be some excess of water in the system. Other authors have indicated that for some systems, it is not necessary to attain exactly the PIT temperature, and the small droplet formation can be satisfied at the vicinity of the PIT.[120]

A schematic representation of the PIT method to form O/W nanoemulsions can be observed in Figure 8.15.[117,126,128] The PIT method is applicable only to surfactants sensitive to temperature changes, i.e., nonionic polyethoxylated surfactants in which changes in temperature induce a change in the hydration of the polar head, and a consequent change of its curvature.[95,144] Through the change in HLD_N (vertical trajectory), the $HLD_N = 0$ line is crossed, and the phase behavior changes from WI (negative HLD_N values) to WII (positive HLD_N values), or vice versa.[28,37,92] In other words, the system continuously changes from an external water phase, through a microemulsion, to an external oil phase and vice versa through a mass transfer process. However, it is worth noting that in most cases these nanodroplets would coalesce since the region near zero HLD_N is intrinsically associated with extremely unstable emulsions.[36,145]

If the phase behavior near the optimum formulation is monophasic (WIV), all droplets formed originate from the exudation of excess phases, i.e., from the transformation of microemulsion domains into droplets, hence all droplets formed are nanometric in size.[92,117] Conversely, if the phase behavior is Winsor III in the optimum formulation, two types of droplets form during the transition. The first category of nanodroplets are the droplets exuded from the microemulsion as the formulation deviates from the optimum, and the droplets that form due to the agitation of the ultralow tension system.[92,146] The second category of droplets (larger size) occurs because the low agitation energy is ineffective in producing sub-micrometric droplets. This leads to a polydisperse droplet system, sometimes with a bimodal distribution.

Roger et al.[120] have revealed that below the PIT, the emulsion may remain homogeneous, irrespective of the traversed zone is WIII or WIV at equilibrium. Moreover, nanoemulsions were obtained even at temperatures lower than PIT. This observation led to the identification of a 'clear boundary' temperature, below the PIT, where

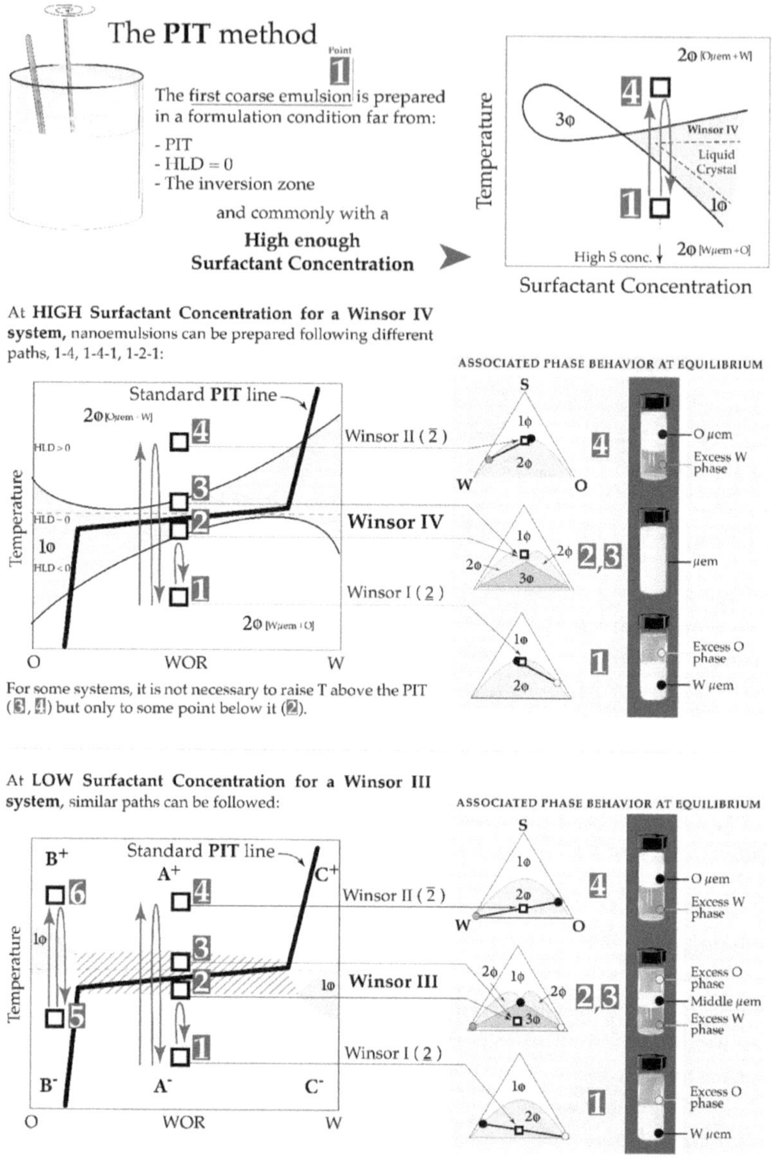

FIGURE 8.15 Schematic representation of the PIT method. At high surfactant concentration, the stirred system is heated so that the temperature change induces the phase transition from O/W (1) to W/O (4), passing through a single-phase (1ϕ , Winsor IV) bicontinuous microemulsion or a lamellar liquid crystal phase. Depending on the desired final nanoemulsion type, the system could be returned to O/W domains. *HLD* < 0. At low surfactant concentration, similar paths can be followed with some regard about solubilization to overcome the instability. At high oil content, a single-phase region can be crossed through (5) to (6), and nanoemulsions are also formed.

nanoemulsions exhibit similar sizes to those formed at or above the PIT when cooled from this clear boundary, termed the sub-PIT method. On the other hand, the temperature vs. surfactant concentration 'fish' diagrams have been particularly useful to determine the optimal condition for PIT emulsification.[40,145,147] In recent years, the PIT method has been successfully used to produce essential oil nanoemulsions with nonionic surfactants, but also to produce nanoemulsions in the presence of cationic surfactant.[148,149] The PIT emulsification technique has also been used for the preparation of nanoemulsions containing monomer droplets, as precursors for polymeric latex particles after an additional polymerization step, among other applications.[150] Besides temperature, other formulation variable used in nanoemulsions formulation is change the nature of the surfactant, by adding a hydrophilic surfactant to a W/O emulsion.[151]

8.3.5 Dilution Methods: The Phase Inversion Composition (PIC) Method

Dilution methods, known as phase inversion composition or emulsion inversion point (EIP),[94,102,152] are based on the gradual dilution of one phase containing a surfactant into another immiscible phase, involving an indirect change on the HLD_N caused by a change in the water/oil ratio (WOR).[92,117] As previously explained, partition of different monomers of technical-grade surfactants produces formulation changes with both the surfactant concentration and the water fraction (see Figures 8.4 and 8.12).

The PIC method was first described by Lin et al. and Dickinson at the end of the 1970s and the beginning of 1980s.[153–155] Traditionally, the PIC method entails the addition of an aqueous phase into an oil phase that already contains a dispersed surfactant, as depicted in top and medium schemes in Figure 8.16. Subsequently, Marszall and Sagitani [156–158] utilized what was named at the time EIP method, in which the phase inversion of water-in-oil (W/O) emulsions to oil-in-water (O/W) emulsions occurs by adding water to systems consisting of water, surfactant, and hydrocarbon, without the need for strong agitation and at a constant temperature. Further studies explain that by adding water to a surfactant solution in oil, a W/O emulsion forms, and by adding more water, the surfactant diffuses to the aqueous phase, forming the O/W emulsion.[102,105,131] Sagitani showed that formation of a lamellar liquid crystal phase during water addition is crucial to obtain O/W emulsions of low polydispersity and small droplet size.[157,158]

Several works have shown that the transition from water-in-oil (WII) microemulsion to bicontinuous microemulsion or lamellar liquid crystalline structures occurs as the polar moieties become hydrated.[94] This transition is marked by a shift in the interfacial spontaneous curvature from negative to zero, corresponding to these structural transitions. Finally, the curvature of the surfactant adsorption layer is inverted, thereby promoting the formation of oil in water droplets, which may be of nanometer size according to the applied conditions.[92,117] The formation of a bicontinuous (cubic or sponge) or lamellar (LαLC) phase has been mentioned as a critical step for nanoemulsification using this technique.[121] The formation of transient viscous phases during emulsification necessitates careful control of the rate of water addition, and

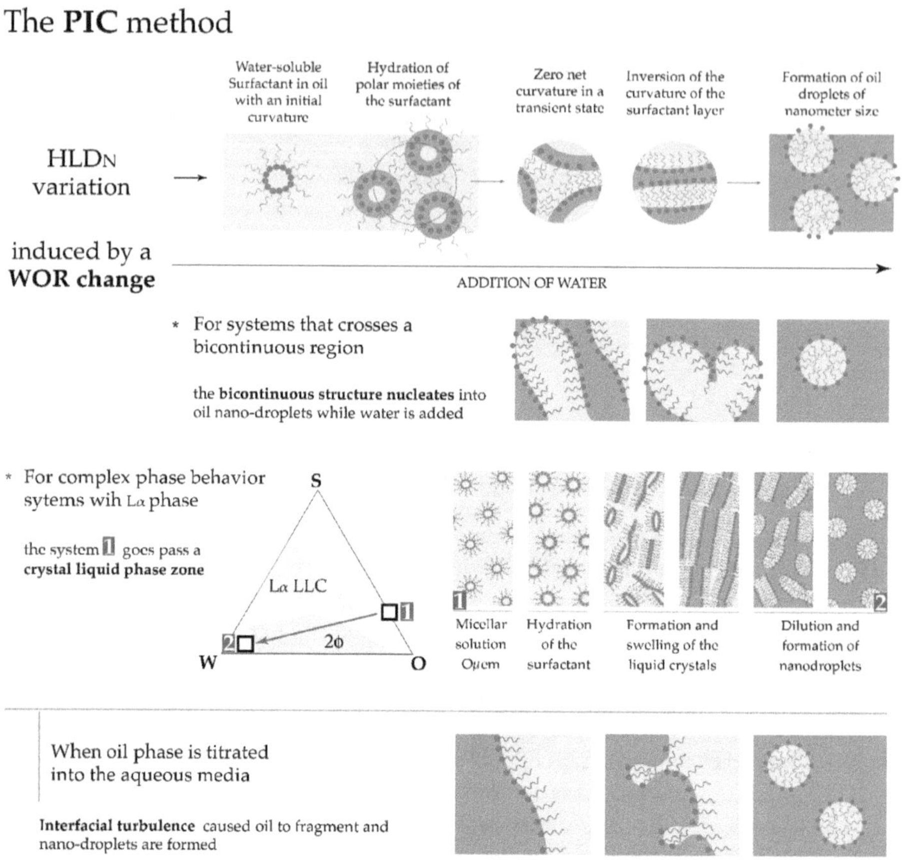

FIGURE 8.16 *Top panel:* Schematic representation of the PIC method. Formulation of O/W nanodroplets: general approach showing the hydration upon adding water changing the surfactant packing parameter and the preferred curvature. *Medium panel:* Specific structures when continuous water addition result in bicontinuous or lamellar phase. *Bottom panel:* Formulation of W/O nanodroplets when adding nonpolar oil into surfactant–water systems.

the intensity of shearing to achieve homogenization. The formation of highly viscous cubic liquid crystal phases can lead to either monomodal or bimodal size distributions in the resulting nanoemulsions, depending on the process parameters.[92]

Water-in-oil nanoemulsions have also been formulated by this method.[159] The occurrence of lamellar phase or bicontinuous structures during the oil addition has been pointed out at the origin of W/O nanoemulsions, analogous to the described O/W systems. However, Cholakova et al.[117] indicate that when the oil phase is nonpolar and titrated into the aqueous media (bottom scheme in Figure 8.16), the main mechanism described to cause W/O nanoemulsion formation is the presence of interfacial turbulence. This interfacial turbulence results in the fragmentation of oil into nanodroplets through the interfacial instability, driven by both surfactant transfer across the interface and Marangoni effects.[117,160]

Emulsifying essential oils or triglycerides with unsaturated fatty acid chains presents challenges due to their polar nature and the generation of steric hindrance, which leads to a bulky configuration.[161,162] Phospholipids or other food-grade surfactants, such as sorbitan esters, and also sucroesters have been utilized to obtain nanoemulsions.[92,163–167] Additionally, triglycerides with shorter fatty acid chains, including medium-chain triglycerides (MCT), have been employed in various formulations. Soy lecithin, with its negative *SCP* and hydrophilic structure, has been demonstrated to work effectively with the PIC method in a mixture of several phosphatidylcholines. The formation of a liquid crystalline phase in SOW systems has been achieved using triglyceride oils by employing different surfactants, among them sorbitan ester surfactants such as Tween and Span. In the HLD_N perspective, to formulate O/W nanoemulsion, the surfactant mixture must have an HLD_N less than 0. This mixture should be at some distance from the optimum but not too far from $HLD_N = 0$, facilitating the creation of a stable O/W emulsion.[123,131,168,169] This emulsion is purportedly stabilized by a liquid crystal layer encapsulating the droplet. The emulsification process primarily depends on two factors: achieving a formulation at $HLD_N < 0$, and close to $HLD_N = 0$,[166,167] which allows for the formation of a hydrophilic surfactant in oil system (point 1 in the ternary diagram in Figure 8.16) that evolves to a zero-curvature lamellar liquid upon slight water addition, and, upon further water addition, the composition shifts toward the W corner, causing the LC to swell into layers that break under low shear to form small, elongated aggregates or nanodroplets.[25,102,166,168,170–172] An additional observation is that diluting the nonionic surfactant solution, hence reducing its concentration, increases the hydrophilicity of the interfacially adsorbed ethoxylated mixture. Consequently, the system's HLD_N decreases, where stability is enhanced. Additionally, this type of emulsification allows obtaining functionalized systems by encapsulating substances with a specific property, e.g., antioxidant, anesthetic, analgesic, solar filter, etc., when formulation principles are applied properly.[108,169,173–175]

8.3.6 Spontaneous Emulsification

Spontaneous emulsification (schematized in Figure 8.17), also known as self-emulsification, has been extensively utilized across various applications, including agricultural formulations, cosmetics, and cleansing products.[123,170,176] This process is triggered by the rapid diffusion of surfactant and/or solvent molecules from the dispersed phase to the continuous phase, facilitating the formation of nanoemulsions without requiring external energy input.[136] It occurs when small droplets of one liquid spontaneously disperse into another liquid without requiring mechanical energy. In this method, the surfactant is first dissolved in the phase where its solubility is lower.[177] Upon contact with another liquid, spontaneous surfactant transfer occurs, which, under suitable conditions, may lead to the formation of nanoemulsions.[170,176,178] Various factors, such as temperature, pH, salt concentration, and the structure and concentration of the surfactant, have been employed to modify the solubility of the surfactant in the involved liquids.[123]

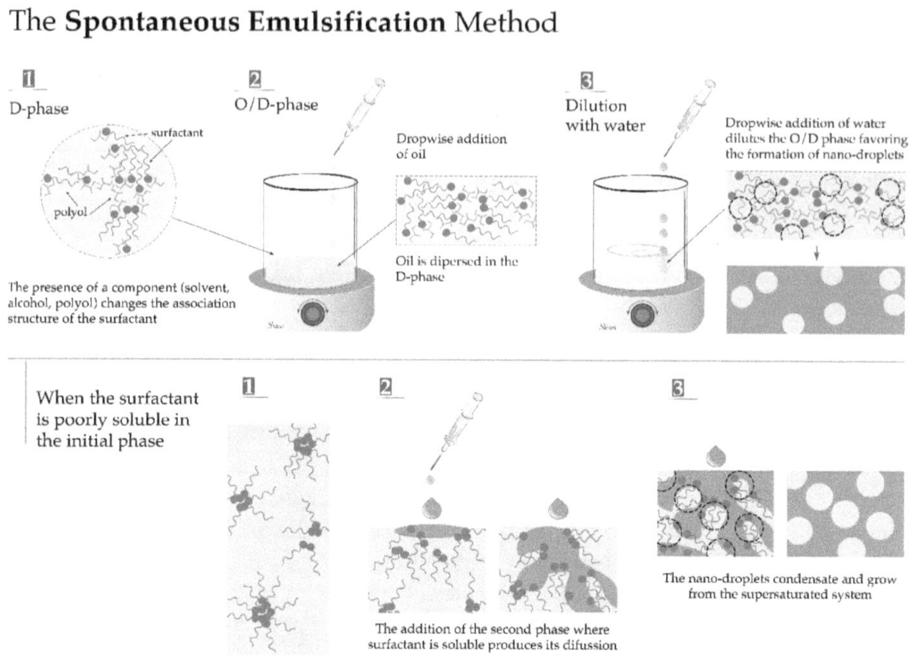

FIGURE 8.17 Scheme of the potential mechanism for formation of nanosized oil droplets through the spontaneous emulsification method.

The concept was first introduced in 1978, focusing on systems in which oils containing fatty acids 'spontaneously' form emulsions upon introduction to alkaline water.[179] In these systems, a substance dissolved in one phase, but more soluble in the other, migrates across the interface, carrying with it a portion of the original solvent into the second liquid, resulting in dispersion in an emulsified form (see Figure 8.17). This process might lead to a significant decrease in interfacial tension, depending on whether the diffusion is facilitated by a surfactant or merely a cosolvent.[92,123] The underlying phenomena include the oil-phase diffusion into the aqueous phase, facilitated by the surfactant migrating into the contacted aqueous phase. Oil droplets emerge from the interface due to interfacial turbulence, potentially accompanied by low interfacial tension. Stabilization primarily occurs through the aggregation and growth of droplets in the supersaturated system, aided by surfactants, or through the formation of lamellar liquid crystal phases when surfactant and oil mixtures contact water. These structures eventually break down as the aqueous content increases, leading to the formation of nanodroplets.[123,178] Figure 8.17 shows a potential mechanism for the spontaneous emulsification method. In this mechanism, a bicontinuous microemulsion phase forms at the barrier where the organic and aqueous phases meet, leading to the spontaneous formation of nanodroplets when the bicontinuous microemulsion phase breaks up.

Finally, it is important to point out the so-called 'ouzo effect', even if this phenomenon doesn't need a surfactant and it is beyond the scope of the chapter. In some ternary systems, a cosolvent coexist with a polar oil and water results from a significant supersaturation of the oil, facilitating the nucleation of oil droplets when combined with water. Consequently, instantaneous diffusion of the oil to the closest droplet occurs, reducing the supersaturation to prevent any further nucleation.[180,181] This method has been applied for different applications, including cleansing formulations, cosmetics, and even generation of membranes by polymerization of a monomer oil.

8.3.7 The Cold-Bursting Method

The cold-bursting method introduced by Cholakova et al.[182,183,117,184] utilizes the *polymorphic* phase transitions of lipids to create nanodroplets from larger lipid droplets, initially up to 100 μm in size. This technique involves cooling and heating of lipid dispersions around the oil's melting temperature or maintaining the dispersion at a temperature slightly below the melting point for extended periods. Through a series of cooling and heating cycles, it is possible to produce nanoemulsions or nanoparticles with diameters of 20 nm. During cooling, lipids molecules tend to initially form in the less stable α-phase, due to its lower nucleation barrier. This phase then transitions into the more stable β′- or β-phase, or undergoes a solid-state transition, particularly after long-term storage at temperatures below the α-phase's melting point.

In these lipid systems, the mass density is lower for the α-phase compared to the β-phase. This leads to the shrinking of crystalline domains during the transition from α-phase to β-phase, creating nanovoids at the grain boundaries within the frozen triglyceride particles (the so-called 'negative pressure effect').[117,184] Initially, a coarse O/W emulsion is formulated and then freeze into a polycrystalline structure composed of α-phase domains. Upon heating at a specific slow rate above the α-phase's melting temperature, the transition from α-phase to β-phase occurs. This transition creates a structure of β-domains separated by nanopores within the large lipid particles. The negative pressure effect then draws the surrounding aqueous surfactant solution into the particle, causing it to swell and ultimately fragmenting into individual β-crystalline nanoparticles that finally evolve to nanodroplets.[117,184] Hence, researchers have successfully identified suitable combinations of oils (monoacid and natural triglycerides, diglycerides, alkanes) and surfactants (nonionics and ionics) that produced nanoemulsions by the cold-busting method.[182–184] If conditions are modified (i.e., heating/cooling rate, initial phase of the surfactant, nature of surfactant), the final system can be a W/O/W multiple emulsion and not a nanoemulsion.[184]

8.4 CONCLUSIONS AND PERSPECTIVES

The hydrophilic–lipophilic deviation semiempirical equation represents adimensionally the formulation variables for SOW systems at equilibrium. Changing the phase behavior from WI to WII, through WIII or WIV zones, is possible by utilizing unidimensional scans, taking into account the variables in the equation. Obtaining the numerical value of the parameters to achieve this remains the principal challenge for formulators, and efforts must be undertaken to characterize more surfactants and oils using the HLD_N approach. This is crucial to describe the behavior of biobased surfactants that aim to partially replace petrochemical-based molecules. The influence of formulation variables on emulsion formulation is undeniable, and the know-how presented using the formulation–composition maps to adjust the morphology, droplet size, viscosity, and stability of emulsions point out the importance of HLD_N and the water/oil proportion, following the Bancroft and Ostwald rules. Transitional and catastrophic phase inversions are well-established methods with industrial applications, and formulation engineering has allowed to obtain emulsions with specific properties using these tools. The improved stability of nanoemulsions and their potential applications in several domains have stimulated a lot of research in recent decades. In the context of energy transition, their formulation with low-energy methods is particularly relevant, and once again, using emulsion phase inversion allows, in some cases, to obtain these systems. The PIT and/or the sub-PIT method, the PIC or EIP, spontaneous emulsification, and the recent cold-bursting method are some of the tools that allow formulators to obtain nanoemulsions. However, research efforts must still be made to fully understand the intrinsic mechanisms that govern some of these methods and to understand the relationship between the occurrence of nanoemulsions, the chemical nature of the surfactants and oils, and the process conditions.

ACKNOWLEDGMENTS

The authors would like to express their gratitude to their colleagues and former coauthors from research centers in Venezuela and France for insightful and enlightening discussions related to the formulation of micro-, nano-, and macroemulsions. They are Jean Louis Salager, Johnny Bullón, and Ana Forgiarini (Venezuela), Valérie Molinier, Dominique Langevin, Véronique Nardello-Rataj, Christel Pierlot, and Jean-Marie Aubry (France). The authors are thankful to Professor Laura Tolosa for contributing the graphical art in the nanoemulsion section of the chapter. The authors also wish to thank all the Ph.D. students and researchers from the FIRP Laboratory (University of Los Andes), and the CISCO team (University of Lille), who have contributed to the body of knowledge on the subject.

REFERENCES

1. Schulman, J.H.; Stoeckenius, W.; Prince, L.M. Mechanism of formation and structure of micro emulsions by electron microscopy. *J. Phys. Chem.* **1959**, *63*, 1677–1680.
2. Stubenrauch, C. *Microemulsions: Background, New Concepts, Applications, Perspectives*, Stubenrauch, C., Ed.; John Wiley & Sons Ltd: Oxford, UK, 2008.
3. Langevin, D. *Emulsions, Microemulsions and Foams*, Springer: Gewerbestr, Switzerland, 2020.
4. Salager, J.-L. Emulsion Properties and related know-how to attain them. In *Pharmaceutical Emulsilons and Suspensions*, Nielloud, F., Marti-Mestres, G., Eds.; Marcel Dekker: New York, NY, USA, 2000; pp. 73–125.

5. Rosen, M.J.; Kunjappu, J.T. *Surfactants and Interfacial Phenomena*, John Wiley & Sons: New York, USA, 2012.
6. Lindman, B.; Olsson, U.; Soderman, O. Characterization of microemulsions by NMR. In *Handbook of Microemulsion Science and Technology*, Kumar, P., Mittal, K.L., Eds.; CRC Press: Boca Ratón, FL, USA, 1999; pp. 309–356.
7. Tartaro, G.; Gentile, L.; Palazzo, G. Characteristic length and curvature of the AOT/Brine/Squalane "sponge" L3 phases. *JCIS Open* **2023**, *9*, 100077.
8. Gradzielski, M.; Duvail, M.; de Molina, P.M.; Simon, M.; Talmon, Y.; Zemb, T. Using microemulsions: Formulation based on knowledge of their mesostructure. *Chem. Re.* **2021**, *121*, 5671–5740.
9. Dufrêche, J.F.; Zemb, Th. Bending: From thin interfaces to molecular films in microemulsions. *Curr. Opin. Colloid Interface Sci.* **2020**, *49*, 133–147.
10. Scriven, L.E. Equilibrium bicontinuous structure. *Nature* **1976**, *263*, 123–125.
11. Hyde, S.; Ninham, B.; Andersson, S.; Larsson, K.; Landh, T.; Blum, Z.; Lidin, S. *The Language of Shape: The Role of Curvature in Condensed Matter: Physics, Chemistry and Biology*, Elsevier Science B.V.: Amsterdam, The Netherlands, 1997.
12. Wennerstroem, H.; Olsson, U. On the flexible surface model of sponge phases and microemulsions. *Langmuir* **1993**, *9*, 365–368.
13. Strey, R. Microemulsion microstructure and interfacial curvature. *Colloids Polym. Sci.* **1994**, *272*, 1005–1018.
14. Salager, J.-L.; Marquez, R.; Rondón, M.; Bullón, J.; Graciaa, A. Review on some confusion produced by the bicontinuous microemulsion terminology and its domains microcurvature: A simple spatiotemporal model at optimum formulation of surfactant-oil-water systems. *ACS Omega* **2023**, *8*, 9040–9057.
15. Griffin, W. Calculation of HLB values of non-ionic surfactants. *J. Soc. Cosmet. Chem.* **1954**, *5*, 248.
16. Griffin, W. Classification of surface-active agents by "HLB". *J. Soc. Cosmet. Chem.* **1949**, *1*, 311–326.
17. Winsor, P. *Solvent Properties of Amphiphilic Compounds*, Butterworths Scientific Publications: London, UK, 1954.
18. Winsor, P. Binary and multicomponent solutions of amphiphilic compounds. solubilization and the formation, structure, and theoretical significance of liquid crystalline solutions. *Chem. Rev.* **1968**, *68*, 1–40.
19. Salager, J.-L.; Forgiarini, A.M.; Márquez, L.; Manchego, L.; Bullón, J.; Marquez, L.; Manchego, L.; Bullon, J. How to attain an ultralow interfacial tension and a three-phase behavior with a surfactant formulation for enhanced oil recovery: A review. Part 2. Performance improvement trends from Winsor's premise to currently proposed inter- and intra-molecular mixture. *J. Surfactants Deterg.* **2013**, *16*, 631–663.
20. Salager, J.-L.; Marquez, R.; Ontiveros, J.F. How to use in practice a simplified HLDN linear equation for surfactant mixtures. In *Proceedings of the 2022 AOCS annual meeting*; AOCS Press: Atlanta, GA, USA, 2022; p. 3166.
21. Davies, J.T. A quantitative kinetic theory of emulsion type. I. Physical chemistry of the emulsifying agent. *Proceedings of the Second International Congress on Surface Activity;* Butterworth: London, UK, 1957; p. 426.
22. Shinoda, K. The correlation between the dissolution state of nonionic surfactant and the type of dispersion stabilized with the surfactant. *J. Colloid Interface Sci.* **1967**, *24*, 4–8.
23. Salager, J.-L. Quantifying the concept of physico-chemical formulation in surfactant-oil-water systems—state of the art. *Progr. Colloid Polym. Sci.* **1996**, *100*, 137–142.
24. Marquez, N.; Anton, R.E.; Graciaa, A.; Lachaise, J.; Salager, J.L. Partitioning of ethoxylated alkylphenol surfactants in microemulsion-oil-water systems. *Colloids Surf. A* **1995**, *100*, 225–231.

25. Salager, J.-L.; Antón, R.E.; Bullón, J.; Forgiarini, A.; Marquez, R. How to use the normalized Hydrophilic-Lipophilic Deviation (HLDN) concept for the formulation of equilibrated and emulsified surfactant-oil-water systems for cosmetics and pharmaceutical products. *Cosmetics* **2020**, *7*, 57.
26. Salager, J.-L.; Marquez, N.; Graciaa, A.; Lachaise, J. Partitioning of ethoxylated octylphenol surfactants in microemulsion-oil-water systems: Influence of temperature and relation between partitioning coefficient and physicochemical formulation. *Langmuir* **2000**, *16*, 13, 5534–5538.
27. Salager, J.-L. *Physico-Chemical Properties of Surfactant-Water-Oil Mixtures: Phase Behavior, Micro-Emulsion Formation and Interfacial Tension*, University of Texas at Austin: Austin, TX, USA, Ph.D. Dissertation, 1977.
28. Aubry, J.M.; Ontiveros, J.F.; Salager, J.-L.; Nardello-Rataj, V. Use of the normalized Hydrophilic-Lipophilic-Deviation (HLDN) Equation for Determining the Equivalent Alkane Carbon Number (EACN) of oils and the Preferred Alkane Carbon Number (PACN) of nonionic surfactants by the Fish-Tail Method (FTM). *Adv. Colloid Interface Sci.* **2020**, *276*, 102098.
29. Sottmann, T.; Strey, R. Ultralow interfacial tensions in water–n-alkane–surfactant systems. *J. Chem. Phys.* **1997**, *106*, 8606–8615.
30. Cayias, J.L.; Schechter, R.S.; Wade, W.H. Modeling crude oils for low interfacial tension. *Soc. Petrol. Eng. J.* **1976**, *16*, 351–357.
31. Marquez, R.; Ontiveros, J.F.; Barrios, N.; Tolosa, L.; Palazzo, G.; Nardello-Rataj, V.; Salager, J.L. Advantages and limitations of different methods to determine the optimum formulation in surfactant–oil–water systems: A review. *J. Surfactants Deterg.* **2024**, *27*, 5–36.
32. Huh, C. Interfacial tensions and solubilizing ability of a microemulsion phase that coexists with oil and brine. *J. Colloid Interface Sci.* **1979**, *71*, 408–426.
33. Rico, F.; Mazabel, A.; Egurrola, G.; Pulido, J.; Barrios, N.; Marquez, R.; García, J. Meta-analysis and analytical methods in cosmetics formulation: A review. *Cosmetics* **2024**, *11*, 1.
34. Lemahieu, G.; Aguilhon, J.; Strub, H.; Molinier, V.; Ontiveros, J.F.; Aubry, J.-M. Hexahydrofarnesyl as an original bio-sourced alkyl chain for the preparation of glycosides surfactants with enhanced physicochemical properties. *RSC Adv.* **2020**, *10*, 16377–16388.
35. Salager, J.-L.; Antón, R.E.; Anderez, J.M.; Aubry, J.-M. Formulation des micro-émulsions par la methode HLD. *Techniques de l'Ingénieur. Génie des Procédés* **2001**, *J2*, 157.1–157.20.
36. Delforce, L.; Ontiveros, J.F.; Nardello-Rataj, V.; Aubry, J.-M. Rational design of O/W nanoemulsions based on the surfactant dodecyldiglyceryl ether using the normalized HLD concept and the formulation-composition map. *Colloids Surf. A* **2023**, *671*, 131678.
37. Salager, J.-L.; Marquez, R.; Bullon, J.; Forgiarini, A. Formulation in surfactant systems: From-Winsor-to-HLDN. *Encyclopedia* **2022**, *2*, 778–838.
38. Cayias, J.L.; Schechter, R.S.; Wade, W.H. The utilization of petroleum sulfonates for producing low interfacial tensions between hydrocarbons and water. *J. Colloid Interface Sci.* **1977**, *59*, 31–38.
39. Cash, L.; Cayias, J.L.; Fournier, G.; Macallister, D.; Schares, T.; Schechter, R.S.; Wade, W.H. The application of low interfacial tension scaling rules to binary hydrocarbon mixtures. *J. Colloid Interface Sci.* **1977**, *59*, 39–44.
40. Queste, S.; Salager, J.L.; Strey, R.; Aubry, J.M. The EACN scale for oil classification revisited thanks to fish diagrams. *J. Colloid Interface Sci.* **2007**, *312*, 98–107.

41. Graciaa, A.; Lachaise, J.; Sayous, J.G.; Grenier, P.; Yiv, S.; Schechter, R.S.; Wade, W.H. The partitioning of complex surfactant mixtures between oil/water/microemulsion phases at high surfactant concentrations. *J. Colloid Interface Sci.* **1983**, *93*, 474–486.
42. Ben Ghoulam, M.; Moatadid, N.; Graciaa, A.; Lachaise, J. Quantitative effect of nonionic surfactant partitioning on the hydrophile–lipophile balance temperature. *Langmuir* **2004**, *20*, 2584–2588.
43. Graciaa, A.; Lachaise, J.; Bourrel, M.; Osborne-Lee, I.; Schechter, R.S.; Wade, W.H. Partitioning of nonionic and ionic surfactant mixtures between oil-microemulsion-water phases. *SPE Reserv. Eng.* **1984**, *10*, 305.
44. Graciaa, A.; Andérez, J.; Bracho, C.; Lachaise, J.; Salager, J.-L.; Tolosa, L.; Ysambertt, F. The selective partitioning of the oligomers of polyethoxylated surfactant mixtures between interface and oil and water bulk phases. *Adv. Colloid Interface Sci.* **2006**, *123–126*, 63–73.
45. Anton, R.E.; Andérez, J.M.; Bracho, C.; Vejar, F.; Salager, J.L. Practical surfactant mixing rules based on the attainment of microemulsion–oil–water three-phase behavior systems. In *Interfacial Processes and Molecular Aggregation of Surfactants*, Narayanan, R., Ed.; Advances in Polymer Science; Springer: Berlin, Germany, 2008; Volume 218, pp. 83–113.
46. Engelskirchen, S.; Elsner, N.; Sottmann, T.; Strey, R. Triacylglycerol microemulsions stabilized by alkyl ethoxylate surfactants—a basic study. *J. Colloid Interface Sci.* **2007**, *312*, 114–121.
47. Burauer, S.; Sottmann, T.; Strey, R. Nonionic microemulsions with cyclic oils. *Tenside Surfactants Deterg.* **2000**, *37*, 8–16.
48. Graciaa, A.; Lachaise, J.; Cucuphat, C.; Bourrel, M.; Salager, J.L. Interfacial segregation of an ethyl oleate/hexadecane oil mixture in microemulsion systems. *Langmuir* **1993**, *9*, 1473–1478.
49. Szekeres, E.; Acosta, E.; Sabatini, D.A.; Harwell, J.H. A two-state model for selective solubilization of benzene–limonene mixtures in sodium dihexyl sulfosuccinate microemulsions. *Langmuir* **2004**, *20*, 6560–6568.
50. Kiran, S.K.; Acosta, E.J.; Moran, K. Evaluating the hydrophilic–lipophilic nature of asphaltenic oils and naphthenic amphiphiles using microemulsion models. *J. Colloid Interface Sci.* **2009**, *336*, 304–313.
51. Ghayour, A.; Acosta, E. Characterizing the oil-like and surfactant-like behavior of polar oils. *Langmuir* **2019**, *35*, 15038–15050.
52. Chen, C.; Shen, H.; Harwell, J.H.; Shiau, B.-J. Characterizing oil mixture and surfactant mixture via Hydrophilic-Lipophilic Deviation (HLD) principle: An insight in consumer products development. *Colloids Surf. A* **2022**, *634*, 127598.
53. Salager, J.L.; Loaiza-Maldonado, I.; Minana-Perez, M.; Silva, F. Surfactant-oil-water systems near the affinity inversion part I: Relationship between equilibrium phase behavior and emulsion type and stability. *J. Disper. Sci. Technol.* **1982**, *3*, 279–292.
54. Pickering, S.U. CXCVI.—Emulsions. *J. Chem. Soc. Trans.* **1907**, *91*, 2001–2021.
55. Esquena, J. Water-in-water (W/W) emulsions. *Curr. Opin. Colloid Interface Sci.* **2016**, *25*, 109–118.
56. Zia, A.; Pentzer, E.; Thickett, S.; Kempe, K. Advances and opportunities of oil-in-oil emulsions. *ACS Appl. Mat. Interfaces* **2020**, *12*, 38845–38861.
57. Ostwald, W. Beiträge Zur Kenntnis Der Emulsionen. *Zeitschrift für Chemie und Industrie der Kolloide* **1910**, *6*, 103–108.
58. Bancroft, W.D. The theory of emulsification, V. *J. Phys. Chem.* **1913**, *17*, 501–518.
59. Marquez, R.; Forgiarini, A.; Langevin, D.; Salager, J.-L. Breaking of water-in-crude oil emulsions. Part 8. New interfacial rheology characteristics measured using a spinning drop rheometer at optimum formulation. *Energ. Fuels* **2019**, *33*, 8151–8164.

60. Firouzi, M.; Kovalchuk, V.I.; Loglio, G.; Miller, R. Salt effects on the dilational viscoelasticity of surfactant adsorption layers. *Curr. Opin. Colloid Interface. Sci.* **2022**, *57*, 101538.
61. Miller, R.; Liggieri, L. *Interfacial Rheology,* CRC Press: Boca Raton, FL, USA, 2008.
62. Marquez, R.; Forgiarini, A.M.; Langevin, D.; Salager, J.-L. Instability of emulsions made with surfactant–oil–water systems at optimum formulation with ultralow interfacial tension. *Langmuir* **2018**, *34*, 9252–9263.
63. Bruggeman, V.D. Berechnung Verschiedener Physikalischer Konstanten von Heterogenen Substanzen. I. Dielektrizitätskonstanten Und Leitfähigkeiten Der Mischkörper Aus Isotropen Substanzen. *Ann. Phys.* **1935**, *416*, 636–664.
64. Lemahieu, G.; Ontiveros, J.F.; Terra Telles Souza, N.; Molinier, V.; Aubry, J.-M. Fast and accurate selection of surfactants for enhanced oil recovery by dynamic Salinity-Phase-Inversion (SPI). *Fuel* **2021**, *289*, 119928.
65. Clausse, M.; Peyrelasse, J.; Heil, J.; Boned, C.; Lagourette, B. Bicontinuous structure zones in microemulsions. *Nature* **1981**, *293*, 636–638.
66. Hadamard, M.J. Mouvement Permanent Lent d'une Sphere Liquide et Visqueuse Dans Un Liquid Visqueux. *Compt. Rend. Acad. Sci.* **1911**, *152*, 1735–1738.
67. Rybczynski, W. Uber Die Fortschreitende Bewegung Einer Flussigen Kugel in Einem Zahen Medium. *Bull. Acad. Sci. Cracovie A* **1911**, *1*, 40–46.
68. Leal-Calderon, F.; Bibette, J.; Schmitt, V. *Emulsion Science: Basic Principles*, Springer: New York, NY, USA, 2007.
69. Kabalnov, A.; Wennerström, H. Macroemulsion stability: The oriented wedge theory revisited. *Langmuir* **1996**, *12*, 276–292.
70. Salager, J.L. Emulsion properties and related know-how to attain them. In *Pharmaceutical emulsions and suspensions*, Nielloud, F., Marti-Mestres, G., Eds.; CRC Press: Boca Raton, FL, USA, 2000; pp. 73–125.
71. Bourrel, M.; Schechter, R.S. *Microemulsions and Related Systems: Formulation, Solvency, and Physical Properties*, M. Dekker: New York, NY, USA, 1988.
72. Zamora, J.M.; Marquez, R.; Forgiarini, A.M.; Langevin, D.; Salager, J.-L. Interfacial rheology of low interfacial tension systems using a new oscillating spinning drop method. *J. Colloid Interface Sci.* **2018**, *519*, 27–37.
73. Rondón, M.; Bouriat, P.; Lachaise, J.; Salager, J.-L. Breaking of water-in-crude oil emulsions. 1. Physicochemical phenomenology of demulsifier action. *Energ. Fuels* **2006**, *20*, 1600–1604.
74. IUPAC Compendium of Chemical Terminology, 3rd Ed. International Union of Pure and Applied Chemistry, 2018.
75. Kolmogorov, A. On the breakage of drops in a turbulent flow. In *Dokl. Akad. Navk. SSSR*; 1949; Volume 66, pp. 825–828.
76. Hinze, J.O. Fundamentals of the hydrodynamic mechanism of splitting in dispersion processes. *AIChE J.* **1955**, *1*, 289–295.
77. L'Estimé, M.; Schindler, M.; Shahidzadeh, N.; Bonn, D. Droplet size distribution in emulsions. *Langmuir* **2024**, *40*, 275–281.
78. Tolosa, L.-I.; Forgiarini, A.; Moreno, P.; Salager, J.-L. Combined effects of formulation and stirring on emulsion drop size in the vicinity of three-phase behavior of surfactant–oil water systems. *Ind. Eng. Chem. Res.* **2006**, *45*, 3810–3814.
79. Gilbert, L.; Picard, C.; Savary, G.; Grisel, M. Rheological and textural characterization of cosmetic emulsions containing natural and synthetic polymers: Relationships between both data. *Colloids Surf. A* **2013**, *421*, 150–163.
80. Briceño, M.I. Rheology of suspensions and emulsions. In *Pharmaceutical Emulsions and Suspensions*, Nielloud, F., Marti-Mestres, G., Eds.; CRC Press: Boca Raton, FL, USA, 2000; pp. 557–607.

81. Einstein, A. A new determination of molecular dimensions. *Ann. Phys.* **1906**, *19*, 289–306.
82. Derkach, S.R. Rheology of emulsions. *Adv. Colloid Interface Sci.* **2009**, *151*, 1–23.
83. Pal, R. Effect of droplet size on the rheology of emulsions. *AIChE J.* **1996**, *42*, 3181–3190.
84. Salager, J.L.; Miñana-Pérez, M.; Andérez, J.M.; Grosso, J.L.; Rojas, C.I.; Layrisse, I.; Intevep, S.A. Surfactant-oil-water systems near the affinity inversion part II: Viscosity of emulsified systems. *J. Disp. Sci. Technol.* **1983**, *4*, 161–173.
85. Pierlot, C.; Ontiveros, J.F.; Catté, M.; Salager, J.-L.; Aubry, J.-M. Cone–plate rheometer as reactor and viscosity probe for the detection of transitional phase inversion of brij30–isopropyl myristate–water model emulsion. *Ind. Eng. Chem. Res.* **2016**, *55*, 3990–3998.
86. Salager, J.L.; Miñana-Pérez, M.; Perez-Sanchez, M.; Ramfrez-Gouveia, M.; Rojas, C.I. Surfactant-oil-water systems near the affinity inversion part III: The two kinds of emulsion inversion. *J. Disp. Sci. Technol.* **1983**, *4*, 313–328.
87. Antón, R.E.; Castillo, P.; Salager, J.L. Surfactant-oil-water systems near the affinity inversion part IV: Emulsion inversion temperature. *J. Disp. Sci. Technol.* **1986**, *7*, 319–328.
88. Salager, J.-L.; Forgiarini, A.; Márquez, L.; Peña, A.; Pizzino, A.; Rodriguez, M.P.; Rondón-González, M. Using emulsion inversion in industrial processes. *Adv. Colloid Interface Sci.* **2004**, *108–109*, 259–272.
89. Zambrano, N.; Tyrode, E.; Mira, I.; Márquez, L.; Rodríguez, M.-P.; Salager, J.-L. Emulsion catastrophic inversion from abnormal to normal morphology. 1. Effect of the water-to-oil ratio rate of change on the dynamic inversion frontier. *Ind. Eng. Chem. Res.* **2003**, *42*, 50–56.
90. Mira, I.; Zambrano, N.; Tyrode, E.; Márquez, L.; Peña, A.A.; Pizzino, A.; Salager, J.-L. Emulsion catastrophic inversion from abnormal to normal morphology. 2. Effect of the stirring intensity on the dynamic inversion frontier. *Ind. Eng. Chem. Res.* **2003**, *42*, 57–61.
91. Pierlot, C.; Ontiveros, J.F.; Royer, M.; Catté, M.; Salager, J.-L. Emulsification of viscous alkyd resin by catastrophic phase inversion with nonionic surfactant. *Colloids Surf. A* **2018**, *536*, 113–124.
92. Forgiarini, A.; Marquez, L.; Celis, M.T.; Salager, J.-L. Nanoemulsiones—Formación Con Baja Energía. In *Nanopartículas: Fundamentos y Aplicaciones*, Larez-Velasquez, C.L., Koteich, S., Lopez, F.; Universidad de Los Andes: Mérida, Venezuela, 2015; pp. 273–293.
93. Solans, C.; Esquena, J.; Forgiarini, A.M.; Uson, N.; Morales, D.; Izquierdo, P.; Azemar, N.; Garcia-Celma, M.J. Nano-emulsions: Formation, properties, and applications. In *Adsorption and Aggregation of Surfactants in Solution*, Mittal, K.L., Shah, D.O., Eds.; CRC Press: Boca Raton, FL, USA, 2003; pp. 525–554.
94. Feng, J.; Rodríguez-Abreu, C.; Esquena, J.; Solans, C. A concise review on nano-emulsion formation by the Phase Inversion Composition (PIC) method. *J. Surfactants Deterg.* **2020**, *2*, 677–685.
95. Miñana-Pérez, M.; Gutron, M.C.; Zundel, C.; Andérez, J.; Salager, J.-L. Miniemulsion formation by transitional inversion. *J. Dispersion Sci. Technol.* **1999**, *20*, 893.
96. Solans, C.; Izquierdo, P.; Nolla, J.; Azemar, N.; Garcia-Celma, M.J. Nano-emulsions. *Curr. Opin. Colloid Interface Sci.* **2005**, *10* (3–4), 102–110. https://doi.org/10.1016/j.cocis.2005.06.004.
97. McClements, D.J. Nanoemulsions versus microemulsions: Terminology, differences, and similarities. *Soft Matter* **2012**, *8*, 1719–1728.
98. Gupta, A.; Eral, H.B.; Hatton, T.A.; Doyle, P.S. Nanoemulsions: Formation, properties and applications. *Soft Matter* **2016**, *12*, 2826–2841.

99. de Oca-Ávalos, J.M.M.; Candal, R.J.; Herrera, M.L. Nanoemulsions: Stability and physical properties. *Curr. Opin. Food Sci.* **2017**, *16*, 1–6.
100. Tadros, T.; Izquierdo, P.; Esquena, J.; Solans, C. Formation and stability of nano-emulsions. *Adv. Colloid Interface Sci.* **2004**, *108–109*, 303–318.
101. Tadros, T. *Encyclopedia of Colloid and Interface Science*, Springer-Verlag: Berlin, Germany, 2013.
102. Forgiarini, A.; Esquena, J.; González, C.; Solans, C. Formation of nano-emulsions by low-energy emulsification methods at constant temperature. *Langmuir* **2001**, *17*, 2076–2083.
103. Komaiko, J.; McClements, D.J. Low-energy formation of edible nanoemulsions by spontaneous emulsification: Factors influencing particle size. *J. Food Eng.* **2015**, *146*, 122–128.
104. Sajjadi, S. Effect of mixing protocol on formation of fine emulsions. *Chem. Eng. Sci.* **2006**, *61*, 3009–3017.
105. Solans, C.; Solé, I. Nano-emulsions: Formation by low-energy methods. *Curr. Opin. Colloid Interface Sci.* **2012**, *17*, 246–254.
106. McClements, D.J.; Rao, J. Food-grade nanoemulsions: Formulation, fabrication, properties, performance, biological fate, and potential toxicity. *Crit. Rev. Food Sci. Nut.* **2011**, *51*, 285–330.
107. Yang, Y.; Marshall-Breton, C.; Leser, M.E.; Sher, A.A.; McClements, D.J. Fabrication of ultrafine edible emulsions: Comparison of high-energy and low-energy homogenization methods. *Food Hydrocolloids* **2012**, *29*, 398–406.
108. Thakur, N.; Garg, G.; Sharma, P.K.; Kumar, N. Nanoemulsions: A review on various pharmaceutical application. *Global J. Pharm.* **2012**, *6*, 222–225.
109. Márquez, R.; Bullón, J.; Márquez, L.; Cárdenas, A.; Briceño, M.I.I.; Forgiarini, A. Rheological changes of parenteral emulsions during phase-inversion emulsification. *J. Disp. Sci. Technol.* **2008**, *29*, 621–627.
110. McClements, D.J. Edible nanoemulsions: Fabrication, properties, and functional performance. *Soft Matter* **2011**, *7*, 2297–2316.
111. Salager, J.-L. Formulation concepts for the emulsion maker. In *Pharmaceutical Emulsions and Suspensions*, Nielloud, F., Marti-Mestres, G., Eds.; Marcel Dekker: New York, NY, USA, 2000; pp. 18–71.
112. Mira, I.; Marquez, L.; Tyrode, E.; Salager, J.-L.; Zambrano, N.; Pena, A. Principles of emulsion formulation engineering. In *Adsorption and Aggregation of Surfactants in Solution*, Mittal, K.L., Shah, D.O., Eds.; CRC Press: Boca Raton, FL, USA, 2003; pp. 501–523.
113. Fryd, M.M.; Mason, T.G. Advanced nanoemulsions. *Ann. Rev. Phys. Chem.* **2012**, *63*, 493–518.
114. Cedeno, Y.; Segovia, L.; Davila, R.; Perez, A.; Bullon, J.; Rennola, L.; Marquez, R. Formulation of a eugenol-based O / W emulsion for application as a topical and oral anesthetic by low-energy emulsification. *Ciencia e Ingenieria* **2022**, *43*, 157–164.
115. Rodríguez-Abreu, C.; Vila, A. Nano-droplet systems by surfactant self-assembly and applications in the pharmaceutical industry. *Curr. Top. Medicin. Chem.* **2014**, *14*, 747–765.
116. Harada, T.; Yokomizo, K. Demulsification of oil-in-water emulsion under freezing conditions: Effect of crystal structure modifier. *J. Am. Oil Sci. Soc.* **2000**, *77*, 859–864.
117. Cholakova, D.; Vinarov, Z.; Tcholakova, S.; Denkov, N.D. Self-emulsification in chemical and pharmaceutical technologies. *Curr. Opin. Colloid Interface Sci.* **2022**, *59*, 101576.
118. Santana, R.C.; Perrechil, F.A.; Cunha, R.L. High- and low-energy emulsifications for food applications: A focus on process parameters. *Food Eng. Rev.* **2013**, *5*, 107–122.

119. Förster, T.; Von Rybinski, W.; Wadle, A. Influence of microemulsion phases on the preparation of fine-disperse emulsions. *Adv. Colloid Interface Sci.* **1995**, *58*, 119–148.
120. Roger, K. Nanoemulsification in the vicinity of phase inversion: Disruption of bicontinuous structures in oil/surfactant/water systems. *Curr. Opin. Colloid Interface Sci.* **2016**, *25*, 120–128.
121. Roger, K.; Cabane, B.; Olsson, U. Emulsification through surfactant hydration: The PIC process revisited. *Langmuir* **2011**, *27*, 604–611. https://doi.org/10.1021/la1042603.
122. Perazzo, A.; Preziosi, V.; Guido, S. Phase inversion emulsification: Current understanding and applications. *Adv. Colloid Interface Sci.* **2015**, *222*, 581–598.
123. Salager, J.-L.; Forgiarini, A.; Lopez, J.C.; Marfisi, S.; Alvarez, G. Dynamics of near-zero energy emulsification. In *6th World Surfactant Congress CESIO*, Berlin, Germany, 2004; Volume 203, pp. 1–11.
124. Anton, N.; Gayet, P.; Benoit, J.-P.P.; Saulnier, P. Nano-emulsions and nanocapsules by the PIT method: An investigation on the role of the temperature cycling on the emulsion phase inversion. *Int. J. Pharm.* **2007**, *344*, 44–52.
125. Corkery, R.W.; Blute, I.A.; Friberg, S.E.; Guo, R. Emulsion inversion in the PIT range: Quantitative phase variations in a two-phase emulsion. *J. Chem. Eng. Data* **2010**, *55* (10), 4471–4475. https://doi.org/10.1021/je100625e.
126. Förster, T.; Schambil, F.; von Rybinski, W. Production of fine disperse and long-term stable oil-in-water emulsions by the phase inversion temperature method. *J. Disp. Sci. Technol.* **1992**, *13*, 183–193.
127. Izquierdo, P.; Esquena, J.; Tadros, T.F.; Dederen, C.; Garcia, M.J.; Azemar, N.; Solans, C. Formation and stability of nano-emulsions prepared using the phase inversion temperature method. *Langmuir* **2002**, *18*, 26–30.
128. Izquierdo, P.; Esquena, J.; Tadros, T.F.; Dederen, J.C.; Feng, J.; Garcia-Celma, M.J.; Azemar, N.; Solans, C. Phase behavior and nano-emulsion formation by the phase inversion temperature method. *Langmuir* **2004**, *20*, 6594–6598.
129. Shinoda, K.; Saito, H. The stability of O/W type emulsions as functions of temperature and the HLB of emulsifiers: The emulsification by PIT-method. *J. Colloid Interface Sci.* **1969**, *30*, 258–263.
130. Elgammal, M.; Schneider, R.; Gradzielski, M. Preparation of latex nanoparticles using nanoemulsions obtained by the Phase Inversion Composition (PIC) method and their application in textile printing. *Colloids Surf. A* **2015**, *470*, 70–78.
131. Forgiarini, A.; Esquena, J.; González, C.; Solans, C. Formation and stability of nano-emulsions in mixed nonionic surfactant systems. *Progr. Colloid Polym. Sci.* **2001**, *118*, 184–188.
132. Morales, D.; Gutiérrez, J.M.; García-Celma, M.J.; Solans, Y.C. A study of the relation between bicontinuous microemulsions and oil/water nano-emulsion formation. *Langmuir* **2003**, *19*, 7196–7200.
133. Bouchemal, K.; Briançon, S.; Perrier, E.; Fessi, H. Nano-emulsion formulation using spontaneous emulsification: solvent, oil and surfactant optimisation. *Int. J. Pharm.* **2004**, *280*, 241–251.
134. Friberg, S.E. Potential correlation between spontaneous emulsification and the plait point in the system water/benzene/ethanol: A phase diagram approach. *Soft* **2013**, *02*, 1–6.
135. Li, Z.; Wu, H.; Yang, M.; Jiang, J.; Xu, D.; Feng, H.; Lu, Y.; Kang, W.; Bai, B.; Hou, J. Spontaneous emulsification via once bottom-up cycle for the crude oil in low-permeability reservoirs. *Energ. Fuels* **2018**, *32*, 3119–3126.
136. López-Montilla, J.C.; Herrera-Morales, P.E.; Pandey, S.; Shah, D.O.; Lapez-Montilla, J.C.; Herrera-Morales, P.E.; Pandey, S.; Shah, D.O. Spontaneous emulsification: Mechanisms, physicochemical aspects, modeling, and applications. *J. Disp. Sci. Technol.* **2002**, *23*, 219–268.

137. López-Montilla, J.C.; James, M.A.; Crisalle, O.D.; Shah, D.O. Surfactants and protocols to induce spontaneous emulsification and enhance detergency. *J. Surfactants Deterg.* **2005**, *8*, 45–53.
138. Miller, C.A. Spontaneous emulsification: Recent developments with emphasis on self-emulsification. In *Emulsions and Emulsion Stability*, Sjoblom, J., Ed.; CRC Press: Boca Raton, FL, USA, 2005; pp.107–126.
139. Shinoda, K.; Saito, H. The effect of temperature on the phase behavior and the types of dispersions of the ternary system composed of water, cyclohexane and nonionic surfactant. *J. Colloid Interface Sci.* **1968**, *26*, 70. https://doi.org/10.1016/0021-9797(68)90273-7.
140. Förster, T. Principles of emulsion formation. In *Surfactants in Cosmetics*, 2nd ed.; Rieger, M., Ed.; CRC Press: Boca Raton, FL, USA, 2017; pp. 105–126.
141. Sonneville-Aubrun, O.; Babayan, D.; Bordeaux, D.; Lindner, P.; Rata, G.; Cabane, B. Phase transition pathways for the production of 100 nm oil-in-water emulsions. *Phys. Chem. Chem. Phys.* **2009**, *11*, 101–110.
142. Sajjadi, S. Nanoemulsion formation by phase inversion emulsification: On the nature of inversion. *Langmuir* **2006**, *22*, 5597–5603.
143. Kunieda, H.; Ishikawa, N. Evaluation of the Hydrophile-Lipophile Balance (HLB) of nonionic surfactants. II. Commercial-surfactant systems. *J. Colloid Interface Sci.* **1985**, *107*, 122–128.
144. Ontiveros, J.F.; Pierlot, C.; Catté, M.; Salager, J.L.; Aubry, J.M. Determining the Preferred Alkane Carbon Number (PACN) of nonionic surfactants using the PIT-slope method. *Colloids Surf. A* **2018**, *536*, 30–37.
145. Pizzino, A.; Molinier, V.; Catté, M.; Ontiveros, J.F.; Salager, J.-L.; Aubry, J.-M.M. Relationship between phase behavior and emulsion inversion for a well-defined surfactant (C10E4)/n-octane/water ternary system at different temperatures and water/oil ratios. *Ind. Eng. Chem. Res.* **2013**, *52*, 4527–4538.
146. Sing, A.J.F.; Graciaa, A.; Lachaise, J.; Brochette, P.; Salager, J.L. Interactions and coalescence of nanodroplets in translucent O/W emulsions. *Colloids Surf. A* **1999**, *152*, 31–38.
147. Ontiveros, J.F.; Pierlot, C.; Catté, M.; Molinier, V.; Pizzino, A.; Salager, J.-L.; Aubry, J.M. Classification of ester oils according to their Equivalent Alkane Carbon Number (EACN) and asymmetry of fish diagrams of C10E4/ester oil/water systems. *J. Colloid Interface Sci.* **2013**, *403*, 67–76.
148. Dario, M.F.; Santos, M.S.C.S.; Viana, A.S.; Arêas, E.P.G.; Bou-Chacra, N.A.; Oliveira, M.C.; da Piedade, M.E.M.; Baby, A.R.; Velasco, M.V.R. A high loaded cationic nanoemulsion for quercetin delivery obtained by sub-PIT method. *Colloids Surf. A* **2016**, *489*, 256–264.
149. Ozogul, Y.; Karsli, G.T.; Durmuş, M.; Yazgan, H.; Oztop, H.M.; McClements, D.J.; Ozogul, F. Recent developments in industrial applications of nanoemulsions. *Adv. Colloid Interface Sci.* **2022**, *304*, 102685.
150. Boscán, F.; Paulis, M.; Barandiaran, M.J. Towards the production of high performance lauryl methacrylate based polymers through emulsion polymerization. *Eur. Polym. J.* **2017**, *93*, 44–52.
151. Miller, D.J.; Henning, T.; Grünbein, W. Phase inversion of W/O emulsions by adding hydrophilic surfactant—a technique for making cosmetics products. *Colloids Surf. A* **2001**, *183–185*, 681–688.
152. Solè, I.; Pey, C.M.; Maestro, A.; González, C.; Porras, M.; Solans, C.; Gutiérrez, J.M. Nano-emulsions prepared by the phase inversion composition method: Preparation variables and scale up. *J. Colloid Interface Sci.* **2010**, *344*, 417–423.
153. Dickinson, E. Thermodynamic aspects of emulsion phase inversion. *J. Colloid Interface Sci.* **1982**, *87*, 416–423.

154. Lin, T. Low-surfactant emulsification. *J. Soc. Cosmetic Chem.* **1978**, *28*, 273–295.
155. Lin, T.J.; Shen, Y.F.; Chemical, H.T. Low-energy part VI : Applications in phase emulsions. *J. Soc. Cosmetic Chem.* **1984**, *368*, 357–368.
156. Marszall, L. Adsorption of nonionic surfactants at the oil-water interface and emulsion inversion point. *Colloid and Polymer Science Kolloid Zeitschrift & Zeitschrift für Polymere* **1976**, *254*, 674–675.
157. Sagitani, H. Making homogeneous and fine droplet o/w emulsions using nonionic surfactants. *J. Am. Oil Chem. Soc.* **1981**, *58*, 738–743.
158. Sagitani, H. Formation of o/w emulsions by surfactant phase emulsification and the solution behavior of nonionic surfactant system in the emulsification process. *J. Disp. Sci. Technol.* **1988**, *9*, 115–128.
159. Uson, N.; Garcia, M.J.; Solans, C. Formation of water-in-oil (w/o) nano-emulsions in a water/mixed non-ionic surfactant/oil systems prepared by a low-energy emulsification method. *Colloids Surf. A* **2004**, *250*, 415–421.
160. Sternling, C.V.; Scriven, L.E. Interfacial turbulence: Hydrodynamic instability and the Marangoni effect. *AIChE J.* **1959**, *5*, 514–523.
161. Artiga-Artigas, M.; Guerra-Rosas, M.I.; Morales-Castro, J.; Salvia-Trujillo, L.; Martín-Belloso, O. Influence of essential oils and pectin on nanoemulsion formulation: A ternary phase experimental approach. *Food Hydrocolloids* **2018**, *81*, 209–218.
162. Prakash, A.; Baskaran, R.; Paramasivam, N.; Vadivel, V. Essential oil based nanoemulsions to improve the microbial quality of minimally processed fruits and vegetables : A review. *Food Res. Int.* **2018**, *111*, 509–523.
163. Erramreddy, V.V.; Ghosh, S. Influence of droplet size on repulsive and attractive nanoemulsion gelation. *Colloids Surf. A* **2015**, *484*, 144–152.
164. Zhou, H.; Yue, Y.; Liu, G.; Li, Y.; Zhang, J.; Gong, Q.; Yan, Z.; Duan, M. Preparation and characterization of a lecithin nanoemulsion as a topical delivery system. *Nanoscale Res. Lett.* **2010**, *5*, 224–230.
165. Klaus, A.; Tiddy, G.J.T.; Solans, C.; Harrar, A.; Touraud, D.; Kunz, W. Effect of salts on the phase behavior and the stability of nano-emulsions with rapeseed oil and an extended surfactant. *Langmuir* **2012**, *28*, 8318–8328.
166. Forgiarini, A.; Pietrangeli, G.; Arandia, M.; Guédez, V.; Celis, M.-T.; Márquez, L.; Ciencia, R. Influencia Del Tipo de Alcohol Sobre La Formación de Nanoemulsiones de Aceite de Soja-En-Agua. *Ciencia e Ingeniería* **2009**, *30*, 115–120.
167. Bullón, J.; Molina, J.; Márquez, R.; Véjar, F.; Scorzza, C.; Forgiarini, A. Nano-Emulsionación de Aceites Triglicéridos Para Uso Parenteral Mediante Un Método de Baja Energía. *Revista Tecnica de la Facultad de Ingenieria Universidad del Zulia* **2007**, *30*, 428–437.
168. Forgiarini, A.; Esquena, J.; González, C.; Solans, C. Studies of the relation between phase behavior and emulsification methods with nanoemulsion formation. *Prog. Colloid Polym. Sci.* **2000**, *115*, 36–38.
169. Villarreal, A.M.; Fernandez, C.; Forgiarini, A.; Marquez, L.; Nielloud, F.; Salager, J. Nanoencapsulation de Filtres Solaires via Nanoémulsions. In *Procédés et formulations au service de la santé*, Durand, A., Canselier, J.-P., Eds.; EDP Sciences: Les Ulies, France, 2011; pp. 1–14.
170. Solans, C.; Morales, D.; Homs, M. Spontaneous emulsification. *Curr. Opin. Colloid Interface Sci.* **2016**, *22*, 88–93.
171. Rao, J.; McClements, D.J. Stabilization of phase inversion temperature nanoemulsions by surfactant displacement. *J. Agric. Food Chem.* **2010**, *58*, 7059–7066.
172. Mohlin, K.; Holmberg, K.; Esquena, J.; Solans, C. Study of low energy emulsification of alkyl ketene dimer related to the phase behavior of the system. *Colloids Surf. A* **2003**, *218*, 189–200.

173. Collins-Gold, L.C.; Lyons, R.T.; Bartholow, L.C. Parenteral emulsions for drug delivery. *Adv. Drug Deliv. Rev.* **1990**, *5*, 189–208.
174. Le Guenic, S.; Chaveriat, L.; Lequart, V.; Joly, N.; Martin, P. Renewable surfactants for biochemical applications and nanotechnology. *J. Surfactants Deterg.* **2019**, *22*, 5–21.
175. Márquez, R.; Bullón, J.; Márquez, L.; Cárdenas, A.; Briceño, M.I.I.; Forgiarini, A. Rheological changes of parenteral emulsions during phase-inversion emulsification. *J. Disp. Sci. Technol.* **2008**, *29*, 621–627.
176. Ahmad, N.; Ahmad, F.J.; Bedi, S.; Sharma, S.; Umar, S.; Ansari, M.A. A novel nanoformulation development of eugenol and their treatment in inflammation and periodontitis. *Saudi Pharm. J.* **2019**, *27*, 778–790.
177. Fernandez, P.; André, V.; Rieger, J.; Kühnle, A. Nano-emulsion formation by emulsion phase inversion. *Colloids Surf. A* **2004**, *251*, 53–58.
178. Miller, C.A. Spontaneous emulsification produced by diffusion - A review. *Colloids and Surfaces* **1988**, *29*, 89–102.
179. Friberg, S.; Solans, C. Emulsification and the HLB-temperature. *J. Colloid Interface Sci.* **1978**, *66*, 367.
180. Ganachaud, F.; Katz, J.L. Nanoparticles and nanocapsules created using the ouzo effect: Spontaneous emulsification as an alternative to ultrasonic and high-shear devices. *ChemPhysChem* **2005**, *6*, 209–216.
181. Vitale, S.A.; Katz, J.L. Liquid droplet dispersions formed by homogeneous liquid-liquid nucleation: "The ouzo effect." *Langmuir* **2003**, *19*, 4105–4110.
182. Tcholakova, S.; Valkova, Z.; Cholakova, D.; Vinarov, Z.; Lesov, I.; Denkov, N.; Smoukov, S.K. Efficient self-emulsification via cooling-heating cycles. *Nat. Commun.* **2017**, *8*, 15012. https://doi.org/10.1038/ncomms15012.
183. Cholakova, D.; Glushkova, D.; Tcholakova, S.; Denkov, N. Nanopore and nanoparticle formation with lipids undergoing polymorphic phase transitions. *ACS Nano* **2020**, *14*, 8594–8604.
184. Cholakova, D.; Glushkova, D.; Pantov, M.; Tcholakova, S.; Denkov, N. Triglyceride mixtures: Cold-bursting and double emulsion formation. *Colloids Surf. A* **2023**, *668*, 131438.

9 Particle vs. Surfactant-Stabilized Emulsions

Armando Maestro, Francisco Ortega, Ramón G. Rubio, and Eduardo Guzmán

9.1 INTRODUCTION

Emulsions are colloidal systems widely used in different industries, including food, cosmetics, pharmaceuticals, and oil production. They consist of two immiscible liquids, usually a polar phase (called the aqueous phase) and a nonpolar phase (commonly called the oil phase), with one of these liquid phases dispersed as droplets (the dispersed phase) in the second liquid (the continuous phase). The stabilization of the droplet dispersion, and hence the emulsions, requires a third component that adsorbs to the droplet-continuous phase interface and at least partially prevents or hinders the phenomena that lead to phase separation. Avoiding the latter is critical as it affects the shelf life, physical properties, and overall performance of the emulsions.[1,2] Therefore, the preparation of stable emulsions requires a critical analysis of the different strategies used to prevent destabilization phenomena and, in particular, the type of substances used as stabilizers, or emulsifiers, and their mechanism of action.[3] Generally, emulsifiers can be divided into three categories: small molecular emulsifiers (surfactant or lipids), macromolecular emulsifiers, and solid emulsifiers (colloidal particles).[4,5] Table 9.1 summarizes some examples of each category of emulsifiers together with examples of their potential uses.

The differences and similarities between particle-stabilized (the so-called Pickering emulsions) and surfactant-stabilized emulsions (generally stabilized by small molecular or macromolecular emulsifiers) are of great interest to researchers and industry professionals. Understanding these aspects is essential for tailoring emulsion formulations to specific applications and optimizing their performance. The first difference between the two families of emulsions emerges from the stabilization mechanism. In fact, surfactant-stabilized emulsions are stabilized by surfactant molecules which, when adsorb to the droplet-continuous phase interface, lower the interfacial tension between the oil and water phases and form a protective layer. This increases the energy barrier to droplet coalescence and delays phase separation. On the other hand, Pickering emulsions are stabilized by solid particles that accumulate at the oil-water interface, although their role in reducing the surface tension can be negligible without compromising emulsion stability, giving higher stability and biocompatibility compared to surfactant-stabilized emulsions.[6–8] Pickering emulsions show advantages such as high resistance to coalescence, long-term stability,

DOI: 10.1201/9781032636184-11

TABLE 9.1
Examples of Each Family of Emulsifiers and Their Potential Applications

Category	Material	Example	Applications
Small molecular emulsifiers	Simple surfactants	Anionic surfactants, cationic surfactants, zwitterionic surfactants, and nonionic surfactants	Widely existing in various emulsifying systems, e.g., petroleum, organic synthesis, materials, biological medicine, electrochemistry, food industry
Macromolecular emulsifiers	Biological macromolecules	Lipids, proteins, polysaccharides	Medicine, food industry, cosmetic industry
	Polymeric surfactants	Random copolymers, block copolymers, branched polymers	Organic synthesis, materials, biological medicine
Solid emulsifiers	Inorganic solid particles	Silicon dioxide, titanium dioxide, ferric oxide, montmorillonite, laponite, layered bimetallic hydroxide	Pharmaceutical industry, oil and gas industry, aerospace industry
	Organic solid particles	Poly (*N*-isopropylacrylamide) gel particles, polyethylene microspheres	
	Janus particles	Polymeric Janus particles	

Source: Adapted from Tian et al.,[5] with permission under Creative Common Attribution License 4.0 [CC-BY 4.0].

and reduced cytotoxicity, making them suitable for various applications, including drug delivery and manufacturing.[7] The presence of solid particles in Pickering emulsions forms a robust barrier that prevents coalescence and maintains stability over time.[9] Figure 9.1 displays a schematic illustration of the structures resulting from the formation of interfacial layers as a result of the adsorption of the different categories of emulsifiers.

This chapter aims to explore the characteristics, stabilization mechanisms, benefits, and limitations of particle and surfactant-stabilized emulsions. To this end, the following sections will explore the unique characteristics and challenges associated with particle-stabilized emulsions, followed by an in-depth analysis of surfactant-stabilized emulsions. By critically examining the strengths and weaknesses of particle and surfactant stabilization, this discussion aims to provide valuable insights into the selection of appropriate stabilization methods for specific emulsion applications. Understanding these differences is essential not only for optimizing product performance and shelf life but also for addressing environmental concerns associated with emulsion formulations.

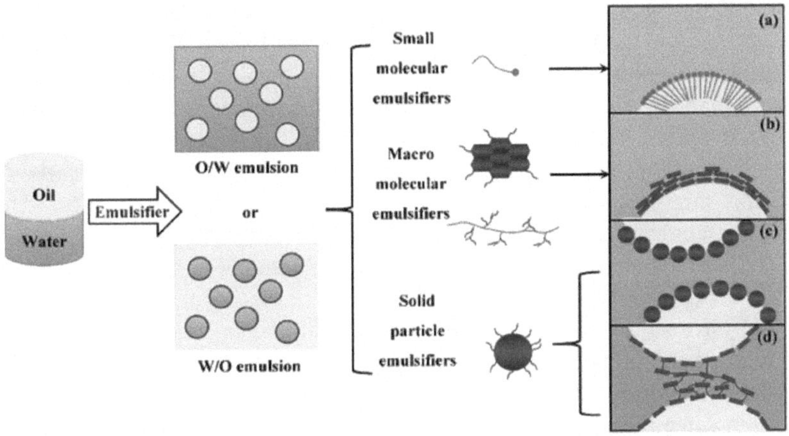

FIGURE 9.1 Structure of the interfacial layer at the droplet-continuous phase interface as a result of the adsorption of emulsifiers belonging to the different families. (a) Small molecular emulsifiers. (b) Macromolecular emulsifiers. (c) Solid particles. (d) 3D network. Adapted from Tian et al.,[5] with permission under Creative Common Attribution License 4.0 [CC-BY 4.0].

9.2 PARTICLE-STABILIZED EMULSIONS

Pickering emulsions rely on the use of solid particles to stabilize the interface between immiscible liquids, i.e., oily and aqueous phases, in contrast to the surfactants more commonly used in conventional emulsions.[10] This feature was first observed by Ramsden[11] and later extended by Pickering,[12] marking a significant departure from conventional emulsion stabilization methods. The current understanding of the stabilization mechanisms of Pickering emulsions is based on two prevailing theories. The mechanical barrier theory proposes that solid particles adsorb at the oil-water interface, forming a protective layer that prevents coalescence by acting as a barrier between the dispersed droplets.[13] On the other hand, the three-dimensional viscoelastic particle network theory suggests that these particles form a network-like structure in the continuous phase, imparting stability to the emulsion by physically immobilizing the droplets within the network.[14] Understanding these theories and the unique properties of Pickering emulsions is crucial to advancing their applications in various industries. Their surfactant-free nature and enhanced stability make them attractive for applications in pharmaceuticals, cosmetics, food, and materials science.[15,16]

9.2.1 Stabilization of Pickering Emulsions

The stabilization of Pickering emulsions requires the adsorption of the solid particles at the oil-water interface, which is only possible when there is a partial wetting of the particles by both phases. The wettability of solid particles significantly influences their functionality at the oil-water interface and can be quantified through contact angle measurements between the particle and the interface. Determination of the emulsion type (oil-in-water, O/W, or water-in-oil, W/O) and stability relies on the particle wettability in both the oil and water phases. The continuous phase of the emulsion tends to be the one in which the particle surface experiences the most significant wetting, with the contact angle serving as a crucial parameter (see Figure 9.2).[7]

The adsorption of particles with contact angles less than 90°, favoring the formation of O/W emulsions due to their hydrophilic nature, contrasts sharply with those with contact angles exceeding 90°, which tend to form W/O emulsions as they display hydrophobic characteristics.[17] Therefore, contact angle underscores the distinct affinities of particles toward water or oil phases, dictating their immersion preferences and ultimately influencing the curvature of the oil-water interface, thereby determining the emulsion type.[18] Altering particle wettability is achievable through the adsorption of compatible surfactants, occasionally resulting in emulsion phase inversion in certain cases.[19] The efficacy of solid substances in stabilizing emulsions

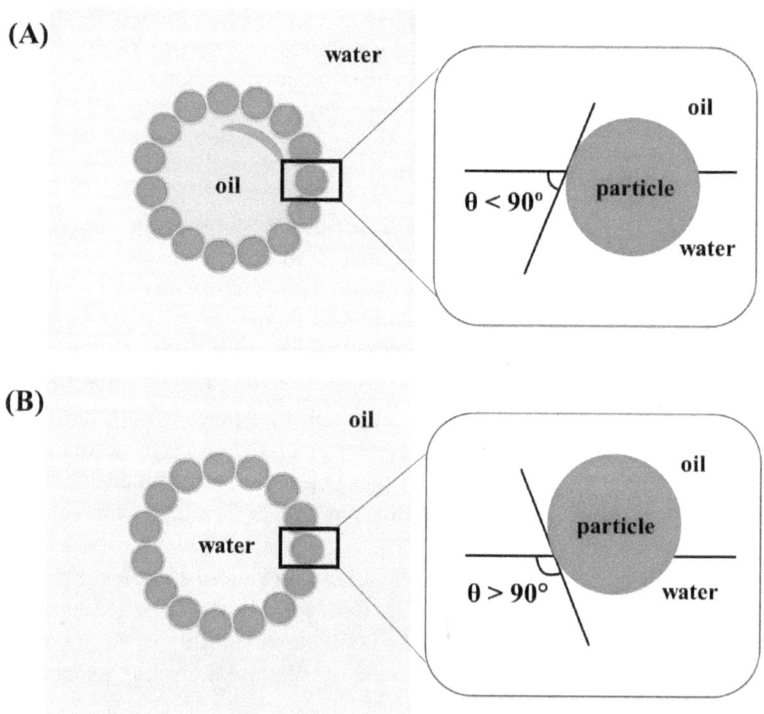

FIGURE 9.2 (a) Sketch of the position of a solid particle at the interface between the drop and continuous phases, displaying a contact angle of less than 90°, indicative of the formation of an oil-in-water (O/W) emulsion. (b) Sketch of the position of a solid particle at the interface between the drop and continuous phases, characterized by a contact angle exceeding 90°, indicative of the formation of a water-in-oil (W/O) emulsion. Adapted from de Caravalho-Guimaraes et al.,[7] with permission under Creative Common Attribution License 4.0 (CC-BY).

depends on various factors, including but not limited to particle size, shape, concentration, wettability, and the interactions occurring between these particles.[13]

The mechanism underpinning the formation of Pickering emulsions involves the homogenization of the aqueous and organic phases. These emulsions achieve stability through the construction of a robust network comprising solid particles within the droplet-continuous phase interface. This network effectively confines the dispersed phase, impeding coalescence, sedimentation, or creaming phenomena.[20] However, the vulnerability of Pickering emulsions lies in their susceptibility to breakdown when defects or fractures occur in the particle layer enveloping droplets.[21,22]

Traditionally, the literature has highlighted that most particles are effective stabilizers for w-o emulsions. However, there is also the possibility to obtain o-w emulsions in specific cases. These emulsions often exhibit larger droplet sizes in comparison to those stabilized by surfactants, primarily due to the significantly larger size of the particles compared to the nanometer-size surfactant molecules.[23]

Beyond particle wettability, there are other parameters, such as particle concentration, particle shape, and size, that influence the stabilization of emulsions by using particles.[13,24] For instance, the concentration of particles in the continuous phase is a key determinant of the stability of Pickering emulsions. In the case of molecular surfactants, achieving optimum emulsion stability requires maintaining a concentration above the critical micelle concentration (CMC). In contrast to molecular surfactants, solid particles acting as colloidal surfactants exhibit a different behavior. The irreversible adsorption of solid particles onto droplet surfaces initially involves coverage of these surfaces by the particles present in the continuous phase. As the particle concentration is further increased, the excess particles begin to form a network structure that envelops each droplet. This phenomenon not only increases emulsion stability but also contributes to an escalation in the viscosity of the emulsion system. An example of this phenomenon can be seen in toluene-in-water emulsions stabilized with varying concentrations of silanized fumed silica nanoparticles, as shown and modeled in Figure 9.3.[25] At lower particle concentrations, the emulsions exhibit instability, with droplets partially coated by silica nanoparticles and a lack of excess nanoparticles in the continuous phase. However, as the nanoparticle concentration increases, the excess nanoparticles in the continuous phase transform from compact flocs to more dispersed networks. This structural change keeps the droplets adequately spaced apart, contributing significantly to improved emulsion stability.

The effect of particle concentration on Pickering emulsions is manifested in three key dimensions: particle density at the droplet surface, droplet size, and the viscosity of the emulsion system. As the particle concentration increases, the particle density at the droplet surface increases until the droplets are completely covered by particles. In systems where the oil-water ratio remains constant, the total droplet surface area increases as the droplet size decreases. Consequently, as the particle concentration increases, the total surface area that can be completely enveloped by particles increases, facilitating a reduction in droplet size to a minimum constrained only by shear forces. However, a subsequent increase in particle concentration results in an excess of particles within the continuous phase. This excess can lead to multilayer adsorption on the droplet surface or induce particle aggregation, culminating in the

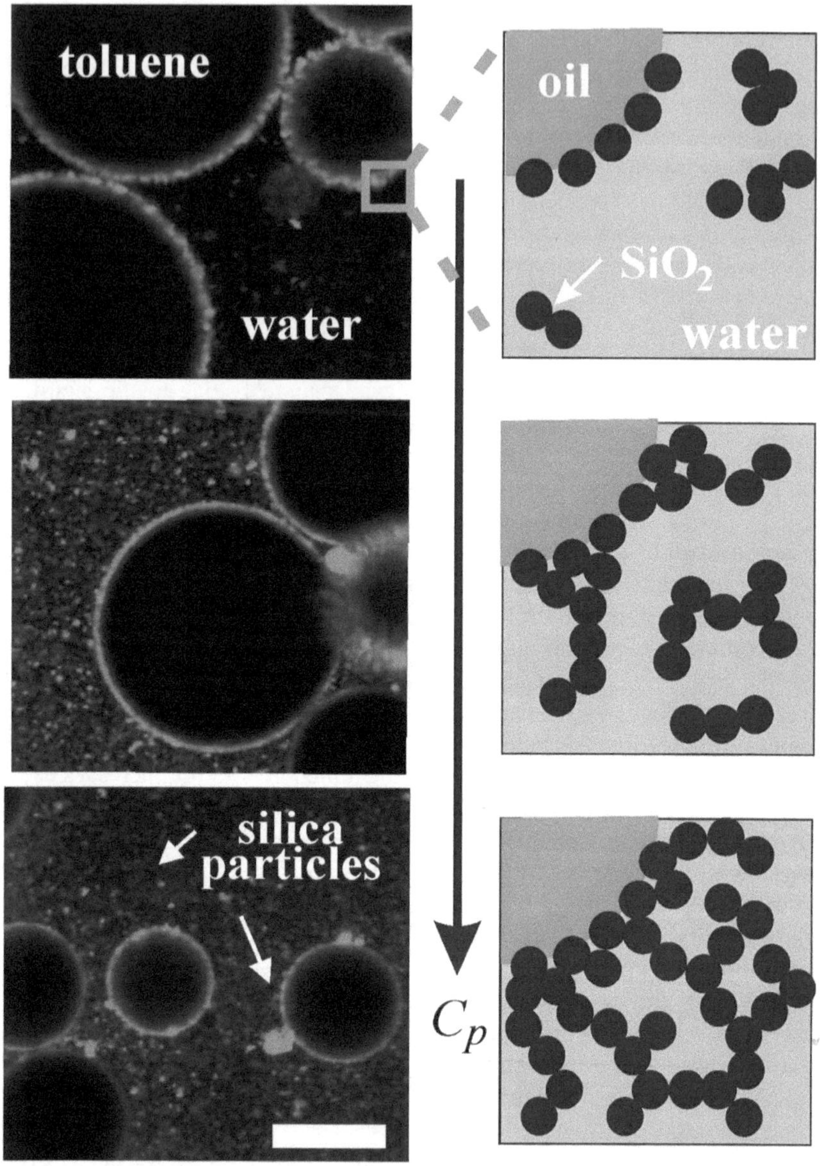

FIGURE 9.3 Confocal fluorescence microscope images showing emulsions with increasing particle concentration from top to bottom. The scale bar corresponds to a measurement of 30 μm. In addition, a schematic representation of the concentration-dependent network structure within the emulsions is shown in the right panel. This diagram represents a conceptual representation in which the emulsion droplets are shown as being coated by a monolayer of particles. This simplification has been adopted for the sake of clarity in visualizing the particle concentration-induced network arrangement within the emulsions. Reprinted from Juárez and Whitby,[25] Copyright (2013), with permission from Elsevier.

formation of a cement-like network structure. This phenomenon leads to the formation of emulsion gels, which represent a threshold where excess particles exceed the capacity for surface coverage, potentially altering the structural integrity of the emulsion.[13,26–29]

The dimensions of the particles also significantly influence in determining both their mobility within the continuous phase and the curvature of Pickering emulsions. In principle, smaller nanoparticles have larger diffusion coefficients, implying faster adsorption on the droplet surface.[30,31] Typically, droplet size tends to increase with increasing particle size due to the constraint that larger particles face in conforming to the pronounced curvature of smaller droplets. This can be understood by considering that flexural elasticity, which characterizes the resistance to bending in a densely packed particle monolayer, is directly proportional to the square of the particle radius. Consequently, an increase in particle size will result in a corresponding increase in droplet size.[32]

9.2.2 Type of Solid Particles Used for Pickering Emulsion Stabilization

A wide range of particles can be used as effective stabilizers for Pickering emulsions. In fact, these emulsions are commonly stabilized by colloidal particles on the micro- or nanoscale with many different chemical natures. This broad category includes a spectrum of materials such as silica, clays, and carbon, as well as polymeric colloids such as starch, cellulose, and chitosan particles.[7,8,33] Their diverse nature not only contributes to stabilization but also expands the possibilities for tailored applications across industries. This adaptability and versatility in particle selection and preparation techniques offers promising avenues for expanding the scope of Pickering emulsion-based solutions in various fields. The ability to customize particle characteristics is in line with emerging trends that focus on precision engineering in formulation design, improving stability and functionality, while exploring environmentally friendly and biodegradable materials for sustainable and impactful applications.[8]

Silica particles are widely used in Pickering emulsions due to their widespread use. Extensive experimentation has revealed significant trends in the behavior of silica particles with respect to their effectiveness in stabilizing Pickering emulsions. Unmodified silica tends to show a tendency to stabilize O/W Pickering emulsions, which is attributed to the inherent hydrophilicity resulting from the surface silanol groups. On the other hand, hydrophobically modified silica particles show a preference for stabilizing W/O Pickering emulsions.[34] As a result, much research has been directed toward the production of various forms of modified silica. Among them, fumed silica particles are of particular importance as they are known for their adaptability in being made hydrophobic by surface modification. This hydrophobic modification involves treating the surface silanols of fumed silica particles with organic silanes, resulting in varying degrees of hydrophobicity. The extent of this modification allows the surface characteristics of the particle to be customized, imparting varying degrees of hydrophobicity depending on the specific application requirements.[35,36] Therefore, the use of silica particles offers a broad range of possibilities for the stabilization of emulsions, enabling to obtaining both W/O and O/W emulsions depending on their silanization degree.[35]

The variables affecting silica-stabilized Pickering emulsions, such as pH and salt concentration, have been systematically investigated,[37] showing that pure fumed silica faces challenges in stabilizing Pickering emulsions under basic conditions due to its high hydrophilicity and surface charge. Conversely, at lower pH levels, silica particles tend to aggregate, reducing their stabilizing capacity. To overcome these limitations, suitable molecules are coupled to pure silica to enhance their stabilizing capacity while maintaining a moderate surface charge.[38]

Unlike silica particles, the use of organic latex particles as emulsifiers is limited. This limitation is because latex particles do not readily attach to the water-oil interface under typical conditions. Despite being synthesized and dispersed in water, latex particles tend to be hydrophobic and therefore can stabilize w-o emulsions.[39] Interestingly, this behavior seems to contradict Bancroft's rule, which applies primarily to surfactant stabilized emulsions, but in the case of Pickering emulsions, stability is ensured by the formation of a densely packed adsorption layer of particles, regardless of the phase origin of these particles, creating a robust and protective shell-like layer which is essential for stability.

Clays are also a widely used material in the preparation of Pickering emulsions, mainly due to their nonpolluting nature, cost-effectiveness, and ready availability, unlike conventional surfactants. Given the inherent hydrophilicity of clay surfaces, modification of these surfaces with organic or amphiphilic molecules is essential to increase the propensity of clay particles to adsorb at oil interfaces.[40]

Recently, considerable attention has been paid to the search for bio-derived particles, e.g., modified cellulose, lignin, modified starch granules, zein particles, chitosan particles, or polyphenol crystals, for use in Pickering stabilization.[8,23] However, the inherent properties of these naturally derived particles sometimes require adjustment to optimize their functionality as stabilizers in Pickering emulsions, which has involved their modification by various chemical or physical treatments. These treatments are used to improve their surface properties, such as hydrophilicity or amphiphilicity, to allow better adsorption at the oil-water interface. In addition, the formation of complexes with other compounds such as surfactants, biopolymers, or additional particles serves to enhance their stability-enhancing capabilities. The seeking for bio-derived particles is driven by the desire to harness natural sources and create stabilizers that are in line with sustainable and environmentally friendly practices. The ongoing quest is to strike a balance between preserving the natural properties of these biological particles and enhancing their performance as effective stabilizers. Future efforts are aimed at further understanding the interplay between natural particle properties and modifications to optimize their use in Pickering stabilization without compromising their inherent bio-origin benefits.

Initially, the use of spherical particles dominated the field of Pickering emulsions. However, as research has progressed, there has been a burgeoning interest in the production of nonspherical particles. This recent trend has led to the investigation of the emulsion-stabilizing capabilities of particles of various shapes, such as rods, fibers, and cubes, beyond their conventional spherical counterparts.[41] Madivala et al.[42] carried out a study to investigate the effect of different particle shapes on emulsion stability. Their research involved the use of particles with different aspect ratios to produce Pickering emulsions. At elevated concentrations, particles aligned end-to-end at the interface formed a triangular mesh structure. This mesh structure acted as a core support for the emulsion, preventing droplet coalescence. Conversely, at lower concentrations, the particles adopted a stripe configuration. Comparative analysis showed that ellipsoidal particles were more effective in stabilizing emulsions than spherical particles. Furthermore, the study found a correlation between increased stability and the aspect ratio of the particles, demonstrating that as the aspect ratio of the particles increased, the stability of the emulsion also increased.[43]

9.2.3 Advantages and Disadvantages of Pickering Emulsions

The unique properties of Pickering emulsions have paved the way for innovative solutions and advances in a wide range of industries, creating an exciting landscape for their use and development. In fact, Pickering emulsions offer several advantages over traditional emulsions that have stimulated extensive research and promising applications. Their key advantages include exceptional stability, reduced toxicity, and a wide range of available particle options.

The superior stability of Pickering emulsions is a defining characteristic that has attracted considerable attention. Stabilized by solid particles at the interface of immiscible phases, these emulsions exhibit robust resistance to coalescence and separation, ensuring extended shelf life and improved durability. This stability characteristic has attracted considerable interest from various industries seeking reliable and durable formulations. Another compelling feature of Pickering emulsions is their reduced toxicity compared to traditional emulsifiers. Many of the particles used in Pickering emulsions are derived from natural or benign materials, reducing concerns about potential adverse health effects or environmental hazards associated with conventional surfactants or emulsifiers. In addition, the wide variety of particles available for Pickering stabilization offers flexibility and versatility in formulation design. This diversity allows emulsions to be tailored to specific requirements, incorporating particles from different sources and with different physicochemical properties. The ability to select particles based on specific needs or desired functionalities opens the door to a wide range of potential applications in different industries.[41]

Despite the multiple advantages of Pickering emulsions, they present several drawbacks that have limited their use. These start with the fact that most of the research in Pickering emulsions remains confined to laboratory research and small-scale experiments. Overcoming several hurdles is imperative to transition these studies into viable research for the true application of Pickering emulsions. Moreover, regulating the size of Pickering emulsions presents a more intricate task compared to emulsions stabilized by surfactants, especially concerning high particle contents or larger particle sizes. While the advent of membrane emulsification and microfluidic methods has partially mitigated this issue, challenges persist in achieving large-scale production. The absence of precise size distributions poses complications in both the utilization and evaluation of Pickering emulsions, underscoring the necessity for refinement in this aspect.[41]

9.2.4 Applications of Particle-Stabilized Emulsions

Pickering emulsions, stabilized by solid particles at the oil-water interface, have become versatile systems with a wide range of applications in a variety of industries. Known for their stability and unique interfacial properties, these emulsions are widely used in a variety of fields due to their potential to revolutionize formulations and processes.[7]

In cosmetics and personal care, Pickering emulsions offer a compelling alternative to traditional surfactant-stabilized systems: their enhanced stability and tunable properties make them ideal candidates for the development of skin care products, sunscreen formulations, and make-up. In addition, the ability to encapsulate active ingredients in these emulsions allows for controlled release, improving efficacy and stability in cosmetic formulations.[15,44]

Pickering emulsions have also attracted considerable interest in the pharmaceutical and biomedical sectors as potential drug delivery vehicles. Their controlled-release capabilities, enabled by the formation of a dense shell of solid particles around droplets, provide opportunities for targeted and sustained drug delivery. Researchers are exploring these emulsions for applications in oral, topical, and intravenous drug delivery systems to achieve better therapeutic outcomes.[45,46] Moreover, in the food industry, Pickering emulsions find use in the creation of healthier, structured food products. By utilizing food-grade solid particles, these emulsions offer improved stability in food formulations, enabling the encapsulation of flavors, nutrients, or bioactive compounds, and enhancing their delivery and bioavailability upon consumption.[16,47] In addition, the use of Pickering emulsions in the food industry can also help in the control of the texture and taste of the products and the control of fats and calories.[48]

Pickering emulsification has become a well-established method for creating smart products by incorporating raw materials with reactive properties. This technique exploits the inherent advantages of Pickering emulsions, where solid particles stabilize emulsions at the interface between immiscible phases. Incorporating stimuli-responsive materials into Pickering emulsions allows the development of novel and tailored products with stimuli- or environmental-responsive behavior, capable of undergoing significant changes triggered by external factors such as temperature, pH changes, ionic strength, electric or magnetic fields, and light. The use of Pickering emulsification to create smart products thus provides a platform for developing materials that respond intelligently to external stimuli, allowing precise control and modulation of their properties. This capability holds great promise for advancing technologies for drug delivery, bioseparation processes, and other biomedical applications, and is driving research and development of functional materials in a wide range of fields.[49–51]

Nano- or micromaterials derived from Pickering emulsions have also attracted much attention in the field of fine chemistry, particularly catalysis.[52–54] The interest of these materials lies in their ability to provide larger interfacial areas, a property that promises to significantly improve catalytic efficiency. The large interfacial areas presented by these materials serve as fertile ground for catalytic reactions, promoting improved efficiency in fine chemical processes. In addition, the unique properties of Pickering emulsions offer distinct advantages to catalysts over traditional emulsions. One such advantage is the facilitation of easier separation and recovery after reaction. This ease of separation is due to the robust stabilization of the emulsions by solid particles, allowing easy and efficient recovery of catalysts, thereby streamlining the process, and minimizing waste. In addition, the novel properties of Pickering emulsions extend their advantage to selective catalysis, enabling targeted and specific catalytic activities. This selectivity is critical in fine chemistry applications where precise control of reactions and product formation is paramount. The above characteristics offer a promising future for the use of Pickering emulsion-derived materials in catalytic applications within fine chemical synthesis, contributing to the development of more efficient and selective chemical processes.

The versatility of Pickering emulsions extends their application beyond the above examples, spanning environmental applications, material science, and enhanced oil recovery, showcasing their potential to innovate and redefine various industrial processes and product formulations.[41]

9.3 SURFACTANT-STABILIZED EMULSIONS

9.3.1 Surfactant in Emulsion Stabilization

Surfactants are amphiphilic molecules consisting of hydrophilic (water-loving) and hydrophobic (water-repellent) moieties. These unique properties allow surfactants to accumulate at the oil-water interface and form a protective interfacial layer. Among surfactants can be included both small molecular and macromolecular emulsifiers. In general, surfactants are classified into four main categories based on their charge: anionic, cationic, nonionic, and amphoteric.[55]

Anionic surfactants, such as sodium dodecyl sulfate (SDS), possess a negatively charged hydrophilic head group. They are commonly used in emulsion stabilization due to their strong adsorption at the oil-water interface, forming stable interfacial films. Cationic surfactants, on the other hand, have a positively charged hydrophilic head group and are often employed in specific applications where anionic systems are not suitable. Nonionic surfactants have uncharged hydrophilic head groups and are widely utilized in various industries due to their compatibility with a wide range of formulations. Amphoteric surfactants contain both positively and negatively charged groups and exhibit versatile properties depending on the pH of the system.

Over time, low molecular weight surfactants have demonstrated their efficacy in ensuring the kinetic stability of various emulsions. However, contemporary research has increasingly focused on advancing and comprehensively grasping the stability aspects provided by what are termed ‘clean label’ emulsifiers. This shift is notably characterized by a keen interest in exploring emulsifying agents derived from biopolymers (macromolecular-like emulsifiers), which encompass a spectrum of substances such as proteins, starches, and non-starch polysaccharides. The pursuit of these 'clean label' emulsifiers stems from a growing emphasis on natural and sustainable alternatives, especially in formulations intended for food, cosmetic, and pharmaceutical applications. The utilization of biopolymers as emulsifiers, either in isolation or in combination with traditional surfactants, represents a progressive step toward achieving stable emulsions with reduced reliance on synthetic additives. It should be noted that this shift not only addresses stability concerns but also aligns with the broader trajectory toward sustainable and environmentally conscious product development.[23]

Proteins inherently possess both hydrophilic and hydrophobic regions along their molecular structure, while natural polysaccharides, in contrast, typically exhibit polar backbones with limited nonpolar side chains attached.[3] Some polysaccharide chains, primarily hydrophilic in nature, may incidentally include trace amounts of protein 'impurities' that confer surface activity.[56] Despite extensive research focusing on the stabilization of O/W emulsions using proteins and polysaccharides, to date, there remains a noticeable absence of studies demonstrating the exclusive stabilization of W/O emulsions solely through a biopolymer interfacial layer. This gap in research highlights an intriguing aspect of emulsion stabilization, particularly in the realm of W/O emulsions, where the role of biopolymers in forming standalone interfacial layers remains unexplored.

9.3.2 Stabilization of Emulsions by Surfactants

When mixtures of two immiscible liquids, i.e., oil and water, are considered, the hydrophilic-lipophilic balance (HLB) emerges as a critical determinant for the disposition of surfactant molecules.[57] The arrangement of the surfactant at the oil-water interface is governed by its packing parameter, which influences whether the surfactant monolayer curves toward water, oil, or maintains a relatively flat structure. This configuration is dictated by the structural geometry of the surfactant molecules, where one side of the monolayer interacts with water, while the other side interacts with oil. In the case of hydrophilic surfactants (nonionic with high degrees of ethoxylation or ionics without salt), the area per head group surpasses that of the chain, causing monolayers to curve around oil, thus fostering the formation O/W emulsions. Conversely, for more lipophilic surfactants (nonionic with low degrees of ethoxylation or ionics in the presence of adequate salt), the area per chain exceeds that of the head group. This leads to the water becoming the dispersed phase, resulting in the creation of W/O emulsions.[13]

Surfactant-stabilized emulsions can take advantage of different stabilization mechanisms to prevent droplet coalescence and maintain long-term stability. These mechanisms include steric hindrance, electrostatic repulsion, and interfacial elasticity. Steric hindrance occurs when surfactant molecules adsorb at the oil-water interface, forming a densely packed layer that physically hinders the approaching droplets. The hydrophilic head groups extend into the continuous phase, while the hydrophobic tails face the oil droplets, creating a repulsive barrier. Electrostatic repulsion is prevalent in emulsions stabilized by charged surfactants or through the addition of electrolytes. The charged surfactant molecules adsorb at the interface, creating an electrical double layer that results in repulsive forces between droplets of the same charge. Interfacial elasticity arises from the formation of surfactant monolayers at the interface. These monolayers possess viscoelastic properties, allowing them to deform and absorb energy upon droplet collision, thereby preventing coalescence.[58]

9.3.3 Advantages and Limitations of Surfactant-Stabilized Emulsions

Surfactant-stabilized emulsions offer several advantages in various applications. They provide excellent droplet size control, allowing the production of emulsions with precise and uniform droplet sizes.[59,60] Surfactants also enhance the stability of emulsions under various environmental conditions, such as temperature and pH fluctuations.[61] Furthermore, surfactants can impart desirable rheological properties, influencing the viscosity and texture of the emulsion system.[62] However, despite their remarkable attributes, surfactant-stabilized emulsions are not without limitations. Concerns regarding the potential toxicity induced by surfactant molecules, notably in pharmaceutical and cosmetic formulations, necessitate meticulous scrutiny and selection of surfactant types, particularly in pharmaceutical and cosmetic formulations, necessitating careful selection and evaluation of surfactant types.[63] Moreover, surfactant monolayers at the interface can be susceptible to displacement or degradation in the presence of external factors, leading to a decrease in emulsion stability over time.[64] Additionally, the concentration of surfactants required for effective stabilization can significantly influence production costs and formulation design, posing economic considerations.[65]

As versatile as surfactant-stabilized emulsions are in providing stability, controlling droplet size, and influencing rheological properties, these advantages come hand in hand with certain constraints that require careful attention and strategic management, especially in industries where safety, stability, and cost-effectiveness are of paramount importance.

9.3.4 Applications and Industrial Relevance

Surfactant-stabilized emulsions are an indispensable asset in a wide range of industries, with a profound impact and versatility that permeates diverse applications. Their widespread use spans numerous sectors, each of which benefits from the distinctive advantages these emulsions offer. The role of these emulsions is far-reaching, encompassing crucial functions in the creation of various consumables within the food industry. From the creamy consistency of mayonnaise to the smooth texture of salad dressings and the stable homogeneity of dairy drinks, the careful application of surfactant-stabilized emulsions forms the basis of these culinary essentials, ensuring uniformity, palatability, and extended shelf life.[66]

Beyond the food industry, the pharmaceutical field benefits from the contribution of surfactant-stabilized emulsions. These emulsions serve as central vehicles in pharmaceutical formulations, enabling the encapsulation and delivery of therapeutic agents. Their role in creating stable drug formulations for oral or topical delivery enhances drug bioavailability, facilitates controlled release, and optimizes therapeutic efficacy. These encapsulation capabilities enhance patient compliance, promote targeted drug delivery, and support a variety of drug delivery systems, driving advances in pharmacotherapy.[67,68]

In cosmetics and personal care, the fusion of science and aesthetics is evident in the formulation of creams, lotions, shampoos, and a range of cosmetic products. This is where surfactant-stabilized emulsions come into their own, providing the ideal base for these products. They contribute significantly to the sensory properties of the products, their stability, and ease of use. By imparting desirable textures and stability, emulsions not only improve product efficacy but also enhance the user experience, playing a key role in meeting consumer expectations for efficacy and sensory satisfaction.[69–71] In addition, surfactant-stabilized emulsions are indispensable in the field of enhanced oil recovery techniques. Their ability to manipulate interfacial properties plays a key role in the mobilization and displacement of oil from reservoirs. By altering the interfacial tension between oil and water, these emulsions facilitate improved oil recovery, mitigating the challenges associated with oil production and optimizing production efficiency in the petroleum industry.[72–74]

In summary, the many benefits of surfactant-stabilized emulsions underline their key role in a wide range of industries. Their ability to control droplet size, stability under variable conditions, and influence rheological properties is driving their widespread adoption. However, critical considerations regarding surfactant toxicity, long-term stability, and production costs require careful evaluation. Nevertheless, the wide range of applications in the food, pharmaceutical, cosmetic, and oil industries underline the industrial relevance of surfactant-stabilized emulsions. Their versatility and key role in improving product performance, formulation efficacy, and industrial processes cements their status as indispensable components in many sectors, promising continued innovation and application in the years ahead.

9.4 MAKING THE DIFFERENCE: A COMPARATIVE ANALYSIS BETWEEN PARTICLE-STABILIZED EMULSIONS AND SURFACTANT-STABILIZED EMULSIONS

The above sections have introduced so far the main aspects of emulsions stabilized by particles and by surfactants. However, to deepen the understanding of the different roles of the stabilizing agents, it is required to perform a comparative analysis between both types of stabilization. In the following, a comparison of the most relevant issues associated with each family of emulsions is included.

The role of surfactants in the stabilization of dispersed systems such as emulsions is linked to their ability to produce a significant decrease in the interfacial tension between the liquid phases, effectively preventing structural coarsening. Nevertheless, the system retains a fluid-like nature, with the resulting emulsion exhibiting a viscosity similar to that of its constituent liquids. Conversely, the solid particles introduce an additional elastic force that stops the structural coarsening while giving the emulsion solid-like properties characterized by a finite modulus of elasticity and a discernible yield stress.[73] The above-mentioned differences significantly influence the long-term stability of the final emulsions as was demonstrated by Chaudhari et al.[75] They demonstrated that while the size of the droplets of particle-stabilized emulsions does not undergo any significant modification under heating or cooling processes, a fast destabilization was found for surfactant-stabilized systems.

Despite that the most accepted framework considers particle-stabilized systems as the paradigm of ultra-stable systems, very recently Dekker et al.[76] have suggested that there is no clear distinction between the physicochemical aspects that govern the stabilization of both types of systems. In fact, they found that independently of the nature of the particles/molecules acting as stabilizing agents, a noticeable decrease in interfacial tension can occur as the concentration of stabilized increases. Surprisingly, high adsorption densities emerge for the particle-based systems, mirroring the behavior observed in the surfactant system. The decrease in interfacial tension for particle systems appears to be a result of numerous particles congregating at the interface, each exhibiting an adsorption energy of a few k_BT. Dynamic measurements of interfacial tension corroborate that these systems are in a state of equilibrium. Notably, the characteristic timescale for adsorption in particle-based systems significantly exceeds that observed in surfactants, aligning with their inherent size disparities. Moreover, emulsions stabilized by particles demonstrate lower resistance against coalescence compared to their surfactant-stabilized counterpart.

9.5 CONCLUDING REMARKS

In emulsion science and technology, the choice of stabilizer is crucial and has a major influence on the stability, properties, and performance of the resulting emulsions. In this chapter, particle-stabilized and surfactant-stabilized emulsions have been extensively reviewed, and their stabilization mechanisms and comparative properties have been analyzed. Particle-stabilized emulsions, orchestrated by Pickering stabilization, show robustness due to particles irreversibly adsorbed at the oil-water interface. In contrast, surfactant-stabilized emulsions rely on the reduction of interfacial tension and excel in kinetic stability, but face challenges in long-term stability. Understanding these mechanisms is critical to tailoring emulsions for specific applications. The choice between particle and surfactant stabilizers therefore has a significant impact on emulsion properties. Particle-stabilized emulsions often exhibit broader droplet size distributions, while surfactant-stabilized counterparts offer finer droplet sizes and improved control. These differences have a profound impact on applications ranging from food formulations to pharmaceutical delivery systems, providing tailor-made solutions for diverse industrial needs. Both types of stabilizers offer unique benefits and challenges. Particle-stabilized emulsions may face hurdles in producing uniform particles or scaling up production, while surfactant-stabilized systems face issues of long-term stability and potential toxicity. Overcoming these challenges depends on continued research and innovation, with recent advances in hybrid stabilizer systems showing promise in synergizing strengths and mitigating limitations.

The horizon of emulsion science is teeming with the potential for revolutionary innovation: novel approaches involving smart materials, bio-derived stabilizers, and cutting-edge characterization techniques promise to improve stability, droplet size control, and overall emulsion properties. Molecularly engineered stabilizers have the potential to redefine emulsion formulations, offering a high degree of control over stability and functionality.

The choice between particle-stabilized and surfactant-stabilized emulsions is not a one-size-fits-all proposition. Each type of stabilizer has its strengths and limitations and is tailored to specific application requirements. Embracing this diversity and understanding the interplay between stabilizer choice and emulsion properties is fundamental to creating tailor-made solutions for different industrial needs. In the ever-evolving landscape of emulsion science, collaborative interdisciplinary efforts will drive innovative formulations that push existing boundaries. These advances will open new applications across industries, unlocking the vast potential of stable and functional emulsions in our interconnected world. The quest for stable, functional emulsions remains a dynamic journey that promises endless opportunities to improve products and processes across multiple industries.

ACKNOWLEDGMENTS

The team in Madrid was funded by grants PID2023-147156NB-I00 funded by MCIN/AEI/10.13039/501100011033 (Spain) and PR12/24-31566 (Ayudas para la Financiación de Proyectos de Investigación UCM 2023), and the European Innovative Training Network-Marie Sklodowska-Curie Action NanoPaInt (grant agreement 955612), funded by the E.U. A.M. was funded by MCIN/AEI/10.13039/501100011033 (Spain) under grant PID2021-129054NA-I00 and the IKUR Strategy under the collaboration agreement between the Ikerbasque Foundation and Materials Physics Center on behalf of the Department of Education of the Basque Government.

REFERENCES

1. Kale, S.N.; Deore, S.L. Emulsion micro emulsion and nano emulsion: A review. *Sys. Rev. Pharm.* **2016**, *8*, 39–47.
2. Israelachvili, J. The science and applications of emulsions — an overview. *Colloids Surf. A* **1994**, *91*, 1–9.
3. McClements, D.J. *Food emulsions*, CRC Press: Boca Raton, FL, USA, 2015.
4. Zembyla, M.; Murray, B.S.; Sarkar, A. Water-in-oil emulsions stabilized by surfactants, biopolymers and/or particles: A review. *Trends Food Sci. Technol.* **2020**, *104*, 49–59.
5. Tian, Y.; Zhou, J.; He, C.; He, L.; Li, X.; Sui, H. The formation, stabilization and separation of oil–water emulsions: A review. *Processes* **2022**, *10*, 739.
6. Deng, W.; Li, Y.; Wu, L.; Chen, S. Pickering emulsions stabilized by polysaccharides particles and their applications: A review. *Food Sci. Technol.* **2022**, *42*, e24722.
7. de Carvalho-Guimarães, F.B.; Correa, K.L.; de Souza, T.P.; Rodríguez Amado, J.R.; Ribeiro-Costa, R.M.; Silva-Júnior, J.O.C. A review of pickering emulsions: Perspectives and applications. *Pharmaceuticals* **2022**, *15*, 1413.
8. Yang, Y.; Fang, Z.; Chen, X.; Zhang, W.; Xie, Y.; Chen, Y.; Liu, Z.; Yuan, W. An overview of pickering emulsions: Solid-particle materials, classification, morphology, and applications. *Front. Pharm.* **2017**, *8*, 00287.

9. Dekker, R.I.; Velandia, S.F.; Kibbelaar, H.V.M.; Morcy, A.; Sadtler, V.; Roques-Carmes, T.; Groenewold, J.; Kegel, W.K.; Velikov, K.P.; Bonn, D. Is there a difference between surfactant-stabilised and pickering emulsions? *Soft Matter* **2023** *19*, 1941–1951.
10. Binks, B.P. Particles as surfactants—similarities and differences. *Curr. Opin. Colloid Interface Sci.* **2002**, *7*, 21–41.
11. Ramsden, W. Separation of solids in the surface-layers of solutions and 'suspensions' (Observations on surface-membranes, bubbles, emulsions, and mechanical coagulation).—preliminary account. *Proc. Roy. Soc. Lond.* **1904**, *72*, 156–164.
12. Pickering, S.U. CXCVI.—emulsions. *J. Chem. Soc., Trans.* **1907**, *91*, 2001–2021.
13. Aveyard, R.; Binks, B.P.; Clint, J.H. Emulsions stabilised solely by colloidal particles. *Adv. Colloid Interface Sci.* **2003**, *100–102*, 503–546.
14. Lagaly, G.; Reese, M.; Abend, S. Smectites as colloidal stabilizers of emulsions. *Appl. Clay Sci.* **1999**, *14*, 83–103.
15. Guzmán, E.; Ortega, F.; Rubio, R.G. Pickering emulsions: A novel tool for cosmetic formulators. *Cosmetics* **2022**, *9*, 69.
16. Chen, L.; Ao, F.; Ge, X.; Shen, W. Food-grade pickering emulsions: Preparation, stabilization and applications. *Molecules* **2020**, *25*, 3202.
17. Maestro, A.; Guzmán, E.; Ortega, F.; Rubio, R.G. Contact angle of micro- and nanoparticles at fluid interfaces. *Curr. Opin. Colloid Interface Sci.* **2014**, *19*, 355–367.
18. Honciuc, A.; Negru, O.-I. Role of surface energy of nanoparticle stabilizers in the synthesis of microspheres via pickering emulsion polymerization. *Nanomaterials* **2022**, *12*, 995.
19. González-González, A.; Sánchez-Arribas, N.; Santini, E.; Rodríguez-Villafuerte, J.L.; Carbone, C.; Ravera, F.; Ortega, F.; Liggieri, L.; Rubio, R.G.; Guzmán, E. Effects of oil phase on the inversion of pickering emulsions stabilized by palmitic acid decorated silica nanoparticles. *Colloids and Interfaces* **2022**, *6*, 27.
20. Chen, J.; Vogel, R.; Werner, S.; Heinrich, G.; Clausse, D.; Dutschk, V. Influence of the particle type on the rheological behavior of pickering emulsions. *Colloids Surf. A* **2011**, *382*, 238–245.
21. Denkov, N.D.; Ivanov, I.B.; Kralchevsky, P.A.; Wasan, D.T. A possible mechanism of stabilization of emulsions by solid particles. *J. Colloid Interface Sci.* **1992**, *150*, 589–593.
22. Kumar, A.; Li, S.; Cheng, C.-M.; Lee, D. Recent developments in phase inversion emulsification. *Ind. Eng. Chem. Res.* **2015**, *54*, 8375–8396.
23. Zembyla, M.; Murray, B.S.; Sarkar, A. Water-in-oil emulsions stabilized by surfactants, biopolymers and/or particles: A review. *Trend Food Sci. Technol.* **2020**, *104*, 49–59.
24. Sun, Z.; Yan, X.; Xiao, Y.; Hu, L.; Eggersdorfer, M.; Chen, D.; Yang, Z.; Weitz, D.A. Pickering emulsions stabilized by colloidal surfactants: Role of solid particles. *Particuology* **2022**, *64*, 153–163.
25. Juárez, J.A.; Whitby, C.P. Oil-in-water pickering emulsion destabilisation at low particle concentrations. *J. Colloid Interface Sci.* **2012**, *368*, 319–325.
26. Arditty, S.; Whitby, C.P.; Binks, B.P.; Schmitt, V.; Leal-Calderon, F. Some general features of limited coalescence in solid-stabilized emulsions. *Eur. Phys. J. E* **2003**, *11*, 273–281.
27. Binks, B.P.; Whitby, C.P. Silica particle-stabilized emulsions of silicone oil and water: aspects of emulsification. *Langmuir* **2004**, *20*, 1130–1137.
28. Binks, B.P.; Philip, J.; Rodrigues, J.A. Inversion of silica-stabilized emulsions induced by particle concentration. *Langmuir* **2005**, *21*, 3296–3302.
29. Chen, Q.-H.; Zheng, J.; Xu, Y.-T.; Yin, S.-W.; Liu, F.; Tang, C.-H. Surface modification improves fabrication of pickering high internal phase emulsions stabilized by cellulose nanocrystals. *Food Hydrocoll* **2018**, *75*, 125–130.

30. Einstein, A. Über Die von Der Molekularkinetischen Theorie Der Wärme Geforderte Bewegung von in Ruhenden Flüssigkeiten Suspendierten Teilchen. *Ann.Phys.* **1905**, *322*, 549–560.
31. Lin, Y.; Skaff, H.; Emrick, T.; Dinsmore, A.D.; Russell, T.P. Nanoparticle assembly and transport at liquid-liquid interfaces. *Science (1979)* **2003**, *299*, 226–229.
32. Kralchevsky, P.A.; Ivanov, I.B.; Ananthapadmanabhan, K.P.; Lips, A. On the thermodynamics of particle-stabilized emulsions: Curvature effects and catastrophic phase inversion. *Langmuir* **2005**, *21*, 50–63.
33. Asfour, M.H.; Elmotasem, H.; Mostafa, D.M.; Salama, A.A.A. Chitosan based pickering emulsion as a promising approach for topical application of rutin in a solubilized form intended for wound healing: In vitro and in vivo study. *Int. J. Pharm.* **2017**, *534*, 325–339.
34. Binks, B.P.; Lumsdon, S.O. Transitional phase inversion of solid-stabilized emulsions using particle mixtures. *Langmuir* **2000**, *16*, 3748–3756.
35. Drelich, A.; Gomez, F.; Clausse, D.; Pezron, I. Evolution of water-in-oil emulsions stabilized with solid particles. *Colloids Surf. A* **2010**, *365*, 171–177.
36. Yan, N.; Gray, M.R.; Masliyah, J.H. On water-in-oil emulsions stabilized by fine solids. *Colloids Surf. A* **2001**, *193*, 97–107.
37. Binks, B.P.; Lumsdon, S.O. Stability of oil-in-water emulsions stabilised by silica particles. *Phys. Chem. Chem. Phys.* **1999**, *1*, 3007–3016.
38. Maestro, A. Tailoring the interfacial assembly of colloidal particles by engineering the mechanical properties of the interface. *Curr. Opin. Colloid Interface Sci.* **2019**, *39*, 232–250.
39. Binks, B.P.; Lumsdon, S.O. Pickering emulsions stabilized by monodisperse latex particles: Effects of particle size. *Langmuir* **2001**, *17*, 4540–4547.
40. Guillot, S.; Bergaya, F.; de Azevedo, C.; Warmont, F.; Tranchant, J.-F. Internally structured pickering emulsions stabilized by clay mineral particles. *J. Colloid Interface Sci.* **2009**, *333*, 563–569.
41. Wu, J.; Ma, G. Recent studies of pickering emulsions: Particles make the difference. *Small* **2016**, *12*, 4633–4649.
42. Madivala, B.; Fransaer, J.; Vermant, J. Self-assembly and rheology of ellipsoidal particles at interfaces. *Langmuir* **2009**, *25*, 2718–2729.
43. Dugyala, V.R.; Daware, S.V.; Basavaraj, M.G. Shape anisotropic colloids: Synthesis, packing behavior, evaporation driven assembly, and their application in emulsion stabilization. *Soft Matter* **2013**, *9*, 6711.
44. Venkataramani, D.; Tsulaia, A.; Amin, S. Fundamentals and applications of particle stabilized emulsions in cosmetic formulations. *Adv. Colloid Interface Sci.* **2020**, *283*, 102234.
45. Marto, J.; Gouveia, L.; Jorge, I.M.; Duarte, A.; Gonçalves, L.M.; Silva, S.M.C.; Antunes, F.; Pais, A.A.C.C.; Oliveira, E.; Almeida, A.J.; Ribeiro, H.M. Starch-based pickering emulsions for topical drug delivery: A QbD approach. *Colloids Surf. B* **2015**, *135*, 183–192.
46. Frelichowska, J.; Bolzinger, M.-A.; Valour, J.-P.; Mouaziz, H.; Pelletier, J.; Chevalier, Y. Pickering w/o emulsions: Drug release and topical delivery. *Int. J. Pharm.* **2009**, *368*, 7–15.
47. Murray, B.S. Pickering emulsions for food and drinks. *Curr. Opin. Food Sci.* **2019**, *27*, 57–63.
48. Skelhon, T.S.; Grossiord, N.; Morgan, A.R.; Bon, S.A.F. Quiescent water-in-oil pickering emulsions as a route toward healthier fruit juice infused chocolate confectionary. *J. Mat. Chem.* **2012**, *22*, 19289.

49. Zhang, K.; Wu, W.; Guo, K.; Chen, J.; Zhang, P. Synthesis of temperature-responsive poly (*N*-Isopropyl Acrylamide)/poly(Methyl Methacrylate)/silica hybrid capsules from inverse pickering emulsion polymerization and their application in controlled drug release. *Langmuir* **2010**, *26*, 7971–7980.
50. Piao, S.H.; Kwon, S.H.; Zhang, W.L.; Choi, H.J. Celebrating soft matter's 10th anniversary: Stimuli-responsive pickering emulsion polymerized smart fluids. *Soft Matter* **2015**, *11*, 646–654.
51. Yao, L.; Wang, R.; Peng, S.; Liu, Z.; Li, H.; Xu, D.; Hu, L.; Mo, H. A "smart-sensing" bactericidal protein-based pickering emulsion. *J. Food Eng.* **2023**, *350*, 111491.
52. Chang, F.; Vis, C.M.; Ciptonugroho, W.; Bruijnincx, P.C.A. Recent developments in catalysis with pickering emulsions. *Green Chem.* **2021**, *23*, 2575–2594.
53. Rodriguez, A.M.B.; Binks, B.P. Catalysis in pickering emulsions. *Soft Matter* **2020**, *16*, 10221–10243.
54. Ni, L.; Yu, C.; Wei, Q.; Liu, D.; Qiu, J. Pickering emulsion catalysis: Interfacial chemistry, catalyst design, challenges, and perspectives. *Angew. Chem. Int. Ed.* **2022**, *134, e202115885.*
55. Palanuwech, J.; Coupland, J.N. Effect of surfactant type on the stability of oil-in-water emulsions to dispersed phase crystallization. *Colloids Surf. A* **2003**, *223*, 251–262.
56. Osano, J.P.; Hosseini-Parvar, S.H.; Matia-Merino, L.; Golding, M. Emulsifying properties of a novel polysaccharide extracted from basil seed (Ocimum Bacilicum L.): Effect of polysaccharide and protein content. *Food Hydrocoll* **2014**, *37*, 40–49.
57. Binks, B.P. *Modern Aspects of Emulsion Science*, The Royal Society of Chemistry: London, UK, 1999.
58. Zhu, Q.; Pan, Y.; Jia, X.; Li, J.; Zhang, M.; Yin, L. Review on the stability mechanism and application of water-in-oil emulsions encapsulating various additives. *Compr. Rev. Food Sci. Food Saf.* **2019**, *18*, 1660–1675.
59. Yuan, Q.; Williams, R.A.; Biggs, S. Surfactant selection for accurate size control of microcapsules using membrane emulsification. *Colloids Surf. A* **2009**, *347*, 97–103.
60. Jeong, W.-C.; Lim, J.-M.; Choi, J.-H.; Kim, J.-H.; Lee, Y.-J.; Kim, S.-H.; Lee, G.; Kim, J.-D.; Yi, G.-R.; Yang, S.-M. Controlled generation of submicron emulsion droplets via highly stable tip-streaming mode in microfluidic devices. *Lab Chip* **2012**, *12*, 1446.
61. Goodarzi, F.; Zendehboudi, S. A comprehensive review on emulsions and emulsion stability in chemical and energy industries. *Can. J. Chem. Eng.* **2019**, *97*, 281–309.
62. Zhu, Y.; Gao, H.; Liu, W.; Zou, L.; McClements, D.J. A review of the rheological properties of dilute and concentrated food emulsions. *J. Texture Stud.* **2020**, *51*, 45–55.
63. Santos, A.C.; Morais, F.; Simões, A.; Pereira, I.; Sequeira, J.A.D.; Pereira-Silva, M.; Veiga, F.; Ribeiro, A. Nanotechnology for the development of new cosmetic formulations. *Exp. Opin. Drug Deliv.* **2019**, *16*, 313–330.
64. McClements, D.J.; Jafari, S.M. Improving emulsion formation, stability and performance using mixed emulsifiers: A review. *Adv. Colloid Interface Sci.* **2018**, *251*, 55–79.
65. Kumar, N.; Ali, S.; Kumar, A.; Mandal, A. Design and formulation of surfactant stabilized o/w emulsion for application in enhanced oil recovery: Effect of PH, salinity and temperature. *Oil Gas Sci. Technol. – Rev. IFP Energies nouvelles* **2020**, *75*, 72.
66. Tan, C.; McClements, D.J. Application of advanced emulsion technology in the food industry: A review and critical evaluation. *Foods* **2021**, *10*, 812.
67. Callender, S.P.; Mathews, J.A.; Kobernyk, K.; Wettig, S.D. Microemulsion utility in pharmaceuticals: Implications for multi-drug delivery. *Int. J. Pharm.* **2017**, *526*, 425–442.
68. Wilson, R.J.; Li, Y.; Yang, G.; Zhao, C.-X. Nanoemulsions for drug delivery. *Particuology* **2022**, *64*, 85–97.

69. Fernández-Peña, L.; Mojahid, B.Z. El; Guzmán, E.; Ortega, F.; Rubio, R.G. Performance of oleic acid and soybean oil in the preparation of oil-in-water microemulsions for encapsulating a highly hydrophobic molecule. *Colloids and Interfaces* **2021**, *5*, 50.
70. Guzmán, E.; Fernández-Peña, L.; Rossi, L.; Bouvier, M.; Ortega, F.; Rubio, R.G. Nanoemulsions for the encapsulation of hydrophobic actives. *Cosmetics* **2021**, *8*, 45.
71. Dubuisson, P.; Picard, C.; Grisel, M.; Savary, G. How does composition influence the texture of cosmetic emulsions? *Colloids Surf. A* **2018**, *536*, 38–46.
72. Mandal, A.; Samanta, A.; Bera, A.; Ojha, K. Characterization of oil–water emulsion and its use in enhanced oil recovery. *Ind. Eng. Chem. Res.* **2010**, *49*, 12756–12761.
73. Faisal, W.; Almomani, F. A Critical review of the development and demulsification processes applied for oil recovery from oil in water emulsions. *Chemosphere* **2022**, *291*, 133099.
74. Zhou, Y.; Yin, D.; Chen, W.; Liu, B.; Zhang, X. A comprehensive review of emulsion and its field application for enhanced oil recovery. *Energy Sci. Eng.* **2019**, *7*, 1046–1059.
75. Chaudhari, A.; Pan, Y.; Nitin, N. Beverage emulsions: Comparison among nanoparticle stabilized emulsion with starch and surfactant stabilized emulsions. *Food Res. Int.* **2015**, *69*, 156–163.
76. Dekker, R.I.; Velandia, S.F.; Kibbelaar, H.V.M.; Morcy, A.; Sadtler, V.; Roques-Carmes, T.; Groenewold, J.; Kegel, W.K.; Velikov, K.P.; Bonn, D. Is there a difference between surfactant-stabilised and pickering emulsions? *Soft Matter* **2023**, *19*, 1941–1951.

10 Recent Advances in Controlling Emulsion Stability

Catherine P. Whitby

10.1 INTRODUCTION

This chapter reviews recent investigations into how emulsions respond to stress and break down over time. An emulsion is a colloidal dispersion of two immiscible liquids, typically oil and water. One of the liquids is dispersed as spherical droplets in the other liquid. This is typically achieved by mechanically mixing, or homogenizing: mixing fragments of one of the liquids into droplets of microscopic or colloidal size that are surrounded by molecules of other liquid. The large interfacial area between the two liquids, and hence the large surface energy, makes emulsions thermodynamically unstable. Emulsions made from pure, immiscible liquids separate rapidly back into the two bulk liquids. Emulsifiers such as surfactants or particles that accumulate at the interface between the two liquid phases, and reduce the surface energy, are incorporated into emulsions to impart kinetic stability.

Emulsions are used in a variety of products, including cosmetics, creams and lotions, paints and coatings, and foods. Products like mayonnaise take advantage of how emulsions yield and flow, their flavor, and visual appearance. The physicochemical properties of emulsions evolve due to physical and chemical processes that occur over time. The stability of an emulsion refers to its ability to resist changes to its properties and hence changes in how it functions.[1,2] The main processes that govern how emulsions age are coalescence, Ostwald ripening, flocculation, and gravity-induced separation. Chemical reactions that alter the composition of an emulsion can also impact emulsion performance. Recent insights into the mechanisms of physical and chemical instabilities in emulsions are described in the first section of this chapter.

Controlling the stability of emulsion-based products during their use is challenging.[3,4] Advances in our understanding of how emulsions break have led to the development of strategies for manipulating emulsion stability. The second section of this chapter discusses the state of the art in approaches to destabilizing emulsions. A common feature of many strategies is the addition of a demulsifier, a surface-active species that disrupts the interfacial layer. The application of chemical stressors that enhance drop coalescence, flocculation, or Ostwald ripening is discussed in turn. This section also highlights investigations into the response of emulsions to physical stresses, and how that response can be controlled to break emulsions on demand.

DOI: 10.1201/9781032636184-12

Examples of new materials prepared by breaking emulsions are described in the last section of the chapter. Emulsions have long been used as templates for fabricating materials ranging from capsules to porous solids and to gels.[4,5] Research into some materials has matured to the extent of harnessing the destabilization processes such as coalescence and flocculation that occur during their formation. The most well-known of these materials are the porous, interconnected solids being used, for example, as scaffolds for growing tissue. Emulsion gels consisting of flocculated drops being formulated as inks for three-dimensional printing are also considered. Finally, the advances made in fabricating oleogels as fat replacements in food products are explored.

10.2 DESTABILIZATION MECHANISMS

10.2.1 Coalescence

Coalescence occurs when two drops merge into a larger drop.[6,7] Coalescence occurs in several stages. It starts with the drops coming into close contact. Regions of each drop surface that are close together flatten and trap a thin film of the suspending liquid in between them. The liquid in the two drops combines into a single droplet when the film ruptures.

Coalescence reduces the total interfacial area in an emulsion. The average size of the drops in an emulsion increases as the drops coalesce, until a bulk layer of liquid forms outside of the emulsion. This section reviews recent advances in understanding the mechanisms of drop coalescence, and the correlations with bulk emulsion stability. The strategies used to destabilize an emulsion by causing the drops to merge are discussed in Section 10.3.

Coalescence is controlled by changes that occur at the microscopic scale of thin films, and by changes that occur at the mesoscopic scale between droplets. So, this section discusses recent investigations of thin film drainage in free-standing thin films, and between pairs of drops brought into contact in, for example, a microfluidic channel. It also describes studies of the impact of drop coalescence on the structure and properties of bulk emulsions.

While it is thought that coalescence is dominated by thin film dynamics, comparisons between behavior observed on the different length scales remain challenging. It is unclear whether the conditions in quiescent films or in a microfluidic channel in which pairs of droplets come into contact, are comparable to conditions on the macroscopic scale, where swarms of droplets interact under complex hydrodynamic flow patterns.[8] This means is not yet possible to predict the behavior that will occur on different length scales as an emulsion destabilizes due to coalescence.

10.2.1.1 Thin Films

Traditionally, coalescence was studied by following the rate of thin film drainage in free-standing thin films.[9] The lifetime of the thin film of suspending liquid that is squashed between the interfaces depends on the attractive and repulsive forces acting on it. The source of the attractive forces is van der Waals interactions. The repulsive forces can be electrostatic interactions if ionic surfactants (or polymers) are present in the suspending liquid, or steric interactions due to nonionic species. The sum of the forces per unit area of the film is called the disjoining pressure.

Recent experiments on thin films were typically done by pushing two liquid interfaces together to cause them to deform and rupture.[10,11] For example, the behavior of water drops held against an oil–water interface by gravitational forces was visualized using an optical microscope, as shown in Figure 10.1a. The lifetime of the thin film separating the drop from the interface was determined by measuring the time taken for the drop to coalesce with the interface. The lifetime of a thin film was found to be less than a few hundred seconds in the absence, or at low concentrations, of surfactant.[12] This is roughly the time it takes for all the liquid to drain from the film. Stable, equilibrium thin films formed at higher concentrations of surfactant.[13] The lifetime of surfactant-containing films depended on the time taken for rupture, rather than drainage.

The causes of rupture in thin liquid films of surfactant remain controversial. This was discussed in a recent review.[14] Fluctuations in surfactant concentration are thought to lower the concentration in regions of the film sufficiently for the repulsive forces to disappear.[15] The area depleted of surfactant is sometimes called a 'hole'. Films rupture when the hole becomes comparable in size to the thickness of the thin film. High-speed, laser-induced fluorescence was used to probe the spatiotemporal distribution of fluorescent surfactants in thin films, for the first time.[16] The surfactant concentration measured[16] along the surface of a drop coalescing with a planar interface confirmed theoretical predictions of surfactant distribution in thin films as they rupture.[17]

10.2.1.2 Droplets in Contact: From Pairs to Swarms

Coalescence is investigated on the mesoscopic scale by bringing drops into close contact using techniques like the atomic force microscope, or microfluidic devices. The drainage of the thin films of suspending liquid between the drops is followed by visualizing the changes in the droplet morphology, or by measuring the forces acting between droplets as they collide together.

The atomic force microscope was used in several recent investigations to measure the interaction forces between two drops as they were pushed together. This technique involves attaching one drop to the tip of a cantilevered beam. The drop is positioned above another droplet sitting on a substrate, as shown in Figure 10.1b. The cantilever is repeatedly lowered to bring the upper droplet into contact with the drop sitting on the substrate and then lifted to separate the two droplets. The vertical position of the cantilever tip and its deflection are measured and used to calculate the interaction forces between the two droplets.

The measured forces are typically compared with the forces predicted using the Stokes–Reynolds–Young–Laplace model. The model quantitatively predicts the forces due to the hydrodynamic pressure resulting from the thin film drainage, and the disjoining pressure generated by the surface forces.[18] The focus of recent studies was on the influence of adsorbed species on droplet interactions.[18–20] For example, the interactions between water drops coated with a food-grade surfactant

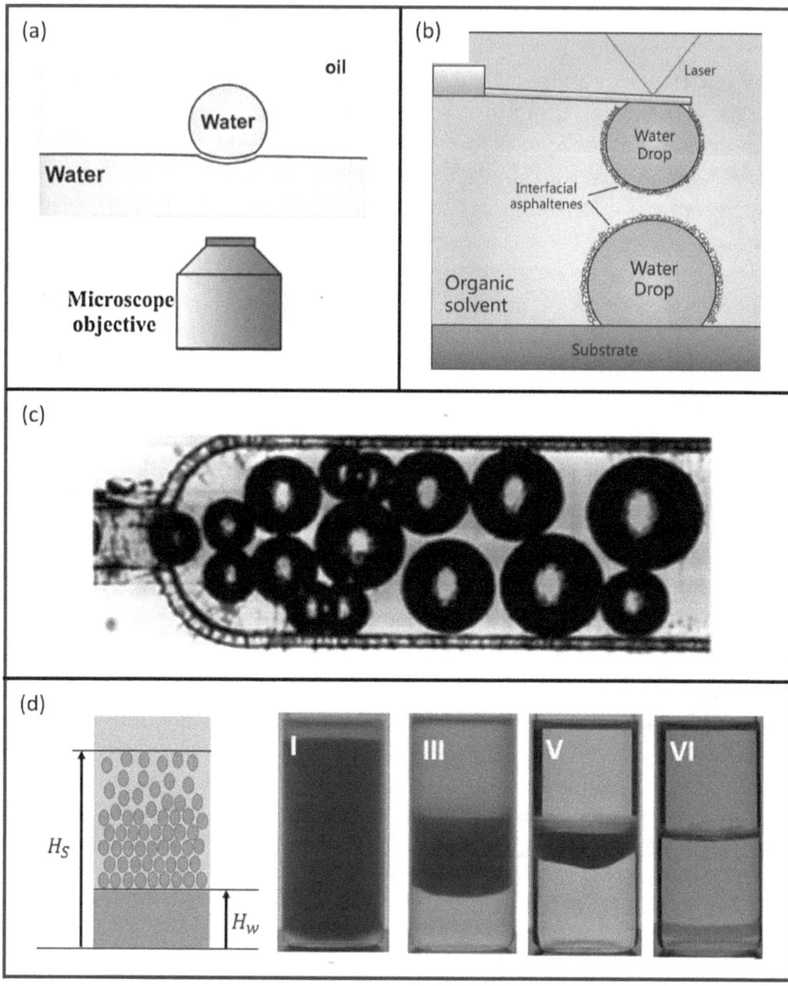

FIGURE 10.1 (a) Schematic illustration of the experimental setup for measuring the lifetime of the thin film of oil between a water drop and a planar oil–water interface. The drop is being held in contact with the planar interface by gravity. Reprinted from Politova et al.,[12] Copyright (2017), with permission from Elsevier. (b) Schematic illustration of the experimental setup for using the atomic force microscope to measure the interaction forces between two water droplets brought into contact in an organic solvent. Adapted with permission from Shi et al.,[19] Copyright (2017), American Chemical Society. (c) Photograph of the expansion channel in a microfluidic device used to bring collections of droplets into contact so that multiple coalescence events could be visualized. Reprinted from Gao et al.,[38] Copyright (2022), with permission from Elsevier. (d) Schematic illustration of the interface between a water-in-oil emulsion layer and the layer of coalesced water in a bulk emulsion. Also shown are photographs of the change in position of the interface in an emulsion destabilizing over time. Adapted with permission from Dinh et al.,[40] with permission from the Royal Chemical Society.

(polyglycerol polyricinoleate) were measured in canola oil.[18] Coalescence between drops incompletely coated by surfactant was linked to surfactant molecules bridging the droplet surfaces and causing a short-range attractive interaction.[18]

Molecular simulations were developed to test hypotheses about the impact of changes on the atomic scale on destabilization at the mesoscopic scale. For example, the coalescence of water droplets coated with asphaltenes and nonionic surfactants was simulated in oil.[21,22] The interaction forces derived from calculations of the free energy profiles were compared to the measured forces. The findings suggested that the magnitude of the energy barrier to coalescence is correlated with the structure and thickness of the interfacial layer.[21,22]

The challenge for modeling the interactions between a pair of drops is describing behavior that occurs over length scales that span from the micrometer scale of the droplets, down to the nanometer scale of liquid film thickness. Computational fluid dynamics simulations were developed to predict the outcome of droplet collisions.[23–27] The simulations revealed that surfactant transport at the interface slows the rate of bubble coalescence.[27] Surfactant accumulates at the neck that forms between bubbles when the film between them ruptures. At the same time, the surfactant is dragged away to the back of the drops by Marangoni forces. These studies were limited by the simplifying assumptions used. For example, the interfaces are coated with a uniform layer of insoluble surfactant.[27]

An alternative to direct numerical simulations is to use phenomenological models. Droplet coalescence models tend to focus on the droplet approach and film thinning,[28,29] rather than film rupture. The former processes are slower, and hence determine the rate of coalescence. Models of liquid film drainage were used, for example, to investigate the effects of surface viscoelasticity on coalescence.[29] Extending this work to deformable interfaces[30] highlighted that the mobility of a viscoelastic interface changes during the drainage process. Thin films between deformable, viscoelastic interfaces were found to initially drain as though the surfaces have negligible viscosity. The delay before the interfaces began to respond like Newtonian interfaces was linked to the Weissenberg number. This is the ratio of the elastic to viscous forces. Thus, it is important to account for the change in interfacial rheology during drainage when estimating the coalescence time.[30]

Many phenomenological models assume that droplets approach each other with a constant velocity, or that they interact under a constant force. This requires that the film separating the drops thin continuously until its thickness is small enough to indicate that the drops coalesced. The models do not account for the effect of droplet inertia, and hence the potential for droplets to rebound after coming into contact. Using a force balance instead allowed the approach velocity, and the interaction forces, to change with time. This led to the development of models capable of simulating droplet collisions that lead to rebounding, or coalescence.[26,31,32]

Several experimental investigations also tackled the challenge of how to correlate processes at the droplet level to those observed in thin films.[33] It is unclear whether measurements of thin film drainage between droplets interacting together in complicated velocity fields can be linked to drainage from a thin film between two interfaces under quiescent conditions. Comparisons between the coalescence of thin films[34] and drops of the same composition colliding in a hydrodynamic microfluidic trap found that the behavior observed on the microscopic and mesoscopic scales were consistent.[35] This suggests that coalescence is dominated by the dynamics in the thin films between drops in close contact.

Another new development was the move to studying coalescence events in larger groups of droplets. This was typically achieved using microfluidic devices which consisted of a structure (like a T-junction) for forming monodisperse drops, and an expansion structure where the drops are brought into close contact. In the example shown in Figure 10.1c, drops move from a narrow channel, through which they were flowing in a single file, into a larger chamber where they slow down and pack closely together. The continuous phase in between the droplets flows out toward the channel walls. Pairs of drops coalesce together when the thin film of liquid separating them ruptures. Anisotropic flows generated in the suspending liquid as pairs of drops merge can cause more drops in the chamber to coalesce. This is called cascade coalescence.[36]

The dynamic processes of cascade coalescence were visualized for a linear series of drops trapped in a sudden expansion microchannel.[37] It was argued that capillary waves on the drop surfaces caused a neck to form between two drops. Growth of the neck generated unstable flow fields inside the droplets that caused the formation of a neck between the second drop and the third one. The increasing instability of the flow field led to the second neck rupturing.[37] The probability of coalescence occurring in swarms of drops trapped in a rectangular expansion structure was linked to the time required for the thin films between the drops to drain.[38] In swarms of highly deformable bubbles in a quasi-two-dimensional vertical cell, it was found that smaller bubbles coalesced rapidly. Larger bubbles spent more time deforming to the shape of each other, before merging.[39]

10.2.1.3 Bulk Emulsions

The classical method for characterizing coalescence processes in bulk emulsions involved visual observations of the time taken for a coalesced layer to separate from the emulsion. The limitation of this approach is that only the release of (relatively) large volumes of liquid is detected by eye. Phase separation can now be detected earlier due to advances in high-speed photography and image processing techniques.[40,41] Liquid separation from emulsions is detected by analyzing digital photographs of the emulsions to extract the position of the interface between the emulsion and the pure liquid layer, as shown in Figure 10.1d. The kinetics of phase separation can be probed by monitoring the change in the position of the interface over time.

Understanding the bulk stability of emulsions (and foams) and its relationship to the behavior observed in thin films, in between pairs of drops, and in swarms of drops, remains challenging.[14] There are two central questions. The first is how to characterize the changes in chemical composition as the emulsion microstructure evolves. The second is how to link bulk stability to interfacial properties.

Novel techniques for probing the composition in situ in dilute emulsions were developed in the last ten years.[42,43] Time-of-flight secondary ion mass spectrometry was used, for example, to identify the locations of surfactant species in emulsions containing a commercial detergent. The emulsions were unstable to coalescence, with the average drop diameter increasing from 2 μm to 3 μm over one day. The mass spectrometry measurements were done by raster scanning over 2 μm × 2 μm areas in the emulsion. The depth of the analysis was less than 10 nm.[43] The data suggested that drops coalesced because larger surfactant species (such as D-tocopheryl polyethylene glycol succinate) took up to 24 h after emulsion formation to diffuse to the droplet surfaces.[43]

Another approach to characterizing changes in emulsion composition during destabilization was to use microsensors to measure the distribution of species, such as hydrogen cations and dissolved oxygen.[42] Measurements were made at 100 μm intervals over depths of 10,000 μm into surfactant-stabilized emulsions, at various times after emulsion formation. The average diameter of the drops in the emulsions was about 10 μm initially. A proportion of the drops coalesced to form a bulk layer of oil during the measurements. The pH in the emulsions increased over time. The rate of increase in pH depended on the type of surfactant present. Measurements at a model, planar oil–water interface indicated that the pH was higher at the interface than in the bulk aqueous phase. It was speculated that components in the oil that are partially miscible in water diffuse into the aqueous phase. It was argued that the rate of diffusion is affected by the charge on the surfactant species present at the interface, and this affects emulsion stability.[42]

There were also significant advances in correlating emulsion stability with interfacial properties, like the capillary pressure in particle-stabilized thin films[44] and the solid-like rheology of surfactant-stabilized interfaces.[45] Using a microscopic model based on the theory of hole nucleation, it was hypothesized that the probability of coalescence in an emulsion is proportional to the area of the thin films separating the drops.[40] A relationship between the proportionality constant (or coalescence frequency) and physicochemical properties of the system, like the surfactant concentration and surface excess, was theorized. The model successfully predicted the initial volume of emulsion formed and the separation kinetics.[40]

10.2.2 Ostwald Ripening

Ostwald ripening occurs when larger droplets grow at the expense of smaller droplets. It is caused by the difference in Laplace pressure between drops of different sizes.[46] Mass transfer between droplets occurs by molecular diffusion, or via micellar or nanodroplet carriers.

The mass transfer was studied by modeling and simulations in recent investigations of Ostwald ripening.[47–50] Ripening was typically modeled in systems consisting of two droplets of different sizes. The focus of several studies was on the impact of surfactants on the rate of Ostwald ripening. For example, the transfer between two benzene drops in water was studied by molecular simulations at a coarse-grained level, as shown in Figure 10.2a.[47] The droplets used in the simulations were made up of beads which represent a cluster of molecules. Figure 10.2b shows how the number of beads per droplet, and hence the size of the two droplets, changed over time. In the absence of surfactants, the smaller drop (called droplet 1) shrank during the simulations, while the larger drop (droplet 2) increased in size. The rate of change in the

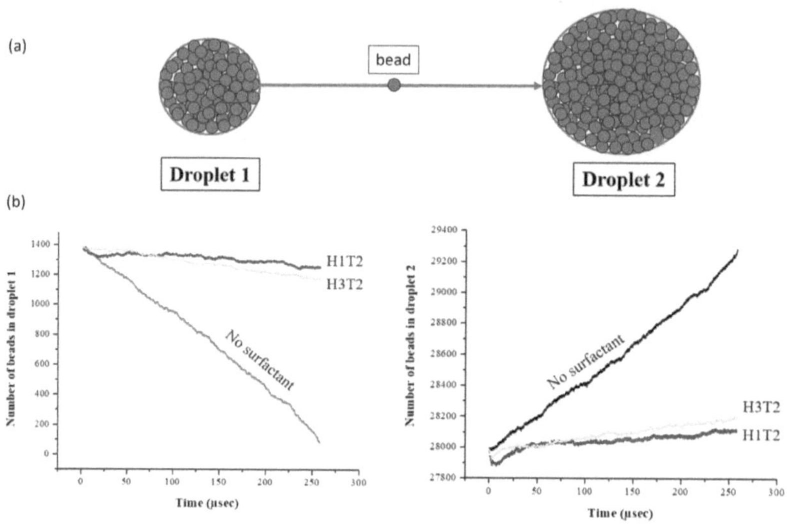

FIGURE 10.2 (a) Schematic illustration of the simulation of Ostwald ripening as the exchange of groups of molecules (called beads) between a small droplet (number 1) and a larger droplet (number 2). (b) Change in the number of beads in two benzene droplets over time during a simulation. In the absence of surfactant, the number of beads in droplet 1 decreases, while the number of beads in droplet 2 increases. There is minimal change in the number of beads in drops coated with surfactant (2000 H1T2 or 1300 H3T2). Adapted with permission from Khedr and Striolo,[47] Copyright (2019), American Chemical Society.

drop size decreased significantly in simulations of drops coated with surfactant. The results indicated that coating the drops with surfactants that form dense layers is an effective way to slow ripening.[47]

In emulsions undergoing ripening, drops move around due to Brownian motion, and encounter many other drops with a range of different sizes. It was argued that a moving droplet could gain and lose mass simultaneously.[48] This is because as it moves around, it encounters smaller drops that are sources of dissolved molecules, and larger drops that act as sinks for dissolved molecules. The evolution in size of a single droplet, and of a collection of 1,000 droplets, was modeled by coupling a Brownian dynamics algorithm to a stochastic equation for mass exchange.[48] Although anomalous diffusive behavior by individual droplets was observed, the average motion of collections of drops was normal. [48] The predicted rate of Ostwald ripening was consistent with experimentally measured rates for surfactant-stabilized emulsions. This included emulsions where the surfactant layer coating the droplet surfaces had a viscous, or elastic response, to applied stress.[48]

10.2.3 Flocculation

Flocculation occurs when two (or more) drops adhere together, but do not merge. This happens when the intermolecular forces between two drops in close contact are sufficiently repulsive to keep the drops separated and prevent the thin film between them from rupturing. An emulsion is described as flocculated when a substantial proportion of the drops exist in clusters (or 'flocs'), rather than being well dispersed in the continuous phase.[51] Flocculation enhances the rate at which the two liquid phases separate. The presence of flocs also affects the appearance of an emulsion, and how it flows.

Experimental investigations of flocculation typically follow changes in the drop size distribution over time. Many techniques used to measure drop size distributions in an emulsion are restricted to very dilute emulsions (with drop volume fractions less than 0.01). In a recent advance, photon density wave spectroscopy was applied to directly probe drop size distributions in concentrated and polydisperse emulsions.[52] This new light scattering technique was successfully used to follow flocculation in an emulsion with a drop volume fraction of 0.1.

Flocculating drops form an extended network throughout the continuous phase that eventually collapses under its own weight. Predicting when a flocculated emulsion will become inhomogeneous remains a challenge.[53] While strongly flocculated emulsions separate quickly, destabilization is slower in weakly flocculated emulsions. The separation caused by gravity (or by contraction of the drop network, which is called syneresis) is thought to be delayed while the strands in the network associate and dissociate.

Multiple static light scattering measurements were used to probe how long before a weakly flocculated emulsion separated.[53] The kinetics of the rearrangements in the droplet networks were followed by measuring the backscattered light intensity. The initial rate of change of the intensity of the backscattered light was found to correlate with the time when the emulsion collapsed. The faster an emulsion structure changed, the sooner it separated. To quantify the rate of change, the photon flux was used to calculate the photon transport path length. The path length depends on the structure factor, that is, on how the drops are arranged in space. The time until an emulsion destabilized was found to correlate with the magnitude of the change in the photon transport path length over one day, and hence with the extent of change in the spatial distribution of the drops for one day.[53]

10.2.4 Creaming and Sedimentation

Creaming occurs because liquid phases of different densities tend to separate due to gravity.[54] Drops that are lighter than the continuous phase tend to rise to the top of the emulsion and form a densely packed layer, called a cream. Heavier dispersed phases sink (sediment) to the bottom. In addition to causing an emulsion to become inhomogeneous, creaming can enhance destabilization processes like coalescence. This is because the drops in a cream (or sediment) are in close contact with each other.

Recent investigations focused on monitoring creaming in situ and modeling this behavior so that the changes in the spatial distribution of drops in an emulsion caused by creaming could be predicted. Separation was followed by visualizing the appearance of the dense layer,[55,56] or by measuring the drop volume fraction profile in emulsions,[55,57] as a function of storage time. For example, Raman spectroscopy was used

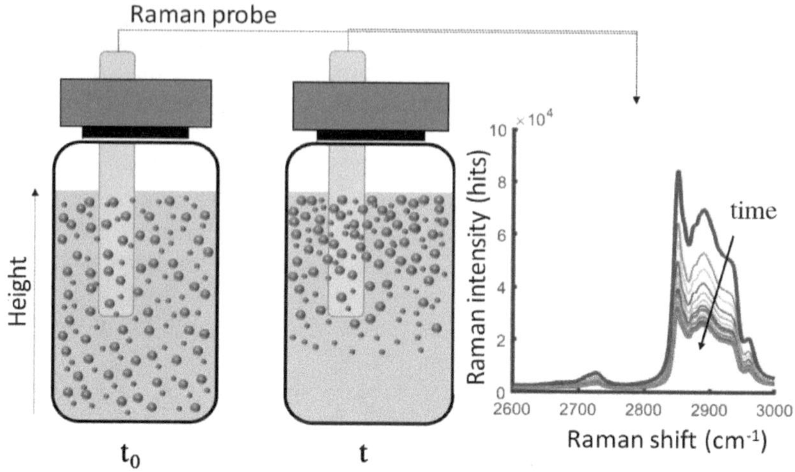

FIGURE 10.3 Schematic illustration of the change in the distribution of drops in an emulsion over time due to creaming. Also shown are examples of Raman spectra measured in hexadecane-in-water emulsions at different times by a Raman probe immersed in the emulsion. Adapted with permission from Deb et al.,[55] Copyright (2022), American Chemical Society.

to monitor creaming in situ in a hexadecane-in-water emulsion. A Raman probe was immersed in the emulsion to a fixed depth and Raman spectra were recorded over time. The intensity of the peaks in the spectra due to characteristic vibrations in hexadecane decreased as the droplets creamed, as shown in Figure 10.3.

The challenges for modeling creaming center around describing the formation and growth of the densely packed layer of drops. Assumptions made about the droplet velocity tend to lead to unrealistic predictions of the drop volume fraction in the layer. Adding a diffusion-type equation to the model enabled predictions that were consistent with the creaming behavior of emulsions over a wide range of drop volume fractions (0.2–0.6) and size distributions.[57]

10.2.5 Chemical Reactions

There are two main types of reactions that alter the chemical structure of the oil phase and hence destabilize oil–water emulsions. The reactions are lipid oxidation and lipid hydrolysis. Reactions between unsaturated fatty acids and active oxygen species cause oxidative degradation of lipids in emulsions. Oxidation occurs mainly at the oil–water interface, due to the contact between unsaturated lipids in the oil and oxidizing species in the aqueous phase. It causes a deterioration in the appearance (color) and taste (flavor and texture) of food emulsions. There have been several excellent reviews of this topic recently.[58–61]

Emulsions destabilize due to chemical reactions during digestion. Lipase enzymes break the triglycerides down into fatty acids and glycerols that are adsorbed into the bloodstream. Lipases are proteins that catalyze reactions at the interface between oil and water phases. Digestion involves a complex interplay between lipase, colipase, and bile salts at the interface. The rate of the reaction is affected by the size of the oil droplets, and hence the interfacial area, as well as the composition of the interface. There have been several outstanding reviews on this topic recently.[62–64]

10.3 MANIPULATING EMULSION STABILITY

10.3.1 Coalescing Drops

The investigations of emulsion stability described in Section 10.2.1 highlight the critical role of the interfacial layer in coalescence. The lifetime of the thin film of liquid in between two drops that have collided together, for example, depends on the attractive and repulsive forces generated by the chemical species located at the drop surfaces. Thus, many of the demulsification strategies developed recently promote droplet coalescence by adding species that interfere with species coating the drops and hence disrupt the interfacial layer. Other approaches to manipulating emulsion stability involve physical disruptions like forcing emulsions through membranes or applying electrical or magnetic fields that enhance the rate of droplet collisions and hence the rate of droplet coalescence.

10.3.1.1 Chemical Changes to Interfacial Layer

10.3.1.1.1 Displacing Surfactants and Polymers

One approach to altering the stability of an emulsion is to displace the emulsifiers with species that have a higher interfacial activity. There was an excellent review of recent investigations of protein–protein displacement in emulsions containing mixtures of proteins.[65] Drop coalescence is enhanced, if the new species replacing adsorbed species in the original layer alters the mechanical properties of the layer from elastic (or solid-like) to viscous (liquid-like).

There have been several studies recently on how surfactants displace asphaltenes from interfaces in water-in-crude oil emulsions.[66] Experimental studies revealed that there is an optimum concentration of surfactant required to break emulsions.[67] Molecular dynamic simulations probed the effect of adsorption of polymeric surfactants on the interfacial layer structure, and the interactions between water drops coated with the layers. The results indicated that surfactant adsorption disrupts the asphaltene film and reduces the repulsive forces between drops, because the surfactant can be located closer to the water.[22] The thin liquid film technique was used to visualize the disruption of asphaltene layers by polymeric surfactants. Interfacial films of asphaltenes were observed to be rigid, thick, and inhomogeneous, due to aggregation of the asphaltenes. Surfactant adsorption in the films transformed them into thinner, fluid-like layers.[68]

Simply removing emulsifiers from the interface can also destabilize an emulsion. In some cases, this was achieved by adding colloidal particles to create a fresh interfacial area available for emulsifiers to occupy in an emulsion. Adding hydrophobic nanoparticles to surfactant-stabilized water-in-oil emulsions, for example, caused drop coalescence. This was attributed to the depletion of surfactant from the liquid interface, as the surfactant molecules were redistributed between the solid–oil and oil–water interfaces.[69]

10.3.1.1.2 Biological Demulsifiers

Biological (or naturally derived) demulsifiers were shown to destabilize emulsions by disrupting the interfacial layer. Biosurfactants are secreted by microorganisms, like bacteria cells, either outside the cell or on the cell surface. Extracellular biosurfactants isolated from bacteria cells were found to promote coalescence in emulsions. For example, the lipopeptides and glycolipid compounds excreted by *Acinetobacter calcoaceticus* were separated from the cells.[70] Adding them to surfactant-stabilized water-in-oil emulsions caused drop coalescence. This was attributed to the compounds displacing surfactants from the interface.[70]

Emulsions were also destabilized by adding bacteria cells that attached to the oil–water interface. For example, mixing surfactant-stabilized water-in-oil emulsions with suspensions of the bacterial strain *Alcaligenes* caused the drops to coalesce.[71] Biosurfactants bound to the bacteria cell walls optimized the cell surface hydrophobicity for partitioning to the interface in emulsions. Cryo-electron microscopy imaging of the emulsion microstructure revealed that the cells were located at the drop surfaces, and clustered in uneven distributions. It was argued that in addition to displacing surfactant molecules from the interface, clusters of bacteria cells enhanced coalescence by bridging the oil films between water drops.[71]

10.3.1.1.3 Detaching Particles

There were studies on the effect of adding surface-active species on the stability of Pickering emulsions.[72] Adding nonionic surfactant to water-in-oil emulsions stabilized by fumed silica particles, for example, was found to displace hydrophobic particles from the interface. On the other hand, nonionic surfactant adsorbed onto partially hydrophobic particles at the interface, rather than displacing them from the interface.[73] Adding hydrophobic fumed silica particles to oil-in-water Pickering emulsions caused the drops to coalesce.[74] It was argued that particles were not displaced from the interface. Rather, the oil drops preferentially wetted the hydrophobic particles added to the emulsion, and hence coalesced.[74] Most experiments were done under low-shear conditions. There was sufficient mixing to expose the particle-coated drops to the new surface-active species, but not enough to increase the frequency of drop collisions and hence induce bulk coalescence.

Shearing emulsions tended to enhance particle displacement from the interface. For example, the combination of adding fumed silica particles and shearing was found to disrupt water-in-oil emulsions stabilized by bacteria.[75] Hydrophilic particles were found to cross the interface, to become fully wetted by the water, and hence located within the water drops. This suggested that the shear stress applied during the experiments was sufficient to induce drop coalescence. Adding hydrophobic particles resulted in phase inversion to a multiple water-in-oil-in-water or a simple oil-in-water emulsion, with the bacteria mainly located in the oil phase and the particles mostly at the interfaces.[75] This confirmed that particle detachment from the interface was due to the shear stress causing the oil drops to coalesce together and entrain water drops within the merging oil drops.

10.3.1.2 Physical Disruptions

10.3.1.2.1 Membranes

Membranes[76–78] and porous solids[79,80] were used to promote coalescence in oily wastewater, which is a dilute oil-in-water emulsion. Membranes (and porous solids) act like sieves and prevent the flow of drops that are larger than the sizes of the pores. The wettability of membranes is also optimized to separate oil from water. The surface chemistry, pore size, and structure of membranes are designed so that high pressures can drive water to wet and pass through the pores, while oil is repelled from the membrane surface.[81]

Recent developments centered on the application of microscopy techniques for visualizing, in situ, droplet behavior on membranes.[82] The efficiency of membrane separations decreases if the pores become blocked.[83,84] Visualization of droplet wetting and spreading and penetration into membranes in real time enhanced understanding of the mechanisms of membrane fouling.[82]

10.3.1.2.2 Electric and Magnetic Fields

The rate of drop coalescence was manipulated by applying magnetic or electrical fields. Magnetic nanoparticles were added to emulsions so that the oil and water could be separated using magnetic fields.[85] The surface chemistry of the particles was designed to optimize the particle wettability so that they attach to the oil–water interface.[86,87] Modifying the particles so that there are electrostatic attractions between the drops and particles was shown to cause drop flocculation, which enhanced the separation.[88] The particles, and hence the drops, move when a magnetic field is applied to the emulsion. The drops are driven together which causes coalescence.

Applying an electric field to a water-in-oil emulsion polarizes the drops and causes them to move. The drops adopt spheroidal shapes with induced positive and negative charges at opposite ends. Attractive interactions between the induced dipoles on different drops pull them close together. The drops align in chains parallel to the direction of the external field. The drops coalesce when the film of oil separating them is squeezed out. A recent review highlighted the advances made using low-field nuclear magnetic resonance to follow drop size evolution, and the gradient in the water concentration that develops as drops sediment and coalesce.[89]

Recently, electrocoalescence between pairs of water drops sitting in air (or oil) was investigated.[90] The drops were sitting on an insulating layer on top of an electrode. They spread out over the substrate and encountered each other when an electric field was applied. A high-speed camera was used to visualize the formation of the neck (or bridge) between the drops as they coalesced. A universal scaling law was discovered for the growth of the neck height h with time t. At short times (typically <1 ms), the neck height increased as $h \sim t^{-1/3}$, while at longer times, it varied linearly with time.[90]

The microscopic processes of electrocoalescence were also studied by simulation. Dissipative particle dynamics were used to probe the influence of surfactants and demulsifiers on the rupture of thin oil films separating water drops in close contact in an electric field.[89] A key finding was that attractive interactions between demulsifiers and surfactants leads to aggregation of these molecules. This leaves areas of the oil film exposed, leading to hole formation in the films and hence film rupture.[91] Molecular dynamic simulations were used to study the effect of electric fields on the adsorption of model asphaltene molecules at a water–toluene interface.[92,93] In the absence of an electric field, the asphaltenes accumulated at the interface and aggregated, forming π–π stacks. Electric fields were predicted to disrupt the π–π stacks and increase asphaltene mobility.[92] This was consistent with experimental observations of enhanced asphaltene dynamics at liquid interfaces under an electric field.[93] The molecular dynamics simulations also predicted that high electric fields would destabilize the interface, by causing a hole to open and the water to form a pillar through the oil phase.[92]

10.3.1.2.3 Temperature

Heating an emulsion enhances the rate of collisions between drops. It can cause desorption of surfactants. This destabilizes interfacial films and leads to an increase in the coalescence frequency. The effectiveness of heating by microwave and radio frequency radiation were investigated.[94,95] The interfacial tension of alkane–water interfaces was probed, in situ, during microwave radiation, to investigate the response of nonionic surfactant layers.[96,97] The interfacial tension was found to increase sharply during irradiation, before decreasing to a stable value. It was argued that this is due to disruption of the surfactant and hydrogen bonding network in the interfacial layer by the switching electromagnetic field of the microwave radiation, rather than localized heating effects.[96]

Cooling and heating emulsions of crystallizable oil cause partial coalescence. Drops do not completely merge because the elastic stress generated by the crystal network acts against the interfacial stress. This behavior was investigated in bulk emulsions[98,99] and between pairs of drops.[100,101] A new approach involved confining the emulsion in a rectangular microfluidic channel, so partial coalescence in ensembles of closely packed drops could be visualized.[102] This revealed how the network of connections between drops propagates partial coalescence by enabling pairs, or small clusters, of partially coalesced drops to merge into larger-scale structures.[103]

10.3.1.2.4 Ultrasound

Ultrasound can fragment drops or cause them to coalesce. The pressure wave generated by applying ultrasound causes movement and mixing in an emulsion. At sufficient intensity, the local pressure in the expansion phase of the cycle will fall below the vapor pressure of the liquid. This causes any tiny bubbles that are present to grow and then collapse. The implosion produces microjets, with high pressure and high temperature, that fragment drops and favors emulsification. But providing the ultrasound intensity remains less than that required for cavitation, studies showed that drop coalescence is enhanced in bulk emulsions.[104–106] A high-speed camera was used to visualize droplet coalescence near cavitation bubbles,[105] and droplet migration and aggregation into bands at nodal planes.[107] It was argued that applying ultrasound enhances drop coalescence by increasing the frequency of drop collisions and disrupting interfacial surfactant layers in emulsions.[105]

The effects of ultrasound on a planar toluene–water interface coated with asphaltenes was investigated by measuring the Brownian movement of probe particles positioned at the interface.[108] The inertial effect of ultrasound on the subphase was characterized using the Reynolds number. The particles were motionless before the application of ultrasound, as they were trapped in asphaltene networks. The particle motion increased as the Reynolds number of the flow caused by the ultrasound increased. Bubbles formed at the interface. Calculation of the viscoelastic surface moduli confirmed that the asphaltene network was weaker and had a shorter relaxation time. Drops coalesced in water-in-toluene emulsions under the same ultrasonic fields. It was argued that the interfacial network of asphaltenes was disrupted by the pressure waves and bubble cavitation, and this led to coalescence when drops collided.[108]

10.3.1.2.5 Shear

Applying shear destabilizes an emulsion by increasing the frequency of collisions between droplets, and hence the rate of coalescence. A review of microfluidic studies from the last 15 years highlighted investigations into the effects of shear flow on coalescence between pairs of drops, or ensembles of drops.[109] Using microfluidic channels had the advantage that coalescence was studied under highly controlled conditions. One limitation was that the flow could be confined to a two-dimensional configuration, which promoted coalescence. This reduced the dependence of the coalescence time (the time required for the thin film separating drops to drain and rupture) on the droplet approach velocity. A stronger correlation between the coalescence time and the approach velocity was found when a drop was sheared past a stationary drop in a three-dimensional channel.[110]

Recent studies focused on probing coalescence in bulk emulsions being sheared under industrially relevant conditions.[111–114] For example, emulsions used in foods, skin, or hair care products are sheared in confined spaces. The drops become entrained as emulsions are forced through a narrow gap. Thus, shear is applied in combination with other stresses, like heat or pressure. The behavior of oil-in-water emulsions sheared in a rolling/sliding soft-tribological contact was investigated.[111] Silicone substrates were used to mimic a soft and compliant surface. This work highlighted the importance of the dynamic oil–water contact angle in determining whether entrained oil drops remain in the contact region long enough to coalesce and wet the substrate.[111] Another advance was the fabrication of substrates with a topography, deformability, and wettability like a human tongue for investigating the lubrication mechanisms of emulsion-based foods.[115]

10.3.1.3 Phase Inversion

Coalescence was investigated as part of studies of phase inversion. The type of fluid that is the continuous (or external) phase of an emulsion changes during phase inversion. For example, a dispersion of oil droplets in water may invert to a dispersion of water droplets in oil. The change is caused by altering the relative proportions of the two immiscible liquids, which is called catastrophic phase inversion, or by changing the affinity of the emulsifier for the two liquids (transitional phase inversion).

Recent studies of catastrophic phase inversion focused on particle-stabilized emulsions.[116,117] Increasing the dispersed phase volume fraction above that required for droplet close packing favors coalescence during the mixing of particle-stabilized emulsions. The mixing time and shear rate, the particle concentration, and the emulsion flow behavior were shown to influence whether complete coalescence and hence inversion occurs. For example, prolonged mixing was required to invert concentrated water-in-limonene oil emulsions containing low concentrations of salinized fumed silica particles.[117] Visualizing the emulsion microstructure revealed that mixing initially produced two populations of drops. There were many small drops densely coated with particles, and a few larger drops incompletely coated by particles. The large droplets were easily deformed by shearing, while the smaller droplets were sufficiently close-packed to resist flowing. Continued mixing caused the larger drops to coalesce around the smaller drops and form water-in-oil-in-water emulsions.[117]

Investigations of transitional phase inversion found that coalescence contributes to the inversion mechanism for emulsions stabilized by more than one emulsifier.[118–120] Transitional inversion is achieved by manipulating the properties of the species stabilizing the drops. The relative proportions of the liquid phases remain the same. The drops do not coalesce as the inversion point is approached unless different emulsifiers that each stabilize a different interfacial curvature are present. In this case, varying the proportions of the different emulsifiers promotes phase inversion.

For example, inversion occurred during the formation of emulsions stabilized by an interfacial bilayer of silica particles of opposite charges and wettability, as the composition of the layer was altered.[119] Layers rich in hydrophobic particles stabilized water drops in oil. Increasing the proportion of hydrophilic particles caused the average size of the water drops to increase until the interfacial bilayer became preferentially wetted by water, and oil-in-water emulsions formed.[119]

Transitional inversion was induced by adding oleic acid to emulsions of decane drops coated with hematite ellipsoids and remixing the emulsions.[120] Adding oleic acid also affected the droplet morphology. Spherical oil or water drops formed at low and high concentrations of oleic acid, respectively. Larger, nonspherical oil (or water) drops formed at oleic acid concentrations just below (or above) that required for phase inversion. High-resolution optical microscopy revealed that the spherical drops were coated with closely packed particles orientated parallel to the interface. The nonspherical drops were partially covered by particles in orientations ranging from parallel to perpendicular to the interface. It was proposed that the negatively charged oleic acid molecules adsorbed on the positively charged particles and modified their wettability. Presumably, the particles were incompletely modified at concentrations of oleic acid insufficient to occupy all the positively charged surface sites on the particles. The formation of nonspherical drops was attributed to arrested coalescence of drops partially coated by particles with a heterogeneous surface chemistry.[120]

10.3.2 Transferring Mass between Drops

The studies of the mechanisms of Ostwald ripening described in Section 10.2.2 highlighted the role of surfactants coating the drops. Investigations into manipulating ripening in oil-in-water emulsions focused on the impact of adding a second oil (called a co-stabilizer) that is poorly soluble in the water. The presence of the co-stabilizer causes compositional ripening that opposes Ostwald ripening. Movement of the (relatively) water-soluble, essential oil molecules from smaller to large droplets increases their concentration in the larger drops. This generates a concentration gradient that drives the transport of oil molecules from the larger to the smaller oil droplets. Droplet growth is halted once the compositional ripening counterbalances Ostwald ripening.

Several studies were done on emulsions of essential oils (like orange oil)-in-water. [121–124]The concentration of the co-stabilizer required varied with the emulsifier used in orange oil-in-water emulsions.[123,124] It was argued that smaller concentrations are required if the emulsifier formed densely packed interfacial layers that slowed the dissolution of oil molecules into the aqueous phase.

The structure of the co-stabilizer also had an effect. Medium-chain length triglycerides were more effective at preventing ripening than the same mole concentration of longer chain triglycerides.[121] It was speculated that the incorporation of medium chain length triglycerides in the emulsifier layer improved its resistance to deformation and hence dissolution of oil. The hydrophobic interactions between the hydrocarbon tails of the emulsifiers and the saturated fatty acids of the medium-chain triglycerides were presumed to be stronger than with the unsaturated fatty acids of longer chain triglycerides.[121]

Experimental studies showed that mass transfer between drops can be arrested by coating drops with sufficient particles to form a percolated network.[125,126] The particle network generates an interfacial yield stress, so that the interface does not flow unless a minimum amount of stress is applied. This prevents droplets from shrinking over a range of pressure gradients.[125,126]

On the other hand, Pickering emulsions were destabilized by the interfacial transport of small molecules that were soluble in the oil and water phases.[127] For example, adding ethanol to the aqueous phase of alkane-in-water Pickering emulsions and gently mixing the emulsion caused complete phase separation, providing ethanol was soluble in the alkane. Phase separation occurred when ethanol was gently added to water-in-alkane Pickering emulsions. A range of particles (spherical and nonspherical iron oxide particles, silica, and latex particles) were used as emulsifiers. The addition of ethanol did not reduce the oil–water contact angle of the particles (or the interfacial tension) sufficiently for the particles to be detached from the interface. Instead, it was proposed that turbulence (Marangoni instability) generated by the mutually soluble and surface-active molecules moving across the interface caused the emulsion destabilization.[127]

10.3.3 Flocculating Drops

Flocculation was a common strategy for destabilizing emulsions that contaminate wastewater.[128] A range of flocculants were investigated. Flocculants based on naturally derived polymers like chitosan were proposed to accumulate at the surfaces of oil drops and cause drop flocculation by charge screening[129] or bridging (associative) interactions.[130] The use of oppositely charged surfactants and polyelectrolytes that form insoluble precipitates (complexes of polymer and surfactant) was also explored. Adding anionic surfactant and then cationic polyelectrolyte to oily wastewater caused the drops to be enmeshed within the complexes as they precipitated. The flocs were separated from the water by dissolved air flotation, providing the ratio of surfactant to polyelectrolyte was optimized to form flocs that withstood the turbulence generated during flotation.[131,132]

Drop flocculation was also identified as the main mechanism for separations of emulsions of crude oil-in-water triggered by adding graphene oxide nanosheets. It was originally thought that the drops coalesced because asphaltenes present at the interface were removed by adsorption onto the graphene oxide.[133] This did not account for microscopy observations suggesting the nanosheets were located at the drop surfaces,[134] or the increase in dilational modulus of pendant drops of crude oil in water when graphene oxide was added to the aqueous phase. Molecular dynamics simulations revealed that the graphene oxide nanosheets accumulated at the drop surfaces, interacting with the asphaltenes to form a mixed interfacial layer.[130,135] It was speculated that attractive interactions between layers coating different drops encourage flocculation, adhesion, and ultimately, coalescence.

10.3.4 Concentrating Drops

Centrifugal forces were used to enhance gravitational forces on emulsions, and hence phase separation by creaming or sedimentation.[136] Another strategy was to incorporate micrometer-sized particles into the dispersed phase to increase the density difference between the drops and the continuous phases.[137] Concentrations of about 10 g of magnetite particles per gram of dispersed phase were gently mixed with emulsions. The mechanism of incorporation depended on the particle wettability. Particles that were partially wet by both liquid phases aggregated with the drops forming multiple layers around the drops. Particles that were preferentially wet by the dispersed phase were absorbed by the drops, if they were smaller than the drops. Larger particles (or flocs of particles) absorbed the drops. Sedimentation occurred once the dispersed phase was sufficiently heavy.[137]

10.3.5 Modulating Lipid Digestion

There were extensive investigations into modifying interfacial properties to manipulate emulsion stability to lipid digestion. One strategy was to coat interfaces with particles that hinder the adsorption of enzymes and bile salts.[138] In addition to experimental studies, there were significant efforts to develop mathematical models to probe the differences between digestion at surfactant- and particle-stabilized interfaces.[139] Another strategy for modulating digestion was to coat the interfaces with polymers that can bind lipases and chelate bile salts. An important advance was the rationalization of the multiple functions of polymers at the interfaces in food emulsions.[140] Conjugates (or complexes) of proteins and polysaccharides were identified as the best candidates for achieving these functions.

Polysaccharides were added to food emulsions to promote drop flocculation and hence control the rate of lipid digestion.[141–144] Aggregation of drops into clusters limited the lipid–water surface area available to the digestive enzymes. This reduced the rate and extent of lipid digestion. The magnitude of the reduction depended on the flocculation mechanism.[142] Depletion flocculation produced clusters of drops surrounded by polymer. On the other hand, bridging flocculation generated clusters of drops that contained polymer within them. It had a greater effect on lipid digestion. It was speculated that the clusters produced by depletion flocculation were more likely to be disrupted by the dilution and shearing that occurs during digestion.[142]

10.4 HARNESSING DESTABILIZATION TO FABRICATE USEFUL MATERIALS

10.4.1 Flocculating Emulsions to Make Gels

Emulsion gels are materials that consist mainly of liquids but behave like soft solids. Their applications include replacing some of the fat in foods, encapsulating active ingredients in foods and cosmetics, or being used as inks for printing three-dimensional objects. There were excellent reviews recently of the different methods used to transform emulsions into gels.[140,145,146] The focus of this section is on gels formed by destabilizing emulsions. Weakly flocculating drops form clusters that increase in size, until the largest clusters assemble into a percolating network. The continuous phase becomes trapped in the interconnected pores of the network. It does not flow readily when stress is applied. The solid-like response of a gelled emulsion is controlled by the number of stress-bearing connections (sometimes called bonds) between the drops.

Drop flocculation was induced by coating oil drops with particles, or globular proteins, and manipulating the electrostatic interactions between the drops in water. For example, mixing oppositely charged droplets coated by two kinds of potato protein caused the cationic and anionic drops to flocculate into space-filling networks due to electrostatic attractions.[147] Protein-coated droplets were also flocculated by adding oppositely charged pectin molecules. It was argued that the network formation was promoted by the polysaccharide bridging the drops.[148] In another example, droplets coated by chitosan particles were flocculated at a pH higher than the pK_a of chitosan.[149] The low surface charge on the particles reduced the electrostatic repulsions between the particle-coated drops sufficiently for them to flocculate together. It was proposed that the emulsion elasticity was enhanced by the presence of excess chitosan particles which aggregated together in the water and formed an interconnected network with the drop clusters.[149]

Another strategy for gelling particle-stabilized oil-in-water emulsions involved adding (nonadsorbing) polysaccharides to cause depletion interactions between the drops.[150] The polysaccharide molecules were excluded from the space between drops when they approached each other. The difference in osmotic pressure between this space and the bulk continuous phase generated an attractive interaction between the drops. The depth of the attractive interaction depended on the polysaccharide concentration. Moderate depletion interactions promoted drop flocculation into microscopic clusters that formed the strands of a network spanning the volume of the emulsion.[150] Stronger depletion interactions generated weaker gels because the drops flocculated into macroscopic clusters. The universality of the strategy was demonstrated for different organic and inorganic particle emulsifiers and various vegetable oils.[150]

10.4.1.1 Extrudable Gels for 3D Printing

Emulsion gels consisting of flocculated drops were primarily developed for use in three-dimensional extrusion printers.[151–154] Extrusion printing builds an object from layers of ink deposited on top of each other.[155] Emulsions used as inks must withstand extrusion and recover to form self-supporting materials. During extrusion, the emulsion is pushed through a nozzle. It should be possible to disrupt the network structure so that the emulsion viscosity decreases as the shear rate increases, and it can flow out of the nozzle. The internal network structure should self-heal after deposition, so that the emulsion viscosity and mechanical strength are recovered. Finally, the printed layer should not flow under the weight of layers deposited on top of it, so that the shape being printed is retained.

The smallest size of features that can be printed, and hence the resolution of 3D printed objects, is limited by the flow behavior of the ink. Improving printing resolution involves reducing the nozzle size to reduce the size of the features printed. This relies on the emulsion flowing through the smaller nozzle without clogging, and preserving its printed geometry after the pressure is removed. Bridging and depletion interactions between drops were exploited to improve the printing performance of emulsion gels. For example, the bridging interactions in protein-stabilized emulsion gels were enhanced by adding chitosan.[152] This ensured objects with high shape fidelity were printed, as shown in Figure 10.4a.

10.4.2 Controlling Coalescence during Dehydration

10.4.2.1 Porous Solids

Dehydration is a common route to making materials like porous or powdered solids and oleogels by destabilizing emulsions. The first application of emulsion templating was to fabricate hydrophobic polymer monoliths.[156] Concentrated emulsions called high internal phase emulsions (HIPE) that are stabilized by polymeric surfactants are used to form porous solids known as polymerized HIPE (or PolyHIPE). The drop volume fraction in the emulsion templates is high enough so that the drops pack closely together. Solids are produced from water-in-oil HIPE where the oil phase

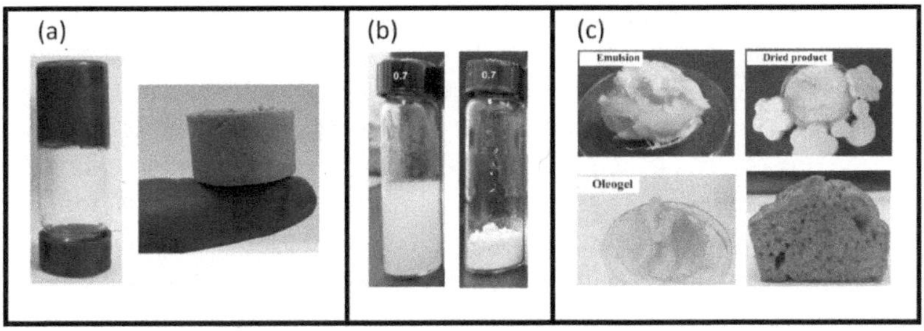

FIGURE 10.4 (a) Photographs of a 25 vol% 1,2 dichloroethane-in-water emulsion stabilized by mixtures of cellulose nanofibrils and graphene oxide (left) and the porous solid produced by freeze-drying the emulsion (right). Reprinted from Liao et al.,[159] Copyright (2021), with permission from Elsevier. (b) Photographs of a 9 vol% hempseed oil-in-water emulsion stabilized by cellulose nanocrystals (left) and the powdered solid produced by spray drying the emulsion (right). Reprinted from Esparza et al.,[165] Copyright (2020), with permission from Elsevier. (c) Photographs of a 60 vol% camelia oil-in-water emulsion stabilized by a mixture of tea polyphenol-palmitate particles and citrus pectin before (upper left) and after freeze-drying (upper right). Also shown are photographs of the oleogel formed by shearing the freeze-dried emulsion (lower left) and a sponge cake baked from a batter using the oleogel as a substitute for butter. Reprinted from Luo et al.,[174] Copyright (2019), with permission from Elsevier.

contains a hydrophobic monomer. Free radical polymerization of monomers in the external phase of the emulsion transforms it into a solid material. Evaporation of the water drops leaves behind air-filled voids, called pores, in the solid. This gives the solid a highly porous and permeable structure.

Porous solids made from a wide range of materials have since been produced, as described in recent reviews.[157,158] Sintering may be used rather than polymerization to transform a liquid emulsion into a solid material. Porous carbons are obtained by carbonization of PolyHIPE. Porous inorganic solids are fabricated by replacing the surfactant stabilizers in HIPE with inorganic particles. Particle-stabilized HIPE can also be used to fabricate porous polymer composites. Figure 10.4a shows a dichloroethane in-water emulsion stabilized by mixtures of graphene oxide and cellulose nanofibrils. The dispersed phase contained polymethyl methacrylate. Freeze-drying removed the volatile liquid phases, leaving behind a porous composite material made of the polymer and particles.[159] Porous solids fabricated from emulsion templates were used as substrates for applications ranging from bone tissue regeneration[160] to catalytic reactions.[161]

One of the key factors that led to the increase in the types of materials used in the emulsion templates was the recent developments in understanding the role of emulsion destabilization in structure formation in porous solids. Surfactant-stabilized emulsion templates tend to produce solids with relatively small pores (<50 μm in diameter) that are interconnected by holes in the wall of the pores. Particle-stabilized emulsion templates, on the other hand, tend to produce solids with larger pores and fewer interconnections. The holes (or pore throats) that connect pores were thought to form when the thin films between drops in close contact shrank during polymerization and ruptured.[156] It was assumed that the rigid particle shell around Pickering emulsion droplets increased the viscoelasticity of the continuous film sufficiently to resist rupture during polymerization. It is now recognized that coalescence and Ostwald ripening occur during polymerization and drying. The interactions between the drops, and the rheology of the interfacial films coating the drops, influence the size of the pore throats.[162] Thus, highly interconnected porous solids were produced using Pickering emulsions that underwent coalescence to a limited extent during polymerization.[163]

10.4.2.2 Powdered Emulsions

Oil-in-water emulsions were dried to form oil-in-air powders by spray drying.[164] The drops do not coalesce if there is sufficient material encapsulating the drops to withstand the stresses caused by drying.[165] Removing the aqueous phase produces a powder of micrometer-sized particles made from the oil droplets. The particles flow freely and do not stick together providing coalescence is minimized. Figure 10.4b shows an emulsion of hempseed oil in water stabilized by sufficient concentrations of cellulose nanocrystals to withstand spray drying and transform into powdered particles.[165] Oil powders are easier to store and transport than emulsions. They are less susceptible to degradation by chemical reactions. Edible oils, for example, were converted into powders to prevent oxidation and loss of flavor. The powders were used as ingredients in foods.[166,167]

Microencapsulation of edible oils by drying aqueous emulsions of the oils was the topic of a recent review.[168] One focus of recent research was on optimizing the amount of oil encapsulated in the powder (the encapsulation efficiency) by adding polysaccharides and proteins that form a protective 'wall' around the droplet cores during drying. Researchers also highlighted the role of wall materials in producing powders that can be reconstituted into emulsions by dispersing the powder in water.[169,170]

10.4.2.3 Oleogels as Fat Substitutes

Oleogels consist of a continuous phase of oil immobilized in a network of polymers or particles. The presence of the network transforms the oil from a viscous liquid into a self-supporting material, with an elastic character. Gels made from edible oils are potential replacements for the fat in a range of food products. Recent reviews have summarized the range of approaches to imparting structure to liquid oils.[171,172] The focus of this section is on oleogels made by drying oil-in-water emulsions stabilized by polymers. This is often described as the emulsion-templated approach to fabricating oleogels. Removing the aqueous phase from an emulsion causes the oil droplets to pack together. They deform into polyhedral shapes with flattened areas of contact and rounded edges and corners. The oil is trapped by the network that forms as the interfacial films of polymer are compressed between the drops.

Recent studies of the emulsion-templated approach to fabricated oleogels involved dehydrating oil-in-water emulsions to form soft, oil-based solids by freeze-drying or oven-drying. The emulsions were stabilized by mixtures of polymeric species to minimize drop coalescence and oil leakage during the water removal. For example, emulsions stabilized by mixtures of cellulose nanocrystals and sodium caseinate dried to produce solids made of individual oil droplets separated by a network of polysaccharides and proteins.[173]

In many cases, it was possible to dilute the dried emulsion residues in water and form an emulsion. This indicated that the polymeric networks separating the drops in the dried solids were sufficiently hydrated to enable water to wet and penetrate the networks and redisperse the drops. Spray- or freeze-dried emulsions were converted into oil-continuous gels by shearing.[171] Figure 10.4c illustrates the dried solid and oleogel obtained from a camelia oil-in-water emulsion stabilized by tea polyphenol-palmitate particles and citrus pectin.[174] Shearing disrupted the networks sufficiently for some of the drops to coalesce into a continuous phase that surrounded the remaining drops.

Oleogels were mainly investigated as substitutes for solid fats (like butter or margarine). As shown in Figure 10.4c, the gels prepared from camelia oil-in-water emulsion stabilized by tea polyphenol-palmitate particles and citrus pectin were used as substitutes for butter in cake batters.[174] Oleogels were also incorporated into pastry doughs.[175]

A key limitation hindering the use of oleogels as substitutes for fat in food products is their susceptibility to oxidation. Oleogels formed using edible oils containing large proportions of unsaturated fatty acids are prone to oxidation during the heat processing that occurs in food preparation. Several studies focused on improving the stability of gels to oxidation by incorporating antioxidants, like polyphenols, into the emulsions.[176,177] Oxidation was reduced in oleogels formed, for example, from emulsions stabilized by colloidal complexes of polyphenols with polysaccharides.[176] It was argued that polyphenols located at the drop surfaces limited the exposure of the unsaturated fatty acids to oxygen and hence slowed the rate of oxidation.

10.5 CONCLUSIONS

Emulsions are thermodynamically unstable systems that undergo various physical and chemical destabilization processes over time. Over the last five years, there have been exciting advances in the experimental techniques and modeling tools used to probe this behavior across the range of length scales that are relevant to emulsion systems. While predicting the lifetime of an emulsion remains challenging, the links between the stability of bulk emulsions and their interfacial properties are starting to be quantified. These developments are aiding the design of demulsification strategies that target specific destabilization processes. They are also informing the formulation of new functional materials from emulsion templates.

REFERENCES

1. McClements, D.J. *Food Emulsions*, CRC Press: Boca Raton, FL, USA, 2015.
2. Tadros, T.F. *Emulsions: Formation, Stability, Industrial Applications*, De Gruyter: Berlin, Germany, 2016.

3. Ghosh, S.; Rousseau, D. Emulsion breakdown in foods and beverages. In *Chemical Deterioration and Physical Instability of Food and Beverages*, Skibsted, L.H., Risbo, J., Andersen, M.L., Eds.; Woodhead Publishing: Sawston, UK, 2010; pp. 260–295.
4. Leal-Calderon, F.; Bibette, J.; Schmitt, V. *Emulsion Science*, Springer: New York NY, USA, 2007.
5. Sjoblom, J. *Emulsions and Emulsion Stability*, CRC Press: Boca Raton, FL, USA, 2005.
6. Langevin, D. Recent advances on emulsion and foam stability. *Langmuir* **2023**, *39*, 3821–3828.
7. Kamp, J.; Villwock, J.; Kraume, M. Drop coalescence in technical liquid/liquid applications: A review on experimental techniques and modeling approaches. *Rev. Chem. Eng.* **2017**, *33*, 1–47.
8. Braun, L.; Kühnhammer, M.; von Klitzing, R. Stability of aqueous foam films and foams containing polymers: Discrepancies between different length scales. *Curr. Opin. Colloid Interface Sci.* **2020**, *50*, 101379.
9. Claesson, P.M.; Ederth, T.; Bergeron, V.; Rutland, M.W. Techniques for measuring surface forces. *Adv. Colloid Interface Sci.* **1996**, *67*, 119–183.
10. Langevin, D. On the rupture of thin films made from aqueous surfactant solutions. *Adv. Colloid Interface Sci.* **2020**, *275*, 102075.
11. Chatzigiannakis, E.; Veenstra, P.; ten Bosch, D.; Vermant, J. Mimicking coalescence using a pressure-controlled dynamic thin film balance. *Soft Matter* **2020**, *16*, 9410–9422.
12. Politova, N.I.; Tcholakova, S.; Tsibranska, S.; Denkov, N.D.; Muelheims, K. Coalescence stability of water-in-oil drops: Effects of drop size and surfactant concentration. *Colloids Surf. A* **2017**, *531*, 32–39.
13. Politova, N.; Tcholakova, S.; Denkov, N.D. Factors affecting the stability of water-oil-water emulsion films. *Colloids Surf. A* **2017**, *522*, 608–620.
14. Wierenga, P.A.; Basheva, E.S.; Delahaije, R.J.B.M. Variations in foam collapse and thin film stability with constant interfacial and bulk properties. *Adv. Colloid Interface Sci.* **2023**, *312*, 102845.
15. Liu, B.; Manica, R.; Liu, Q.; Klaseboer, E.; Xu, Z.; Xie, G. Coalescence of bubbles with mobile interfaces in water. *Phys. Rev. Lett.* **2019**, *122*, 194501.
16. Dong, T.; Weheliye, W.H.; Angeli, P. Laser induced fluorescence studies on the distribution of surfactants during drop/interface coalescence. *Phys. Fluids* **2019**, *31* 012106.
17. Burrill, K.A.; Woods, D.R. Film shapes for deformable drops at liquid-liquid interfaces. II. The mechanisms of film drainage. *J. Colloid Interface Sci.* **1973**, *42*, 15–34.
18. Mettu, S.; Wu, C.; Dagastine, R.R. Dynamic forces between emulsified water drops coated with Poly-Glycerol-Poly-Ricinoleate (PGPR) in canola oil. *J. Colloid Interface Sci.* **2018**, *517*, 166–175.
19. Shi, C.; Zhang, L.; Xie, L.; Lu, X.; Liu, Q.; He, J.; Mantilla, C.A.; Van den berg, F.G.A.; Zeng, H. Surface interaction of water-in-oil emulsion droplets with interfacially active asphaltenes. *Langmuir* **2017**, *33*, 1265–1274.
20. Xie, L.; Lu, Q.; Tan, X.; Liu, Q.; Tang, T.; Zeng, H. Interfacial behavior and interaction mechanism of pentol/water interface stabilized with asphaltenes. *J. Colloid Interface Sci.* **2019**, *553*, 341–349.
21. Sun, X.; Yang, D.; Zhang, H.; Zeng, H.; Tang, T. Unraveling the interaction of water-in-oil emulsion droplets via molecular simulations and surface force measurements. *J. Phys. Chem. B* **2021**, *125*, 7556–7567.
22. Sun, X.; Zeng, H.; Tang, T. Molecular simulations on the coalescence of water-in-oil emulsion droplets with non-ionic surfactant and model asphaltene. *Langmuir* **2023**, *39*, 2233–2245.

23. Liu, B.; Manica, R.; Liu, Q.; Klaseboer, E.; Xu, Z. Coalescence or bounce? How surfactant adsorption in milliseconds affects bubble collision. *J. Phys. Chem. Lett.* **2019**, *10*, 5662–5666.
24. Liu, B.; Manica, R.; Liu, Q.; Xu, Z.; Klaseboer, E.; Yang, Q. Nanoscale transport during liquid film thinning inhibits bubble coalescing behavior in electrolyte solutions. *Phys. Rev. Lett.* **2023**, *131*, 104003.
25. Liu, B.; Manica, R.; Xu, Z.; Liu, Q. The boundary condition at the air–liquid interface and its effect on film drainage between colliding bubbles. *Curr. Opin. Colloid Interface Sci.* **2020**, *50*, 101374.
26. Vakarelski, I.U.; Yang, F.; Tian, Y.S.; Li, E.Q.; Chan, D.Y.C.; Thoroddsen, S.T. Mobile-surface bubbles and droplets coalesce faster but bounce stronger. *Sci. Adv.* **2019**, *5*, 4292.
27. Lu, J.; Corvalan, C.M.; Chew, Y.M.J.; Huang, J.-Y. Coalescence of small bubbles with surfactants. *Chem. Eng. Sci.* **2019**, *196*, 493–500.
28. Ozan, S.C.; Jakobsen, H.A. On the effect of the approach velocity on the coalescence of fluid particles. *Int. J. Multiphase Flow* **2019**, *119*, 223–236.
29. Ozan, S.C.; Jakobsen, H.A. On the role of the surface rheology in film drainage between fluid particles. *Int. J. Multiphase Flow* **2019**, *120*, 103103.
30. Ozan, S.C.; Jakobsen, H.A. Effect of surface viscoelasticity on the film drainage and the interfacial mobility. *Int. J. Multiphase Flow* **2020**, *130*, 103377.
31. Ozan, S.C.; Hosen, H.F.; Jakobsen, H.A. On the prediction of coalescence and rebound of fluid particles: A film drainage study. *Int. J. Multiphase Flow* **2021**, *135*, 103521.
32. Song, R.; Zhang, L.; Yi, Z.; Zhou, Y.; Yuan, H.; Han, L. On the role of the macroscopic deformation in liquid film drainage between bubbles. *AIChE J.* **2023**, *69*, e18044.
33. Narayan, S.; Metaxas, A.E.; Bachnak, R.; Neumiller, T.; Dutcher, C.S. Zooming in on the role of surfactants in droplet coalescence at the macroscale and microscale. *Curr. Opin. Colloid Interface Sci.* **2020**, *50*, 101385.
34. Chatzigiannakis, E.; Chen, Y.; Bachnak, R.; Dutcher, C.S.; Vermant, J. Studying coalescence at different lengthscales: From films to droplets. *Rheol. Acta* **2022**, *61*, 745–759.
35. Narayan, S.; Makhnenko, I.; Moravec, D.B.; Hauser, B.G.; Dallas, A.J.; Dutcher, C.S. Insights into the microscale coalescence behavior of surfactant-stabilized droplets using a microfluidic hydrodynamic trap. *Langmuir* **2020**, *36*, 9827–9842.
36. Bremond, N.; Thiam, A.R.; Bibette, J. Decompressing emulsion droplets favors coalescence. *Phys. Rev. Lett.* **2008**, *100*, 024501.
37. Yi, H.; Fu, T.; Zhu, C.; Ma, Y. Cascade coalescence of droplets in a sudden expansion microchannel. *Chem. Eng. J.* **2022**, *442*, 136240.
38. Gao, C.; Ling, S.; Chen, Z.; Wang, Y.; Xu, J. Coalescence law of microdroplet swarms in microchannels. *Chem. Eng. Sci.* **2022**, *262*, 118055.
39. Ruiz-Rus, J.; Ern, P.; Roig, V.; Martínez-Bazán, C. Coalescence of bubbles in a high reynolds number confined swarm. *J. Fluid Mech.* **2022**, *944*, A13.
40. Dinh, H.-H.-Q.; Santanach-Carreras, E.; Schmitt, V.; Lequeux, F. Coalescence in concentrated emulsions: Theoretical predictions and comparison with experimental bottle test behaviour. *Soft Matter* **2020**, *16*, 10301–10309.
41. Diaz, D.; Church, J.; Willner, M.R.; Sarnyai, S.; Lundin, J.G.; Paynter, D.M.; Lee, W.H. Evaluation of bilgewater emulsion stability using nondestructive analytical methods. *Ind. Eng. Chem. Res.* **2021**, *60*, 1014–1025.
42. Church, J.; Paynter, D.M.; Lee, W.H. *In Situ* Characterization of oil-in-water emulsions stabilized by surfactant and salt using microsensors. *Langmuir* **2017**, *33*, 9731–9739.
43. Shen, Y.; Yao, J.; Son, J.; Zhu, Z.; Yu, X.-Y. Liquid ToF-SIMS revealing the oil, water, and surfactant interface evolution. *Phys. Chem. Chem. Phys.* **2020**, *22*, 11771–11782.

44. Kaganyuk, M.; Mohraz, A. Impact of particle size on droplet coalescence in solid-stabilized high internal phase emulsions. *Langmuir* **2019**, *35*, 12807–12816.
45. Botti, T.C.; Hutin, A.; Quintella, E.; Carvalho, M.S. Effect of interfacial rheology on drop coalescence in water–oil emulsion. *Soft Matter* **2022**, *18*, 1423–1434.
46. Koroleva, M. Yu.; Yurtov, E.V. Ostwald ripening in macro- and nanoemulsions. *Russ. Chem. Rev.* **2021**, *90*, 293–323.
47. Khedr, A.; Striolo, A. Quantification of Ostwald ripening in emulsions via coarse-grained simulations. *J. Chem. Theory Comput.* **2019**, *15*, 5058–5068.
48. Rodriguez-Lopez, G.; O'Neil Williams, Y.; Toro-Mendoza, J. Individual and collective behavior of emulsion droplets undergoing Ostwald ripening. *Langmuir* **2019**, *35*, 5316–5323.
49. Mwasame, P.M.; Wagner, N.J.; Beris, A.N. On the macroscopic modeling of the rheology and Ostwald ripening of dilute stabilized emulsions. *Phys. Fluids* **2019**, *31*, 021206.
50. Reyes, Y.; Hamzehlou, S.; Leiza, J.R. Ostwald ripening in nano/miniemulsions in the presence of two costabilizers as revealed by molecular dynamics simulations. *J. Mol. Liquids* **2021**, *335*, 116152.
51. Dickinson, E. Strategies to control and inhibit the flocculation of protein-stabilized oil-in-water emulsions. *Food Hydrocoll* **2019**, *96*, 209–223.
52. Bressel, K.; Müller, W.; Leser, M.E.; Reich, O.; Hass, R.; Wooster, T.J. Depletion-induced flocculation of concentrated emulsions probed by photon density wave spectroscopy. *Langmuir* **2020**, *36*, 3504–3513.
53. Tisserand, C.; Brambilla, G.; Meunier, G.; Parker, A. Predicting the long-term stability of depletion-flocculated emulsions by Static Multiple Light Scattering (SMLS). *J. Dispers. Sci. Technol.* **2020**, *41*, 648–655.
54. Pal, R. Modeling of sedimentation and creaming in suspensions and pickering emulsions. *Fluids* **2019**, *4*, 186.
55. Deb, T.K.; Lebaz, N.; Ozdemir, M.S.; Govoreanu, R.; Mhamdi, A.; Sin, G.; Sheibat-Othman, N. Monitoring and modeling of creaming in oil-in-water emulsions. *Ind. Eng. Chem. Res.* **2022**, *61*, 4638–4647.
56. Daglish, J.; Blacker, A.J.; de Boer, G.; Crampton, A.; Hose, D.R.J.; Parsons, A.R.; Kapur, N. Determining phase separation dynamics with an automated image processing algorithm. *Org. Process Res. Dev.* **2023**, *27*, 627–639.
57. Assar, M.; Simon, S.; Sørland, G.H.; Grimes, B.A. A theoretical and experimental investigation of batch oil-water gravity separation. *Chem. Eng. Res. Design* **2023**, *194*, 136–150.
58. Aslam, A.; Schroën, K. Lipid oxidation in food emulsions: A review dedicated to the role of the interfacial area. *Curr. Opin. Food Sci.* **2023**, *51*, 101009.
59. Villeneuve, P.; Bourlieu-Lacanal, C.; Durand, E.; Lecomte, J.; McClements, D.J.; Decker, E.A. Lipid oxidation in emulsions and bulk oils: A review of the importance of micelles. *Crit. Rev. Food Sci. Nutr.* **2023**, *63*, 4687–4727.
60. Oh, W.Y.; Kim, M.-J.; Lee, J. Approaches of lipid oxidation mechanisms in oil matrices using association colloids and analysis methods for the lipid oxidation. *Food Sci. Biotechnol.* **2023**, *32*, 1805–1819.
61. Miyagawa, Y.; Adachi, S. Dispersion and oxidative stability of o/w emulsions and oxidation of microencapsulated oil. *Biosci. Biotechnol. Biochem.* **2017**, *81*, 625–633.
62. Bertsch, P.; Bergfreund, J.; Windhab, E.J.; Fischer, P. Physiological fluid interfaces: Functional microenvironments, drug delivery targets, and first line of defense. *Acta Biomaterial.* **2021**, *130*, 32–53.
63. Maldonado-Valderrama, J. Probing in vitro digestion at oil–water interfaces. *Curr, Opin. Colloid Interface Sci.* **2019**, *39*, 51–60.

64. Koziolek, M.; Carrière, F.; Porter, C.J.H. Lipids in the stomach – implications for the evaluation of food effects on oral drug absorption. *Pharm. Res.* **2018**, *35*, 55.
65. Hinderink, E.B.A.; Meinders, M.B.J.; Miller, R.; Sagis, L.; Schroën, K.; Berton-Carabin, C.C. Interfacial protein-protein displacement at fluid interfaces. *Adv. Colloid Interface Sci.* **2022**, *305*, 102691.
66. Abdulredha, M.M.; Hussain, S.A.; Abdullah, L.C.; Hong, T.L. Water-in-oil emulsion stability and demulsification via surface-active compounds: A review. *J. Petrol. Sci. Eng.* **2022**, *209*, 109848.
67. Salager, J.-L.; Marquez, R.; Delgado-Linares, J.G.; Rondon, M.; Forgiarini, A. Fundamental basis for action of a chemical demulsifier revisited after 30 years: HLD_N as the primary criterion for water-in-crude oil emulsion breaking. *Energy Fuels* **2022**, *36*, 711–730.
68. Yang, F.; Tchoukov, P.; Qiao, P.; Ma, X.; Pensini, E.; Dabros, T.; Czarnecki, J.; Xu, Z. Studying demulsification mechanisms of water-in-crude oil emulsions using a modified thin liquid film technique. *Colloids Surf. A* **2018**, *540*, 215–223.
69. Yegya Raman, A.K.; Aichele, C.P. Demulsification of surfactant-stabilized water-in-oil (Cyclohexane) emulsions using silica nanoparticles. *Energy Fuels* **2018**, *32*, 8121–8130.
70. Akbari, N.; Biria, D. Investigation of the activity of *Acinetobacter calcoaceticus* biodemulsifier to break stable water in oil emulsions. *J. Env. Chem. Eng.* **2018**, *6*, 4144–4150.
71. Peng, K.; Liu, W.; Xiong, Y.; Lu, L.; Liu, J.; Huang, X. Emulsion microstructural evolution with the action of environmentally friendly demulsifying bacteria. *Colloids Surf. A* **2018**, *553*, 528–538.
72. Zhao, X.; Huang, B.; El-Aooiti, M.; Rousseau, D. Demulsification to control solute release from pickering crystal-stabilized water-in-oil emulsions. *J. Colloid Interface Sci.* **2018**, *509*, 360–368.
73. Yegya Raman, A.K.; Aichele, C.P. Influence of non-ionic surfactant addition on the stability and rheology of particle-stabilized emulsions. *Colloids Surf. A* **2020**, *585*, 124084.
74. Griffith, C.; Daigle, H. Destabilizing pickering emulsions using fumed silica particles with different wettabilities. *J. Colloid Interface Sci.* **2019**, *547*, 117–126.
75. Xie, H.; Zhao, W.; Zhang, X.; Wang, Z. Demulsification of bacteria-stabilized pickering emulsions using modified silica nanoparticles. *ACS Appl. Mat. Interfaces* **2022**, *14*, 24102–24112.
76. Chen, C.; Weng, D.; Mahmood, A.; Chen, S.; Wang, J. Separation mechanism and construction of surfaces with special wettability for oil/water separation. *ACS Appl. Mat. Interfaces.* **2019**, *11*, 11006–11027.
77. Deng, Y.; Dai, M.; Wu, Y.; Peng, C. Emulsion system, demulsification and membrane technology in oil–water emulsion separation: A comprehensive review. *Crit. Rev. Env. Sci. Technol.* **2023**, *53*, 1254–1278.
78. Rasouli, S.; Rezaei, N.; Hamedi, H.; Zendehboudi, S.; Duan, X. Superhydrophobic and superoleophilic membranes for oil-water separation application: A comprehensive review. *Mat. Design* **2021**, *204*, 109599.
79. Wu, P.; Luo, Q.; Zhang, X.; He, J.; Liu, C.; Jiang, W. Universal rapid demulsification by vacuum suction using superamphiphilic and underliquid superamphiphobic polyurethane/diatomite composites. *ACS Appl. Mat. Interfaces* **2022**, *14*, 24775–24786.
80. Wang, Y.; Zhu, Y.; Yang, C.; Liu, J.; Jiang, W.; Liang, B. Facile two-step strategy for the construction of a mechanically stable three-dimensional superhydrophobic structure for continuous oil–water separation. *ACS Appl. Mat. Interfaces* **2018**, *10*, 24149–24156.
81. Xiang, B.; Sun, Q.; Zhong, Q.; Mu, P.; Li, J. Current research situation and future prospect of superwetting smart oil/water separation materials. *J. Mat. Chem. A* **2022**, *10*, 20190–20217.

82. Zhang, J.; Peng, K.; Xu, Z.-K.; Xiong, Y.; Liu, J.; Cai, C.; Huang, X. A comprehensive review on the behavior and evolution of oil droplets during oil/water separation by membranes. *Adv. Colloid Interface Sci.* **2023**, *319*, 102971.
83. Sutrisna, P.D.; Kurnia, K.A.; Siagian, U.W.R.; Ismadji, S.; Wenten, I.G. Membrane fouling and fouling mitigation in oil–water separation: A review. *J. Env. Chem. Eng.* **2022**, *10*, 107532.
84. Tummons, E.; Han, Q.; Tanudjaja, H.J.; Hejase, C.A.; Chew, J.W.; Tarabara, V.V. Membrane fouling by emulsified oil: A review. *Sep. Purif. Technol.* **2020**, *248*, 116919.
85. Simonsen, G.; Strand, M.; Øye, G. Potential applications of magnetic nanoparticles within separation in the petroleum industry. *J. Petrol. Sci. Eng.* **2018**, *165*, 488–495.
86. Ali, N.; Bilal, M.; Khan, A.; Ali, F.; Yang, Y.; Khan, M.; Adil, S.F.; Iqbal, H.M.N. Dynamics of oil-water interface demulsification using multifunctional magnetic hybrid and assembly materials. *J. Mol. Liquids* **2020**, *312*, 113434.
87. Hamedi, H.; Rezaei, N.; Zendehboudi, S. A comprehensive review on demulsification using functionalized magnetic nanoparticles. *J. Clean. Product.* **2022**, *380*, 134868.
88. Lü, T.; Zhang, S.; Qi, D.; Zhang, D.; Zhao, H. Enhanced demulsification from aqueous media by using magnetic chitosan-based flocculant. *J. Colloid Interface Sci.* **2018**, *518*, 76–83.
89. Sjöblom, J.; Mhatre, S.; Simon, S.; Skartlien, R.; Sørland, G. Emulsions in external electric fields. *Adv. Colloid Interface Sci.* **2021**, *294*, 102455.
90. Quintero, J.S.M.; Majhy, B.; Caesar, M.; Waghmare, P.R. Electrowetting-induced coalescence of sessile droplets in viscous medium. *Langmuir* **2023**, *39*, 4917–4923.
91. Skartlien, R.; Simon, S.; Sjöblom, J. Electrocoalescence of water in oil emulsions: A DPD simulation study and a novel application of electroporation theory. *RSC Adv.* **2019**, *9*, 34172–34183.
92. Li, W.; Zeng, H.; Tang, T. Molecular dynamics simulation on water/oil interface with model asphaltene subjected to electric field. *J. Colloid Interface Sci.* **2022**, *628*, 924–934.
93. Mhatre, S.; Simon, S.; Sjöblom, J. Experimental evidence of enhanced adsorption dynamics at liquid–liquid interfaces under an electric field. *Anal. Chem.* **2020**, *92*, 12860–12870.
94. Lv, X.; Song, Z.; Yu, J.; Su, Y.; Zhao, X.; Sun, J.; Mao, Y.; Wang, W. Study on the demulsification of refinery oily sludge enhanced by microwave irradiation. *Fuel* **2020**, *279*, 118417.
95. Kovaleva, L.; Zinnatullin, R.; Musin, A.; Gabdrafikov, A.; Sultanguzhin, R.; Kireev, V. Influence of radio-frequency and microwave electromagnetic treatment on water-in-oil emulsion separation. *Colloids Surf. A* **2021**, *614*, 126081.
96. Hyde, A.; Saiuchi, K.; Sonobe, S.; Shibata, Y.; Asakuma, Y.; Phan, C. Influence of microwave pulsing patterns on oil/water interfacial behavior. *Colloids Surf. A* **2021**, *631*, 127660.
97. Shibata, Y.; Hyde, A.; Asakuma, Y.; Phan, C. Thermal response of a non-ionic surfactant layer at the water/oil interface during microwave heating. *Colloids Surf. A* **2018**, *556*, 127–133.
98. Goibier, L.; Lecomte, S.; Leal-Calderon, F.; Faure, C. The effect of surfactant crystallization on partial coalescence in o/w emulsions. *J. Colloid Interface Sci.* **2017**, *500*, 304–314.
99. MacWilliams, S.V.; Sebben, D.A.; Clulow, A.J.; Ferri, J.K.; Gillies, G.; Golding, M.; Boyd, B.J.; Beattie, D.A.; Krasowska, M. The effect of emulsifier type on the secondary crystallisation of monoacylglycerol and triacylglycerols in model dairy emulsions. *J. Colloid Interface Sci.* **2022**, *608*, 2839–2848.

100. Dahiya, P.; DeBenedictis, A.; Atherton, T.J.; Caggioni, M.; Prescott, S.W.; Hartel, R.W.; Spicer, P.T. Arrested coalescence of viscoelastic droplets: Triplet shape and restructuring. *Soft Matter* **2017**, *13*, 2686–2697.
101. Otazo, M.R.; Ward, R.; Gillies, G.; Osborne, R.S.; Golding, M.; Williams, M.A.K. Aggregation and coalescence of partially crystalline emulsion drops investigated using optical tweezers. *Soft Matter* **2019**, *15*, 6383–6391.
102. Abedi, S.; Suteria, N.S.; Chen, C.-C.; Vanapalli, S.A. Microfluidic production of size-tunable hexadecane-in-water emulsions: Effect of droplet size on destabilization of two-dimensional emulsions due to partial coalescence. *J. Colloid Interface Sci.* **2019**, *533*, 59–70.
103. Abedi, S.; Chen, C.-C.; Vanapalli, S.A. Catastrophic thermal destabilization of two-dimensional close-packed emulsions due to synchronous coalescence initiation. *Soft Matter* **2020**, *16*, 6032–6037.
104. Antes, F.G.; Diehl, L.O.; Pereira, J.S.F.; Guimarães, R.C.L.; Guarnieri, R.A.; Ferreira, B.M.S.; Flores, E.M.M. Effect of ultrasonic frequency on separation of water from heavy crude oil emulsion using ultrasonic baths. *Ultrason. Sonochem.* **2017**, *35*, 541–546.
105. Luo, X.; Gong, H.; Yin, H.; He, Z.; He, L. Optimization of acoustic parameters for ultrasonic separation of emulsions with different physical properties. *Ultrason. Sonochem.* **2020**, *68*, 105221.
106. Pedrotti, M.F.; Enders, M.S.P.; Pereira, L.S.F.; Mesko, M.F.; Flores, E.M.M.; Bizzi, C.A. Intensification of ultrasonic-assisted crude oil demulsification based on acoustic field distribution data. *Ultrason. Sonochem.* **2018**, *40*, 53–59.
107. Luo, X.; Cao, J.; Yin, H.; Yan, H.; He, L. Droplets banding characteristics of water-in-oil emulsion under ultrasonic standing waves. *Ultrason. Sonochem.* **2018**, *41*, 319–326.
108. Khalesi Moghaddam, R.; Yarranton, H.W.; Natale, G. Ultrasound treated asphaltene laden w/o interface: Insights into emulsion destabilization mechanism using a microrheology approach. *Chem. Eng. Sci.* **2023**, *280*, 119065.
109. Porto Santos, T.; Cejas, C.M.; Cunha, R.L. Microfluidics as a tool to assess and induce emulsion destabilization. *Soft Matter* **2022**, *18*, 698–710.
110. Tian, Y.S.; Li, E.Q.; Elsaadawy, E.; Zhang, J.M.; Vakarelski, I.U.; Thoroddsen, S.T. Coalescence time of water-in-oil emulsions under shear. *Chem. Eng. Sci.* **2022**, *250*, 117257.
111. Lim, M.Y.; Stokes, J.R. Lubrication of non-ionic surfactant stabilised emulsions in soft contacts. *Biotribology* **2021**, *28*, 100199.
112. Lim, M.Y.; Xu, Y.; Shewan, H.M.; Stokes, J.R. Entrainment mechanism of viscoplastic fat particles and tribofilm formation in soft contact tribology. *Biotribology* **2022**, *32*, 100220.
113. Haghighat, A.K.; Olsen, M.G.; Vigil, R.D.; Sarkar, A. Droplet coalescence and phase separation in a topical ointment: Effects of fluid shear and temperature. *Int. J. Pharm.* **2020**, *591*, 119872.
114. Nadeem, H.; Vigil, R.D.; Samuel, A.; Sarkar, A.; Yeoh, T.; Olsen, M.G. Coalescence-induced phase separation of an oil in water emulsion under controlled shear and temperature conditions. *Chem. Eng. Res. Desing* **2022**, *182*, 517–524.
115. Soltanahmadi, S.; Bryant, M.; Sarkar, A. Insights into the multiscale lubrication mechanism of edible phase change materials. *ACS Appl. Mat. Interfaces* **2023**, *15*, 3699–3712.
116. Fernandes Barros, F.M.; Chassenieux, C.; de Souza Lima, M.M.; Benyahia, L. Structure and rheology during catastrophic phase inversion of pickering emulsions stabilized with fumed silica particles. *Colloids Surf. A* **2020**, *593*, 124630.
117. Sawiak, L.; Bailes, K.; Harbottle, D.; Clegg, P.S. Mixing time, inversion and multiple emulsion formation in a limonene and water pickering emulsion. *Front. Chem.* **2018**, *6*, 00132.

118. Barros, F.M.F.; Pelegrini, B.L.; Chassenieux, C.; de Souza Lima, M.M.; Benyahia, L. Rheology and structure of pickering emulsions undergoing transitional phase inversion using a mixture of hydrophilic and hydrophobic silica particles. *Colloids Surf. A* **2022**, *644*, 128801.
119. Sun, G.; Guo, T.; Luo, J.; Liu, R.; Ngai, T.; Binks, B.P. Phase inversion of pickering emulsions induced by interfacial electrostatic attraction. *Langmuir* **2023**, *39*, 1386–1393.
120. Kumar, H.; Dugyala, V.R.; Basavaraj, M.G. Phase inversion of ellipsoid-stabilized emulsions. *Langmuir* **2021**, *37*, 7295–7304.
121. Park, S.H.; Hong, C.R.; Choi, S.J. Prevention of Ostwald ripening in orange oil emulsions: Impact of surfactant type and Ostwald ripening inhibitor type. *LWT* **2020**, *134*, 110180.
122. Park, J.; Lee, J.; McClements, D.J.; Choi, S.J. Inhibition of droplet growth in model beverage emulsions stabilized using poly (Ethylene Glycol) alkyl ether surfactants having various hydrophilic head sizes: Impact of ester gum. *Appl. Sci.* **2020**, *10*, 5588.
123. Jang, Y.; Park, J.; Song, H.Y.; Choi, S.J. Ostwald ripening rate of orange oil emulsions: Effects of molecular structure of emulsifiers and their oil composition. *J. Food Sci.* **2019**, *84*, 440–447.
124. Han, S.W.; Song, H.Y.; Moon, T.W.; Choi, S.J. Influence of emulsion interfacial membrane characteristics on Ostwald ripening in a model emulsion. *Food Chem.* **2018**, *242*, 91–97.
125. Beltramo, P.J.; Gupta, M.; Alicke, A.; Liascukiene, I.; Gunes, D.Z.; Baroud, C.N.; Vermant, J. Arresting dissolution by interfacial rheology design. *Proc. Nat. Acad. Sci. USA* **2017**, *114*, 10373–10378.
126. Thompson, K.L.; Giakoumatos, E.C.; Ata, S.; Webber, G.B.; Armes, S.P.; Wanless, E.J. Direct observation of giant pickering emulsion and colloidosome droplet interaction and stability. *Langmuir* **2012**, *28*, 16501–16511.
127. Kumar, H.; Upendar, S.; Mani, E.; Madivala G. Basavaraj. Destabilization of pickering emulsions by interfacial transport of mutually soluble solute. *J. Colloid Interface Sci.* **2023**, *633*, 166–176.
128. Zhao, C.; Zhou, J.; Yan, Y.; Yang, L.; Xing, G.; Li, H.; Wu, P.; Wang, M.; Zheng, H. Application of coagulation/flocculation in oily wastewater treatment: A review. *Sci. Total Env.* **2021**, *765*, 142795.
129. Ma, J.; Xia, W.; Zhang, R.; Ding, L.; Kong, Y.; Zhang, H.; Fu, K. Flocculation of emulsified oily wastewater by using functional grafting modified chitosan: The effect of cationic and hydrophobic structure. *Journal of Hazard. Mat.* **2021**, *403*, 123690.
130. Lamanna, L.; Giacoia, G.; Friuli, M.; Leone, G.; Carlucci, N.; Russo, F.; Sannino, A.; Demitri, C. Oil–water emulsion flocculation through chitosan desolubilization driven by PH variation. *ACS Omega* **2023**, *8*, 20708–20713.
131. Scharnberg, A.; Oliveira, H.; Weschenfelder, S.; Rubio, J.; Azevedo, A. Flocculation of emulsified oil-in-water with dodecylbenzene sulfonate and polyacrylamide and floc separation by dissolved air flotation. *Colloids Surf. A* **2023**, *669*, 131496.
132. Santos, A.S.; Oliveira, L.F.S.; Marques, A.M.T.; Silva, D.C.A.; Mansur, C.R.E. Evaluation of the efficiency of polyethylenimine as flocculants in the removal of oil present in produced water. *Colloids Surf. A* **2018**, *558*, 200–210.
133. Liu, J.; Li, X.; Jia, W.; Li, Z.; Zhao, Y.; Ren, S. Demulsification of crude oil-in-water emulsions driven by graphene oxide nanosheets. *Energy Fuels* **2015**, *29*, 4644–4653.
134. Fang, S.; Chen, T.; Wang, R.; Xiong, Y.; Chen, B.; Duan, M. Assembly of graphene oxide at the crude oil/water interface: A new approach to efficient demulsification. *Energy Fuels* **2016**, *30*, 3355–3364.
135. Lan, T.; Zeng, H.; Tang, T. Molecular dynamics study on the mechanism of graphene oxide to destabilize oil/water emulsion. *J. Phys. Chem. C* **2019**, *123*, 22989–22999.

136. Al-Jumaily, A.M.; Meshkinzar, A.; Torres, L.M.P. On the development of emulsion destabilization technologies for dairy industry. *Food Eng. Rev.* **2023**, *15*, 215–229.
137. Ghanbari, M.; Esmaeilzadeh, F. Demulsification by increasing the gravitational force acting upon the dispersed phase owing to the adsorption/absorption of the magnetite particles. *J. Disp. Sci. Technol.* **2019**, *40*, 1581–1590.
138. Chen, S.; Dima, C.; Kharazmi, M.S.; Yin, L.; Liu, B.; Jafari, S.M.; Li, Y. The colloid and interface strategies to inhibit lipid digestion for designing low-calorie food. *Adv. Colloid Interface Sci.* **2023**, *321*, 103011.
139. Sarkar, A.; Zhang, S.; Holmes, M.; Ettelaie, R. Colloidal aspects of digestion of pickering emulsions: Experiments and theoretical models of lipid digestion kinetics. *Adv. Colloid Interface Sci.* **2019**, *263*, 195–211.
140. Murray, B.S.; Ettelaie, R.; Sarkar, A.; Mackie, A.R.; Dickinson, E. The perfect hydrocolloid stabilizer: Imagination versus reality. *Food Hydrocoll* **2021**, *117*, 106696.
141. Zhai, H.; Gunness, P.; Gidley, M.J. Barley β-Glucan effects on emulsification and in vitro lipolysis of canola oil are modulated by molecular size, mixing method, and emulsifier type. *Food Hydrocoll* **2020**, *103*, 105643.
142. Zhai, H.; Gunness, P.; Gidley, M.J. Depletion and bridging flocculation of oil droplets in the presence of β-glucan, arabinoxylan and pectin polymers: Effects on lipolysis. *Carbohydrate Polym.* **2021**, *255*, 117491.
143. Wang, X.; Lin, Q.; Ye, A.; Han, J.; Singh, H. Flocculation of oil-in-water emulsions stabilised by milk protein ingredients under gastric conditions: Impact on in vitro intestinal lipid digestion. *Food Hydrocoll* **2019**, *88*, 272–282.
144. Grundy, M.M.L.; McClements, D.J.; Ballance, S.; Wilde, P.J. Influence of oat components on lipid digestion using an in vitro model: Impact of viscosity and depletion flocculation mechanism. *Food Hydrocoll.* **2018**, *83*, 253–264.
145. Wan, C.; Cheng, Q.; Zeng, M.; Huang, C. Recent progress in emulsion gels: From fundamentals to applications. *Soft Matter* **2023**, *19*, 1282–1292.
146. Lin, D.; Kelly, A.L.; Miao, S. Preparation, structure-property relationships and applications of different emulsion gels: Bulk emulsion gels, emulsion gel particles, and fluid emulsion gels. *Trends Food Sci. Technol.* **2020**, *102*, 123–137.
147. Li, S.; McClements, D.J. Controlling textural attributes of plant-based emulsions using heteroaggregation of cationic and anionic potato protein-coated oil droplets. *Food Hydrocoll* **2023**, *145*, 109126.
148. Simo, O.K.; Mao, Y.; Tokle, T.; Decker, E.A.; McClements, D.J. Novel strategies for fabricating reduced fat foods: heteroaggregation of lipid droplets with polysaccharides. *Food Res. Int.* **2012**, *48*, 337–345.
149. Wang, X.-Y.; Wang, J.; Rousseau, D.; Tang, C.-H. Chitosan-stabilized emulsion gels via Ph-induced droplet flocculation. *Food Hydrocoll* **2020**, *105*, 105811.
150. Tong, Q.; Yi, Z.; Ma, L.; Tan, Y.; Chen, X.; Cao, X.; Liu, D.; Li, X. Polysaccharide-dependent depletion strategy to fabricate pickering emulsion gels. *Food Hydrocoll* **2023**, *145*, 109175.
151. Zhang, Y.; Yu, D. High internal phase pickering emulsions stabilized by ODSA-functionalized cyclodextrin metal-organic frameworks for extrusion 3d printing application. *LWT* **2023**, *182*, 114800.
152. Wang, C.; Yan, R.; Li, X.; Sang, S.; McClements, D.J.; Chen, L.; Long, J.; Jiao, A.; Wang, J.; Qiu, C.; Jin, Z. Development of emulsion-based edible inks for 3d printing applications: Pickering emulsion gels. *Food Hydrocoll* **2023**, *138*, 108482.
153. Feng, G.; Wang, G.; Li, Q.; Han, C.; Wei, D.; Guo, J.; Wan, Z.; Wang, J.; Yang, X. Depletion attraction driven formation of spirulina emulsion gels for 3d printing. *Food Hydrocoll* **2023**, *141*, 108691.

154. Johannesson, J.; Pathare, M.M.; Johansson, M.; Bergström, C.A.S.; Teleki, A. Synergistic stabilization of emulsion gel by nanoparticles and surfactant enables 3d printing of lipid-rich solid oral dosage forms. *J. Colloid Interface Sci.* **2023**, *650*, 1253–1264.
155. Cui, L.; Guo, J.; Meng, Z. A review on food-grade-polymer-based o/w emulsion gels: Stabilization mechanism and 3d printing application. *Food Hydrocoll.* **2023**, *139*, 108588.
156. Cameron, N.R. High internal phase emulsion templating as a route to well-defined porous polymers. *Polymer* **2005**, *46*, 1439–1449.
157. Bago Rodriguez, A.M.; Binks, B.P. High internal phase pickering emulsions. *Curr. Opin. Colloid Interface Sci.* **2022**, *57*, 101556.
158. Zhang, T.; Sanguramath, R.A.; Israel, S.; Silverstein, M.S. Emulsion templating: Porous polymers and beyond. *Macromolecules* **2019**, *52*, 5445–5479.
159. Liao, D.; Guan, Y.; He, Y.; Li, S.; Wang, Y.; Liu, H.; Zhou, L.; Wei, C.; Yu, C.; Chen, Y. Pickering emulsion strategy for high compressive carbon aerogel as lightweight electromagnetic interference shielding material and flexible pressure sensor. *Cer. Int.* **2021**, *47*, 23433–23443.
160. Yang, T.; Hu, Y.; Wang, C.; Binks, B.P. Fabrication of hierarchical macroporous biocompatible scaffolds by combining pickering high internal phase emulsion templates with three-dimensional printing. *ACS Appl. Mat. Interfaces* **2017**, *9*, 22950–22958.
161. Lee, J.; Chang, J.Y. A hierarchically porous catalytic monolith prepared from a pickering high internal phase emulsion stabilized by microporous organic polymer particles. *Chem. Eng. J.* **2020**, *381*, 122767.
162. Foudazi, R. HIPEs to PolyHIPEs. *React. Funct. Polym.* **2021**, *164*, 104917.
163. Durgut, E.; Sherborne, C.; Aldemir Dikici, B.; Reilly, G.C.; Claeyssens, F. Preparation of interconnected pickering polymerized high internal phase emulsions by arrested coalescence. *Langmuir* **2022**, *38*, 10953–10962. h
164. Massicotte, M.; Cranston, E.D. Comparison of techniques for drying cellulose nanocrystal pickering emulsions into oil powders. *ACS Sustain. Chem. Eng.* **2022**, *10*, 14914–14925.
165. Esparza, Y.; Ngo, T.-D.; Boluk, Y. Preparation of powdered oil particles by spray drying of cellulose nanocrystals stabilized pickering hempseed oil emulsions. *Colloids Surf. A* **2020**, *598*, 124823.
166. Amaya Cano, J.S.; Segura Pacheco, S.; Salcedo Galán, F.; Arenas Bustos, I.; Rincón Durán, C.; Hernández Carrión, M. Formulation of a responsive in vitro digestion wall material, sensory and market analyses for chia seed oil capsules. *J. Food Eng.* **2021**, *296*, 110460.
167. Srivastava, S.; Mishra, H.N. Development of microencapsulated vegetable oil powder based cookies and study of its physicochemical properties and storage stability. *LWT* **2021**, *152*, 112364.
168. Sandhya, K.; Leena, M.M.; Moses, J.A.; Anandharamakrishnan, C. Edible oil to powder technologies: Concepts and advances. *Food Biosci.* **2023**, *53*, 102567.
169. Lu, W.; Maidannyk, V.; Kelly, A.L.; Miao, S. Fabrication and characterization of highly re-dispersible dry emulsions. *Food Hydrocoll* **2020**, *102*, 105617.
170. Li, K.-Y.; Yu, W.-K.; Xiao, J.-X.; Huang, G.-Q. Addition of gelatin increased the spray drying performance of the pickering emulsion stabilized by the ovalbumin – gum Arabic electrostatic complex. *Colloids Surf. A* **2023**, *671*, 131640.
171. Scholten, E. Edible Oleogels: How suitable are proteins as a structurant? *Curr. Opin. Food Sci.* **2019**, *27*, 36–42.
172. Patel, A.R. Structuring edible oils with hydrocolloids: Where do we stand? *Food Biophys.* **2018**, *13*, 113–115.

173. Urbánková, L.; Sedláček, T.; Kašpárková, V.; Bordes, R. Formation of oleogels based on emulsions stabilized with cellulose nanocrystals and sodium caseinate. *J. Colloid Interface Sci.* **2021**, *596*, 245–256.
174. Luo, S.-Z.; Hu, X.-F.; Jia, Y.-J.; Pan, L.-H.; Zheng, Z.; Zhao, Y.-Y.; Mu, D.-D.; Zhong, X.-Y.; Jiang, S.-T. Camellia oil-based oleogels structuring with tea polyphenol-palmitate particles and citrus pectin by emulsion-templated method: Preparation, characterization and potential application. *Food Hydrocoll* **2019**, *95*, 76–87.
175. Tavernier, I.; Doan, C.D.; Van der Meeren, P.; Heyman, B.; Dewettinck, K. The Potential of waxes to alter the microstructural properties of emulsion-templated oleogels. *Eur. J. Lipid Sci. Technol.* **2018**, *120*, 1700393.
176. Pan, H.; Xu, X.; Qian, Z.; Cheng, H.; Shen, X.; Chen, S.; Ye, X. Xanthan gum-assisted fabrication of stable emulsion-based oleogel structured with gelatin and proanthocyanidins. *Food Hydrocoll* **2021**, *115*, 106596.
177. Hong, S.J.; Shin, G.H.; Kim, J.T. Vitamin E-fortified emulsion-templated oleogels structured with xanthan gum and soybean lecithin and their application in pound cakes. *Food Biosci.* **2024**, *57*, 103505.

11 Phase Behavior of Aqueous Two-Phase Systems and Stabilization of Water-in-Water Emulsions

Jordi Esquena

11.1 INTRODUCTION

Liquid–liquid phase separation (LLPS) in two immiscible aqueous solutions is quite common in mixtures of water-soluble polymers or salts and occurs in the absence of oil or surfactant. For example, phase separation in two aqueous solutions occurs inside eukaryotic cells, forming membraneless organelles (MLO) that can appear spontaneously in biological systems by mechanisms of phase separation in thermodynamic equilibrium. Aqueous two-phase systems (ATPS) can be formed by mixing two water-soluble polar components, with phase separation occurring through two distinct mechanisms: (a) attractive interactions between chemical species, leading to association or self-aggregation; or (b) repulsive interactions in segregative systems.[1]

Figure 11.1 shows schematically the different cases that may be observed in aqueous mixtures of hydrophilic polymers, depending on whether two polymers interact by attractive or repulsive intermolecular forces. Five situations may be observed: (a) formation of kinetically trapped coacervate particles or droplets that may remain in colloidally stable suspensions; (b) coprecipitation that produces an insoluble hydrated solid; (c) homogeneous mixture of the two polymers in a single-phase aqueous solution; (d) phase separation in two nonstructured aqueous phases; and (e) emulsification and formation of droplets. The first two cases are associative systems caused by attractive interactions between the two polymers; the third case is a noninteracting system; and the last two cases are segregative systems produced by repulsive interactions.

In associative systems, self-aggregation of two macromolecules frequently arises from electrostatic attractions between oppositely charged biopolymers. This phenomenon, of ionic complexation, can give rise to the formation of particles or a precipitate with high water content and liquid appearance. However, in segregative

 DOI: 10.1201/9781032636184-13

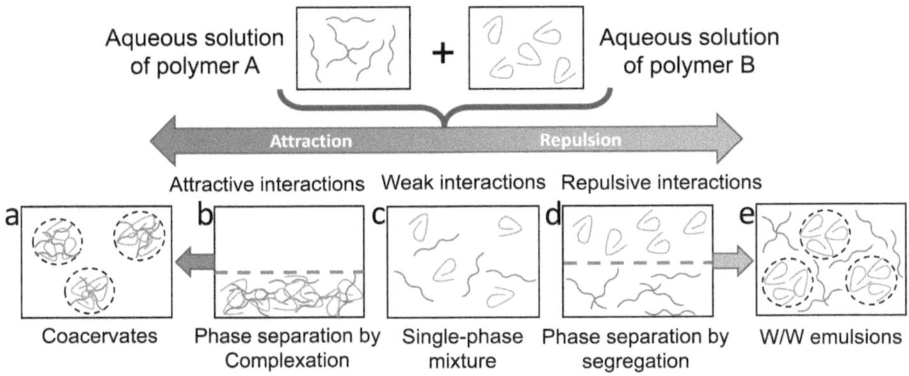

FIGURE 11.1 Schematic representation of the phase behavior of aqueous mixtures of two polymers, ranging from attractive to repulsive interactions. Coacervation occurs because of attractive interactions and coacervate droplets or particles may remain as colloidally stable suspensions (a) or coprecipitate and sediment (b); weakly or noninteracting systems, commonly found in diluted systems, form homogeneous single-phase solutions (c); and segregative interactions lead to the formation of two equilibrium aqueous solutions (d). In this last case, water-in-water (W/W) emulsions that consist of two liquid equilibrium phases can be prepared (e).

systems, phase separation occurs due to the immiscibility of the two solutes, producing the separation of two liquid phases; each phase becomes enriched with one of the two solutes, while it is saturated with the other solute. Molecular association and segregation can be distinguished by their respective phase behavior (Figure 11.1). Association commonly results in the formation of a coacervate that contains a mixture of the two components in equilibrium with almost pure water (Figures 11.1a and 11.1b), whereas segregation leads to the formation of two aqueous solutions, each one enriched in one component and depleted in the other component (Figures 11.1d and 11.1e), as reported by Piculell and Lindman.[2] Consequently, the ternary phase diagrams of associative and segregative systems are different (Figure 11.2).

The phrase 'water-in-water emulsions', abbreviated as W/W emulsions, refers to colloidal dispersions in which an aqueous solution forms droplets within another aqueous solution, the compositions of the two phases being in thermodynamic equilibrium. This is a broad definition that includes a very wide variety of different systems, in which the phase separation can be due to association (coacervation), segregation (immiscibility), or a combination of both.

Traditionally, formation of viscous droplets by association, such as ionic complexation, is denoted as coacervation and coacervates are the microgels that constitute the dispersed phase. The term coacervate was coined by Bungenberg de Jong and de Vries.[3–5] They studied a wide variety of systems that formed by ionic complexation, obtained in aqueous mixtures of biomolecules with opposed electric charges. Complex coacervation refers to aggregation of two molecules with opposite charges, while simple coacervation refers to self-aggregation of a single component. Therefore, coacervates are highly hydrated condensates of matter induced

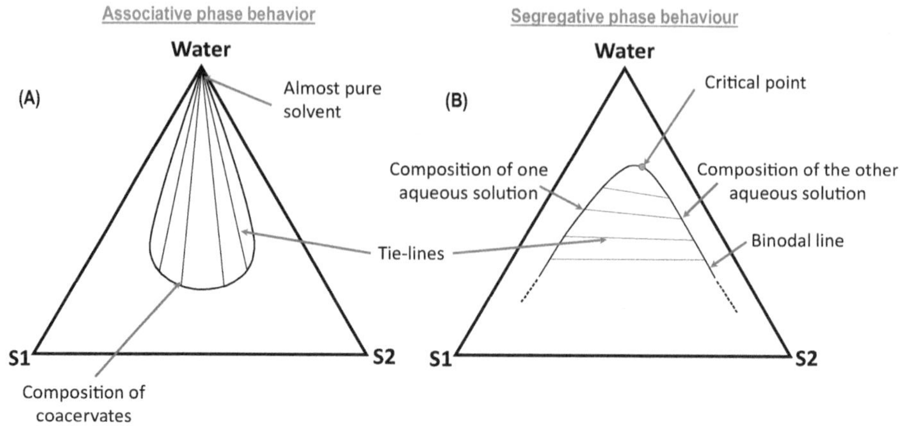

FIGURE 11.2 Schematic representation of the ternary diagrams of aqueous mixtures of two solutes (S1 and S2) that can interact either by association (A) or by segregation (B). The tie-lines relate the compositions that are in mutual thermodynamic equilibrium. In associative phase behavior (a), the two solutes condense in a coacervate in equilibrium with excess water. In the case of segregative phase behavior (b), two liquid phases can be found within the boundary of the binodal line. The tie-lines converge in a critical point, where the two phases reach the same composition, and the interface vanishes. Therefore, the two liquid phases merge into one single phase at the critical point. Adapted from Piculell and Lindman,[2] Copyright (1992), with permission from Elsevier.

by aggregation of components, usually charged polyelectrolytes. Often, coacervates remain in suspension as colloidally stable microgels or liquid-like droplets.

Nowadays, the phrase 'water-in-water emulsions' is mostly used to describe colloidal liquid–liquid dispersions in segregative systems,[6–8] despite the fact that colloidal dispersions of ionic complexes (coacervates) can also be considered to be W/W emulsions. Coacervates and segregative W/W emulsions have in common that they consist of aqueous droplets, albeit with different degrees of viscosity, since coacervate droplets are usually very viscous and are considered microgels, while droplets of segregative W/W emulsions consist of liquids with lower viscosities. In both cases, the interfaces may have no membrane and/or monolayers, since water–water interfaces have very low interfacial tensions, and then adsorption of molecules is very weak at these interfaces.

'Membraneless organelles', abbreviated as MLOs, is the phrase used in cell biology to describe condensates of proteins and other components that appear as liquid droplets in eukaryotic cells, either in the cytoplasm or inside the nucleus. These structures typically form through liquid–liquid phase separation, often driven by ionic complexation and coacervation.[9, 10] MLOs enable the organization of diverse cellular components in a compartmentalized manner, increasing local concentrations and facilitating a broad spectrum of cellular functions. MLOs are coacervates that have the capacity to increase the velocity of enzymatic reactions and are actively investigated for their potential role as protocells.[11] Consequently, coacervates are recognized as simplified synthetic models of artificial cells and MLOs.[12–16]

There is some terminological confusion caused by calling 'water-in-water emulsions' to all fully aqueous systems,[17] including those with dispersed phases that are nanostructured liquids or liquid crystals. Certainly, these can be completely aqueous colloidal dispersions, but one can use more specific names such as cubosomes, tactoids, etc., referring to self-assembled colloidal nanostructures. Thus, the phrase 'water-in-water emulsions' would be applied to colloidal dispersions in aqueous two-phase systems (ATPS) without organization inside droplets. This definition would not encompass dispersions of liquid crystals, even though they are sometimes referred to in the literature as 'W/W emulsions'.[17] A recent review of colloidal dispersions of aqueous liquid crystalline phases can be found elsewhere.[18] Vesicles and liposomes, which are dispersions of lamellar liquid crystals, should not be considered 'W/W emulsions' since the internal and external aqueous solutions are not immiscible liquids.

Therefore, the phrase 'water-in-water emulsions' (generally abbreviated as W/W emulsions) denotes colloidal dispersions of low-viscosity aqueous immiscible phases. This concept is simple and should not be confused with 'complex coacervates',[18–20] 'simple coacervates', or 'liquid coacervates',[21] which refer to colloidal dispersions of hydrogels (microgels) in associative systems. Usually, 'water-in-water emulsions' highlight the liquid nature with relatively low viscosity of the two liquid phases, albeit simple or complex coacervates can also have low viscosity and thus sometimes denoted as 'liquid coacervates'.[22] These coacervates of low viscosity can also be considered as 'water-in-water emulsions'. Coacervates with higher viscosity can be classified as 'microgels'.

This chapter centers on water-in-water (W/W) emulsions formed in segregative systems characterized by the incompatibility between solutes in aqueous mixtures. In these systems, the two solutions share water as the common solvent and consist of two non-mixable hydrophilic solutes, resulting in the formation of two liquid phases that are in thermodynamic equilibrium.[8,23,24] A good critical review has been published recently.[25] Aqueous biphasic systems (ABS)[26,27] are most often denoted as 'aqueous two-phase systems'.[24,28] The latter terminology is widely accepted in the scientific community and it is used throughout this chapter. It is important to note that this chapter does not cover the study of coacervation (either complex or simple), nor does it address research on membraneless organelles (MLOs) or artificial organelles. While these topics are related to ATPS, they fall outside the scope of the current discussion.

Water-in-water emulsions have a long history, starting with the pioneering studies of Beijerinck in 1896.[29] He was a microbiologist, considered the founder or virology, who observed the presence of droplets in culture media composed of partly hydrolyzed starch, gelatin, and water. He observed droplet formation across a wide range of starch/gelatin ratios and observed a shift in the emulsion characteristics, ranging from starch-in-gelatin at elevated gelatin concentrations to gelatin-in-starch at increased starch contents. At intermediate starch/gelatin ratios, he noted the presence of multiple emulsions. His publication in 1896[29] is considered the first description of W/W emulsions, although they were not yet known by this name, which began to be used in the 1980s. Otto Bütschli[30] studied Beijerink experiments in 1898, preparing emulsion droplets in aqueous mixtures of hydrolyzed starch and gelatin, and studying the effects of various preparation parameters. Beijerinck[31] noted a similar behavior in the starch–agar system in 1910, leading him to believe that the formation of emulsions in aqueous mixtures was a characteristic of 'certain gelatinizing colloids'.

A significant interest in ATPS emerged with Albertsson's groundbreaking findings in 1958–1960, which unveiled the potential of ATPS systems for the extraction and concentration of proteins[32] and viruses.[33,34] He discovered that, by using the adsorption of the influenza virus at water/water interfaces, the virus could be concentrated up to 200 times more using aqueous mixtures of dextran with methylcellulose or polyethylene glycol, which are two water-soluble polymers with lower hydrophilicity than dextran.[33]

Nowadays, it is widely recognized that ATPS in segregative systems have significant importance owing to their extensive applications in the extraction of biomolecules and cells.[35] W/W emulsions could be prepared in a very great variety of different systems. Multiphasic aqueous mixtures focus a great attention on a wide variety of areas, including the food industry, chemical processes, and biomedical engineering. Currently, widespread industrial use of ATPS occurs in the purification of proteins and antibodies, along with the extraction of water-soluble macromolecules. Important scientific advances are currently taking place in the understanding of water/–water interfaces, the controlled formation of W/W emulsions, and the development of biomimetic systems that can be considered as artificial cells.[25, 36]

11.2 AQUEOUS MIXTURES WITH SEGREGATIVE PHASE SEPARATION

There is a broad variety of different aqueous mixtures capable of forming two immiscible liquid solutions through phase segregation. Typically, ATPS can appear in polymer–polymer, polymer–salt, and salt–salt aqueous mixtures, but it can also be observed in many other mixtures of water-soluble components.[35,37–42] All the possible aqueous mixtures for segregative-phase separation are comprehensively summarized and reviewed by Pereira and Coutinho.[38,40] These systems can be categorized as follows: (i) highly hydrated polymers with less hydrated polymers; (ii) water-soluble polymers with kosmotropic salts; (iii) mixtures of chaotropic salts with kosmotropic salts; (iv) polar small organic molecules in the presence of salts; (v) mixtures in the presence of surfactant micelles; (vi) deep eutectic solvents (DES) in the presence of water-soluble polymers or organic molecules, and (vii) other systems in the presence of weakly hydrated molecules or polymers. The following section discusses these various systems.

11.2.1 Highly Hydrated Polymers with Less Hydrated Polymers

The aqueous mixtures of two water-soluble polymers[38,40,43] are the ATPS that have undergone the most extensive studies, tracing back to the early investigations by Beijerinck.[29,31] Segregation in two aqueous phases occurs when one polymer is more hydrophilic, and thus more hydrated, than the other polymer causing volume exclusions due to hydration differences.[42, 44] The phase separation in aqueous polymer mixtures is a common phenomenon that manifests in the absence of opposing charges that could lead to complexation instead.

The phase behavior of ATPS was deeply studied from the 1950s, when Albertsson et al. demonstrated the effectiveness of ATPS for extracting macromolecules, viruses, and bacteria.[32–34, 45] In 1997, a review by Grinberg and Tolstoguzov[44] listed approximately 100 different combinations of polysaccharides and proteins that had shown segregative-phase separation. Currently, the most widely known systems to form W/W emulsions are composed of proteins and polysaccharides.

Immiscibility of two polymers in aqueous mixtures can be explained in terms of the Flory–Huggins theory, which might offer straightforward analytical expressions that quantify the hydration differences that influence phase behavior in aqueous two-phase systems.[42,46] Flory–Huggins theory has some limitations, mainly because this theory does not specifically include charge effects, but nevertheless this theory allows explaining quite precisely the phase behavior of ATPS, providing correct predictions of phase diagrams.[42] Moreover, this theory can also explain that increasing the molecular weight of the polymers leads to an entropic contribution that favors demixing and formation of ATPS.

ATPS can be obtained in a wide variety of different combinations of two polymers: (a) two nonionic polymers; (b) one nonionic and one ionic polymer; and (c) two ionic polymers. A comprehensive table summarizing many diverse combinations of two polymers can be found in a review by Pereira and Coutinho.[38]

In combinations of two nonionic polymers, it is typically necessary for one of them to be more polar than the other. This is evident in mixtures of polyethylene glycol (PEG) and a polysaccharide, where PEG has significantly lower ability to form hydrogen bonds.[47] An illustrative example is dextran–PEG aqueous mixtures, which have been deeply studied and phase diagrams are available in the literature.[47–50] PEG exhibits a reduced capacity to establish hydrogen bonds with water, compared to polysaccharides. Consequently, PEG can form aqueous two-phase systems with virtually all nonionic polysaccharides.

In this context, the dextran–PEG system has been widely used as a model for the formation and stabilization of water-in-water emulsions.[51–53] Nicolai and coworkers[7,54–57] used the dextran–PEG system to investigate mechanisms and techniques for stabilizing emulsions, employing either particles or polymers as stabilizers. This system has also found extensive application in the formation of water-in-water emulsions through microfluidics.[25] Additionally, Ravaine and coworkers[50] achieved the production of thermosensitive dextran–PEG emulsions, stabilized by microgels based on pNIPAM. Their reports will be explained in more detail in section 11.4.2 of this chapter.

Separation into distinct aqueous phases can also take place when two nonionic polysaccharides are combined. The guar–amylopectin system[58] exemplifies polysaccharide incompatibility in aqueous solutions. Guar, a linear polysaccharide composed of alternating D-galactose and D-mannose units, and amylopectin, a highly branched polymer of D-glucose units, exhibit mutual immiscibility in solution. This incompatibility arises from volume exclusion effects, leading to spontaneous phase separation and demixing. Analogous phase behavior is observed in mixtures of other nonionic polysaccharides such as maltodextrin–dextran[59] and pullulan-amylopectin.[60] In these examples, maltodextrin and pullulan are linear polysaccharides, while dextran and amylopectin exhibit branched structures. The structural differences between these polysaccharides contribute to their incompatibility in solution, leading to the formation of aqueous two-phase systems. One intriguing example is the phase separation observed in xyloglucan–amylopectin aqueous mixtures, which involves two polysaccharides that are both nonionic and branched.[61,62] The mechanism that produces this phase separation is not yet fully understood.

When combining an ionic polymer with a nonionic one, a well-documented example in the literature is gelatin–maltodextrin.[46,63,64–71,72] Bovine or porcine gelatin, derived from the partial hydrolysis of collagen, is highly soluble and offers the advantage of forming gels at temperatures below body temperature. This characteristic allows the formation of microgels by cooling down gelatin-in-maltodextrin emulsion droplets.[69] Alternatively, fish gelatin, which does not form gels, provides a model system for studying ATPS at lower temperatures.[73] Other interesting systems involve proteins in the presence of a nonionic polymer. For example, monoclonal antibodies have a tendency to undergo phase separation when combined with nonionic polymers like PEG. The phase behavior of monoclonal antibodies with PEG has been investigated by Rogers et al. [74]

Significant scientific focus has been directed toward investigating phase separation in aqueous mixtures of two hydrophilic polymers, leading to the formation of ATPS. Although this area has been extensively documented, the scientific understanding in this topic is rapidly advancing. Interested readers can find many repo rts.[19,23,25,44,57,66,72,75]

11.2.2 Water-Soluble Polymers with Kosmotropic Salts

Another category of ATPS involves combinations of a water-soluble polymer with a kosmotropic salt, such as mixtures of PEG with Na_2SO_4[76,77] or PEG with alkylmethyl imidazolium chlorides.[77,78] These ATPS in polymer–salt aqueous mixtures are well known, with numerous systems extensively documented in the literature.[35,37,79,80] Typically, the salt is composed of highly hydrated ions that induce a dehydration of the polymer.[81] Therefore, ATPS occurs with polymers that are less hydrated than the salts, requiring the use of kosmotropic salts. These polymer–salt mixtures result in the segregation into two liquid phases, at concentrations higher than those of the binodal line. The phase with lower density is enriched in polymer, while the phase with higher density has a higher salt concentration.

Kosmotropic electrolytes reduce the solubility of polymers in aqueous solutions, inducing crystallization through a phenomenon known as salting-out. Conversely, chaotropic electrolytes increase polymer solubility, an effect termed salting-in. This distinction arises because kosmotropes establish strong hydrogen bonds with water, resulting in ion–water interactions that are stronger than water-water hydrogen bonds. Conversely, chaotropes are weakly hydrated electrolytes, whose ion–water interactions are weaker than those between water molecules.[82–84]

Na_2SO_4 and NaSCN serve as archetypical examples of kosmotropic and chaotropic salts, respectively. Frank Hofmeister ordered ions based on their capacity to either solubilize or precipitate proteins, establishing the well-known series that bear his name.[85,86] The 'salting-out' capacity follows the order $SO_4^{2-} > F^- > I^- > SCN^-$, and therefore, high-charge-density anions behave as kosmotropes and low-charge-density anions act as chaotropes. Regarding cations, the order might be reversed, with low charge density species showing pronounced salting-out properties. Therefore, in terms of salting-out, the $N(CH_3)_4N^+ > Cs^+ > K^+ > Na^+ > Li^+$ sequence might be observed, ranking from stronger to weaker effects of cations. Readers interested in historical articles can find German to English translations of Hofmeister's two most important original articles.[85]

It should be noted that the order of the ions in the Hofmeister series depends on the specific adsorption effects as well as the hydrophilicity/hydrophobicity of the substrates, and then Hofmeister series can be altered or even completely reversed.[84,87] For this reason, different versions of Hofmeister series can be found in the scientific literature. Additionally, it is noteworthy that the kosmotropic/chaotropic effects are considerably more pronounced in anions than in cations. As a result, the type of anion is the primary criterion for selecting salts in particular applications. Detailed insights into recent discoveries regarding the Hofmeister series can be found in excellent review articles.[83, 84]

In ATPS, kosmotropes typically promote separation, while chaotropes facilitate mixing in one single liquid phase. As a result, combinations of a weakly hydrated polymer with a kosmotrope lead to phase separation. For example, this occurs in aqueous blends of acrylate polymers with trisodium citrate, which is a strong kosmotrope.[88] The ATPS is formed because of the salting-out effects of the kosmotrope that induce a partial dehydration of the polymer.[38,89]

The capacity of a specific salt to induce phase separation in the presence of a polymer is primarily determined by the salt dehydrating ability.[38,90–93] Rogers and Bauer[76] showed that the salting-out effectiveness of ions on PEG is closely linked to the Gibbs free energy of ion hydration.

Salts with kosmotropic properties often possess bulky cations that frequently lead to low melting points; these salts can be categorized as ionic liquids (ILs) if the melting temperature is lower than 100°C. For example, 1-*n*-butyl-3-methylimidazolium chloride (abbreviated as C_4mimCl) is a well-known IL with a melting point of approximately 65°C.[94] Many commercial ILs are kosmotropes,[82] and therefore these ILs enhance phase segregation and formation of ATPS.

Following the research led by Rogers and coworkers,[95,96] significant attention has been directed toward employing ILs for obtaining ATPS, particularly with polysaccharides, short-chain alcohols, or other nonionic compounds. Many interesting ATPS systems incorporating ILs have been reported, as comprehensively reviewed by Freire et al.[97] Polymer–salt mixtures in water have the advantage, compared to aqueous polymer–polymer systems, that the two immiscible phases have greater differences in their macroscopic properties (refractive indices, dielectric constants, etc.)

which implies that polymer–salt systems allow better yields in extraction processes based on liquid–liquid separation.[97, 98]

Pereira, Coutinho, and their collaborators documented a wide range of ILs capable of producing ATPS when combined with polymers.[78,99] A commonly used polymer is PEG, since it is a water-soluble polymer with a low degree of hydration, and thus, its miscibility in water decreases with the presence of salting-out electrolytes (kosmotropes) and consequently it can produce ATPS with a wide range of different ILs.[38, 99–105] These ATPS based on ILS can have very interesting applications in extraction or purification processes.[97]

Rogers and coworkers[77] showed that differential scanning calorimetry can be used to study the interactions between kosmotropic and chaotropic electrolytes in ATPS. This study demonstrated that aqueous mixtures of PEG and $(NH_4)_2SO_4$ exhibit a transition to vitrified states upon supercooling. Subsequent heating of these glassy states results in devitrification, producing eutectic mixtures.[77]

It should be remarked that ILs in the past were labeled as *green solvents*, because of their very low vapor pressure. However, many ILs are not environmentally benign and furthermore they are toxic. These problematic issues could be solved by using ILs derived from biocompatible anions (e.g., acetate, sorbate, or ascorbate) and cations (e.g., choline). However, fully biocompatible ATPS based on nontoxic ILs have only been sparsely studied. This could become a topic of interest in the future.

11.2.3 Mixtures of Chaotropic Salts with Kosmostropic Salts

ATPS are formed when combining two ions situated on opposing ends of the Hofmeister series. Rogers and coworkers[77,95,106] deeply studied mixtures of kosmotropic and chaotropic salts. For instance, phase separation occurs in mixtures of K_3PO_4 (a strong kosmotrope) with C_4mim Cl acting as a mild chaotrope. This results in the formation of two aqueous phases in thermodynamic equilibrium: a less dense phase rich in C_4mimCl and a denser phase rich in K_3PO_4.[107] These ATPS offer the advantage, compared to other polymer-based ATPS, that the two liquid solutions exhibit quite low viscosities. Consequently, phase separation is fast, enabling rapid extraction processes.[97]

Many mixtures of chaotropic and kosmotropic salts have been used to generate ATPS, with numerous phase diagrams available in the literature.[97,98] Chaotropic salts may incorporate tetrafluoroborates, sulfonates, and alkanoate anions, among others situated in one side of the Hofmeister series. Conversely, kosmotropic salts might include anions like sulfates, phosphates, or carbonates. The greater the separation between distinct ions within the Hofmeister series, the lower the minimum concentrations required for inducing phase separation.[95,108] The effectiveness of phase separation is linked to variations in the free energy of hydration of the ionic species.[95,106] Thus, the position of an ionic species within the Hofmeister series holds crucial significance, demonstrated by systematic studies varying cations and anions.[108–111] A classic illustration is the mixing of ethylmethylimidazolium tetrafluoroborate with phosphate salts.[107] Tetrafluoroborate (BF_4^-) serves as a potent chaotropic anion, while phosphates act as robust kosmotropes. As mentioned before, the kaotropic/

kosmotropic behavior of anions is more pronounced than that of cations. Anyway, it should be noted that the mixtures are not biocompatible, since powerful chaotropes are cytotoxic due to their capability to denature proteins.

Ionic liquids often consist of low-charge-density anions, leading to weak anion–water hydrogen bonds that induce chaotropic behavior (salting-in). Consequently, phase separation may occur upon the addition of kosmotropic high-charge-density anions to aqueous solutions of ionic liquids.[106] Often, these ATPS based on ionic liquids are referred to as 'ionic liquid/salt' aqueous mixtures.[38,111] Nevertheless, it is worth noting that not all chaotropic salts have melting points below 100°C,[112] and, consequently, two-phase mixtures of kosmotropes and chaotropes can be formulated in the absence of ionic liquids.[113] For this reason, the phrase *chaotropic salt/kosmotropic salt* is preferred here instead of *ionic liquid/salt*, the latter being a particular case of the former. In addition, the phrase *chaotropic salt/kosmotropic salt* is considered more precise than *salt/salt*. Regarding the classification of salts, this chapter avoids using the phrase *hydrophobic salt/hydrophilic salt*, preferring the phrase *chaotropic salt/kosmotropic salt* that is related to their free energies of hydration.

In an ATPS composed of two salts, the system has the potential to include two distinct cations and two distinct anions. The distribution of these ionic species between the two aqueous phases is determined by the principle of electroneutrality, which dictates that each phase must maintain an equal number of positive and negative charges. Bridges et al.[106] demonstrated that in an ATPS, the ions with pronounced chaotropic properties tend to concentrate in one phase, while those with prominent kosmotropic characteristics appear in the other phase. For ions with intermediate properties, like Na^+, they are distributed more evenly between the two phases, maintaining electroneutrality in each phase.

11.2.4 Polar Small Organic Molecules in the Presence of Salts

ATPS can also be prepared by mixing polar small organic molecules with salts. The nature of the salt (kosmotropic or chaotropic) depends on the hydration characteristics of the organic molecules. For instance, short-chain alcohols like ethanol or 1-propanol, which are weakly hydrated in aqueous solutions, exhibit dehydration in the presence of strong kosmotropic salts like $(NH_4)_2SO_4$, resulting in the formation of two separated aqueous phases.[114]

ATPS made of alcohol–salt mixtures can be applied in the extraction of active components. An illustrative example is the retrieval of vanillin and L-ascorbic acid from wastewater generated during food processing.[115] This extraction process employs ATPS composed of short-chain alcohols (methanol, ethanol, 1-propanol, and 2-propanol) with K_2HPO_4 or K_3PO_4. The results indicated a preferential partitioning of vanillin to the alcohol-rich aqueous phase, while L-ascorbic acid concentrated into the phosphate-rich aqueous phase.[115]

ATPS can also be obtained by mixing carbohydrates with salts. Mono- and disaccharides are highly hydrated molecules that might phase separate by adding chaotropes. One example is the formation of ATPS by mixing sucrose or fructose with an ionic liquid such as $C_4mim^+BF_4^-$ (1-butyl-3-methylimidazolium tetrafluoroborate), which is a strong chaotrope.[116] Most likely, carbohydrates play the role of weak kosmotropes (hydrated salting-out agents), leading to immiscibility when mixed with chaotropes (less hydrated salting-in agents).[117] Additionally, ionic liquids based on tetraalkylphosphonium or tetraalkylammonium cations, paired with Br^- or short-chain carboxylates as counterions, can also generate ATPS in the presence of carbohydrates.[118] These ionic liquids may also act as chaotropes. Another study showed that the C_4mim^+ $CF_3SO_3^-$ ionic liquid produces ATPS with short-chain saccharides[119] because $CF_3SO_3^-$ is a potent chaotrope.

Amino acids (AAs) can also form aqueous two-phase systems when combined with some ILs. For example, aqueous mixtures of AAs and ILs based on tetralkylphosphonium or tetraalkylammonium cations, which are typically identified as kosmotropes in the Hofmeister series, produce ATPS with AAs.[120] Aromatic and aliphatic AAs showed different partitioning. Aromatic AAs (L-tyrosine, L-tryptophan, and L-phenylalanine) display a marked preference for partitioning into the IL-rich aqueous solution. Conversely, aliphatic AAs (L-lysine and L-proline) exhibit a higher inclination to remain at the IL-depleted aqueous phase. Consequently, this different partitioning offers a viable strategy for the selective separation of aromatic and aliphatic AAs.[120]

11.2.5 Mixtures in the Presence of Micelles

Aqueous micellar two-phase systems (abbreviated as AMPTS) frequently manifest in solutions of surfactants with limited solubility. It is a well-known phenomenon that aqueous dilutions of ethoxylated nonionic surfactants can phase separate into two coexisting liquid phases. The lighter phase becomes enriched with micellar aggregates, while the denser phase experiences a depletion in the surfactant concentration.[121] The temperature at which phase separation occurs is the cloud point of the surfactant. The hydrophilicity of nonionic ethoxylated surfactants decreases with temperature, and thus, these surfactants often have a cloud point.

AMPTS can form in the presence of zwitterionic[122] or nonionic surfactants.[122,123] Nonionic surfactant, in particular, may exhibit a lower critical solution temperature (LCST), above which their solubility in water decreases significantly.[124] At temperatures above the LCST, phase separation occurs, observing cloudiness (at the cloud point or cloud temperature) when the two-phase region is reached.

Nonionic surfactants with poly(ethylene oxide) (PEO) chains typically have a LCST, exhibiting a cloud point upon heating. Above the LCST, the PEO chains become less hydrated, leading to decreased surfactant solubility and phase separation.[124] The LCST can be controlled through the addition of other components. For example, the cloud point of an ethoxylated triglyceride surfactant (Kolliphor ELP), around 80°C, is lowered by the presence of gelatin. This decrease of the cloud point allows the formation of two aqueous phases, one enriched in gelatin and the other in surfactant, enabling the preparation of gelatin-in-surfactant and surfactant-in-gelatin emulsions.[125]

It is worth noting that LCST is not exclusive to PEO chains. Sugar surfactants, for instance, can also exhibit a cloud point, and the transition temperature is also influenced by surfactant concentration.[126] It is well known that chaotropic salts, such as NaI, increase the cloud temperature of nonionic surfactants, whereas kosmotropes, such as NaF, decrease it, promoting phase separation. Therefore, phase separation of micellar solutions is enhanced by highly hydrated anions, such as F^-.

Nevertheless, the processes governing the phase separation in micellar solutions are very complex, with molecular interactions playing crucial roles. Quite small amounts of ionic surfactants can induce a large increase in the cloud temperature of nonionic surfactants. For example, the addition of 10^{-5} mol/L of sodium lauryl sulfate increases the cloud temperature of Triton X-100 nonionic surfactant by approximately 10°C.[127] This is caused by the incorporation of ionic surfactants into the micelles, which increases the electrical charge density and prevents the formation of concentrated micellar solutions due to the electrostatic repulsion between micelles.

This behavior contrasts with the addition of 1-ethyl-3-methyl imidazolium ethylsulfate, which lowers the cloud point of Triton X-100 solutions.[128] Despite being a relatively hydrophilic ionic liquid that one might expect to act as a kosmotrope, promoting immiscibility, it actually increases the solubility of Triton X-100 in water.[128]

11.2.6 Deep Eutectic Solvents in the Presence of Water-Soluble Polymers or Organic Molecules

Deep eutectic solvents (DESs) also produce aqueous two-phase systems. DES are typically constituted by equimolar mixtures of a hydrogen bond acceptor (HBA) and a hydrogen bond donor (HBD).[129–132] In a study by Yan and coworkers, a DES formed with betaine (hydrogen bond acceptor) and xylitol (hydrogen bond donor) was studied in the presence of short-chain alcohols like 1-propanol, 2-propanol, and *tert*-butanol.[129] The addition of these alcohols induced the formation of two aqueous phases, attributed to the limited miscibility between the betaine–xylitol DES and the alcohols. *tert*-Butanol, being the most hydrophobic water-soluble alcohol, induced the highest degree of phase separation, an observation potentially applicable to various other aqueous two-phase systems founded on DES.

A debate arose concerning the nature of the aqueous two-phase systems in the presence of DES, as it was initially unclear whether the equimolarity of HBA–HBD pairs was maintained after phase separation. Coutinho and colleagues questioned whether the systems with DES, polypropylene glycol (PPG), and water were true ternary systems or should be considered quaternary mixtures.[131] In their investigation of HBA/HBD ratio, specifically with cholinium chloride and carboxylic acids or urea, mixed with PPG, they discovered that the two DES components exhibited a nonstoichiometric distribution between the coexisting phases. Consequently, the HBA/HBD ratio was not preserved in each phase. The authors concluded that the phase separation was not caused by genuine PPG–DES immiscibility.[131] The betaine-xylitol DES system, combined with a short-chain alcohol, also exhibited a similar nonstoichiometric partitioning of DES components between the two coexisting aqueous phases.[129]

Nevertheless, stoichiometric partition of DES components occurs in some systems. In a study of a DES consisting of cholinium chloride (as the hydrogen bond acceptor) and glucose (as the hydrogen bond donor) in the presence of polyethylene glycol and water, Coutinho and coworkers[132] noted that the HBA/HBD molar ratios were conserved in the two coexisting phases. Therefore, this work demonstrated that DES–PPG–water can be considered as ternary systems that forms two aqueous phases in thermodynamic equilibrium.

A drawback of ATPS relying on DES is that the complexes formed through hydrogen bonding are often not stable at low concentrations in aqueous solutions. This challenge has been addressed by incorporating electrolytes into mixtures of HBA and HBD. Farias et al.[133] investigated quaternary systems comprising cholinium chloride acting as HBA, short-chain alcohols or ethylene glycol acting as HBD, K_2HPO_4 (kosmotropic salt), and water. They found that the presence of the salt enabled phase separation to occur at lower concentrations of the other components, and with higher water content.

In more recent works, ATPS based on deep eutectics have been evaluated as extraction media.[130,–134] Marchel et al.[130] prepared a DES by stoichiometrically combining betaine (as HBA) with fructose, glucose, and sucrose of urea (as HBDs). These DES formed aqueous two-phase systems when mixed with polyethylene glycol and water, and these ATPS were used for the extraction of pepsin enzyme. The results showed that the enzyme become concentrated in the DES with a very good extraction efficiency, above 96%. In another example, Zafarani-Moattar et al.[134] used two-phase systems prepared with tetrabutylammonium bromide, citrate salts, and PEG for the extraction of dyes.

11.2.7 Other Systems in the Presence of Weakly Hydrated Molecules or Polymers

Phase separation in aqueous mixtures is also observed in the combination of two polar molecules. An interesting example is that composed of acetonitrile, saccharides, and water.[135] Acetonitrile, while soluble in water, lacks acidity and does not establish powerful hydrogen bonds with water molecules. Consequently, its behavior in aqueous solution leans toward being chaotropic. Conversely, small saccharides, being more hydrated molecules, have relatively strong hydrogen bonds with water, often behaving as kosmotropes. Therefore, the distinct hydration properties of the two solutes lead to phase separation in these aqueous mixtures.

Aqueous solutions of one single polymer can have a cloud temperature, observing phase separation above a certain lower critical solubilization temperature (LCST). Some polymers exhibit a reduction in solubility with increasing temperature, producing the formation of two aqueous phases, one enriched in polymer and another depleted in polymer. In a comprehensive review published in 1993, Galaev and Mattiasson reported over 20 polymers that showed a cloud temperature.[136] Among them, polyethylene oxide (PEO) or polypropylene oxide (PPO) random or block copolymers are typically used, since they are fully soluble at low temperatures but become partially dehydrated at higher temperatures, inducing the formation of two liquid phases. One of the two liquid phases contains polymer micelles, whereas the other liquid phases is almost pure water. Tjerneld and coworkers[137–139] showed that proteins sharply partition between water and copolymer phases, demonstrating that these systems are useful for the extraction of macromolecules of biological interest.[137, 138]

11.3 FORMATION OF W/W EMULSIONS

Agitation is commonly used to generate water-in-water emulsions from aqueous two-phase systems. Unfortunately, traditional agitation methods result in a wide range of droplet sizes, with high polydispersity in droplet size distribution. Microfluidics offers a solution by producing highly uniform droplets, but its limited throughput hinders its use in large-volume industrial settings. W/W emulsions, with small droplet size and low polydispersity, can be obtained through phase transitions that prompt droplet condensation. For instance, gelatin-in-maltodextrin, bicontinuous, or maltodextrin-in-gelatin emulsions can be 'spontaneously' obtained by increasing pH from 3.5 to 5.0.[63] Emulsification occurs because these polymers form a homogeneous single-phase mixture at low pH and a two-phase system at neutral pH. Increasing pH produces an oversaturation of the disperse phase, leading to the formation of more uniform and smaller droplets. Different emulsions, gelatin-in-maltodextrin, bicontinuous, or maltodextrin-in-gelatin emulsions, can be observed, depending on the composition. Typically, the phase with a larger volume fraction becomes the continuous phase, as there is no inherent driving force to direct curvature orientation. It is worth noting that bicontinuous emulsions are inherently unstable, quickly progressing towards complete phase separation.

Water-in-water-in-water (W/W/W) multiple emulsions can be produced through straightforward and cost-effective methods, eliminating the need for microfluidic techniques. Beldengrün et al.[63] demonstrated the formation of maltodextrin-in-gelatin-in-maltodextrin emulsions by reducing the temperature of gelatin-in-maltodextrin dispersions. This occurs because the solubility of maltodextrin in the gelatin solution decreases with decreasing temperature. Therefore, gelatin droplets become oversaturated with maltodextrin, which condenses forming maltodextrin-enriched droplets inside the gelatin-solution droplets. Another intriguing type of emulsion is Janus droplets within aqueous three-phase systems (A3PS). For instance, Shum and coworkers reported the creation of Janus droplets with diverse morphologies (such as dumbbell- or snowman-shaped) by introducing a mixture of poly(2-ethyl-2-oxazoline) and dextran into a concentrated PEG aqueous solution, using coflow microfluidics.[140] The three water-soluble polymers (poly-2-ethyl-2-oxazoline, dextran, and PEG) are mutually immiscible, creating an A3PS that allows the formation of Janus droplets, and their morphology depends on composition parameters.

Porous membranes with homogeneous pore size, for example, Shirasu porous membranes, proved to be an effective strategy for generating W/W emulsions with low polydispersity and controlled droplet size, facilitating the production of homogeneous emulsions. Akamatsu et al. obtained Janus droplets by mixing two separate W/O emulsions previously prepared by membrane emulsification.[141] The droplets of the first emulsion contained PEG and dextran at concentrations below the threshold of their phase separation, while the second W/O emulsion contained NaCl or glucose at high concentrations. In both emulsions, the continuous oil phase was kerosene with 1 wt% tetraglycerol ricinoleate. These two different W/O emulsions were mixed and the difference in osmotic pressure produced the extraction of water from the PEG–dextran droplets, inducing phase separation and forming Janus droplets, since PEG and dextran concentrations had increased above the threshold of phase separation.

When the polymer primarily located in the dispersed phase is cross-linked, microgels are formed. For instance, gelatin microgels can be obtained by cooling gelatin-in-maltodextrin emulsion droplets below the gelation temperature of gelatin, typically around ≈35°C.[69] Gelation and microgel formation occurs because of association of gelatin chains at low temperatures. These gelatin microgels have been utilized for encapsulating enzymes such as lactase (β-galactosidase), and the encapsulation was found to retard enzyme inactivation in simulated gastric fluid.[69] Microgels can also be obtained by covalent cross-linking of the polymers located in the disperse phase of the emulsion. Egito and coworkers[142] showed the formation of microgels in emulsions of xylan (a polysaccharide), dispersed in PEG solutions. The introduction of trisodium trimetaphosphate into the xylan phase induced covalent cross-linking of xylan, further promoted by the addition of $CaCl_2$ into the PEG phase.

11.4 INTERFACIAL PROPERTIES AND EMULSION STABILIZATION

11.4.1 W/W INTERFACES

In W/W emulsions, the interfacial regions are ill-defined, interfaces are relatively thick and fuzzy, and interfacial tensions are very low.[143–148] In the case of polymer–polymer ATPS, interfaces have thickness larger than the chain length of polymers, and thus, they are much thicker than conventional O/W interfaces.

The two liquid phases across the interface have small differences in composition, with the same solvent (water). Consequently, the macroscopic characteristics (such as density or refractive index) exhibit small variation across the interface, especially when compared to conventional oil–water interfaces. Therefore, the interfacial energy in water–water boundaries is remarkably low, resulting in typical interfacial tensions from 10^{-2} to 10^{-4} mN/m.[143,144,147,149,150] The interfacial tension decreases as the difference in composition between the two phases decreases. The compositions of the two phases become virtually identical at the critical point, where the interfacial tension becomes zero and the interface disappears.[143] As systems approach their critical point, interfacial tension decreases dramatically, reaching values below the measurement limit of spinning drop methods. In these cases, rheo-optical techniques are necessary for precise measurements of interfacial tension.[151]

A fascinating area of investigation is the development of ferrofluidic ATPS, described first by Timonen and coworkers.[152] These researchers studied the incorporation of maghemite superparamagnetic nanoparticles into the dextran phase

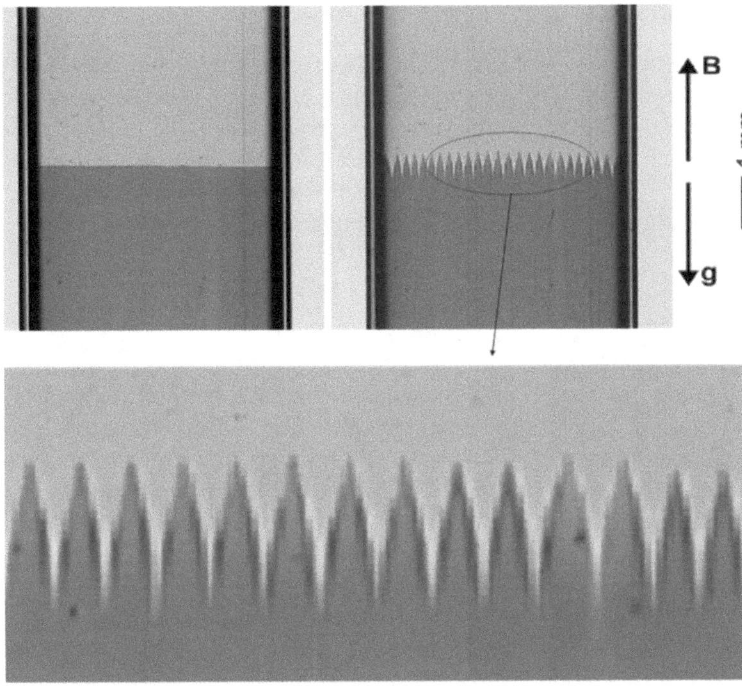

FIGURE 11.3 Dextran–PEG interface, with magnetic nanoparticles in the denser dextran phase. The top images show the interface in the absence (left) and the presence (right) of a magnetic field (B) against gravity (*g*). The inset shows the detail of the interfacial deformations. The scale bar indicates 1 mm. Reproduced from Rigone et al.,[152] with permission under Creative Common Attribution License 4.0 [CC-BY 4.0].

of dextran-in-PEG emulsions. The resultant ATPS exhibited superparamagnetism while preserving the typical phase behavior and distinctive properties of ATPS. This biliquid system, with extremely low interfacial tension, showcased high responsiveness to external magnetic fields. The combination of magnetic field and low tension produced interesting patterns at the interface, with high deformations and curious motions (Figure 11.3). This unique behavior opens up possibilities for the development of magnetically responsive liquid materials with exciting potential applications.[152]

The very low interfacial energy leads to low colloidal stability of W/W emulsions. They can be stable in dilute situations or in the presence of viscous liquid phases, but in most cases coalescence is very fast.[153] Adsorption of molecules is often poor on W/W interfaces, since energy of adsorption is often lower than thermal energy. The low colloidal stability is the main drawback of the use of W/W emulsions in practical applications. Therefore, developing novel methods to improve stability of W/W emulsions is a very important topic, and substantial scientific effort is currently devoted to addressing this challenge.

11.4.2 Stabilization by Particle Adsorption

Over the past ten years, researchers have focused on improving colloidal stability of water-in-water emulsions by adsorbing particles at interfaces. Many examples of using particles as stabilizers have been reported, since the pioneering work of Poortinga[154] and the comprehensive studies conducted by Nicolai and coworkers[7,50,54,55,57,58,153,155,156] The stabilization of emulsions by adsorption of spherical particles on liquid/liquid interfaces is described by the equation proposed by Binks and colleagues.[157,158]

$$\Delta G = \pi R^2 \gamma (1 - \cos\theta)^2, \tag{11.1}$$

where ΔG is the free energy of particle adsorption, R is the particle radius, γ is the interfacial tension, and θ is the contact angle of the adsorbed particle. At water-in-water interfaces, the very low interfacial tension requires a large droplet radius to achieve stability. Moreover, small particles exhibit Brownian motion caused by thermal energy $\approx k_B T$, and adsorption is only secured if its free energy is much higher than thermal energy. Consequently, only relatively large particles may have a significant adsorption energy and stabilize W/W emulsions.

Equation (11.1) would provide the prediction of the minimum particle size for adsorption of particles and stabilization of emulsions. However, this equation is rarely used in the quantitative analysis of W/W emulsions[8,159] for a number of reasons: First, Equation (11.1) is restricted to spherical particles, excluding rod-like[155,160] or flat particles[161] that can impart a much better stabilization because of their large transversal area;[8] second, the contact angle at fuzzy water–water interfaces is often meaningless; and third, specific interactions of chemical species to particle surface play a major role in particle adsorption and such interactions are difficult to generalize and predict.

Nevertheless, stabilization of W/W emulsions by particle adsorption is common and it has been widely studied. Poortinga was the first to describe Pickering water-in-water emulsions, demonstrating that the colloidal stability of emulsions prepared in aqueous mixtures of methylcellulose and dextran could be significantly enhanced by the adsorption of particles like whey proteins or quartz crystals.[154] Subsequently, Nicolai and his collaborators showed that β-lactoglobulin microgel particles formed adsorbed layers around dextran-in-PEG droplets,[6] as well as amylopectin-in-xyloglucan emulsion droplets.[61]

These first studies opened the door to many further contributions. In recent times, a wide range of particles that can stabilize water-in-water emulsions have been extensively documented, as reviewed by Dickinson[8,159] and by Chao and Shum.[25] Examples include oil droplets in caseinate–xanthan gum aqueous mixtures;[162,163] phospholipid liposomes in dextran-PEG mixtures;[164,165] *E. coli* bacteria in the casein–xanthan system;[166] gibbsite clay nanoplates in gelatin–dextran emulsions;[167] polymer microgels or cellulose nanocrystals in the dextran–PEG system;[153,160] cellulose microfibrils to stabilize maltodextrin-in-dextran emulsions;[59] β-lactoglobulin protein particles in emulsions of hydroxypropyl methyl cellulose (HPMC) and dextran;[168]

and starch platelet nanocrystals in PEG-in-dextran emulsions;[169] as well as many other particles.[8,25,159,170]

Whey protein soft microparticles have demonstrated great versatility, as they can stabilize water-in-water emulsions in many different systems. Examples include their use in corn starch/locust bean gum aqueous mixtures[171] and gelatin/carboxymethyl cellulose mixtures.[172] The attachment of latex particles in gelatin–dextran interfaces that have a very low tension (≈0.1 mN/m) was thoroughly studied by Monteux et al.[173] They observed that attachment of gelatin on latex particles played a very important role in facilitating adsorption of particles on the water–water interface.

One example of particle adsorption at the W/W interface[174] is shown in Figure 11.4a, demonstrating that zein particles with a size of several micrometers certainly adsorb at the gelatin–dextran interface. Pickering W/W emulsions can be observed with great detail by confocal microscopy, labeling particles and emulsion constituents with different fluorescence dyes. An example of gelatin-in-maltodextrin emulsion droplet stabilized with mucin particles[63] is shown in Figure 11.4b. Other examples are the use of zein protein particles for the stabilization of bicontinuous emulsions in gelatin–PEG aqueous mixtures (Figure 11.4c),[174] mucin glucoprotein particles,[63] or PEG-functionalized latex particles.[57]

The examples shown in Figure 11.2 illustrate that W/W emulsions can be stabilized by a great variety of different particles able to adsorb on the interface, as summarized in several reviews.[8,25] These particles should have a balanced affinity for the two liquid phases, leading to a contact angle near 90°. Nicolai and coworkers[61] observed that the increase in colloidal stability is strongly related to attachment of

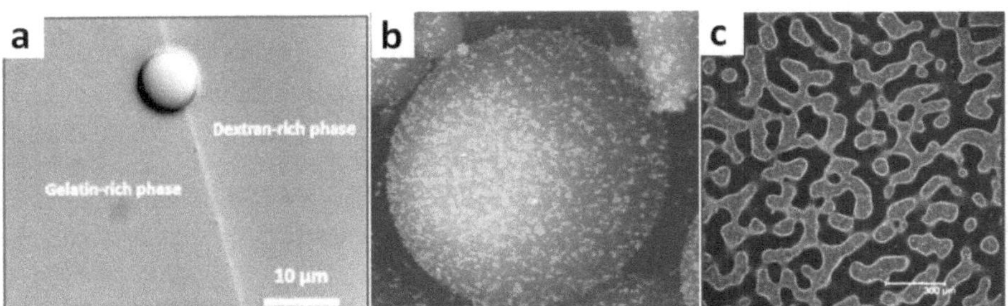

FIGURE 11.4 Stabilization by particle adsorption. (a) Detail of a zein protein particle, adsorbed in the gelatin–dextran interface. Reprinted from Chatsisvili et al.,[174] Copyright (2017), with permission from Elsevier. (b) 3D software-assisted reconstruction of a Pickering-stabilized gelatin-in-maltodextrin W/W emulsion, in which gelatin has been labeled with fluorescein isothiocyanate (gray large droplets) and mucin particles have been labeled with rhodamine isothiocyanate (small white particles at the gelatin/dextran interface). (c) Bicontinuous emulsions obtained by temperature quenching in the region of spinodal decomposition in a gelatin–starch aqueous mixture. The gray regions indicate the presence of COOH-coated latex particles (0.20 μm size) within the gelatin phase; white interfaces spots highlight the presence of PEG-coated latex particles (0.32 μm size) at the gelatin–starch interface; and dark areas correspond to starch-rich domains, which is mostly free of particles. Reprinted from Nicolai and Murray,[57] Copyright (2017), with permission from Elsevier.

polymer molecules on particles. Amylopectin-in-xyloglucan emulsions were stabilized with β-lactoglobulin particles coated with xyloglucan, and this surface modification of the microgels played a major role in enhancing interfacial adsorption and improving the colloidal stability.

An interesting example of particle-stabilized W/W emulsions is that described by Douliez and coworkers.[175] They prepared microgels with particles covalently fixed at the surface, by cross-linking dextran-in-PEG droplets stabilized by amino-functionalized polystyrene (PS) particles. The dextran solution was gelated to produce microgels, which were stabilized by covalently anchored polystyrene particles. These hydrogelled droplets were capable of reversible swelling.

Nicolai and coworkers also studied the stabilization of W/W emulsions using rheological determinations and observations of emulsion droplets under shear with a confocal microscope.[156] The presence of stabilizer showed to inhibit the coalescence of small droplets generated by shear. Particles remained at the interface even when droplet disruption occurred at strong shear. Stopping shear lead to the formation of small droplets, which were stabilized by a surface layer of adsorbed particles.

Stabilization of W/W emulsions by particle adsorption opens great possibilities for biomedical applications. Hao and coworkers studied the rheological and tribological properties of dextran-in-PEG emulsions, and measured friction coefficients. These emulsions were stabilized using collagen nanofibrils, and the researchers also evaluated the potential cytotoxicity and suitability of these emulsions for joint lubrication.[176] The emulsions showed absence of cytotoxicity on EGFP-hMSC-TERT human mesenchymal stem cells (Wang et al. 2022).

The study of tribology of W/W emulsions is a fascinating subject. You et al. have explored the tribological properties of emulsions containing gelatinized corn starch in κ-carrageenan, stabilized by whey protein soft particles.[177] These W/W emulsions showed a reduction in the friction coefficient at higher volume fractions of whey particles. These results might be relevant for controlling the mouthfeel of food formulations based on W/W emulsions.

Effective colloidal stabilization can be attained when employing soft microgel particles that adsorb onto emulsion droplets. In the case of O/W emulsions, it has been demonstrated that microgels are surface active and therefore they are effective as both emulsifiers and stabilizers.[178] Moreover, microgels are soft and deformable, and consequently, they interpenetrate with water and oil at O/W interface, producing thick and elastic adsorbed layers. It is also interesting to remark that swelling/shrinking of hydrogels is sensitive to external stimuli (temperature, pH, salinity, osmotic pressure, etc.) and thus, stability of microgel-stabilized emulsions can be easily controlled.[178]

The use of microgels for stabilizing W/W emulsions provides responsiveness to external stimuli, as shown by Nguyen et al.[153] in the dextran–PEG system. They covalently cross-linked microgels consisting of poly(ethylacrylate-*co*-methacrylic acid-*co*-1,4-butanediol diacrylate). These microgels exhibited significant swelling at pH levels near the pK_a of the polymer, ≈7. The optimal emulsion stability, lasting at least one week, was observed within a narrow pH range, between 7 and 7.5.

Therefore, the stability of the emulsions shifted from stable to unstable with just a small variation of pH.

Schmidt and coworkers highlighted the significant potential presented by temperature-sensitive microgels.[179] They successfully synthesized a water-soluble double-block copolymer composed of two hydrophilic chains, poly(*N*,*N*-dimethylacrylamide)-*b*-poly(*N*,*N*-diethylacrylamide), denoted as PDMA-*b*-PDEA. This block copolymer self-aggregates forming polymer micelles above its cloud point, stabilizing dextran-in-PEG emulsions. They observed that these polymer micelles had a preferential affinity for the PEG phase and consequently the micelles preferably stabilized dextran-in-PEG emulsions.[179]

The main conclusion from their research is that W/W emulsions can be engineered to exhibit significant sensitivity to temperature, enabling the modulation of emulsion stability. In a more recent study, Schmidt and coworkers[180,181] reported the development of pH-sensitive water-in-water emulsions based on pullulan and poly(*N*,*N*-dimethylacrylamide) (PDMA). These emulsions, consisting of pullulan-rich droplets dispersed in PDMA solutions (pull-in-PDMA emulsions), were stabilized by the addition of poly(2-(dimethylamino)ethyl methacrylate)-*b*-poly(oligo(ethyleneglycol) methyl ether methacrylate)) (PDMAEMA-*b*-POEGMA block copolymer). This polymer forms micelles by self-aggregation, in response to pH changes.

Emulsion stability can also be precisely controlled via temperature using poly(N-isopropylacrylamide) (PNIPAM) microgels. Ravaine and coworkers[50] demonstrated this by stabilizing both PEG-in-dextran and dextran-in-PEG emulsions with dextran-modified pNIPAM microgels. This functionalization enabled temperature-responsive behavior, allowing the modulation of emulsion stability through temperature variations. The functionalization enhanced the adsorption of the microgels at the interface while preserving the same transition temperature as pristine pNIPAM. The addition of microgels to dextran-in-PEG emulsions resulted in their stabilization at low temperatures. However, when temperature was higher than the transition point of pNIPAM, phase inversion occurred, shifting from dextran-in-PEG to PEG-in-dextran by increasing temperature.[50] Consequently, these Pickering W/W emulsions were thermosensitive, and the nature of emulsions, dextran-in-PEG or PEG-in-dextran, could be controlled by temperature.

The control of particle adsorption in W/W interfaces can be achieved through various methods. Gonzalez-Jordan et al. demonstrated that the stability of Pickering water-in-water emulsions is strongly influenced by the precise arrangement of particles at interfaces, dictated by the distribution of particles between the two separate aqueous solutions.[55] The adsorption of particles, and their contact angle at the interface, can be adjusted by modifying the surface properties of particles, allowing control over emulsion stability (see illustrative example in Figure 11.5). In this case, polystyrene particles (PS) were coated with different quantities of whey proteins and utilized as stabilizers for both PEG-in-dextran (P/D) and dextran-in-PEG (D/P) emulsions.

Nicolai and coworkers[56] found that the stability of emulsions (P/D or D/P) correlated with the preferential partitioning of surface-modified PS particles. Emulsion stability was significantly enhanced when PS particles were preferentially wetted by the continuous phase. This aligns with the conventional behavior observed

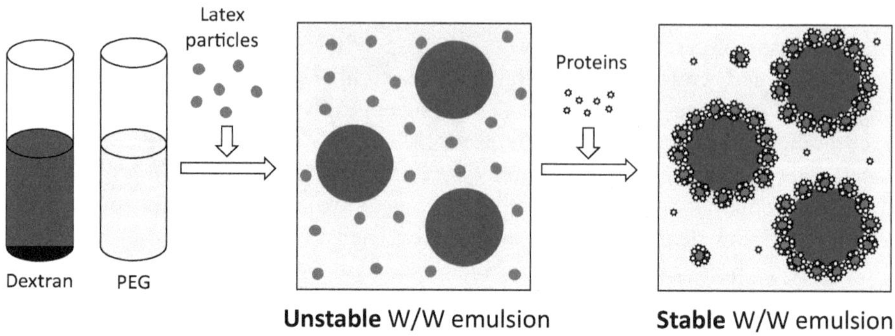

FIGURE 11.5 Scheme of the enhancement of particle adsorption by modification with proteins, resulting in the improvement of the particle stabilization. Adapted from Gonzalez-Jordan et al.,[55] Copyright (2018), with permission from Elsevier.

before in O/W or W/O Pickering emulsions, as described by Binks and coworkers.[157] Consequently, in this aspect, W/W emulsions behave similarly to traditional emulsions with two immiscible solvents, such as water and oil.

Nicolai and coworkers[182] successfully controlled the adsorption of whey protein particles at pullulan–amylopectin interfaces through the addition of small quantities of other polysaccharides, including alginate, xyloglucan, pectin, or κ-carrageenan.

The pullulan–amylopectin system is intriguing due to the occurrence of phase separation, even though both pullulan and amylopectin are nonionic polysaccharides. Pullulan, being a linear molecule, and amylopectin, characterized by high branching, induce phase separation in a nonionic aqueous mixture that remains compatible with the addition of a third polysaccharide. Their immiscibility is caused by a combination of hydration and entropic factors, still not well understood.

The addition of another component can cause whey proteins to migrate from the pullulan-rich liquid phase to the amylopectin-rich liquid phase. Whey protein particle adsorption at the interface was maximized at intermediate concentrations of the added component.[182] The authors inferred that particle adsorption is influenced by both the pullulan–amylopectin interfacial tension and the tensions between the particles and the individual liquid phases. These observations are applicable beyond the studied system, as similar behavior was noted in other water-in-water emulsion systems featuring different types of particles.

In a related study, also using whey proteins to stabilize pullulan–amylopectin emulsions, Nicolai and coworkers[60] explored the role of pH in the presence of anionic polysaccharides. The researchers noted that reducing the pH resulted in a shift in the preferential affinity of the protein microgels from the pullulan phase to the amylopectin phase, causing a transition from amylopectin-in-pullulan emulsions to pullulan-in-amylopectin emulsions. Additionally, the introduction of anionic polysaccharides significantly influenced the system by forming complexes with the protein microgel particles, inducing a partial aggregation of these particles, and changing interfacial tensions.

It is worth noting that most of these investigations focus on W/W emulsions formed in systems containing two water-soluble polymers: a highly hydrated polymer, such as a protein or ionic polysaccharide, and a less hydrated polymer, such as PEG or a nonionic polysaccharide. The most frequently studied system is that composed of dextran and PEG. However, there exists a vast variety of different systems that can be explored. For instance, the study of W/W emulsions based on salts might be the focus of future research works. Interesting examples are PEG-in-Na_2SO_4 and Na_2SO_4-in-PEG emulsions, which were stabilized with silica nanoparticles that had been partially hydrophobized with dichlorodimethylsilane.[183] The degree of particle surface hydrophobization controlled the stabilization of both polymer-in-salt and salt-in-polymer emulsions.

11.4.3 Stabilization by Forming Membranes in Water/Water Interfaces

Frequently, Pickering W/W emulsions lack prolonged colloidal stability, and achieving high stability beyond several weeks is challenging.[182] Although some WW emulsions exhibit colloidal stability longer than one year,[183] this is still insufficient for practical applications such as cosmetics. Therefore, much scientific research is devoted to find new methods for long-term stability.

A potential approach to improve the colloidal stability of water/water emulsions is covalent bonding between particles that remain adsorbed at the interface. Zhang et al.[184] demonstrated this by cross-linking dopamine particles that were adsorbed on dextran-in-PEG droplets. Remarkably, these droplets remained stable even after dilution in water, despite swelling caused by osmotic pressure differences. These results demonstrated that interfacial cross-linking in W/W emulsions allows obtaining capsules that can remain stable after dilution, as also reported by Perro et al.[185]

Improving the colloidal stability of W/W emulsions can be significantly achieved by constructing membranes at the interfaces through adsorption and/or self-assembly processes. Small molecules typically do not adsorb readily at water/water (W/W) interfaces due to their inherently low interfacial tension. Adsorption at these interfaces is mainly characteristic of particles or macromolecules, such as globular proteins or high-molecular-weight polymers.

Therefore, polymers can indeed adsorb on W/W interfaces and improve colloidal stability. One of the first examples was reported by Nicolai and coworkers[7] They evaluated 16 different hydrophilic polymers in dextran-in-PEG or PEG-in-dextran emulsions. Propylene glycol alginate and chitosan were found to be good stabilizers of PEG-in-dextran emulsions.[7] The good stabilization properties of these polymers was attributed to their amphiphilic character, with the simultaneous presence of hydrophobic domains and charged functional groups in the polymer.

More recently, Nicolai and coworkers[186] stabilized emulsions in the dextran–PEG system by addition of xanthan gum, which is a frequently employed thickening agent. The polymer demonstrated a preference for partitioning into the dextran phase, resulting in a high rise in the viscosity of PEG-in-dextran emulsions. This viscosity increase stabilized the emulsions for at least one week. Intriguingly, xanthan gum also exhibited stabilizing properties for dextran-in-PEG emulsions. These findings suggested that xanthan contributed to stability by adsorbing at the interface, forming a protective layer, and, in conjunction with the viscosity effect, enhancing the overall stability of the W/W emulsions.

A very interesting strategy is the use of aqueous three-phase systems (abbreviated as A3PS). Mixtures of three immiscible water-soluble polymers can result in phase separation in three distinct aqueous liquid solutions. This very curious phenomenon can be observed in systems such as fish gelatin/guar/amylopectic[58] or PEG/dextran/poly(2-ethyl-2-oxazoline),[140] which can be considered as versatile multiphasic system with a great potential for structured liquids with confined domains.

In another investigation led by Nicolai and coworkers,[58] they successfully stabilized amylopectin-in-guar and guar-in-amylopectin emulsions by creating a bovine gelatin layer on the interfaces. Amylopectin and guar are immiscible with each other, and at temperatures below approximately 25°C, bovine gelatin has a low solubility in both phases. Therefore, A3PS systems can be prepared in aqueous mixtures of amylopectin, guar, and gelatin. This three-phase mutual immiscibility allows forming a stabilizing layer of gelatin at the interfaces of the emulsions, improving their stability. At high temperatures, amylopectin–guar–gelatin aqueous mixtures were used to prepare W/W emulsions. Subsequent cooling induced further separation of gelatin in a third phase, resulting in the formation of a continuous gelatin layer on the surface of droplets. Consequently, this system comprised three separated aqueous domains that were mutually immiscible. This gelatin layer allowed stabilizing W/W emulsions.

In a more recent work developed by Nicolai and coworkers,[187] they also obtained an A3PSs by mixing PEG, dextran, and gelatin in water. This facilitated the preparation of double (core–shell) emulsions, with the gelatin phase serving as the intermediate phase. Consequently, as the gelatin underwent gelation during cooling, hydrogel shells formed around the droplets, leading to core–shell capsules. The formation of an encapsulating shell, sorrounding the droplets, conferred stability against dilution or agitation. In conclusion, this research introduced a novel approach to W/W emulsion stabilization by gelling a third phase that envelops the emulsion droplets. The formation of core–shell capsules in A3PS is illustrated in Figure 11.6.

Lamellar phases demonstrated efficacy in stabilizing W/W emulsions. Martin and coworkers created bilayers or multilayers at the W/W interface of dextran-in-PEG emulsions, using sodium oleate and *n*-decanol to simulate conventional vesicles.[185,188] These lamellar nanostructures not only improved stability but also showed substantial impermeability to large molecules, opening new possibilities for active component encapsulation.

In a more recent study developed by Hao and coworkers,[189] PEG-in-dextran and dextran-in-PEG emulsions were stabilized using DNA strands. These researchers noted that DNA molecules adhered to the interface, leading to a reduction in droplet size as the length of the DNA strands increased. Interestingly, single-stranded DNA (ssDNA) conferred greater colloidal stability than its double-stranded counterpart (dsDNA), probably because the higher hydrophobicity and flexibility of ssDNA. At high DNA concentrations, the researchers also noted the formation of onion-like multilayer structures (scheme shown in Figure 11.7). The formation of these

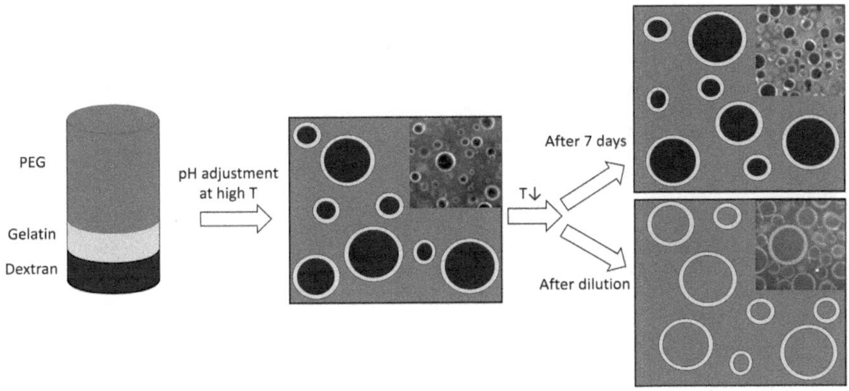

FIGURE 11.6 Scheme of the formation of core–shell droplets in aqueous three-phase systems (A3PS). Polyethylene glycol, PEG (denoted as polyethylene oxide, PEO), dextran, and gelatin are mutually immiscible in aqueous solution and phase-separate in three distinct aqueous layers, depending on pH. The gelatin liquid phase can be gelified by decreasing temperature, obtaining capsules that are stable at least for one week. These capsules are also stable upon dilution with purified water. Reprinted from Meng et al.,[187] Copyright (2022), with permission from Elsevier.

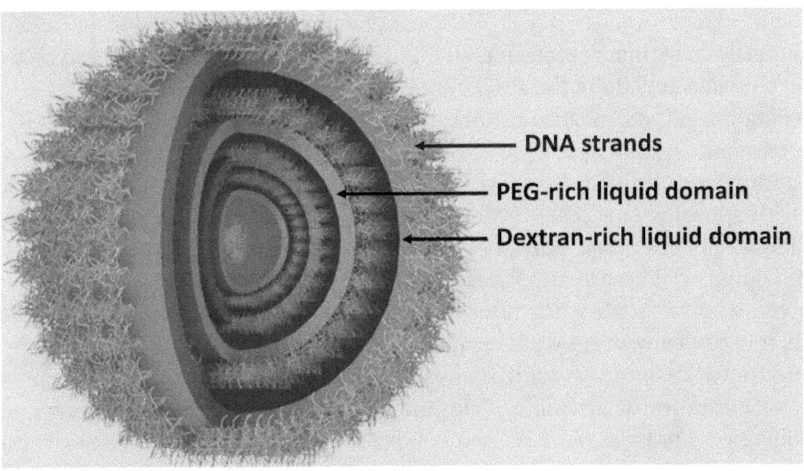

FIGURE 11.7 Scheme of an onion-like multiple droplet, which resembles vesicles. DNA strands intercalate between PEG and dextran liquid domains. Reproduced with permission from Wang et al.,[189] Copyright (2022), American Chemical Society.

multilayers was attributed by authors to nanofilms formed between PEG droplets and the glass slide. The gradient of the chemical potential between the film and the droplet meniscus could produce diffusion of DNA strands from the nanofilm to the meniscus and osmotic pressure differences might thin the nanofilms to produce onion-like morphologies.

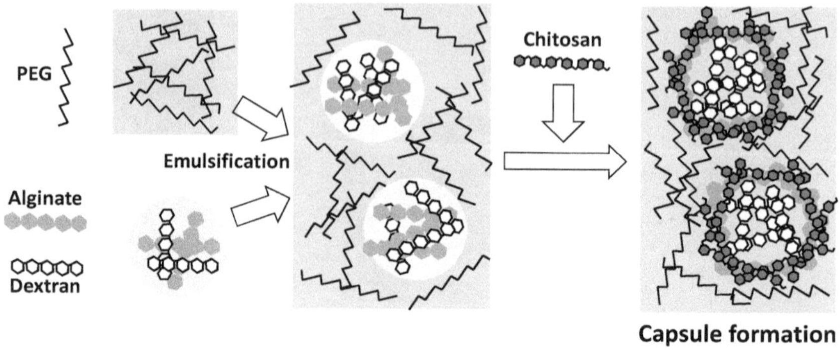

FIGURE 11.8 Scheme of the formation of capsules by ionic complexation at the interface of W/W emulsions. Alginate (as the anionic polyelectrolyte) is incorporated into the dextran disperse phase, whereas chitosan (cationic polyelectrolyte) is added to the PEG continuous phase. Precipitation of chitosan alginate occurs at the interface, obtaining dextran-in-PEG capsules with a rigid chitosan–alginate shell.

Obtaining stable W/W emulsions has shown great promise through the formation of rigid polyelectrolyte complexes at the interface. These complexes, which consist of two oppositely charged polyelectrolytes, can be formed at the W/W interface because of the ionic complexation of a polyanion with a polycation. One example of interfacial ionic complexation in W/W emulsions is the stabilization of dextran-in-PEG emulsions by incorporating sodium alginate in the dextran disperse phase and adding chitosan acetate in the PEG continuous phase (Figure 11.8).[190]

To create rigid microcapsules, Shum and coworkers[190,191] utilized four polymers. This combination comprised the two nonionic polymers, dextran, and PEG, which form an aqueous two-phase system. Additionally, they incorporated two polyelectrolytes with opposing charges: poly(allylamine hydrochloride), abbreviated as PAH, as the cationic polymer, and poly(sodium 4-styrenesulfonate), denoted as PSS, as the anionic polymer. These two polyelectrolytes coprecipitate to form a shell of PAH–PSS ionic complex shell at the interface between dextran and PEG solution, obtaining microcapsules with rigid shells at neutral pH.

This approach based on ionic complexation at the W/W interface presents exciting possibilities for designing and developing stable W/W emulsions with complex morphologies. For example, Lee and colleagues created double PEG-in-dextran-in-PEG emulsions through interfacial complexation of poly(diallyldimethylammonium chloride), PDADMAC, with Ludox silica nanoparticles.[24] The complexation of PDADMAC and SiO_2 formed rigid microcapsule shells at the dextran-in-PEG droplet interfaces. Subsequently, osmotic pressure differences drove water diffusion from the dextran droplets to the external PEG phase, while PEG simultaneously diffused into the droplets. This diffusion induced the condensation of PEG inside the dextran rigid microcapsules and the formation of inner PEG droplets, leading to multiple PEG-in-dextran-in-PEG droplets, in which the external dextran–PEG interface is a rigid polyelectrolyte membrane.[24]

W/W interfaces stabilized by ionic complexation have demonstrated promising applications in 3D printing. A study by Russell and coworkers[193] showcased the formation of tubular structures, using the cationic polymer PDADMAC and the anionic polymer PSS. Before 3D printing, PDADMAC was introduced into the dextran solution, while PSS was introduced into the PEG aqueous solution. The resulting polycation–polyanion ionic complexation between the two solutions formed rigid coacervate films at the dextran–PEG interface. These films served as the structural framework for 3D printing tubular structures (Figure 11.9).

The tubular structures produced by 3D printing were examined as flow reactors. The surfaces of the tubes displayed permeability to small molecules, allowing their application in cascade reactions requiring enzyme confinement. This permeability

FIGURE 11.9 Three-dimensional printing of tubular structures prepared by injecting dextran in a PEG solution, which contained PDADMAC and PSS, respectively. The tubular structures were stabilized by a membrane of PDADMAC–PSS ionic complex. (a) Scheme of fabrication. (b) Image of a printed tube. The scale bar corresponds to 3 mm (c). Observation by confocal microscopy of the tubular structure. White spots indicate the dextran labeled with fluorescein isothiocyanate (FITC), black external areas correspond to rhodamine-labeled PEG, and the bright interface highlights the anthracene-labeled polyanion. The scale bar indicates 100 μm. (d–f) Visual aspect of printed spirals using nozzles with various diameters (d: 0.16 mm, e: 0.26 mm, and f: 1.60 mm). The scale bar corresponds to 3 mm. Reprinted from Xie et al.,[193] Copyright (2019), with permission from Elsevier.

enabled continuous refreshing of reactants and extraction of synthesis products at a considerable rate.[193] This novel approach paves the way for designing functional tubular structures with controlled porosity, offering potential applications as two-phase reactors.

11.4.4 Formation and Stabilization of W/W Emulsions in Segregative Systems by Microfluidic Methods

The generation of emulsions through microfluidic techniques is a broad scientific area that has garnered significant recent interest. Microfluidics encompasses highly versatile techniques applicable to a wide range of different systems.[194] Microfluidics can be used to control compartmentalization in both associative (coacervation) and segregative systems. Recent review articles[194–196] comprehensively detail a wide variety of different microfluidic methods, which can be used on many systems for a wide range of applications.

A short selection of works is presented here, illustrating recent achievements in the microfluidic manipulation of W/W segregative systems and highlighting some works of interest. Readers interested in microfluidics of W/W systems can consult the very comprehensive review articles by Shum and coworkers.[25, 197]

Shum and colleagues have conducted research on the formation of multiple water-in-water-in-oil (W/W/O) emulsions using microfluidics. In their work, the stabilization of the inner W/W interface was achieved through ionic complexation.[192] One of the aqueous phases consisted of dextran solution in the presence of PAH, while the other aqueous phase contained PEG with PSS. The combination of these oppositely charged polymers and the formation of a PAH–PSS ionic complex at the W/W interface allowed for the generation of stable W/W/O multiple emulsion structures. Another example is that described by Zhou et al., [198] who obtained $W_1/O/W_2$ multiple emulsions, in which the inner W_1 phase was an aqueous phase enriched with either dextran or PEG, the oil was in the middle phase, and the W_2 continuous phase was enriched with PEG or dextran, respectively. Depending on the relative interfacial tensions, the most inner droplets could exit the middle phase and form binary emulsions with coexistence of both O/W_2 and W_1/W_2 droplets.

Formation of W/W emulsions by microfluidics requires lower pressures than formation of O/W systems. For example, PEG-in-dextran emulsion droplets are prepared in simple microfluidic chips by passive hydrostatic forces, without applying external pressure.[199,200] Another advantage of fully aqueous systems is that W/W interfaces are often highly permeable and droplets quickly react to osmotic pressure differences. Shum and coworkers obtained high-order multiple emulsions (double, triple, and quadruple emulsions) in a simple microfluidic device, introducing an initial saturated mixture of sodium citrate and PEG into a nonsaturated PEG solution.[201] The osmotic difference produces a continuous removal of water from the droplet core and phase separation occurs repeatedly. During each cycle of phase separation, smaller inner droplets of the new phase spontaneously form and coalesce into a new phase in the core of existing droplets. This osmotic-driven formation of multiple droplets is schematically shown in Figure 11.10.

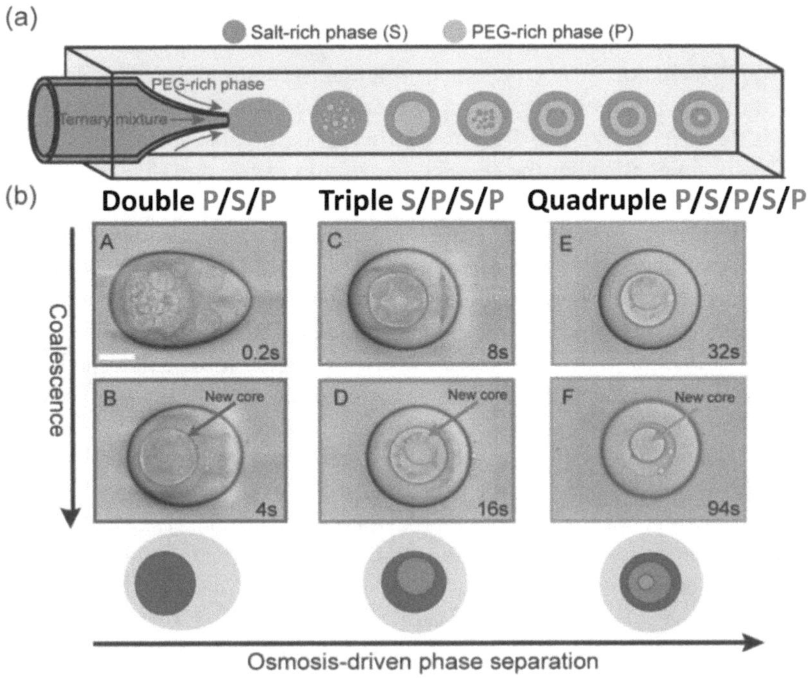

FIGURE 11.10 Scheme of the stepwise formation of multiple emulsions, starting from a simple PEG-in-salt emulsion, and producing double PEG-in-salt-in-PEG (P/S/P) emulsions, triple salt-in-PEG-in-salt-in-PEG (S/P/S/P), and even quadruple PEG-in-salt-in-PEG-in-salt-in-PEG (P/S/P/S/P) emulsions. Reproduced from Chao et al.,[201] with permission under Creative Common Attribution License 4.0 [CC-BY 4.0].

11.5 W/W EMULSIONS IN BIOMEDICINE

Water-in-water emulsions are biocompatible colloidal dispersions, free of oil and surfactants, and can be formulated using macromolecules (proteins, polysaccharides, etc.) of biological interest. Therefore, these emulsions could be used to encapsulate cells. As an example, *Lactobacillus rhamnosus* GG (LGG) probiotic bacteria was encapsulated in carboxymethyl cellulose-in-gelatin emulsions.[202] The viability of LGG bacteria in a simulated gastric fluid was improved after their entrapment in the W/W emulsions.

These emulsions may also have very interesting applications in the preparation of cell organoids or 3D cell cultures.[36] Paunov and coworkers[203] presented an interesting example where cell spheroids were prepared using W/W emulsion droplets as carriers. Cell spheroids were obtained by encapsulating cells in dextran-in-PEG emulsion droplets. Increasing the PEG concentration in the continuous phase generated an osmotic pressure difference, which caused the dextran droplets to shrink and led to the formation of compact cell spheroids. The emulsion was then broken by dilution, and the isolated cell spheroids were transferred to culture media. This method illustrates the capability of W/W emulsions to facilitate the growth and well-controlled organization of cells in 3D cultures.

The stabilization of W/W emulsions, which closely resemble biological systems, has paved the way for the development of innovative biomedical applications. These emulsions demonstrated a low friction coefficient on diverse surfaces, offering significant utility in controlling the sensory perception of food products. In a study by You et al.,[177] the tribological properties of gelatinized corn starch in κ-carrageenan emulsions were investigated, stabilizing these emulsions with whey protein microgel particles. Furthermore, Hao and coworkers[176] investigated the tribology of dextran-in-PEG emulsions stabilized with collagen nanofibrils. They evaluated the potential of these emulsions for joint lubrication and found no cytotoxicity. In in vivo tests on Sprague-Dawley rats, they observed protective effects on articular cartilage, suggesting that these emulsions could serve as an alternative to synthetic synovial fluid in osteoarthritis management.[176]

Certainly, W/W emulsions demonstrate exceptional biocompatibility and have displayed potential in encapsulating living cells. For example, carboxymethyl cellulose-in-gelatin emulsions have been employed for encapsulating *L. rhamnosus* GG (LGG) probiotic bacteria.[202] This encapsulation approach enhanced the viability of LGG bacteria, indicating the potential of W/W emulsions in safeguarding and delivering delicate cells. W/W emulsions are highly appealing for creating controlled microenvironments in organoid formation and 3D cultures.

The research developed by Qin and coworkers[204] has significantly advanced the field of W/W emulsions, particularly in the context of cell encapsulation and organoid formation through the utilization of microfluidic devices. Their work showcases the creation and control of W/W droplets for encapsulating cells, leading to the formation of cell organoids. A very interesting study focused on producing multicore hydrogel capsules, functioning as carriers with multiple compartments for co-encapsulating hepatic and endothelial cells.[204] This method facilitates the formation of multicellular systems for studying cell–cell interactions and tissue development. Qin and coworkers also explored high-throughput generation of dextran-in-PEG emulsion droplets, which they subsequently gelified with calcium alginate.[205] Additionally, Qin and coworkers[206] also engineered hydrogel fibers as carriers for cell organoids, providing a distinctive structure to support the encapsulated cells.

In another investigation, pancreatic endocrine cell spheroids were produced in dextran-in-PEG emulsion droplets that were stabilized through the ionic complexation between alginate anionic polymer and chitosan cationic polymer.[190] The cell spheroids exhibited good cell viability and maintained their capacity to produce insulin. This highlights the potential of W/W emulsions as a versatile method for enclosing and preserving pancreatic cells, with significant implications for diabetes studies and possible treatment approaches. Overall, the study developed by Qin and coworkers[190] showcases the versatility and potential of W/W emulsions in cell encapsulation, organoid generation, and the creation of sophisticated cell-based systems for various biomedical uses.

In a related study, Wang and coworkers[207] presented and alternative approach for encapsulating cells within in W/W droplets. They used dextran-in-PEG microdroplets containing cells, with sodium alginate present in the external PEG solution. These microdroplets were stabilized through the adsorption of particles at the W/W interface and subsequently enveloped within alginate beads by introducing the W/W emulsions into $CaCl_2$ solutions. This process resulted in the creation of calcium alginate shells encompassing microdroplets contanining cells.

11.6 CONCLUDING REMARKS

Water-in-water-type emulsions can be prepared in a wide variety of segregative systems, where hydration phenomena lead to the phase separation of two immiscible aqueous liquid solutions. This phase separation has been known for a very long time, but there are many aspects that remain unknown and should be investigated. W/W emulsions can be produced in a wide variety of different segregative systems, including mixtures of biomolecules of biological interest. Phase separation plays a fundamental role in biological processes, since numerous membrane-free organelles (MLOs) are related to aqueous two-phase systems.

Some systems have been studied and used very frequently for the preparation of W/W emulsions, such as those in the presence of polyethylene glycol. However, there is a wide range of different systems that have only been barely explored, among which we can mention aqueous two-phase systems in mixtures of polysaccharides and proteins, mixtures of two proteins or two different polysaccharides, macromolecules and kosmotropic salts, and mixtures of a chaotropic salt with a kosmotropic salt. Exploring these systems could pave the way for discovering novel emulsion types and finding new applications of W/W emulsions. Probably, in the near future, new research will focus on completely biocompatible systems with biomolecules of biological interest.

The remarkable progress in the field of W/W emulsions can be attributed, in part, to recent advancements in their colloidal stabilization. W/W emulsions inherently suffer from instability and therefore require methods to confer colloidal stability. In recent years, significant scientific attention has been devoted to stabilizing W/W emulsions through the adsorption of particles at the interface, leading to the development of highly stable Pickering W/W emulsions. Furthermore, noteworthy contributions have been made in the stabilization by interfacial membranes, particularly produced through ionic complexation, which have enabled the production of W/W emulsions and/or capsules with enhanced colloidal stability.

A novel application of W/W emulsions is their use for 3D cell cultures, allowing the preparation of living cell clusters and obtaining organoids with specific metabolic functions. Undoubtedly, the development of new 3D cell culture systems will be associated with the use of aqueous two-phase systems. Moreover, W/W emulsions can allow the preparation of synthetic models of MLOs, which in turn will allow a much better understanding of the physical chemistry of cells.

During the last years there has been a great development of segregative W/W systems, thanks to the new tools based on microfluidics, which have allowed enormous and very precise control of the size, structure, and morphology of W/W dispersions. This precise control also allows the development of new applications, such as 3D cell culturing. However, microfluidic methods continue to suffer from certain difficulties, such as the low capacity to obtain emulsions and/or capsules in large quantities at industrial scale, at least in the range of kg/h. Possible improvements in high-throughput preparation methods at large scales could mean the definitive leap from laboratory preparation to mass production.

ACKNOWLEDGMENTS

Jordi Esquena expresses gratitude to various institutions for their financial backing. He acknowledges the support from the Spanish Ministry of Science, Innovation and Universities (MCIU) and the Spanish Research Foundation (Agencia Estatal de Investigación, AEI) for funding CTQ2017-84998-P and PID2021-122187NB-C31 research projects. Additionally, he recognizes the contributions of the European Regional Development Fund (FEDER) and the Generalitat de Catalunya (2017SGR1778 grant). J. Esquena also highlights the 'Pharmaceutical Nanotechnology Group' UB-CSIC Associated R&D Unit. He extends his appreciation to colleagues Carlos Rodríguez-Abreu from the Institute of Advanced Chemistry of Catalonia (IQAC-CSIC) and Maria José Garcia-Celma from the University of Barcelona (UB) for their strong collaboration and many useful discussions. He is also very grateful to Susana Vílchez and Jonathan Miras, who are part of the IQAC-CSIC Surface Chemistry Group, for their assistance and support.

COMPETING INTERESTS

The author declares no conflict of interest.

REFERENCES

1. Matalanis, A.; Jones, O.G.; McClements, D.J. Structured biopolymer-based delivery systems for encapsulation, protection, and release of lipophilic compounds. *Food Hydrocoll.* **2011**, *25*, 1865–1880.
2. Piculell, L.; Lindman, B. Association and segregation in aqueous polymer/polymer, polymer/surfactant, and surfactant/surfactant mixtures: Similarities and differences. *Adv. Colloid Interface Sci.* **1992**, *41*, 149–178.
3. Bungenberg de Jong, H.G.; de Vries, N.F. Orientierende Untersuchungen Ueber Das Sol and Die Koazervate Des Ichthyocolls (Preliminary Studies on the Sol and the Coacervates of the Ichthyocoll). *Recl. des Trav. Chim. des Pays-Bas* **1931**, *50*, 238–247.
4. Bungenberg de Jong, H.G. Colloïden En Biologie (Colloids and Biology). *Ned. Tijdschr. Geneeskd.* **1952**, *96*, 1489–1494.
5. Booij, H.L.; Bungenberg de Jong, H.G. Colloid systems. In *Biocolloids and Their Interactions*, Booij, H.L., Bungenberg de Jong, H.G., Eds.; Springer Verlag: Vienna, Austria, 1956; pp. 8–14.
6. Balakrishnan, G.; Nicolai, T.; Benyahia, L.; Durand, D. Particles trapped at the droplet interface in water-in-water emulsions. *Langmuir* **2012**, *28*, 5921–5926.
7. Tea, L.; Nicolai, T.; Renou, F. Stabilization of water-in-water emulsions by linear homo-polyelectrolytes. *Langmuir* **2019**, *35*, 9029–9036.
8. Dickinson, E. Particle-based stabilization of water-in-water emulsions containing mixed biopolymers. *Trends Food Sci. Technol.* **2019**, *83*, 31–40.

9. Alberti, S.; Gladfelter, A.; Mittag, T. Considerations and challenges in studying liquid-liquid phase separation and biomolecular condensates. *Cell* **2019**, *176*, 419–434.
10. Linsenmeier, M.; Kopp, M.R.G.; Stavrakis, S.; de Mello, A.; Arosio, P. Analysis of Biomolecular condensates and protein phase separation with microfluidic technology. *Biochim. Biophys. Acta - Mol. Cell Res.* **2021**, *1868*, 118823.
11. Drobot, B.; Iglesias-Artola, J.M.; Le Vay, K.; Mayr, V.; Kar, M.; Kreysing, M.; Mutschler, H.; Tang, T.Y.D. Compartmentalised RNA catalysis in membrane-free coacervate protocells. *Nat. Commun.* **2018**, *9*, 1–9.
12. Crowe, C.D.; Keating, C.D. Liquid–liquid phase separation in artificial cells. *Interface Focus* **2018**, *8*, 2018.0032.
13. Martin, N. Dynamic synthetic cells based on liquid–liquid phase separation. *ChemBioChem* **2019**, *20*, 2553–2568.
14. Fanalista, F.; Birnie, A.; Maan, R.; Burla, F.; Charles, K.; Pawlik, G.; Deshpande, S.; Koenderink, G.H.; Dogterom, M.; Dekker, C. Shape and size control of artificial cells for bottom-up biology. *ACS Nano* **2019**, *13*, 5439–5450.
15. Yoshizawa, T.; Nozawa, R.S.; Jia, T.Z.; Saio, T.; Mori, E. Biological phase separation: Cell biology meets biophysics. *Biophys. Rev.* **2020**, *12*, 519–539.
16. Yewdall, N.A.; André, A.A.M.; Lu, T.; Spruijt, E. Coacervates as models of membraneless organelles. *Curr. Opin. Colloid Interface Sci.* **2021**, *52*, 101416.
17. Simon, K.A.; Sejwal, P.; Gerecht, R.B.; Luk, Y.Y. Water-in-water emulsions stabilized by non-amphiphilic interactions: Polymer-dispersed lyotropic liquid crystals. *Langmuir* **2007**, *23*, 1453–1458.
18. Fraccia, T.P.; Zanchetta, G. Liquid–liquid crystalline phase separation in biomolecular solutions. *Curr. Opin. Colloid Interface Sci.* **2021**, *56*, 101500.
19. Doublier, J.L.; Garnier, C.; Renard, D.; Sanchez, C. Protein-polysaccharide interactions. *Curr. Opin. Colloid Interface Sci.* **2000**, *5*, 202–214.
20. Schmitt, C.; Turgeon, S.L. Protein/polysaccharide complexes and coacervates in food systems. *Adv. Colloid Interface Sci.* **2011**, *167*, 63–70.
21. Fang, Y.; Li, L.; Inoue, C.; Lundin, L.; Appelqvist, I. Associative and segregative phase separations of gelatin/κ- carrageenan aqueous mixtures. *Langmuir* **2006**, *22*, 9532–9537.
22. Capasso Palmiero, U.; Paganini, C.; Kopp, M.R.G.; Linsenmeier, M.; Küffner, A.M.; Arosio, P. Programmable zwitterionic droplets as biomolecular sorters and model of membraneless organelles. *Adv. Mater.* **2022**, *34*, 2104837.
23. Esquena, J. Water-in-water (w/w) emulsions. *Curr. Opin. Colloid Interface Sci.* **2016**, *25*, 109–119.
24. Hann, S.D.; Stebe, K.J.; Lee, D. AWE-somes: All water emulsion bodies with permeable shells and selective compartments. *ACS Appl. Mater. Interfaces* **2017**, *9*. 25023–25028.
25. Chao, Y.; Shum, H.C. Emerging aqueous two-phase systems: From fundamentals of interfaces to biomedical applications. *Chem. Soc. Rev.* **2020**, *49*, 114–142.
26. Abraham, M.H.; Zissimos, A.M.; Huddleston, J.G.; Willauer, H.D.; Rogers, R.D.; Acree, W.E. Some novel liquid partitioning systems: Water - ionic liquids and aqueous biphasic systems. *Ind. Eng. Chem. Res.* **2003**, 413–418.
27. Wu, B.; Zhang, Y.M.; Wang, H.P. Aqueous biphasic systems of hydrophilic ionic liquids + sucrose for separation. *J. Chem. Eng. Data* **2008**, *53*, 983–985.
28. Jeyhani, M.; Gnyawali, V.; Abbasi, N.; Hwang, D.K.; Tsai, S.S.H. Microneedle-assisted microfluidic flow focusing for versatile and high throughput water-in-water droplet generation. *J. Colloid Interface Sci.* **2019**, *553*, 382–389.
29. Beijerinck, M.W. Ueber Eine Eigentümlichkeit Der Löslichen Stärke. *Zentralblatt für Bakteriol. Parasitenkd. und Infekt.* **1896**, *2*, 697–699.

30. Bütschli, O. Untersuchungen Über Strukturen: Insbesondere Über Strukturen Nichtzelliger Erzeugnisse Des Organismus Und Über Ihre Beziehungen Zu Strukturen, Welche Ausserhalb Des Organismus Entstehen; Verlag Von Wilhelm Engelmann: Leipzig, Germany, 1898; p. 251.
31. Beijerinck, M.W.W. Ueber Emulsionsbildung Bei Der Vermischung Wässeriger Lösungen Gewisser Gelatinierender Kolloide. *Zeitschr. Chem. Ind. Kolloide (Kolloid Z)* **1910**, *7*, 16.
32. Albertsson, P.Å. Partition of proteins in liquid polymer-polymer two-phase systems. *Nature* **1958**, *182*, 709–711.
33. Wesslen, T.; Albertsson, P.a; Philipson, L. Concentration of animal viruses using two-phase systems of aqueous polymer solutions. *Arch. Gesamte Virusforsch.* **1959**, *9*, 510–520.
34. Norrby, E.C.J.; Albertsson, P.Å. Concentration of poliovirus by an aqueous polymer two-phase system. *Nature* **1960**, *188*, 1047–1048.
35. Grilo, A.L.; Aires-Barros, M.R.; Azevedo, A.M. Partitioning in aqueous two-phase systems: Fundamentals, applications and trends. *Sep. Purif. Rev.* **2016**, *45*, 68–80.
36. Esquena, J. Recent advances on water-in-water emulsions in segregative systems of two water-soluble polymers. *Curr. Opin. Food Sci.* **2023**, *51*, 101010.
37. Hatti-Kaul, R. Aqueous two-phase systems: A general overview. *Mol. Biotechnol.* **2001**, *19*, 269–277.
38. Pereira, J.F.B.; Coutinho, J.A.P. Aqueous two-phase systems. In *Liquid-Phase Extraction*; Poole, C., Ed.; Elsevier: Amsterdam, The Netherlands, 2019; pp. 157–182.
39. Chakraborty, A.; Sen, K. Impact of PH and temperature on phase diagrams of different aqueous biphasic systems. *J. Chromatogr. A* **2016**, *1433*, 41–55.
40. Pereira, J.F.B.; Freire, M.G.; Coutinho, J.A.P. Aqueous two-phase systems: Towards novel and more disruptive applications. *Fluid Phase Equilib.* **2020**, *505*, 112341.
41. Asenjo, J.A.; Andrews, B.A. Aqueous two-phase systems for protein separation: A perspective. *J. Chromatogr. A* **2011**, *1218*, 8826–8835.
42. Johansson, H.-O.; Karlström, G.; Tjerneld, F.; Haynes, C.a. Driving forces for phase separation and partitioning in aqueous two-phase systems. *J. Chromatogr. B Biomed. Sci. Appl.* **1998**, *711*, 3–17.
43. Diamond, A.D.; Hsu, J.T. Aqueous two-phase systems for biomolecule separation. *Adv. Biochem. Eng. Biotechnol.* **2006**, *47*, 89–135.
44. Grinberg, V.Y.; Tolstoguzov, V.B. Thermodynamic incompatibility of proteins and polysaccharides in solutions. *Food Hydrocoll.* **1997**, *11*, 145–158.
45. Albertsson, P.-Å.; Johansson, G.; Tjerneld, F. Aqueous two-phase separations. In *Separation Processes in Biotechnology*, Juan A. Asenjo, Ed.; CRC Press: Boca Raton, FL, USA, 1990; pp. 287–328.
46. Masullo, F.; Beldengrün, Y.; Miras, J.; Mackie, A.D.; Esquena, J.; Avalos, J.B. Phase behavior of gelatin/maltodextrin aqueous mixtures studied from a combined experimental and theoretical approach. *Fluid Phase Equilib.* **2020**, *524*, 112675.
47. Da Silva, L.M.; Meirelles, A.A. Phase equilibrium in polyethylene glycol/maltodextrin aqueous two-phase systems. *Carbohydr. Polym.* **2000**, *42*, 273–278.
48. Machado, F.L.C.; Dos Reis Coimbra, J.S.; Zuniga, A.D.G.; Da Costa, A.R.; Martins, J.P. Equilibrium data of aqueous two-phase systems composed of poly(ethylene glycol) and maltodextrin. *J. Chem. Eng. Data* **2012**, *57*, 1984–1990.
49. Giraldo-Zuniga, A.D.; Dos Reis Coimbra, J.S.; Arquete, D.A.; Minim, L.A.; Da Silva, L.H.M.; Maffia, M.C. Interfacial tension and viscosity for poly(ethylene glycol) + maltodextrin aqueous two-phase systems. *J. Chem. Eng. Data* **2006**, *51*, 1144–1147.

50. Merland, T.; Waldmann, L.; Guignard, O.; Tatry, M.C.; Wirotius, A.L.; Lapeyre, V.; Garrigue, P.; Nicolai, T.; Benyahia, L.; Ravaine, V. Thermo-induced inversion of water-in-water emulsion stability by bis-hydrophilic microgels. *J. Colloid Interface Sci.* **2022**, *608*, 1191–1201.
51. Buzza, D.M.a; Fletcher, P.D.I.; Georgiou, T.K.; Ghasdian, N. Water-in-water emulsions based on incompatible polymers and stabilized by triblock copolymers-templated polymersomes. *Langmuir* **2013**, *29*, 14804–14814.
52. Jeyhani, M.; Thevakumaran, R.; Abbasi, N.; Hwang, D.K.; Tsai, S.S.H. Microfluidic generation of all-aqueous double and triple emulsions. *Small* **2020**, *16*, 1906565.
53. Aydin, D.; Kizilel, S. Water-in-water emulsion based synthesis of hydrogel nanospheres with tunable release kinetics. *J. Miner. Met. Mater. Soc.* **2017**, *69*, 1185–1194.
54. Nguyen, B.T.; Nicolai, T.; Benyahia, L. Stabilization of water-in-water emulsions by addition of protein particles. *Langmuir* **2013**, *29*, 10658–10664.
55. Gonzalez-Jordan, A.; Nicolai, T.; Benyahia, L. Enhancement of the particle stabilization of water-in-water emulsions by modulating the phase preference of the particles. *J. Colloid Interface Sci.* **2018**, *530*, 505–510.
56. Gonzalez-Jordan, A.; Nicolai, T.; Benyahia, L. Influence of the protein particle morphology and partitioning on the behavior of particle-stabilized water-in-water emulsions. *Langmuir* **2016**, *32*, 7189–7197.
57. Nicolai, T.; Murray, B. Particle stabilized water in water emulsions. *Food Hydrocoll.* **2017**, *68*, 157–163.
58. Chen, J.; Guo, J.; Yang, X.; Nicolai, T. Water-in-water-in-water emulsions formed by cooling mixtures of guar, amylopectin and gelatin. *Food Hydrocoll.* **2021**, *118*, 106763.
59. Lei, C.; Xie, Y.; Wu, Y.; Li, Y.; Li, B.; Pei, Y.; Liu, S. Properties and stability of water-in-water emulsions stabilized by microfibrillated bacterial cellulose. *Food Hydrocoll.* **2022**, *130*, 107698.
60. Machado, J.P.E.; Capron, I.; de Freitas, R.A.; Benyahia, L.; Nicolai, T. Stabilization of amylopectin-pullulan water in water emulsions by interacting protein particles. *Food Hydrocoll.* **2022**, *124*, 107320.
61. de Freitas, R.A.; Nicolai, T.; Chassenieux, C.; Benyahia, L. Stabilization of water-in-water emulsions by polysaccharide-coated protein particles. *Langmuir* **2016**, *32*, 1227–1232.
62. Hazt, B.; Bassani, H.P.; Elias-Machado, J.P.; Aldinucci Buzzo, J.L.; Silveira, J.L.M.; de Freitas, R.A. Effect of PH and protein particle shape on the stability of amylopectin–xyloglucan water-in-water emulsions. *Food Hydrocoll.* **2020**, *104*, 105769.
63. Beldengrün, Y.; Dallaris, V.; Jaén, C.; Protat, R.; Miras, J.; Calvo, M.; García-Celma, M.J.J.; Esquena, J. Formation and stabilization of multiple water-in-water-in-water (w/w/w) emulsions. *Food Hydrocoll.* **2020**, *102*, 105588.
64. Lorén, N.; Hermansson, A.M.; Williams, M.A.K.; Lundin, L.; Foster, T.J.; Hubbard, C.D.; Clark, A.H.; Norton, I.T.; Bergström, E.T.; Goodall, D.M. Phase separation induced by conformational ordering of gelatin in gelatin/maltodextrin mixtures. *Macromolecules* **2001**, *34*, 289–297.
65. Borza, A.D.; Annan, N.T.; Moreau, D.L.; Allan-Wojtas, P.M.; Ghanem, A.; Rousseau, D.; Paulson, A.T.; Hansen, L.T. Microencapsulation in genipin cross-linked gelatine-maltodextrin improves survival of *Bifidobacterium adolescentis* during exposure to in vitro gastrointestinal conditions. *J. Microencapsul.* **2010**, *27*, 387–399.
66. Kasapis, S.; Morris, E.R.; Norton, I.T.; Clark, A.H. Phase equilibria and gelation in gelatin/maltodextrin systems – Part I: Gelation of individual components. *Carbohydr. Polym.* **1993**, *21*, 243–248.

67. Hoey, A.; Ryan, J.T.; Fitzsimons, S.M.; Morris, E.R. Segregative interactions in single-phase mixtures of gelling (potato) maltodextrin with other hydrocolloids. In *Gums and Stabilisers for the Food Industry 18 Hydrocolloid Functionality for Affordable and Sustainable Global Food Solutions*, Williams, P.A., Phillips, G., Eds.; RSC: London, UK, 2016; pp. 305–312.
68. Williams, M.; Fabri, D.; Hubbard, C. Kinetics of droplet growth in gelatin/maltodextrin mixtures following thermal quenching. *Langmuir* **2001**, *17*, 3412–3418.
69. Beldengrün, Y.; Aragon, J.; Prazeres, S.; Montalvo, G.; Miras, J.; Esquena, J. Gelatin-maltodextrin water-in-water (w/w) emulsions for the preparation of crosslinked enzyme-loaded microgels. *Langmuir* **2018**, *34,* 9731–9743.
70. Alevisopoulos, S.; Kasapis, S. Molecular weight effects on the gelatin/maltodextrin gel. *Carbohydr. Polym.* **1999**, *40*, 83–87.
71. Butler, M.F.; Heppenstall-Butler, M. Phase separation in gelatin/dextran and gelatin/maltodextrin mixtures. *Food Hydrocoll.* **2003**, *17*, 815–830.
72. Kasapis, S.; Morris, E.R.; Norton, I.T.; Clark, A.H. Phase equilibria and gelation in gelatin/maltodextrin systems – Part IV: Composition-dependence of mixed-gel moduli. *Carbohydr. Polym.* **1993**, *21*, 269–276.
73. Vis, M.; Peters, V.F.D.; Blokhuis, E.M.; Lekkerkerker, H.N.W.; Erné, B.H.; Tromp, R.H. Decreased interfacial tension of demixed aqueous polymer solutions due to charge. *Phys. Rev. Lett.* **2015**, *115*, 078303.
74. Rogers, B.A.; Rembert, K.B.; Poyton, M.F.; Okur, H.I.; Kale, A.R.; Yang, T.; Zhang, J.; Cremer, P.S. A stepwise mechanism for aqueous two-phase system formation in concentrated antibody solutions. *Proc. Natl. Acad. Sci. U.S.A.* **2019**, *116*, 15784–15791.
75. Turgeon, S.L.; Beaulieu, M.; Schmitt, C.; Sanchez, C. Protein-polysaccharide interactions: Phase-ordering kinetics, thermodynamic and structural aspects. *Curr. Opin. Colloid Interface Sci.* **2003**, *8*, 401–414.
76. Rogers, R.D.; Bauer, C.B. Partitioning behavior of group 1 and 2 cations in poly(ethylene glycol)-based aqueous biphasic systems. *J. Chromatogr. B Biomed. Appl.* **1996**, *680*, 237–241
77. Berton, P.; Kelley, S.P.; Bridges, N.J.; Klingshirn, M.A.; Huddleston, J.G.; Willauer, H.D.; Baldwin, J.W.; Moody, M.L.; Rogers, R.D. Water in solutions of chaotropic and kosmotropic salts: A differential scanning calorimetry investigation. *J. Chem. Eng. Data* **2019**, *64*, 4781–4792.
78. Pereira, J.F.B.; Ventura, S.P.M.; E Silva, F.A.; Shahriari, S.; Freire, M.G.; Coutinho, J.A.P. aqueous biphasic systems composed of ionic liquids and polymers: A Platform for the purification of biomolecules. *Sep. Purif. Technol.* **2013**, *113*, 83–89.
79. Iqbal, M.; Tao, Y.; Xie, S.; Zhu, Y.; Chen, D.; Wang, X.; Huang, L.; Peng, D.; Sattar, A.; Shabbir, M.A.B.; et al. Aqueous Two-Phase System (ATPS): An overview and advances in its applications. *Biological Procedures Online* **2016**, *18*, 18.
80. Ruiz-Ruiz, F.; Benavides, J.; Aguilar, O.; Rito-Palomares, M. Aqueous two-phase affinity partitioning systems: Current applications and trends. *J. Chromatography A* **2012**, *1244*, 1–13.
81. Huddleston, J.G.; Willauer, H.D.; Griffin, S.T.; Rogers, R.D. Aqueous polymeric solutions as environmentally benign liquid/ liquid extraction media. *Ind. Eng. Chem. Res.* **1999**, *38*, 2523–2539.
82. Yang, Z. Hofmeister effects: An explanation for the impact of ionic liquids on biocatalysis. *J. Biotechnol.* **2009**, *144*, 12–22.
83. Hyde, A.M.; Zultanski, S.L.; Waldman, J.H.; Zhong, Y.L.; Shevlin, M.; Peng, F. General principles and strategies for salting-out informed by the Hofmeister series. *Org. Process Res. Dev.* **2017**, *21*, 1355–1370.

84. Gregory, K.P.; Elliott, G.R.; Robertson, H.; Kumar, A.; Wanless, E.J.; Webber, G.B.; Craig, V.S.J.; Andersson, G.G.; Page, A.J. Understanding specific ion effects and the Hofmeister series. *Phys. Chem. Chem. Phys.* **2022**, *24*, 12682–12718.
85. Kunz, W.; Henle, J.; Ninham, B.W. "Zur Lehre von Der Wirkung Der Salze" (about the Science of the Effect of Salts): Franz Hofmeister's historical papers. *Curr. Opin. Colloid Interface Sci.* **2004**, *9* (1–2), 19–37. https://doi.org/10.1016/j.cocis.2004.05.005.
86. Hofmeister F.. Zur Lehre von Der Wirkung Der Salze ("About the Science of the Effects of Salts"). *Archiv fuer Exp. Pathol. und Pharmakologie* **1888**, *24*, 247–260.
87. Schwierz, N.; Horinek, D.; Sivan, U.; Netz, R.R. Reversed Hofmeister series – The rule rather than the exception. *Curr. Opin. Colloid Interface Sci.* **2016**, *23*, 10–18.
88. Chen, J.; Ding, Z.; Pan, H.; Cao, X. Development of PH-responsive polymer and citrate aqueous two-phase system for extractive bioconversion of cefprozil. *Talanta* **2017**, *174*, 256–264.
89. Ananthapadmanabhan, K.P.; Goddard, E.D. Aqueous biphase formation in polyethylene oxide-inorganic salt systems. *Langmuir* **1987**, *3*, 25–31.
90. Silvério, S.C.; Gracia, J.; Teixeira, J.A.; Macedo, E.A. Polyethylene glycol 8000+ citrate salts aqueous two-phase systems: Relative hydrophobicity of the equilibrium phases. *Fluid Phase Equilib.* **2015**, *407*, 298–303.
91. Sadeghi, R.; Kahaki, H.B. Thermodynamics of aqueous solutions of poly ethylene glycol di-methyl ethers in the presence or absence of ammonium phosphate salts. *Fluid Phase Equilib.* **2011**, *306*, 219–228.
92. Marcus, Y. Thermodynamics of solvation of ions. Part 5. - Gibbs free energy of hydration at 298.15 K. *J. Chem. Soc. Faraday Trans.* **1991**, *87*, 2995–2999.
93. Park, S.; Barnes, R.; Lin, Y.; Jeon, B. jin; Najafi, S.; Delaney, K.T.; Fredrickson, G.H.; Shea, J.E.; Hwang, D.S.; Han, S. Dehydration entropy drives liquid-liquid phase separation by molecular crowding. *Commun. Chem.* **2020**, *3*, 83.
94. Iolitec (Ionic Liquid Technologies). Iolitec Product Catalogue https://iolitec.de/en/products/ionic_liquids/catalogue (accessed Aug 13, 2022).
95. Gutowski, K.E.; Broker, G.A.; Willauer, H.D.; Huddleston, J.G.; Swatloski, R.P.; Holbrey, J.D.; Rogers, R.D. Controlling the aqueous miscibility of ionic liquids: Aqueous biphasic systems of water-miscible ionic liquids and water-structuring salts for recycle, metathesis, and separations. *J. Am. Chem. Soc.* **2003**, *125*, 6632–6633.
96. Bridges, N.J.; Gutowski, K.E.; Rogers, R.D. Investigation of aqueous biphasic systems formed from solutions of chaotropic salts with kosmotropic salts (Salt-Salt ABS). *Green Chem.* **2007**, *9*, 177–183.
97. Freire, M.G.; Cláudio, A.F.M.; Araújo, J.M.M.; Coutinho, J.A.P.; Marrucho, I.M.; Canongia Lopes, J.N.; Rebelo, L.P.N.; Aqueous biphasic systems: A boost brought about by using ionic liquids (2012) *Chemical Society Rev* **2012**, *41*, 4966–4995.
98. Freire, M.G. *Ionic-Liquid-Based Aqueous Biphasic Systems: Fundamentals and Applications*, Springer: Berlin, Germany, 2016.
99. Pereira, J.F.B.; Kurnia, K.A.; Cojocaru, O.A.; Gurau, G.; Rebelo, L.P.N.; Rogers, R.D.; Freire, M.G.; Coutinho, J.A.P. Molecular interactions in aqueous biphasic systems composed of polyethylene glycol and crystalline vs. liquid cholinium-based salts. *Phys. Chem. Chem. Phys.* **2014**, *16*, 5723–5731.
100. Visak, Z.P.; Canongia Lopes, J.N.; Rebelo, L.P.N. Ionic liquids in polyethylene glycol aqueous solutions: salting-in and salting-out effects. *Monatshefte fur Chemie* **2007**, *138*, 1153–1157.
101. Pereira, J.F.B.; Kurnia, K.A.; Freire, M.G.; Coutinho, J.A.P.; Rogers, R.D. Controlling the formation of ionic-liquid-based aqueous biphasic systems by changing the hydrogen-bonding ability of polyethylene glycol end groups. *ChemPhysChem* **2015**, *16*, 2219–2225.

102. Rodríguez, H.; Francisco, M.; Rahman, M.; Sun, N.; Rogers, R.D. Biphasic liquid mixtures of ionic liquids and polyethylene glycols. *Phys. Chem. Chem. Phys.* **2009**, *11*, 10916–10922.
103. Mourão, T.; Tomé, L.C.; Florindo, C.; Rebelo, L.P.N.; Marrucho, I.M. Understanding the role of cholinium carboxylate ionic liquids in PEG-based aqueous biphasic systems. *ACS Sustain. Chem. Eng.* **2014**, *2*, 2426–2434.
104. Freire, M.G.; Pereira, J.F.B.; Francisco, M.; Rodríguez, H.; Rebelo, L.P.N.; Rogers, R.D.; Coutinho, J.A.P. Insight into the interactions that control the phase behaviour of new aqueous biphasic systems composed of polyethylene glycol polymers and ionic liquids. *Chem. - A Eur. J.* **2012**, *18*, 1831–1839.
105. Neves, C.M.S.S.; Shahriari, S.; Lemus, J.; Pereira, J.F.B.; Freire, M.G.; Coutinho, J.A.P. Aqueous biphasic systems composed of ionic liquids and polypropylene glycol: insights into their liquid-liquid demixing mechanisms. *Phys. Chem. Chem. Phys.* **2016**, *18*, 20571–20582.
106. Bridges, N.J.; Gutowski, K.E.; Rogers, R.D. Investigation of aqueous biphasic systems formed from solutions of chaotropic salts with kosmotropic salts (Salt-Salt ABS). *Green Chem.* **2007**, *9*, 177–183.
107. Bao, X.; Chen, Z.; Han, J.; Wang, Y.; Li, C.; Ni, L. Liquid – liquid equilibrium of imidazolium ionic liquids + phosphate + water aqueous two - phase systems and correlation. *J. Solution Chem.* **2019**, *48*, 1167–1187.
108. Shahriari, S.; Neves, C.M.S.S.; Freire, M.G.; Coutinho, J.A.P. Role of the hofmeister series in the formation of ionic-liquid-based aqueous biphasic systems. *J. Phys. Chem. B* **2012**, *116*, 7252–7258.
109. Zafarani-Moattar, M.T.; Hamzehzadeh, S. Salting-out effect, preferential exclusion, and phase separation in aqueous solutions of chaotropic water-miscible ionic liquids and kosmotropic salts: Effects of temperature, anions, and cations. *J. Chem. Eng. Data* **2010**, *55*, 1598–1610
110. Ventura, S.P.M.; Neves, C.M.S.S.; Freire, M.G.; Marrucho, I.M.; Oliveira, J.; Coutinho, J.A.P. Evaluation of anion influence on the formation and extraction capacity of ionic-liquid-based aqueous biphasic systems. *J. Phys. Chem. B* **2009**, *113*, 9304–9310.
111. Cláudio, A.F.M.; Ferreira, A.M.; Shahriari, S.; Freire, M.G.; Coutinho, J.A.P. Critical assessment of the formation of ionic-liquid-based aqueous two-phase systems in acidic media. *J. Phys. Chem. B* **2011**, *115*, 11145–11153.
112. Wilkes, J.S. A short history of ionic liquids - from molten salts to neoteric solvents. *Green Chem.* **2002**, *4*, 73–80.
113. Shahriari, S.; Tomé, L.C.; Araújo, J.M.M.; Rebelo, L.P.N.; Coutinho, J.A.P.; Marrucho, I.M.; Freire, M.G. Aqueous biphasic systems: A benign route using cholinium-based ionic liquids. *RSC Adv.* **2013**, *3*, 1835–1843
114. Liu, D.; Zou, X.; Gao, M.; Gu, M.; Xiao, H. Hydrophilic organic/salt-containing aqueous two-phase solvent system for counter-current chromatography: A novel technique for separation of polar compounds. *J. Chromatogr. A* **2014**, *1356*, 157–162.
115. Reis, I.A.O.; Santos, S.B.; Santos, L.A.; Oliveira, N.; Freire, M.G.; Pereira, J.F.B.; Ventura, S.P.M.; Coutinho, J.A.P.; Soares, C.M.F.; Lima, Á.S. Increased significance of food wastes: Selective recovery of added-value compounds. *Food Chem.* **2012**, *135*, 2453–2461.
116. Jamehbozorg, B.; Sadeghi, R. Evaluation of the effect of carbohydrates as renewable, none-charged and non-toxic soluting-out agents on the ionic-liquid-based abs implementation. *J. Mol. Liq.* **2018**, *255*, 476–491.
117. Zafarani-Moattar, M.T.; Shekaari, H.; Mazaher Haji Agha, E. Thermodynamic studies on the phase equilibria of ternary {ionic Liquid, 1-Hexyl-3-Methyl Imidazolium Chloride + D-Fructose or Sucrose + Water} systems at 298.15 K. *Fluid Phase Equilib.* **2017**, *436*, 38–46.

118. Quental, M.V.; Pereira, M.M.; Ferreira, A.M.; Pedro, S.N.; Shahriari, S.; Mohamadou, A.; Coutinho, J.A.P.; Freire, M.G. Enhanced separation performance of aqueous biphasic systems formed by carbohydrates and tetraalkylphosphonium- or tetraalkylammonium- based ionic liquids. *Green Chem. Commun.* **2018**, *20*, 2978–2983.
119. Freire, M.G.; Louros, C.L.S.; Rebelo, L.P.N.; Coutinho, J.A.P. Aqueous biphasic systems composed of a water-stable ionic liquid + carbohydrates and their applications. *Green Chem.* **2011**, *13*, 1536–1545.
120. Capela, E.V.E.V.; Quental, M.V.M.V.M.V.; Domingues, P.; Coutinho, J.A.P.P.; Freire, M.G. Effective separation of aromatic and aliphatic amino acid mixtures using ionic-liquid-based aqueous biphasic mixtures. *Green Chem.* **2017**, *19*, 1850–1854.
121. Blankschtein, D.; Thurston, G.M.; Benedek, G.B. Phenomenological theory of equilibrium thermodynamic properties and phase separation of micellar solutions. *J. Chem. Phys.* **1986**, *85*, 7268–7288.
122. Liu, C.L.; Nikas, Y.J.; Blankschtein, D. Novel bioseparations using two-phase aqueous micellar systems. *Biotechnol. Bioeng.* **1996**, *52*, 185–192.
123. Liu, C.; Kamei, D.T.; King, J.A.; Wang, D.I.C.; Blankschtein, D. Separation of proteins and viruses using two-phase aqueous micellar systems. *J. Chromatogr. B Biomed. Sci. Appl.* **1998**, *711*, 127–138.
124. Hinze, W.L.; Pramauro, E. A critical review of surfactant-mediated phase separations (cloud-point extractions): Theory and applications. *Crit. Rev. Anal. Chem.* **1993**, *24*, 133–177.
125. Miras, J.; Vílchez, S.; Bouju, C.; Blanco-Pérez, M.; Ros-Espinal, M.; Rodríguez-Abreu, C.; Esquena, J., Formation of water-in-water emulsions and microgels in nonionic surfactant + gelatin aqueous mixtures. *Journal of Colloid and Interface Science*, **2025**, *684 (1)*, 319–330.
126. Aveyard, R.; Binks, B.P.; Chen, J.; Esquena, J.; Fletcher, P.D.I.; Buscall, R.; Davies, S. Surface and colloid chemistry of systems containing pure sugar surfactant. *Langmuir* **1998**, *14*, 4699–4709.
127. Valaulikar, B.S.; Manohar, C. The mechanism of clouding in triton x-100: The effect of additives. *J. Colloid Interface Sci.* **1985**, *108*, 403–406.
128. Álvarez, M.S.; Rivas, M.; Deive, F.J.; Sanromán, M.A.; Rodríguez, A. Ionic liquids and non-ionic surfactants: A new marriage for aqueous segregation. *RSC Adv.* **2014**, *4*, 32698–32700.
129. Yin, K.; Chen, L.; Liu, F.; Fan, T.; Wu, Y.; Li, S.; Yan, Z. Liquid – liquid equilibrium data for novel aqueous two-phase systems composed of betaine-xylitol deep eutectic solvent + alcohol (1-Propanol / 2-Propanol / *tert*-Butanol) + Water. *J. Chem. Eng. Data* **2022**, *67*, 2475–2485.
130. Marchel, M.; Coroadinha, A.S.; Marrucho, I.M. Novel acidic deep eutectic solvent-based aqueous biphasic systems for efficient extraction of pepsin. *ACS Sustain. Chem. Eng.* **2020**, *8*, 12400–12408.
131. Passos, H.; Tavares, D.J.P.; Ferreira, A.M.; Freire, M.G.; Coutinho, J.A.P. Are aqueous biphasic systems composed of deep eutectic solvents ternary or quaternary systems? *ACS Sustain. Chem. Eng.* **2016**, *4*, 2881–2886.
132. Farias, F.O.; Passos, H.; Lima, Á.S.; Mafra, M.R.; Coutinho, J.A.P. Is it possible to create ternary-like aqueous biphasic systems with deep eutectic solvents? *ACS Sustain. Chem. Eng.* **2017**, *5*, 9402–9411.
133. Farias, F.O.; Passos, H.; Sanglard, M.G.; Igarashi-Mafra, L.; Coutinho, J.A.P.; Mafra, M.R. Designer solvent ability of alcohols in aqueous biphasic systems composed of deep eutectic solvents and potassium phosphate. *Sep. Purif. Technol.* **2018**, *200*, 84–93.
134. Zafarani-Moattar, M.T.; Shekaari, H.; Fattah, S.G.; Mokhtarpour, M. Novel aqueous two-phase systems containing polymer-based deep eutectic solvent and citrate salts for high-performance extraction of dyes. *J. Mol. Liq.* **2022**, *360*, 119475.

135. De Brito Cardoso, G.; Mourão, T.; Pereira, F.M.; Freire, M.G.; Fricks, A.T.; Soares, C.M.F.; Lima, Á.S. Aqueous two-phase systems based on acetonitrile and carbohydrates and their application to the extraction of vanillin. *Sep. Purif. Technol.* **2013**, *104*, 106–113.
136. Galaev, I.Y.; Mattiasson, B. Thermoreactive water-soluble polymers, nonionic surfactants, and hydrogels as reagents in biotechnology. *Enzyme Microb. Technol.* **1993**, *15*, 354–366.
137. Johansson, H.O.; Persson, J.; Tjerneld, F. Thermoseparating water/polymer system: A novel one-polymer aqueous two- phase system for protein purification. *Biotechnol. Bioeng.* **1999**, *66*, 247–257.
138. Persson, J.; Johansson, H.O.; Tjerneld, F. Biomolecule separation using temperature-induced phase separation with recycling of phase-forming polymers. *Ind. Eng. Chem. Res.* **2000**, *39*, 2788–2796.
139. Persson, J.; Johansson, H.-O.; Galaev, I.; Mattiasson, B.; Tjerneld, F. Aqueous polymer two-phase systems formed by new thermoseparating polymers. *Bioseparation* **2000**, *8*, 105–116.
140. Song, Q.; Chao, Y.; Zhang, Y.; Shum, H.C. Controlled formation of all-aqueous Janus droplets by liquid – liquid phase separation of an aqueous three-phase system. *J. Phys. Chem. B* **2021**, *125*, 562–570.
141. Akamatsu, K.; Kurita, R.; Sato, D.; Nakao, S.I. Aqueous two-phase system formation in small droplets by Shirasu porous glass membrane emulsification followed by water extraction. *Langmuir* **2019**, *35*, 9825–9830.
142. Souza, B.S.; Marcelino, H.R.; Alexandrino, F.; Urtiga, S.C.C.; Silva, K.C.H.; Soares, D.C.F.; Egito, E.S.T. Water-in-water emulsion as a new approach to produce mesalamine-loaded xylan-based microparticles. *Appl. Sci.* **2019**, *9*, 3519.
143. Antonov, Y.A.; Van Puyvelde, P.; Moldenaers, P. Interfacial tension of aqueous biopolymer mixtures close to the critical point. *Int. J. Biol. Macromol.* **2004**, *34*, 29–35.
144. Scholten, E.; Visser, J.E.; Sagis, L.M.C.; van der Linden, E. Ultralow interfacial tensions in an aqueous phase-separated gelatin/dextran and gelatin/gum Arabic system: A comparison. *Langmuir* **2004**, *20*, 2292–2297.
145. Lü, Y.; Feng, Y.; Zhao, Y.; Zhang, S.; Shu, S.; Luo, P. Anionic polyacrylamide "water-in-water" emulsion prepared by dispersion polymerization in mixed salt aqueous media. *e-Polymers* **2009,** *9*, 93.
146. Ding, P.; Wolf, B.; Frith, W.J.; Clark, a H.; Norton, I.T.; Pacek, a W. Interfacial tension in phase-separated gelatin/dextran aqueous mixtures. *J. Colloid Interface Sci.* **2002**, *253*, 367–376.
147. Forciniti, D.; Hall, C.K.; Kula, M.R. Interfacial tension of polyethyleneglycol-dextran-water systems: Influence of temperature and polymer molecular weight. *J. Biotechnol.* **1990**, *16*, 279–296.
148. Vis, M.; Erné, B.H.; Tromp, R.H. Chemical physics of water–water interfaces. *Biointerphases* **2016**, *11*, 018904.
149. Ryden, J.; Albertsson, P. Interfacial tension of dextran-polyethylene glycol-water two-phase systems. *J. Colloid Interface Sci.* **1971**, *37*, 219–222.
150. Vis, M.; Blokhuis, E.M.; Erné, B.H.; Tromp, R.H.; Lekkerkerker, H.N.W. Interfacial tension of phase-separated polydisperse mixed polymer solutions. *J. Phys. Chem. B* **2018**, *122*, 3354–3362.
151. Tromp, R.H. Water – water interfaces. In *Soft Matter at Aqueous Interfaces*, Lang, P., Liu, Y., Eds.; Springer: Berlin, Germany, 2016; pp. 159–186.
152. Rigoni, C.; Beaune, G.; Harnist, B.; Sohrabi, F.; Timonen, J.V.I. Ferrofluidic aqueous two-phase system with ultralow interfacial tension and micro-pattern formation. *Commun. Mater.* **2022**, *3*, 26.

153. Nguyen, B.T.; Wang, W.; Saunders, B.R.; Benyahia, L.; Nicolai, T. PH-responsive water-in-water Pickering emulsions. *Langmuir* **2015**, *31*, 3605–3611.
154. Poortinga, A.T. Microcapsules from self-assembled colloidal particles using aqueous phase-separated polymer solutions. *Langmuir* **2008**, *24*, 1644–1647.
155. Peddireddy, K.R.; Nicolai, T.; Benyahia, L.; Capron, I. Stabilization of water-in-water emulsions by nanorods. *ACS Macro Lett.* **2016**, *5*, 283–286.
156. Tea, L.; Nicolai, T.; Benyahia, L.; Renou, F. Viscosity and morphology of water-in-water emulsions: The effect of different biopolymer stabilizers. *Macromolecules* **2020**, *53*, 3914–3922.
157. Aveyard, R.; Binks, B.; Clint, J. Emulsions stabilised solely by colloidal particles. *Adv. Colloid Interface Sci.* **2003**, *100*, 503–546.
158. Binks, B.P. Colloidal particles at a range of fluid-fluid interfaces. *Langmuir* **2017**, *33*, 6947–6963
159. Dickinson, E. Exploring the frontiers of colloidal behaviour where polymers and particles meet. *Food Hydrocoll.* **2016**, *52*, 497–509.
160. Ben Ayed, E.; Cochereau, R.; Dechancé, C.; Capron, I.; Nicolai, T.; Benyahia, L. Water-in-water emulsion gels stabilized by cellulose nanocrystals. *Langmuir* **2018**, *34*, 6887–6893.
161. Ganley, W.J.; Ryan, P.T.; van Duijneveldt, J.S. Stabilisation of water-in-water emulsions by montmorillonite platelets. *J. Colloid Interface Sci.* **2017**, *505*, 139–147.
162. Hanazawa, T.; Murray, B.S. The influence of oil droplets on the phase separation of protein-polysaccharide mixtures. *Food Hydrocoll.* **2012**, *34*, 128–137.
163. Hanazawa, T.; Murray, B.S. Effect of oil droplets and their solid/liquid composition on the phase separation of protein-polysaccharide mixtures. *Langmuir* **2013**, *29*, 9841–9848.
164. Dewey, D.C.; Strulson, C.A.; Cacace, D.N.; Bevilacqua, P.C.; Keating, C.D. Bioreactor droplets from liposome-stabilized all-aqueous emulsions. *Nat. Commun.* **2014**, *5*, 1–9.
165. Cacace, D.N.; Rowland, A.T.; Stapleton, J.J.; Dewey, D.; Keating, C.D. Aqueous emulsion droplets stabilized by lipid vesicles as microcompartments for biomimetic mineralization. *Langmuir* **2015**, *31*, 11329–11338.
166. Hann, S.D.; Goulian, M.; Lee, D.; Stebe, K.J. Trapping and assembly of living colloids at water-water interfaces. *Soft Matter* **2015**, *11*, 1733–1738.
167. Vis, M.; Opdam, J.; Van't Oor, I.S.J.; Soligno, G.; Van Roij, R.; Tromp, R.H.; Erné, B.H. Water-in-water emulsions stabilized by nanoplates. *ACS Macro Lett.* **2015**, *4*, 965–968.
168. Zhang, J.; Mei, L.; Ma, P.; Li, Y.; Yuan, Y.; Zeng, Q.-Z.Z.; Wang, Q. Microgel-stabilized hydroxypropyl methylcellulose and dextran water-in-water emulsion: Influence of PH, ionic strength, and temperature. *Langmuir* **2021**, *37*, 5617–5626.
169. Qian, X.; Peng, G.; Ge, L.; Wu, D. Water-in-water Pickering emulsions stabilized by the starch nanocrystals with various surface modifications. *J. Colloid Interface Sci.* **2022**, *607*, 1613–1624.
170. Ansari, N.; Shekarchizadeh, H. Stabilisation of water in water Pickering emulsion containing gelatin and maltodextrin by bitter vetch protein nanoparticles. *Int. J. Food Sci. Technol.* **2022**, *57*, 3816–2825.
171. Murray, B.S.; Phisarnchananan, N. Whey protein microgel particles as stabilizers of waxy corn starch + locust bean gum water-in-water emulsions. *Food Hydrocoll.* **2016**, *56*, 161–169.
172. Laranjo, M.R.; de Sá Costa, B.; Garcia-Rojas, E.E. Stabilization of gelatin and carboxymethylcellulose water-in-water emulsion by addition of whey protein. *Polimeros* **2020**, *29*, e2019051.
173. Keal, L.; Colosqui, C.E.; Tromp, R.H.; Monteux, C. Colloidal particle adsorption at water-water interfaces with ultralow interfacial tension. *Phys. Rev. Lett.* **2018**, *120*, 208003.

174. Chatsisvili, N.; Philipse, A.P.; Loppinet, B.; Tromp, R.H. Colloidal zein particles at water-water interfaces. *Food Hydrocoll.* **2017**, *65*, 17–23.
175. Douliez, J.-P.; Martin, N.; Beneyton, T.; Eloi, J.-C.; Chapel, J.-P.; Navailles, L.; Baret, J.-C.; Mann, S.; Béven, L. Preparation of swellable hydrogel-containing colloidosomes from aqueous two-phase Pickering emulsion droplets. *Angew. Chemie Int. Ed.* **2018**, *57*, 7780–7784.
176. Wang, Y.; Yuan, J.; Zhao, Y.; Wang, L.; Guo, L.; Feng, L.; Cui, J.; Dong, S.; Wan, S.; Liu, W.; Hoffman, H.; Tieu, K.; Hao, J. Water-in-water emulsions, ultralow interfacial tension, and biolubrication. *CCS Chem.* **2022**, *4*, 2102–2114.
177. You, K.-M.; Murray, B.S.; Sarkar, A. Tribology and rheology of water-in-water emulsions stabilized by whey protein microgels. *Food Hydrocoll.* **2023**, *134*, 108009.
178. Dickinson, E. Microgels – An alternative colloidal ingredient for stabilization of food emulsions. *Trends Food Sci. Technol.* **2015**, *43*, 178–188.
179. Pavlovic, M.; Plucinski, A.; Zeininger, L.; Schmidt, B.V.K.J. Temperature sensitive water-in-water emulsions. *Chem. Commun.* **2020**, *56*, 6814–6817.
180. Plucinski, A.; Schmidt, B.V.K.J. PH Sensitive water-in-water emulsions based on the pullulan and poly(*N,N*-dimethylacrylamide) aqueous two-phase system. *Polym. Chem.* **2022**, *13*, 4170–4177.
181. Plucinski, A.; Schmidt, B.V.K.J. Erratum: PH sensitive water-in-water emulsions based on the pullulan and poly(*N,N*-dimethylacrylamide) aqueous two-phase system (Polymer Chemistry (2022) DOI: 10.1039/D2py00469k). *Polym. Chem.* **2022**, *13*, 4352.
182. Machado, J.P.E.; Benyahia, L.; Nicolai, T. Effect of adding a third polysaccharide on the adsorption of protein microgels at the interface of polysaccharide-based water in water emulsions. *J. Colloid Interface Sci.* **2021**, *603*, 633–640.
183. Binks, B.P.; Shi, H. Phase inversion of silica particle-stabilized water-in-water emulsions. *Langmuir* **2019**, *35*, 4046–4057.
184. Zhang, J.; Hwang, J.; Antonietti, M.; Schmidt, B.V.K.J. Water-in-water Pickering emulsion stabilized by polydopamine particles and cross-linking. *Biomacromolecules* **2019**, *20*, 204–211.
185. Perro, A.; Coudon, N.; Chapel, J.P.; Martin, N.; Béven, L.; Douliez, J.P. Building microcapsules using water-in-water emulsion droplets as templates. *J. Colloid Interface Sci.* **2022**, *613*, 681–696.
186. Meng, Y.; Nicolai, T.; Benyahia, L.; Nicol, E. Utilization of xanthan to stabilize water in water emulsions and modulate their viscosity. *Carbohydr. Polym.* **2022**, *277*, 118812.
187. Meng, Y.; Nicol, E.; Nicolai, T. Exploiting multiple phase separation to stabilize water in water emulsions and form stable microcapsules. *J. Colloid Interface Sci.* **2022**, *617*, 65–72.
188. Coudon, N.; Navailles, L.; Nallet, F.; Ly, I.; Bentaleb, A.; Chapel, J.P.; Béven, L.; Douliez, J.P.; Martin, N. Stabilization of all-aqueous droplets by interfacial self-assembly of fatty acids bilayers. *J. Colloid Interface Sci.* **2022**, *617*, 257–266.
189. Wang, Y.; Yuan, J.; Dong, S.; Hao, J. Multilayer-stabilized water-in-water emulsions. *Langmuir* **2022**, *38*, 4713–4721.
190. Liu, H.; Wang, Y.; Wang, H.; Zhao, M.; Tao, T.; Zhang, X.; Qin, J. A droplet microfluidic system to fabricate hybrid capsules enabling stem cell organoid engineering. *Adv. Sci.* **2020**, *7*, 1903739.
191. Ma, Q.; Yuan, H.; Song, Y.; Chao, Y.; Mak, S.Y.; Shum, H.C. Partitioning-dependent conversion of polyelectrolyte assemblies in an aqueous two-phase system. *Soft Matter* **2018**, *14*, 1552–1558.
192. Nan, L.; Cao, Y.; Yuan, S.; Shum, H.C. Oil-mediated high-throughput generation and sorting of water-in-water droplets. *Microsystems Nanoeng.* **2020**, *6*, 70.
193. Xie, G.; Forth, J.; Chai, Y.; Ashby, P.D.; Helms, B.A.; Russell, T.P. Compartmentalized, all-aqueous flow-through-coordinated reaction systems. *Chem* **2019**, *5*, 2678–2690.

194. Sohrabi, S.; Kassir, N.; Keshavarz Moraveji, M. Droplet microfluidics: Fundamentals and its advanced applications. *RSC Adv.* **2020**, *10*, 27560–27574.
195. Wang, X.; Liu, J.; Wang, P.; deMello, A.; Feng, L.; Zhu, X.; Wen, W.; Kodzius, R.; Gong, X. Synthesis of biomaterials utilizing microfluidic technology. *Genes* **2018**, *9*, 283.
196. Keller, S.; Teora, S.P.; Boujemaa, M.; Wilson, D.A. Exploring new horizons in liquid compartmentalization via microfluidics. *Biomacromolecules* **2021**, *22*, 1759–1769.
197. Ma, Q.; Song, Y.; Sun, W.; Cao, J.; Yuan, H.; Wang, X.; Sun, Y.; Shum, H.C. Cell-inspired all-aqueous microfluidics: From intracellular liquid–liquid phase separation toward advanced biomaterials. *Adv. Sci.* **2020**, *7*, 1903359.
198. Zhou, C.; Zhu, P.; Han, X.; Shi, R.; Tian, Y.; Wang, L. Microfluidic generation of ATPS droplets by transient double emulsion technique. *Lab Chip* **2021**, *21*, 2684–2690.
199. Cheung Shum, A.H.; Varnell, J.; Weitz, D.A. Microfluidic fabrication of water-in-water (w/w) jets and emulsions. *Biomicrofluics* **2012**, *6*, 12808.
200. Choi, D.; Lee, E.; Kim, S.J.; Han, M. Passive droplet generation in aqueous two-phase systems with a variable-width microchannel. *Soft Matter* **2019**, *15*, 4647–4655.
201. Chao, Y.; Mak, S.Y.; Rahman, S.; Zhu, S.; Shum, H.C. Generation of high-order all-aqueous emulsion drops by osmosis-driven phase separation. *Small* **2018**, *14*, 1802107.
202. Singh, P.; Medronho, B.; Miguel, M.G.; Esquena, J. On the encapsulation and viability of probiotic bacteria in edible carboxymethyl cellulose-gelatin water-in-water emulsions. *Food Hydrocoll.* **2018**, *75*, 41–50.
203. Das, A.A.K.; Filby, B.W.; Geddes, D.A.; Legrande, D.; Paunov, V.N. High throughput fabrication of cell spheroids by templating water-in-water Pickering emulsions. *Mater. Horizons* **2017**, *4*, 1196–1200.
204. Wang, H.; Liu, H.; He, F.; Chen, W.; Zhang, X.; Zhao, M.; Wang, L.; Qin, J. Flexible generation of multi-aqueous core hydrogel capsules using microfluidic aqueous two-phase system. *Adv. Mater. Technol.* **2020**, *5*, 02000045.
205. Liu, H.H.T.; Wang, H.; Wei, W.B.; Liu, H.H.T.; Jiang, L.; Qin, J.H. A microfluidic strategy for controllable generation of water-in-water droplets as biocompatible microcarriers. *Small* **2018**, *14*, 1801095.
206. Wang, H.; Liu, H.; Zhang, X.; Wang, Y.; Zhao, M.; Chen, W.; Qin, J. One-step generation of aqueous-droplet-filled hydrogel fibers as organoid carriers using an all-in-water microfluidic system. *ACS Appl. Mater. Interfaces* **2021**, *13*, 3199–3208.
207. Liu, T.; Yi, S.; Liu, G.; Hao, X.; Du, T.; Chen, J.; Meng, T.; Li, P.; Wang, Y. Aqueous two-phase emulsions-templated tailorable porous alginate beads for 3d cell culture. *Carbohydr. Polym.* **2021**, *258*, 117702.

Section III

Emulsion Applications

12 Emulsions in Petroleum Industry and Role of Asphaltenes

Plamen Tchoukov and Jan Czarnecki

12.1 INTRODUCTION

Despite systematic efforts to transition away from fossil fuels to mitigate climate change, crude oil remains an important component of global energy consumption, comprising approximately 30% as of 2021, according to the International Energy Agency (IEA).[1] The significance of crude oil resources prompts research efforts aimed at optimizing its extraction and processing, including with a focus on reducing its environmental footprint. Typically, crude oil reservoirs coexist with water-based brine, and many extraction processes are facilitated by injecting water or steam, thus the formation of different emulsion types during the extraction process is inevitable.[2,3] This fact explains why a significant body of research is devoted to petroleum emulsions-related problems. These emulsions could be undesirable, such as those that occur during extraction and processing, and desirable emulsions that facilitate these processes. Apparently, research goals will be different in these two cases. While for the desirable emulsions the objective will be to stabilize those emulsions, for the undesirable emulsions the aim will be their destabilization and effective separation of aqueous and oil phases. In both cases, we have to answer the following questions: (1) What are the interfacially active materials acting as emulsifiers? (2) What the mechanisms of emulsion stabilization/destabilization are? (3) What are the optimal system conditions that will ensure stabilization or destabilization of the emulsion? Three main types of emulsions occur in the petroleum industry: water-in-crude oil (W/CO), oil-in-water (O/W), and multiple emulsions of more complex structures (e.g., W/CO/W or CO/W/CO). While the crude oil emulsions share many common features with other industrial emulsions, they have some specific features that set them apart. One of them is that crude oils are a complex blend of hydrocarbons that vary between crude oils from different reservoirs, and even between different locations within a reservoir. This specific feature of crude oil along with the large number of processing steps, changing conditions, chemical additives, and large-scale operations create a very challenging task that requires both fundamental scientific and engineering approaches to address emulsion problems in the oil industry. In this chapter, we focus on the specifics of the petroleum emulsions and will leave out some fundamental aspects that are common for all emulsions and covered in other chapters of this book or in the literature, e.g., see Refs 4–6.

DOI: 10.1201/9781032636184-15

12.2 COMPOSITION OF CRUDE OIL AND INTERFACIALLY ACTIVE SPECIES

12.2.1 Composition of Crude Oil – SARA Fractions and Beyond

Crude oil consists of a mixture of hydrocarbons with no well-defined chemical structure. It is commonly accepted in the petroleum industry to separate the crude components to saturates, aromatics, resins, and asphaltenes, known as SARA fractions.[7,8] These fractions are based on their solubility properties and usually separated through chromatographic methods. In this fractionation procedure, asphaltenes are initially precipitated by dilution (40:1 by volume) with *n*-pentane or *n*-heptane. The remaining fraction is called maltenes. Maltenes are further separated by liquid chromatography using solvents with different polarity into aromatics, saturates, and resins. Aromatics consist of molecules containing one or more aromatic rings. The saturates fraction consists of alkenes and cycloalkanes. Resins and asphaltenes are composed of fused aromatic rings with attached paraffinic or polar segments and are differentiated by their solubility in *n*-alkanes (pentane and heptane), molecular weight, and polarizability. Resins have lower molecular weight than asphaltenes and are soluble in *n*-pentane and *n*-heptane, whereas asphaltenes are insoluble in these solvents. The main SARA fractions are shown in Figure 12.1. Large number of studies have identified asphaltenes as a key factor in the stabilization of crude oil emulsions, justifying the research interest in their chemical structure, aggregation, and solubility properties. Most work on dividing asphaltenes into subfractions is based on their precipitation in solvents with increasing aliphatic content.[9] Recently, we suggested sub-fractionation of asphaltenes fraction based on the adsorption affinity of individual species toward the water-oil interface.[10,11] Thus, Qiao et al.[11] introduced the term Extended-SARA (E-SARA) fractionation, which is considered more relevant to the emulsion and interfacial problems related to petroleum systems (see Figure 12.1).

As mentioned earlier, crude oils of different types and sources can vary significantly in their composition and properties. Specific gravity (SG) is used to classify oil as light, medium, or heavy crude using American Petroleum Institute (API)

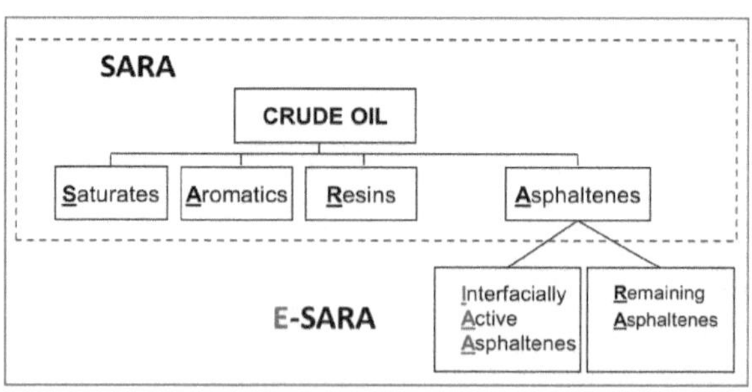

FIGURE 12.1 Presentation of main SARA fractions and Extended-SARA. Adapted from Qiao et al.,[11] Copyright (2017), with permission from American Chemical Society.

gravity scale. In this scale, crude oil gravity is measured in degrees and calculated from its SG using the following formula:

$$°API = 141.5 / (\text{Crude Oil } SG \left(\text{at } 12.6°\text{C}\right) - 131.5.$$

The specific gravity of water is 1 and will correspond to 10° API. Thus, crude oil with gravity above 10° API will be lighter than water, and crude oil with gravity below 10° API will be heavier than water. Based on API gravity the crude oil is light (greater than 31.1° API), medium (22.3°–31.1° API), heavy (10.0°–22.3° API), and extra heavy (less than 10.0° API).[2]

In crude oil, various indigenous interfacially active species such as carboxylic and naphthenic acids adsorb at the water-oil (W/O) interface and can lower significantly W/O interfacial tension,[12] thus stabilizing the emulsions. Additionally, fine inorganic particles, such as clays, silicates, oxides, and others, are present in crude oil. Their surface properties are modified by resins and asphaltenes in such a way that they adsorb at interfaces and introduce Pickering type of stabilization.[13] Sodium naphthenates are also reported to form a multilayer structure in crude oil emulsions and liquid crystalline phase that stabilizes crude oil emulsions[14,15]

12.2.2 Asphaltenes

It is generally believed that petroleum W/O emulsions are stabilized by asphaltenes,[5,16] a solubility class containing a complex mixture of chemical compounds of various compositions, functional groups, molecular weight, and other properties. Asphaltenes are not a specific family of chemicals with common functionality and varying molecular weight. Their only unifying property is insolubility in a specified *n*-alkane.

Carbon to hydrogen ratio for asphaltenes is about 1.2-1.3 and about 40% of carbon is in aromatic rings. About 90% of hydrogen is linked to saturated carbon atoms. Sulfur is mostly present in thiophene and sulfide groups. Almost all nitrogen is present in pyrrole and pyridine aromatic rings (see Table 12.1 for the average asphaltene composition).

Asphaltenes exhibit high propensity to aggregate forming dimers, trimers, and so on, likely via stacking of aromatic and polyaromatic rings. Asphaltenes aggregation eventually leads to precipitation in poor solvents. This aggregation makes

TABLE 12.1
Typical Asphaltene Composition

	Carbon	Hydrogen	Sulfur	Nitrogen	Oxygen
wt% from elemental analysis	8.2	8.5	5.5	1.5	2,5
Moles in 100 g sample	6.8	8.4	0.2	0.1	0.2
Atoms in 750 Da molecule	51	64	1.3	0.8	1.3

estimations of their average molecular mass using osmotic pressure or other colligative methods prone to overestimation of the measured property. Average molecular mass of asphaltenes is now believed to be about 700–800 Da, meaning that most single molecules may contain only one functional group.

Many hypothetical structures were drawn to match the average amounts of nitrogen, oxygen, sulfur, and aromaticity measured for asphaltenes. These average structures do not represent the properties of asphaltenes well. It is also worth noting that all suggested asphaltene structures are not like those of conventional surfactants, which are all composed of a hydrophilic 'head' and a lipophilic 'tail'. Because of their hydrophilic head, surfactants are soluble in water. Surfactants associate into aggregates called micelles above critical micelle concentration (CMC).

Asphaltenes aggregation and their ability to stabilize emulsions led to a mistaken notion of similarity between asphaltenes and surfactants and even to attempts at estimation of their CMC. Because asphaltene aggregates usually contain only a few monomeric molecules, the transition from true solution to one containing both individual molecules and their aggregates takes place over a wide concentration range, without very sharp transition characteristic for surfactants CMC, where micelles contain about a hundred or so of monomers.[17]

Asphaltenes may contain emulsion stabilizers as well as compounds that are not involved in emulsion formation at all. It would be thus interesting and beneficial to identify the chemicals and mechanisms responsible for petroleum emulsion stability. Conventional surfactants can be both emulsion stabilizers and emulsion breakers, based on their hydrophilic-lipophilic balance (HLB). Surfactants have at least some affinities to water. Asphaltenes do not have any affinity to water.

Substantial amounts of work have been done on diluted Athabasca bitumen, a major source of petroleum recovered from oil sands in Canada, which is classified as extra heavy oil crude according to the API gravity scale. The bitumen density is very close to that of water, and it is too viscous to remove emulsified water without prior dilution with a light organic solvent, either aliphatic like natural gas condensate or naphtha, which is a mixture of both aliphatic and aromatic hydrocarbons. There is a 'critical dilution' of bitumen in the diluted bitumen below which, i.e., at high bitumen concentration, the W/O interface is flexible and emulsions are very stable. Above critical dilution (at low bitumen concentration) the interface is rigid, and emulsions are easier to break.[18]

12.2.2.1 How Big Is the Asphaltenes Fraction Involved in W/O Petroleum Emulsions Stabilization?

Xu et al.[19] performed clever experiments aimed at finding out how much material is responsible for W/O emulsion stability. Without going into the details described in the original paper, the idea behind the study was that if some water is emulsified in the studied oil (50 wt% Athabasca bitumen dissolved in toluene) and then centrifuged at high speed, forming a cake at the bottom of the centrifuge cell, the supernatant would be depleted in the emulsion stabilizer. Thus, if one adds more water to the supernatant and agitates the sample again, the secondary emulsion formed should be less stable. And if enough water is added to the oil in the first stage of the procedure, then stable secondary emulsions cannot be formed. Figure 12.2 shows the results

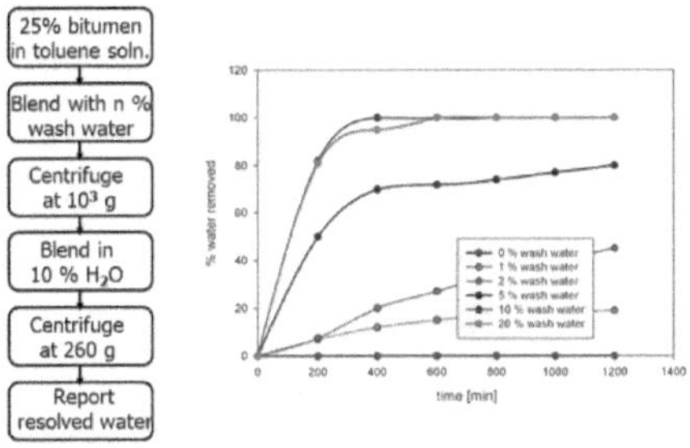

FIGURE 12.2 Washing experiment. Adapted from Xu et al.,[19] Copyright (1999), with permission from Taylor & Francis.

of the washing experiment for 25 wt% bitumen in toluene solution. When no wash was performed (0% wash water), the primary emulsion was stable, and no water was resolved even after 20 h centrifugation at 260 g. Even when the primary emulsion was made with addition of only 1% water, some water could be resolved from the secondary emulsion after long centrifugation at 260 g. Above 10% wash water, all added water was removed from the secondary emulsion after 10 h centrifuging. That indicates that practically all stabilizing materials were removed from the sample.

The cake formed after high-speed centrifugation can be freeze-fracture photographed to determine the primary emulsion droplet sizes which, with knowing how much water was present, allows calculation of the total W/O interface area in the sample. Making conservative assumptions on the surface layer thickness and density, one can estimate the amount of the material involved in emulsion stabilization. Xu et al.[19] found that there is less than 2 wt% of the total oil which contributes to the emulsion stability. With the asphaltene forming about 17 wt% of the bitumen it is clear that only a small fraction of asphaltenes may be responsible for emulsion stability. Or, in other words, the often-repeated general statement that 'asphaltenes are emulsion stabilizers' is simply not true. Based on FTIR spectra, the emulsion-stabilizing fraction is mostly composed of polar compounds such as carbonic acids and other oxygen-containing compounds.

12.2.2.2 What Is the Emulsion-Stabilizing Material?

Although many analytical techniques can be used to characterize the interfacial material, the problem is that only a very small amount of the interfacial material can be isolated, and even then, it is always contaminated by the compounds from the neighboring phases. To overcome that problem, Wu[20] emulsified heavy water (D_2O) in diluted bitumen. This emulsion was then carefully placed above an ordinary water layer in a centrifuge cell and then centrifuged at high speed (see Figure 12.3). Since D_2O has density greater than H_2O, the heavy water emulsion droplets break through W/O interface, forming a cake at the bottom. The heavy water emulsion droplets washed themselves when travelling down the cell but carried the interfacial material on their surfaces to the cake at the bottom. After drying, the cake yields about a few milligrams of the interfacial material, which can now be analyzed.

Wu[20] used solubility, H/C ratio, time-of-flight secondary ion mass spectroscopy (ToF-SIMS), and photoacoustic FT-IR (PA-FTIR) spectroscopy methods to study

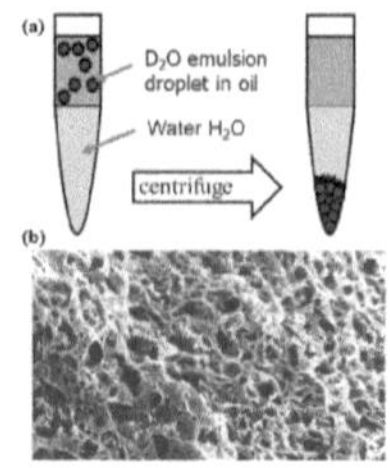

FIGURE 12.3 (a) Collection of interfacial material using D_2O. (b) Freeze-fracture photo of the cake. Reprinted from Wu,[20] Copyright (2003), with permission from American Chemical Society.

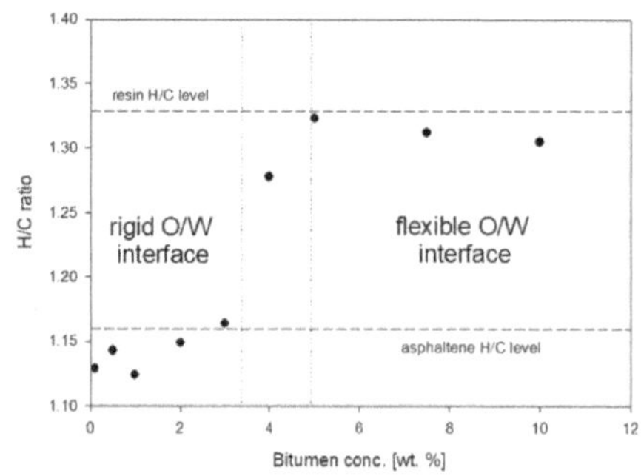

FIGURE 12.4 H/C ratio in the interfacial material separated using heavy water as a function of bitumen concentration in the initial emulsion. Adapted from Wu,[20] Copyright (2003), with permission from American Chemical Society.

the collected interfacial material. As seen in Figure 12.4, at low bitumen content (high dilution) of the oil, the H/C ratio of the interfacial material is about 1.12-1.15, which indicates that the interfacial material is mostly composed of asphaltenes. In that regime, the W/O interface is rigid and crumbles at droplet deflation, and emulsions can be broken. Above about 4 wt% the H/C increases to about 1.34, i.e., to values characteristic for resins. Here, the W/O interface is flexible, and emulsions are very stable. He concluded that below critical dilution, the interfacial material was a mixture of asphaltenes and heavy sodium naphthenates containing more than 20 carbon atoms. Above the critical dilution, the interfacial film was composed mostly of asphaltene alone. The techniques used by Wu were not suitable to determine the elemental composition of the compounds in the analyzed samples.

Consider that ethane C_2H_6 and formaldehyde CH_2O have the same nominal mass of 30 Da, but their exact molecular masses are 30.046950 Da and 30.010565 Da, respectively. Therefore, if one can measure molecular mass with accuracy better than 4 or 5 decimal points, it is possible to find the elemental composition of the molecule. That could be done using high resolution Fourier transform ion cyclotron resonance mass spectrometry (FT-ICR MS). This technique determines mass-to-charge ratio of ions (*m*/*z*) measuring cyclotron frequency of the ions in a fixed high-intensity magnetic field. Electro-spray ionization is used to inject the sample into the instrument to prevent fragmentation of the injected molecules. A sample diluted with

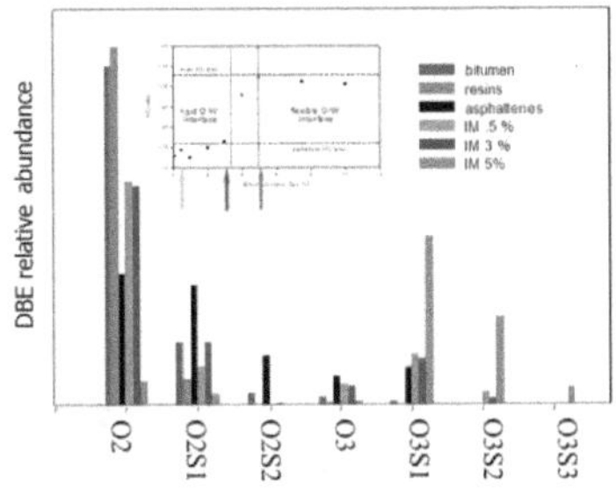

FIGURE 12.5 Relative abundance of interfacial material molecules containing oxygen atoms for Athabasca bitumen, asphaltenes and resins separated from the parent bitumen, and for interfacial material below and above critical dilution. Adapted from Stanford et al.,[21] Copyright (2007), with permission from American Chemical Society.

a volatile solvent is injected through a nozzle with a high electric voltage applied (negative or positive). Each tiny droplets formed is thus charged, and as the solvent evaporates in the high vacuum inside the instrument, the surface charge density on the droplets increases until it causes the droplet to burst into several smaller ones. This process repeats itself several times, resulting in producing single molecules with a single charge. Thus, the measured m/z with $z = 1$ gives molecular mass m with high accuracy, sufficient for estimating elemental composition of the molecule.

The results of FT-ICR MS studies[21] of the material from water in diluted Athabasca bitumen emulsions, collected using Wu's method,[20] for samples below and above critical concentration are shown in Figure 12.5.[21] There, relative abundances of oxygen-containing molecules are shown for materials collected from Athabasca bitumen, asphaltenes, and resins separated from the parent bitumen, and for interfacial material collected from D_2O in diluted bitumen emulsions. The latter samples contained 0.5 wt%, 3.0 wt%, and 5.0 wt% of bitumen dissolved in heptol (heptane/toluene 1:1 mixture by volume). The 0.3% and 3.0% samples are above critical dilution and the 5 wt% sample is below critical dilution.

Each 1 mg sample of interfacial materials for FT-ICR MS analysis was dissolved in 1 mL of 50:50 chloroform to MeOH mixture by volume because chloroform (unlike toluene) is sufficiently polar to dissolve the emulsion interfacial material completely. The sensitivity of FT-ICR MS method varies between molecule classes such that one can only compare relative abundances within a class and not between the classes. If we look at the class of molecules containing two oxygen atoms (O_2) in Figure 12.5 (see Stanford et al. [21] for a color version and additional details), it is clearly visible that the relative abundances of molecules in the interfacial materials for 0.5% and 3.0% emulsions and 5% emulsion are not similar to either the parent bitumen, resins, or asphaltenes. This means that the composition of the emulsion interfacial material is markedly different from either asphaltenes, resins, and/or the parent bitumen. It is also visible that both samples below critical bitumen concentration are very similar to each other but different from the one above critical concentration. The same holds true for all oxygen classes shown in Figure 12.5. Again, the conclusion must be drawn that for only a small sub-fraction the whole asphaltene is emulsion stabilizer.

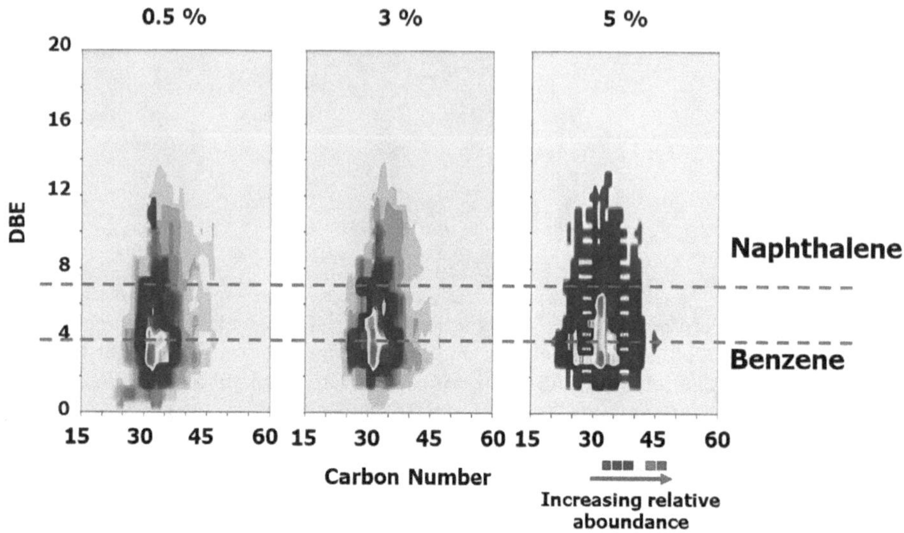

FIGURE 12.6 Double-bond equivalent (DBE) against the number of carbon atoms. Adapted from Stanford et al.,[21] Copyright (2007), with permission from American Chemical Society.

Figure 12.6 (see Stanford et al. [21] for a color version and additional details) shows double bond equivalent (DBE) against the number of carbon atoms in O_2 class molecules from W/O emulsions below and above critical dilutions (see insert to Figure 12.5). In the figure, the values of DBE for benzene (4) and naphthalene (7) are shown as dotted lines. It is important to note that the DBE for some of the interfacial materials is below DBE of 4 and most is below 7, meaning most of this material is not polyaromatic, and some is not even aromatic, thus does not conform to the general notion that asphaltenes are composed of large polyaromatic molecules. Therefore, we must conclude that the interfacial material contains some material that is not asphaltene.

12.3 PETROLEUM EMULSIONS: SIGNIFICANCE AND MECHANISMS OF STABILIZATION

12.3.1 Water-in-Crude Oil Emulsions

The presence of asphaltenes, indigenous surface-active species, and fine particles under agitation during extraction, transport, pumping, or mixing can form very stable W/CO emulsions. This type of emulsion must be resolved during primary separation processes, which removes a large fraction of the emulsified water. A small amount of water (0.1-5%) remains dispersed in the crude oil in the form of micrometer size droplets and is very difficult to remove. This dispersed water phase contains salts contributing to corrosion of the pipelines and other industrial infrastructure. It can also severely affect the quality of the final product and influence transportation and downstream processing operations.

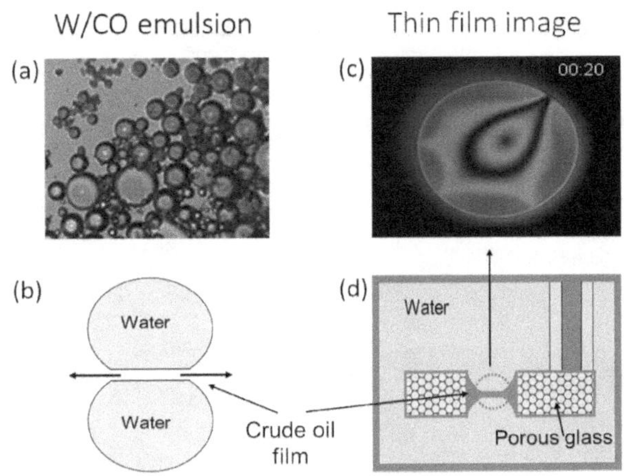

FIGURE 12.7 Schematic of thin emulsion film as a model of emulsion studies. (a) A typical microscopic image of W/O emulsions. (b) Thin oil film formed between two approaching water droplets. (c) Microscopic image of the thin film with a dimple in reflected light. (d) Porous glass holder with hole, where the thin oil film is formed.

While 'bottle' tests are widely used to evaluate emulsion stability and the effect of different system parameters, model microscopic techniques such as micropipette[18] and thin liquid films (TLF)[15,22,23] have proven to be very helpful in revealing the mechanisms of emulsion stabilization. Here, we will discuss selected results from TLF studies, which provide new insights into the stability of W/CO emulsions. A thin oil film sandwiched by two water phases is conveniently formed and studied using Scheludko–Exerowa cell,[22] shown in Figure 12.7. This configuration mimics two approaching water droplets separated by an oil film in W/CO emulsion. The stability of these emulsions can be evaluated by measuring thin film properties such as thickness using optical interferometry, studying film drainage kinetics by following film thickness in time, the lifetimes of films before rupture, and equilibrium thickness and corresponding surface forces in the case of stable films. For more details on TLF technique, refer to Chapter 6.

To facilitate primary water separation and transportation of more viscous types of crude oil, such as heavy oil and bitumen, these are typically diluted with light hydrocarbon to reduce the viscosity and increase the density difference between water droplets and the oil phase. Processing requirement prompted extensive research on the dilution ratio and solvent composition (degree of aromaticity) and their effect on emulsion stability. It should be noted that the solvent type and dilution ratio directly affect the solubility of the asphaltenes and their aggregation behavior.[9]

12.3.1.1 Critical Dilution and Effect of Solvent-to-Bitumen Ratio (*S*/*B*)

Systematic study of water-in-diluted bitumen thin films[23] revealed that film drainage and stability changed dramatically at the critical *S*/*B*, as shown in Figure 12.8.

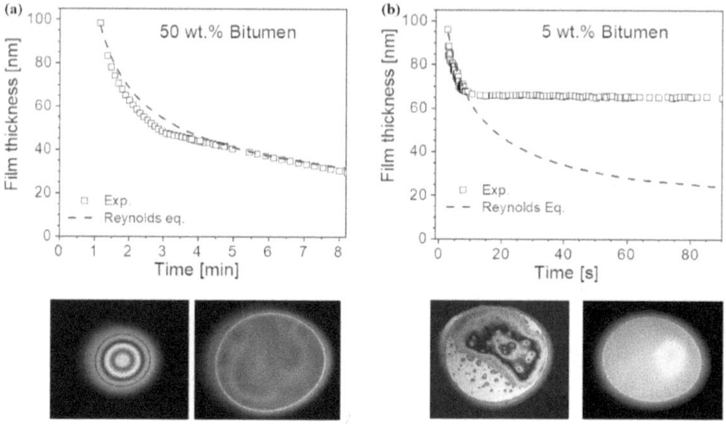

FIGURE 12.8 Film thinning and typical images. (a) 50 wt% bitumen in heptol (80:20) (below critical *S/B*). (b) 5 wt% bitumen in heptol (80:20) (above critical *S/B*). Adapted from Tchoukov et al.,[23] Copyright (2010), with permission from Elsevier.

Below the critical *S/B* (less diluted solutions), the films were homogeneous, reached lower equilibrium thickness, and were less stable overall. Above the critical *S/B*, the films became inhomogeneous with 'rigid' interfaces, drained more slowly, and had higher equilibrium thickness compared to those below the critical *S/B*. The illustrative results shown in Figure 12.8 are for bitumen dissolved in a blend of heptane and toluene at a volume ratio (80:20), referred to as heptol (80:20). The dashed lines in the figures are estimated time dependences of the film thickness calculated using the Reynolds model (Equation (12.1))

$$V_{\mathrm{Re}} = -\frac{\mathrm{d}h}{\mathrm{d}t} = \frac{2h^3 \Delta P}{3\eta R^2}, \tag{12.1}$$

where $\Delta P = P_\gamma - \Pi(h)$, h is the film thickness, η is the bulk viscosity of the film liquid, R is the film radius, P_γ is the capillary pressure due to the curvature of the film meniscus, and $\Pi(h)$ is the disjoining pressure. Disjoining pressure is estimated to be negligible for these relatively thick films. The capillary pressure is estimated from $P_\gamma = 2\gamma / r$ (γ is the water-oil interfacial tension and r is the radius of the hole in the glass plate). The agreement between measured and calculated thinning curves is reasonably good for 50 wt% bitumen (R = 70 μm, r = 400 μm, η = 9.3 mPa·s, γ = 14 mN/m, and $\Pi(h)$ = 0). In the case of 5 wt% bitumen (R = 100 μm, r = 400 μm, η = 0.56 mPa·s, γ = 16 mN/m, and $\Pi(h)$ = 0), a significant divergence between theoretical and measured curves was observed.

At first glance, these results are counterintuitive: films with a higher concentration of the stabilizing agent (bitumen) were less stable and thinner at equilibrium. This observation is contrary to the behavior of films stabilized with amphiphilic molecules but agrees with the observed transition in properties of water-heavy oil interfaces at critical *S/B* ratio from fluid to rigid interfaces.[18,24] It is worth noting that the critical *S/B* depended on the composition of the diluent and coincided with the critical dilution where initial onset of asphaltene precipitation as a separate phase is observed.

12.3.1.2 Bitumen Fraction Responsible for Emulsion Stability – TLF Studies

Most of the studies on the role of asphaltenes in petroleum emulsion stabilization are typically conducted using whole crude oil, which contains all SARA fractions. The presence of all fractions conceals the individual roles of each component and hinders the understanding of the mechanisms by which asphaltenes affect emulsion stability. To highlight the effects of the pure asphaltene fraction, we separated the whole bitumen into asphaltenes and maltenes and studied toluene solutions of these fractions separately using the TLF technique.[25] Film lifetimes, drainage kinetics, and equilibrium film thicknesses were measured and analyzed. Figure 12.9 shows the lifetimes of films with a diameter of approximately 200 μm. The lifetime is measured from the moment of film formation to its rupture. Each film was monitored for 25 min, and if it did not rupture within this period, it was considered stable against coalescence. Asphaltene films at a relatively low concentration of 0.5 g/L demonstrated significant stability with average lifetimes of about 7 min, while films at concentrations above 1 g/L were stable (lifetime >25 min). To achieve similar film lifetimes, concentrations of bitumen and maltenes need to be more than an order of magnitude higher.

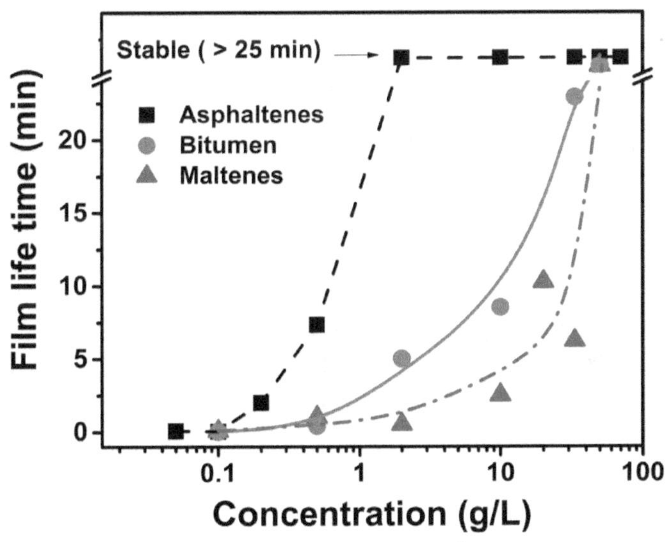

FIGURE 12.9 Film lifetime as a function of concentration for films obtained from asphaltenes, maltenes, and bitumen in toluene solutions. Adapted from Tchoukov et al.,[25] Copyright (2014), with permission from American Chemical Society.

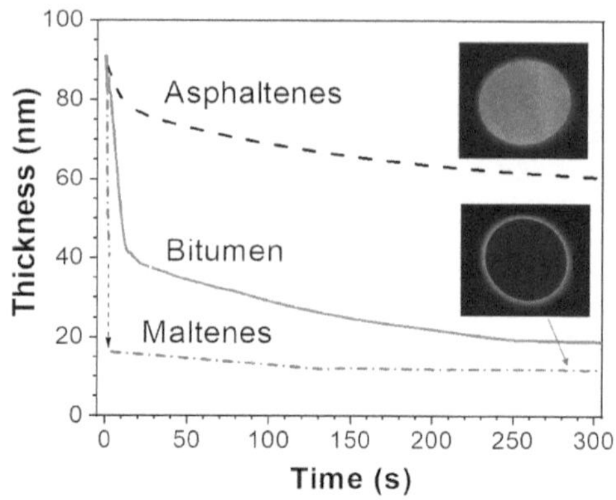

FIGURE 12.10 Drainage kinetics emulsion films from 10 g/L asphaltenes, 59 g/L bitumen, and 50 g/L maltenes in toluene solutions. Adapted from Tchoukov et al.,[25] Copyright (2014), with permission from American Chemical Society.

Emulsions settling tests of Yan et al.[26] and McLean and Kilpatrick[27] also found that asphaltenes are the main stabilizer of W/CO emulsions and the presence of resins and aromatics decreases emulsion stability. Micropipette experiments conducted by Gao et al.[28] similarly concluded that the asphaltene fraction is mainly responsible for the decreased coalescence probability of water droplets in model W/CO emulsions.

TLF studies revealed that asphaltene films drain significantly slower than the films obtained from maltenes, as shown in Figure 12.10. The equilibrium film thicknesses of asphaltene were in the range 40-90 nm, which is significantly higher than those of maltenes films, which were about 11 nm. Bitumen films had a thickness that fell in between these values.[25] The entirely different drainage kinetics and final equilibrium film thicknesses are intriguing findings as they suggest different mechanisms of stabilization for W/CO emulsions by asphaltenes and maltenes, a topic that we will explore further in the discussion on emulsion stabilization mechanisms later in this chapter. These results demonstrate the key role of asphaltenes in stabilizing W/CO emulsions. However, they do not answer the question of asphaltenes as a whole or only a subfraction is responsible for emulsion stabilization.

12.3.1.3 Interfacially Active Subfraction of Asphaltenes

Based on Xu et al.[19] washing experiments discussed above, it was concluded that less than 2% of the crude heavy oil is responsible for emulsion stabilization, which is significantly less than the asphaltene fraction in heavy oil. Yang et al.[10] proposed a new procedure to isolate the most interfacially active material by using only the asphaltene subfraction and not the whole bitumen. In brief, the procedure is as follows: 10% v/v deionized water is emulsified in 10 g/L asphaltene-in-toluene solution using a homogenizer and shaken for 24 h to equilibrate. The stable emulsion is then centrifuged at 20,000 *g* to separate water droplets, at which stage no free water is observed. The supernatant is removed, and the remaining cake is diluted with fresh toluene to wash away the asphaltenes not bound to the interface. The washing steps are repeated until the supernatant appears colorless. The washed water droplets are dried in a vacuum oven at 60°C. The interfacially active asphaltenes (IAA) extracted in this way are compared to the whole asphaltenes (WA) and the remaining fraction (RA). The isolated IAA fraction represents less than 2% of the asphaltene fraction in agreement with the Xu[19] conclusion discussed earlier. Removing this fraction from the asphaltenes dramatically decreased their capacity to stabilize W/CO emulsions, as confirmed with TLF experiments and bottle tests. It was also determined that important features of asphaltenes, such as ageing and the formation of 'rigid' interfaces, are entirely due to the IAA fraction and not observed with the RA fraction.

12.3.1.4 Mechanism of Stabilization of W/CO Emulsions by Asphaltenes, and More Precisely, by Their Most Interfacially Active Subfraction

Since the asphaltenes are not similar to surfactants, they cannot stabilize the emulsions the same way.[29] The average molecular weight of asphaltenes, which is below 1000 Da, makes them significantly different from large molecules like polymers, so related stabilization mechanisms are not applicable. To consistently explain the results with thin emulsions films,[23,25] we proposed a possible stabilization mechanism based on the hierarchical aggregation of IAA fraction and formation of extended 3D network at the W/CO interphase. Such a structure can modify the rheological properties of the film liquid, imparting a gel-like behavior. In Ref. 25, we demonstrated that the Bingham plastic fluid model with a small yield stress (~10^{-2} Pa) would be sufficient to explain the results for asphaltene film thickness and drainage kinetics discussed above and shown in Figure 12.10. Aging effects and the increasing film thickness common for asphaltene-in-toluene or bitumen-in-aliphatic solvent solutions provide experimental evidence of continuous growth of this 3D network.

12.3.2 Crude Oil-in-Water and Multiple Emulsions (Rag Layer)

Crude oil-in-water emulsions have not drawn the same research attention as W/CO emulsions. However, they can also unintendedly occur in the course of crude oil extraction and processing or are deliberately formed to facilitate processing.[30] As noted by Langevin et al.,[31] most indigenous interfacially active species in crude oil are mostly oil soluble because of their predominantly hydrophobic structure and naturally stabilize W/CO emulsions. However, the addition of surfactants with HLB (hydrophilic-lipophilic balance) in the range of 8-16 could stabilize CO/W emulsions. Indigenous for crude oil acids (e.g., carboxylic or naphthenic) at high pH conditions deprotonate and become water soluble and can stabilize CO/W emulsions. The effect of pH is illustrated by Stasser[32] by mixing Venezuelan crude oil (13.2° API) and distilled water at a ratio 50:50 and varying the pH from 3 to 13. He observed stable W/CO emulsions at acidic conditions, and at pH above 11 stable CO/W emulsions were formed.

Enhanced oil recovery (EOR),[33] which is the tertiary recovery stage designed to extract the remaining recoverable oil from the reservoir after the primary (using the natural pressure in the reservoir or gravity) and the secondary recovery (aided by injection of water or gas) involves CO/W emulsions. Another important application of CO/W emulsion is viscosity reduction. Water has a lower viscosity than typical heavy crude oil products, so along with dilution with light hydrocarbons, emulsifying CO in water is one way to facilitate its transportation.[34,35] The reduction of pressure losses per unit length in transport of CO/W emulsions compared to straight crude oil could reach 90-99%, according to Hoshyargar et al.[35]

CO/W emulsions can form accidentally as a result of crude oil spills in aquatic basins. Natural agitation by wind and waves causes the oil slick to break apart and disperse as oil droplets in water. This process depends on crude oil properties like viscosity, specific gravity, and composition. When a mechanical recovery of the spill is pursued using oil skimmers, the formation of CO/W is undesirable, as it leads to contamination of the natural basin. However, in some cases, dispersing the oil in water in form of small droplets can be a practical strategy for oil spill response to protect shoreline or sensitive areas.[36] Hounjet et al.[33] investigated the dispersed crude oil in the subsurface water layer of oil slick and the effects of oil type, water chemistry, and the presence of fine particles in a model lab-scale setting. Consistent with previous results,[37] it was found that lower salinity (lower ionic strength) and higher pH of the water led to a higher concentration of dispersed crude oil (greater CO/W stability). Both factors - lower ionic strength and alkaline pH (which increases the surface charge of water-crude oil interface) - increase the electrostatic stabilization of CO/W emulsions according to classical DLVO theory.[38,39]

During the oil-water separation process, sometimes a highly undesirable intermediate zone of complex fluid called a rag layer forms between the two phases. The rag layer formation represents a major operational issue for oil recovery operations,[40,41] and its composition and mechanisms of formation are not fully understood. According to Czarnecki et al.[42] and Varadaraj et al.,[43] rag layers encompass W/CO emulsions, solids-in-oil dispersions, and oil-in-water-in-oil multiple emulsions in the continuous oil phase. For such complex and inhomogeneous phases, multiple stabilization mechanisms are involved. The formation of rag layers is facilitated by inhomogeneities in the system and specific composition and conditions that might occur in the proximity of the water-oil interface during operations, diverging from the average conditions in the system. Formation of rigid water-oil interfaces, gel-forming asphaltene hierarchical aggregation,[25,44] and presence of oil-wet solids[45] are highlighted as stabilization factors for the occurrence of these very stable and detrimental complex emulsions.

12.4 CONCLUSIONS

Petroleum emulsions continue to attract significant research interest due to their economic importance and the intriguing scientific questions they present. The stability of these emulsions is greatly influenced by indigenous species in crude oil, particularly a subfraction of asphaltenes. Asphaltenes do not have a typical amphiphilic structure (they are not surfactant-like) and their interfacial affinity and ability to stabilize petroleum emulsions depend on their aggregation state and propensity to form extended 3D macrostructures. This extended network can induce non-Newtonian behavior of the crude oil within the nanosized confined geometry of the thin film separating dispersed water droplets, thereby imparting effective emulsion stabilization. One significant result from the research over the last couple of decades reviewed in this chapter is the realization that only a small subfraction of asphaltenes is responsible for emulsion stabilization and that the previous research overlooked this fact. Focusing our research on this asphaltene subfraction could help better understand emulsion problems associated with crude oil recovery and processing and improve the design of more effective chemical demulsifiers. CO/W and multiple emulsions that occur in petroleum have attracted less attention, and more research in this area will be required to address important applications like in situ and enhance oil recovery, and oil spills remediation.

ACKNOWLEDGMENTS

We express our gratitude to our colleagues and friends with whom we had the pleasure of collaborating over the years on the topic of petroleum emulsions: Tadeusz Dabros (passed away 2016), Zhenghe Xu, Jacob Masliyah, Fan Yang, Peiqi Qiao, Johan Sjoblom, and many others.

REFERENCES

1. *Total energy supply (TES) by source, World, 1990-2021.* https://www.iea.org/data-and-statistics/data-tools/energy-statistics-data-browser?country=WORLD&fuel=Energy supply&indicator=TESbySource (accessed 2024-04-01).
2. Masliyah, J.H.; Czarnecki, J.; Xu. Z. *Handbook on Theory and Practice on Bitumen Recovery from Athabasca Oil Sands*, Kinngsley Knowledge Publishing, Canada, 2011; Volume 1.
3. Kokal, S.; Aramco, S. Crude-Oil Emulsions: A state-of-the-art review. *SPE Prod. &* Fac. **2005**, *20*, 5–13.
4. Schramm, L.L. *Emulsions, Foams, and Suspensions: Fundamentals and Applications*, Wiley-VCH: Saskatchewan Research Council, Saskatoon, SK S7N 2X8, 15 Innovation Blvd., Canada, 2006.
5. Tadros, T.F. *Emulsion Science and Technology*, Wiley-VCH: Wokingham, Berkshire RG40 4HE, 89 Nash Grove Lane, United Kingdom, 2009.
6. Sjöblom, J. *Emulsions and Emulsion Stability*, Taylor & Francis: New York, NY, USA, 2006; Volume 45.
7. Mullins, O.C.; Sheu, E.Y.; Hammami, A.; Marshall, A.G. *Asphaltenes, Heavy Oils, and Petroleomics*, Springer Science & Business Media, New York, NY, USA, 2007.
8. Jewell, D.M.; Weber, J.H.; Bunger, J.W.; Plancher, H.; Latham, D.R. Ion-exchange, coordination, and adsorption chromatographic separation of heavy-end petroleum distillates. *Anal. Chem.* **1972**, *44*, 1391–1395.
9. Spiecker, P.M.; Gawrys, K.L.; Kilpatrick, P.K. Aggregation and solubility behavior of asphaltenes and their subfractions. *J. Colloid Interface Sci.* **2003**, *267*, 178–193.
10. Yang, F.; Tchoukov, P.; Pensini, E.; Dabros, T.; Czarnecki, J.; Masliyah, J.; Xu, Z. Asphaltene subfractions responsible for stabilizing water-in-crude oil emulsions. Part 1: Interfacial behaviors. *Energy and Fuels* **2014**, *28*, 6897–6904
11. Qiao, P.; Harbottle, D.; Tchoukov, P.; Masliyah, J.; Sjoblom, J.; Liu, Q.; Xu, Z. Fractionation of asphaltenes in understanding their role in petroleum emulsion stability and fouling. *Energy and Fuels* **2017**, *31*, 3330–3337.

12. Kilpatrick, P.K. Water-in-crude oil emulsion stabilization: Review and unanswered questions. *Energy and Fuels* **2012**, *26* 7, 4017–4026.
13. Liu, J.-X.; Zhu, H.-J.; Wang, P.; Pan, J.-M. Recent studies of Pickering emulsion system in petroleum treatment: The role of particles. *Pet. Sci.* **2021**, *18*, 1551–1563.
14. Taylor, S.D.; Czarnecki, J.; Masliyah, J. Aqueous foam films stabilized by sodium naphthenates. *J. Colloid Interface Sci.* **2006**, *299*, 283–290.
15. Czarnecki, J.; Khristov, K.; Masliyah, J.; Panchev, N.; Taylor, S.D.; Tchoukov, P. Application of Scheludko–Exerowa thin liquid film technique to studies of petroleum w/o emulsions. *Colloids Surf. A* **2017**, *519*, 2–10.
16. McLean J.D.; Speicher, P.; Sullivan, A.P.; Kilpatrick, P.K. *The Role of Petroleum Asphaltenes in The Stabilization of Water-in-Oil Emulsions. Structures and Dynamics of Asphaltenes*, Mullins, O.C., Sheu, E.Y., Eds.; Plenum Press: New York, 1998.
17. Friberg, S. Micellization. In *Asphaltenes, Heavy Oils, and Petroleomics*, Mullins, O., Sheu, E., Hammami, A., Marshall, A., Eds.; Springer: New York, 2007.
18. Yeung, A.; Dabros, T.; Czarnecki, J.; Masliyah, J. On the interfacial properties of micrometre-sized water droplets in crude oil. *Proc. R. Soc. London. Ser. A Math. Phys. Eng. Sci.* **1999**, *455*, 3709–3723.
19. Xu, Y.; Dabros, T.; Hamza, H.; Shefantook, W. Destabilization of water in Bitumen emulsion by washing with water. *Pet. Sci. Technol.* **1999**, *17*, 1051–1070.
20. Wu, X. Investigating the stability mechanism of water-in-diluted Bitumen emulsions through isolation and characterization of the stabilizing materials at the interface. *Energy and Fuels* **2003**, *17*, 179–190.
21. Stanford, L.A.; Rodgers, R.P.; Marshall, A.G.; Czarnecki, J.; Wu, X.A. Compositional characterization of Bitumen/water emulsion films by negative-and positive-ion electrospray ionization and field desorption/ionization fourier transform ion cyclotron resonance mass spectrometry. *Energy & Fuels* **2007**, *21*, 963–972.
22. Sheludko, A. Thin liquid films. *Adv. Colloid Interface Sci.* **1967**, *1*, 391–464.
23. Tchoukov, P.; Czarnecki, J.; Dabros, T. Study of water-in-oil thin liquid films: Implications for the stability of petroleum emulsions. *Colloids Surf. A* **2010**, *372*, 15–21.
24. Zhang, L.Y.; Breen, P.; Xu, Z.; Masliyah, J.H. Asphaltene films at a Toluene/water interface. *Energy & Fuels* **2007**, *21*, 274–285.
25. Tchoukov, P.; Yang, F.; Xu, Z.; Dabros, T.; Czarnecki, J.; Sjöblom, J. Role of asphaltenes in stabilizing thin liquid emulsion films. *Langmuir* **2014**, *30*, 3024–3033.
26. Yan, Z.; Elliott, J.A.W.; Masliyah, J.H. Roles of various Bitumen components in the stability of water-in-diluted-Bitumen emulsions. *J. Colloid Interface Sci.* **1999**, *220*, 329–337.
27. McLean, J.D.; Kilpatrick, P.K. Effects of asphaltene aggregation in model heptane-toluene mixtures on stability of water-in-oil emulsions. *J. Colloid Interface Sci.* **1997**, *196*, 23–34.
28. Gao, S.; Moran, K.; Xu, Z.; Masliyah, J. Role of Bitumen components in stabilizing water-in-diluted oil emulsions. *Energy & Fuels* **2009**, *23*, 2606–2612.
29. Czarnecki, J.; Tchoukov, P.; Dabros, T. Possible role of asphaltenes in the stabilization of water-in-crude oil emulsions. *Energy and Fuels* **2012**, *26*, 5782–5786.
30. Yonguep, E.; Kapiamba, K.F.; Kabamba, K.J.; Chowdhury, M. Formation, stabilization and chemical demulsification of crude oil-in-water emulsions: A review. *Pet. Res.* **2022**, *7*, 459–472.
31. Langevin, D.; Poteau, S.; Hénaut, I.; Argillier, J.F. Crude oil emulsion properties and their application to heavy oil transportation. *Oil Gas Sci. Technol.* **2004**, *59*, 511–521.
32. Strassner, J.E. Effect of PH on interfacial films and stability of crude oil-water emulsions. *J. Pet. Technol.* **1968**, *20*, 303–312.

33. Mandal, A.; Samanta, A.; Bera, A.; Ojha, K. Characterization of oil– water emulsion and its use in enhanced oil recovery. *Ind. Eng. Chem. Res.* **2010**, *49*, 12756–12761.
34. Ahmed, N.S.; Nassar, A.M.; Zaki, N.N.; Gharieb, H.K. Stability and rheology of heavy crude oil-in-water emulsion stabilized by an anionic-nonionic surfactant mixture. *Pet. Sci. Technol.* **1999**, *17*, 553–576.
35. Hoshyargar, V.; Ashrafizadeh, S.N. Optimization of flow parameters of heavy crude oil-in-water emulsions through pipelines. *Ind. Eng. Chem. Res.* **2013**, *52*, 1600–1611.
36. Moldestad, M.Ø.; Leirvik, F.; Johansen, Ø.; Daling, P.S.; Lewis, A. Environmental emulsions: A practical approach. In *Emulsions and Emulsion Stability*, Sjöblom, J., Ed.; CRC Press, 2005; pp. 375–402.
37. Hounjet, L.J.; Stoyanov, S.R.; Chao, D. Distributions of diluted Bitumen and conventional crude oil in a range of water types. *Chemosphere* **2018**, *211*, 1212–1218.
38. Verwey, E.J.W. Theory of the stability of lyophobic colloids. *J. Phys. Chem.* **1947**, *51*, 631–636.
39. Derjaguin, B.; Landau, L. The theory of stability of highly charged lyophobic sols and coalescence of highly charged particles in electrolyte solutions. *Acta Physicochim. URSS* **1941**, *14*, 58.
40. Kiran, S.K.; Acosta, E.J.; Moran, K. Study of solvent– Bitumen– water rag layers. *Energy & Fuels* **2009**, *23*, 3139–3149.
41. Hristova, E.; Stoyanov, S.R.; McFarlane, R.; Nikooyeh, K. Investigation of unresolved interface "rag layer" in Athabasca oil sand bitumen in situ recovery. *ACS omega* **2024**, *9* 29, 32201–32209.
42. Czarnecki, J.; Moran, K.; Yang, X. On the "rag layer" and diluted bitumen froth dewatering. *Can. J. Chem. Eng.* **2007**, *85*, 748–755.
43. Varadaraj, R.; Brons, C. Molecular origins of crude oil interfacial activity part 3: Characterization of the complex fluid rag layer formed at crude oil– water interfaces. *Energy & Fuels* **2007**, *21*, 1617–1621.
44. Hristova, E.; Tchoukov, P.; Stoyanov, S.R.; McFarlane, R. Asphaltene precipitation onsets in relation to the critical dilution of Athabasca bitumen in paraffinic solvents. *Energy and Fuels* **2022**, *36*, 1832–1841.
45. Kupai, M.M.; Yang, F.; Harbottle, D.; Moran, K.; Masliyah, J.; Xu, Z. Characterising rag-forming solids. *Can. J. Chem. Eng.* **2013**, *91*, 1395–1401.

13 Emulsions in Pest Control

Alejandro Lucia, Agustina Fañani, and Eduardo Guzmán

13.1 INTRODUCTION

Pesticides are pure substances or mixtures used to prevent, eliminate, or mitigate pests and their effects. Pesticides are also applied as plant regulators, defoliants, or desiccants. However, inappropriate use of pesticides leads to several problems, including soil and water contamination, development of pest resistance, and ecological and environmental concerns.[1] Several factors contribute to suboptimal pesticide application performance: (i) Many active ingredients (AIs) have poor solubility in water, and conventional pesticide formulations have inadequate foliar retention and adhesion properties. (ii) AIs in conventional formulations are not adequately shielded, making them susceptible to premature degradation, particularly in the case of environmentally sensitive pesticides. (iii) Spray system limitations, including the potential for spray drift, compound the challenges.[2–4]

In recent years, the pesticide industry has seen a noticeable shift to developing innovative, smart, and environmentally friendly water-based pesticide formulations. This trend is primarily motivated by the dual goals of improving pesticide efficacy and ensuring the safety of people involved in the handling and application of pesticides.[3,5] One of the challenges in pesticide formulation arises from the fact that many AIs used in pest control are inherently water-insoluble compounds. As a result, scientists and researchers have actively pursued various strategies to improve the water dispersibility of these essential components. These strategies have been crucial in optimizing the performance of pesticides in various agricultural, horticultural, and pest management contexts.

To address the challenge of the vehiculation of water-insoluble actives, a common and effective strategy is to incorporate these actives into carrier materials. In this approach, the active ingredients are embedded in specially designed carriers that act as delivery vehicles for the active components. This technique improves the solubility and dispersibility of the active ingredients in water-based pesticide formulations. The use of carriers has proven to be a valuable solution, significantly improving the performance and efficacy of pesticide products while ensuring precise and controlled delivery. This method has gained popularity for its ability to overcome the problem of water insolubility. In addition, nanocarrier materials protect AIs from premature degradation and enable their controlled release.[6] This protective function ensures that the AIs remain stable and effective over time, while the controlled release mechanism allows for the gradual and precisely regulated distribution of

 DOI: 10.1201/9781032636184-16

these active components, ultimately improving the performance of pesticides and making them more reliable and efficient.[7,8] The quest for smart and safe water-based pesticide formulations has led to the development of different product forms. These formulations are tailored to meet specific pest control needs and challenges. Some of the notable formulations include capsules,[9–11] spheres,[12] micelles,[13,14] suspensions,[15] emulsions,[16,17] and mesoporous silica.[11,18–20]

Among the many possibilities offered by colloidal systems, oil-in-water (O-W) emulsions have emerged as a promising solution to address key challenges associated with pesticide management, storage, and application.[21,22] This approach exploits the unique ability of emulsions to encapsulate and effectively deliver water-insoluble molecules. The large surface area for interaction and the simplicity of some of the methods used to prepare some emulsions have stimulated extensive research into the use of emulsions for integrated pest management strategies.[23–25] In addition, this often allows for in situ preparation of formulations, reducing costs and minimizing pollutant emissions associated with transporting formulations containing significant amounts of solvent, usually water, from storage facilities to application sites. This approach helps to reduce the overall carbon footprint. In many cases, emulsion-like systems act as vehicles by dissolving the active ingredients in the dispersed phase, typically the oil phase.

Nowadays, emulsions are fundamental to the effectiveness and sustainability of modern pest control strategies. This chapter provides a comprehensive examination of the development of emulsion technology in pest control, tracing its transformative impact on improving the efficiency and precision of pesticide applications. In addition, contemporary applications in agriculture, forestry, and public health highlight the versatility of emulsion-based formulations in addressing diverse pest-related challenges.

13.2 AN OVERVIEW OF EMULSIONS IN PEST CONTROL

Since the 1940s, several pesticides have been subject to formulation improvements to enhance delivery, cost-effectiveness, reduced toxicity, and boosted efficacy. Nonetheless, conventional pesticide formulations entail inherent downsides related to nontargeted organisms and environmental impact. Furthermore, traditional formulations display thermal instability, require specific application methods, and possess other limitations.[20-25] In contrast, pesticide-based emulsion formulations offer significant advantages over conventional pesticides, as shown in Figure 13.1. This alternative formulation has the potential to address several limitations associated with conventional pesticide application.[26]

Pesticide-containing oil-in-water emulsions are formulations characterized by the dispersion of insoluble liquid pesticides or organic solvents, including the pesticide, as small liquid droplets in an aqueous medium by the addition of appropriate molecules to ensure the stabilization of the dispersion and the application of energy.[27] This type of formulation contributes to a significant reduction in the required concentration of molecules generally identified as toxic, such as surfactants and/or

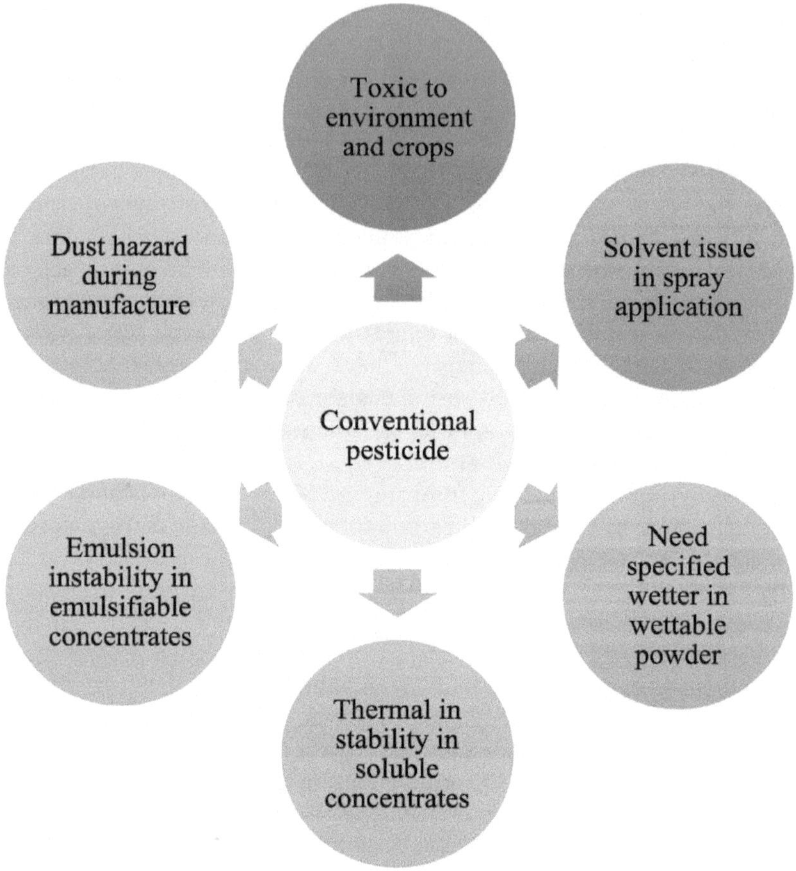

FIGURE 13.1 Constraints associated with the application of traditional pesticide formulations. Reprinted from Rana et al.,[26] Copyright (2023), with permission from Elsevier.

organic solvents (e.g., dimethylbenzene, toluene, or xylene).[28,29] In addition, the use of oil-in-water emulsions not only improves the delivery and penetration efficiency of bioactive ingredients in pesticide delivery systems but also offers a host of other benefits that have driven their increased adoption in pest management. Beyond their role as effective carriers, emulsions have proven to be versatile platforms for the formulation of insecticides, fungicides, and herbicides.[30,31]

One notable aspect contributing to the widespread acceptance of emulsions is their exceptional ability to provide precise control over the distribution of active ingredients.[32,33] This precision is particularly important in the agricultural context, where achieving uniform crop coverage is paramount to the efficacy of pest control measures. By ensuring uniform distribution of pesticides, emulsions not only improve efficacy but also reduce the risks associated with an uneven application, be it over- or underapplication.[34,35] Moreover, emulsions offer a versatile platform adaptable to a range of pest control products. In agriculture, different pests, from insects to fungi and weeds, often require tailored treatments. Emulsions, with their customizable nature, allow the formulation of pesticide solutions to suit the unique characteristics of each pest control agent, thereby extending their applicability across a range of agricultural challenges.[36,37] Finally, the controlled release capability of emulsions is a valuable attribute in pest management strategies. Emulsions can be precisely formulated to release active ingredients gradually, providing extended protection over time. This controlled release not only ensures sustained efficacy but also reduces the frequency of reapplication, offering cost-effective and environmentally friendly solutions. This is particularly beneficial in scenarios involving pests with extended life cycles or in regions characterized by longer growing seasons, where a single application can provide prolonged and consistent protection.[38] The multiple benefits of emulsions thus position them as key players in the evolution of pest control strategies, demonstrating their potential to address multiple challenges in agriculture and environmental management.[16]

It is worth emphasizing that the integration of emulsions into pest control strategies is not a recent development, with a rich history of traditional pesticide formulations rooted in emulsion-based technologies. A prominent example in this area is the category known as suspoemulsions, a formulation that allows formulation chemists to effectively blend active ingredients with significantly different solubility profiles. This family of products combines two different formulations into a single solution by incorporating two separate phases: an oil-in-water emulsion and a suspension concentrate phase.[39,40]

Another widely used emulsion-based formulation in pest control is the emulsifiable concentrate. In this type of formulations, the active ingredient is dissolved in an oil-based solvent and then emulsified with water. The resulting stable emulsion offers the convenience of easy dilution with water for practical application. This category of emulsion-based pesticides exploits the inherent property that most insecticidal active ingredients are oil-soluble and water-insoluble.[40,41] However, the use of emulsifiable concentrates presents several drawbacks, including the potential stability problems within the emulsion as a result of the dilution may lead to. On the other hand, there may be cases of phytotoxicity observed in vegetable crops at application. In addition, there is also a risk of increased dermal toxicity associated with the active ingredient due to the enhanced dermal penetration. Finally, there is also a potential fire hazard and the solvents used may have adverse effects on plastics and rubbers within spray applicators.[42]

It should be noted that, despite the widespread use of suspoemulsions and emulsifiable concentrates in pest control, their use is beginning to be banned by international regulations due to their marked dependence on petroleum-based solvents, resulting in high product toxicity, flammability, and odor.[43] It was, therefore, necessary to find effective alternatives that were easy to use, dust-free, less toxic, and nonflammable, while reducing the carbon footprint associated with their manufacture, transport, and use. In this context, the use of most traditional types of emulsion-like systems, e.g., emulsions, nanoemulsions, microemulsions, or Pickering emulsions, has gained importance in recent years.[16]

13.3 EMULSIFIABLE CONCENTRATE FORMULATIONS

Emulsifiable concentrates are formulations containing a combination of pesticides, emulsifiers, and adjuvants dissolved in a volatile oil. The emulsifiable concentrate formulation is a solvent-rich system (with approximately 10-90 parts of solvent in the formulation) used to deliver oil-soluble pesticides with low melting points or liquid, e.g., organo-phosphorous compounds and pyrethroids.[44] When introduced into the water in the spray tank, the emulsifiable concentrates form a stable, diluted emulsion.[45] Most emulsifiable concentrates typically consist of 60-65% (w/w) herbicide dissolved in a 30-35% organic solvent, together with 3-7% of a suitable emulsifier to facilitate the formation of an oil-in-water emulsion when the concentrate is mixed with water. Occasionally they may include a small amount of emulsion stabilizer and a surfactant,[46] generally of anionic and nonionic species (calcium dodecyl benzenesulfonate, nonylphenolethoxylate-6, nonylphenolethoxylate-30, or Atlas® G-5000), specifically chosen to ensure effective interaction with plant surfaces. Manufacturers often aim for the highest possible concentration, but a common measure is approximately 0.5 kg/L.[47]

Despite their simplicity, emulsifiable concentrate formulations can pose significant environmental risks to the environment. This is due to the elevated levels of harmful organic solvents within their composition that may result in contamination.[48] Therefore, the substitution of toxic organic solvents, generally with aromatic nature, for other compounds with better toxicity profiles is good of the challenges for pesticide formulators. In this context, Chin et al.[48] evaluated the use of biodiesel, a nontoxic and biodegradable alternative, as the organic phase in emulsifiable concentrate formulations to minimize the environmental risks and hazards associated with commonly used aromatic solvents. To this aim, they prepared an emulsifiable concentrate formulation loaded with 40.8% w/w chlorpyrifos by mixing biodiesel, organic cosolvent, and surfactants. This formulation was compared with a commercial product based solely on an aromatic solvent as a carrier. The results of the study showed minimal differences between the biodegradable formulation and the commercial counterpart in stability assessments. In addition, the use of biodiesel helped to reduce the unpleasant odor typically associated with chlorpyrifos due to its lower volatility. The higher flash point of biodiesel also facilitated safer storage, transport, and application of the formulation. Results from residue experiments suggested that the lipophilic nature of biodiesel could increase the amount of active ingredients on plant surfaces, potentially improving the efficacy of the pesticide delivery system against pests. On the other hand, in phytotoxicity assessments, biodiesel as a carrier did not cause any damage to plants at different application rates, indicating a potential reduction in environmental risks. Therefore, for oil-soluble pesticides, biodiesel is a promising alternative carrier for emulsifiable concentrate formulations. In contrast to commercial formulations, the use of biodiesel could lead to dosage reductions, cost-effective food production, and reduced adverse effects. In addition, biodiesel has added value, environmentally friendly properties, and reduced hazards. Further exploring the substitution of aromatic organic solvents for others with lower toxicity profiles, Zhang et al.[49] formulated an emulsifiable concentrate formulation containing lambda-cyhalothrin as the active ingredient using ethylene glycol diacetate (EGDA) as the solvent. This provides an environmentally friendly alternative to the adverse effects associated with aromatic solvents. The study evaluated the physicochemical properties, wettability, phytotoxicity, and bioassays of the EGDA formulation in comparison to a formulation using xylene as a solvent. The results showed that both formulations exhibited excellent emulsion properties and storage stability. In addition, the EGDA formulation showed a significantly higher flash point (96°C), suggesting improved safety during production, storage, and transport. Furthermore, the EGDA formulation exhibited 1.22-1.46 times higher leaf retention compared to the xylene sample. It also showed reduced surface tensions and contact angles, which could reduce drift and improve utilization. In addition, bioassays showed that the EGDA formulation exhibited lower acute toxicity to aquatic organisms, while demonstrating higher efficacy in controlling target insects compared to the xylene formulation. EGDA is therefore emerging as a promising carrier for oil-soluble agrochemicals, offering improved application performance and reduced adverse effects. Continuing with the efforts toward the replacement of organic toxic solvent, Tran et al.[50] proposed the formulation of cypermethrin-loaded emulsifiable concentrate using methyl ester as an environmentally friendly solvent. The study included an evaluation of the physicochemical properties, emulsion properties, and storage stability of the methyl ester emulsifiable concentrate formulations compared with a traditional emulsifiable concentrate using an aromatic compound as a solvent. The results highlighted the exceptional emulsifiability and storage stability of the new formulations compared to those based on aromatic solvents.

Arthur[51] evaluated the efficacy of a novel emulsifiable concentrate formulation containing the pyrethroid deltamethrin for the treatment of brown rice, which was then stored for 12 months. The treated rice was mixed with untreated brown rice at various ratios one day after treatment and every 3 months during the 12-month storage period. Bioassays were conducted by exposing the rice mixtures to 10 adult individuals of *Rhyzopertha dominica* and *Sitotroga cerealella* and assessing progeny production, feeding damage, and weight loss. The results showed that progeny production of *S. cerealella* decreased slightly with the amount of rice treated, with minimal feeding damage or weight loss observed. In contrast, progeny production of *R. dominica* decreased as the proportion of treated rice increased. It should be noted that weight loss and feeding damage showed a correlation with progeny production. These results suggest that the new deltamethrin formulation is promising for protecting brown rice, although *S. cerealella* may be comparatively less susceptible to deltamethrin than *R. dominica*.

Safar et al.[52] prepared a novel emulsifiable concentrate formulation following established guidelines from professional pesticide organizations. Once prepared, the formulation was subjected to rigorous biological testing against adult *Tetranychus urticae* in controlled laboratory conditions. The results showed promising acaricidal activity, with a lethal concentration of 4,548 ppm ensuring the elimination of 50% of the individuals. Further greenhouse trials with *T. urticae* immatures, adult males, females, and egg deposition were conducted and showed a direct relationship between concentration levels and mortality rates at all developmental stages.

Notably, 100% mortality was observed in immatures after seven days of treatment, while both adult males and females showed complete mortality after three days. In addition, the formulation completely stopped egg deposition within 14 days of treatment. These results suggest the potential use of the new formulation for effective control of infestations.

Emulsifiable concentrate formulations are cost-effective, user-friendly products that are easy to produce, dose, and transport. These formulations do not require agitation in the spray tank, which reduces the likelihood of nozzle plugging and prevents active ingredients from settling or segregating during equipment operation. As a result, they are frequently chosen for a variety of pesticides. Moreover, given their concentrated nature, the potential for overapplication increases, thereby presenting a challenge for formulating chemists to reduce or ideally eliminate hazards to machinery and individuals associated with these required toxic formulating agents.[47]

13.4 SUSPOEMULSIONS FORMULATIONS

A suspoemulsion is a water-based product containing both suspended solids and emulsion droplets (similar to an emulsifiable concentrate when diluted). This type of formulation requires suspension aids to prevent the settling of the solids or creaming of the emulsion droplets. Designing such formulations tends to be a significant formulation development challenge. Suspoemulsions have significantly higher viscosities than water. They are designed to disperse rapidly when diluted in water and require thorough agitation after dilution to maintain a uniform dispersion.[53]

Challenges frequently arise when creating suspoemulsions because a dispersant is required to suspend solid particles and an emulsifier is needed for the oil component. Variations in the chemical nature and affinities of dispersants and emulsifiers toward solids and oils can trigger issues with competitive adsorption and displacement. The relative ratios of these agents in the system and their adsorption strength are crucial factors in such occurrences. Even when utilizing the same dispersant for both solids and liquids, there is a possibility of flocculation because of the lateral movement of dispersant molecules on solid particles and oil droplets. This movement generates exposed surface areas that are vulnerable to aggregation. Heteroflocculation can happen when the oil has a stronger attraction to solid particles than the dispersant's hydrophobic portion, causing the oil droplets and solid particles to merge. Comprehending these interactions requires fundamental research that considers the multitude of forces at play. Another difficulty, known as 'phase transfer', occurs when solid particles exhibit some solubility in oil droplets, leading to particle dissolution in the oil and subsequent formation of larger, supersaturated solution-based crystals. Such a phenomenon may cause modifications to crystal habits, often resulting in needle-shaped crystals and thickening of subsequent formulations. The meticulous selection of diluents, emulsifiers, and dispersants is necessary to address these issues in agreement with the results by Webster et al.[54] They found that the maximum solubility of the solid on the droplets controls the transfer and dissolution process. Depending on this parameter, the pesticide formulation can display either an emulsion-like character or a suspoemulsion character. Furthermore, the active ingredient may crystallize within the oil phase, particularly when utilizing a supersaturated solid-in-oil solution to increase the concentration of the active ingredient in the formulation. Temperature fluctuations, prevalent in practical settings, can expedite the crystallization process. Despite the potential mitigation of the phenomenon by crystal growth inhibitors, controlling it remains a challenge.[55]

13.5 MICROEMULSIONS AS PESTICIDES

Pesticide formulations based on microemulsions offer several advantages, such as long-term thermodynamic stability, low viscosity, cost-effectiveness, and remarkable efficacy. Microemulsions, described as homogeneous blends of oil and water, provide convenient, environmentally friendly, and economically feasible solvents. They enhance the availability of poorly soluble compounds and reduce the negative environmental impact of petroleum and solvent-based pesticides through customized formulations. Moreover, these formulations showcase enhanced stability, reduced chemical usage, minimized toxicity, heightened loading capacity of active ingredients, and ease of handling, making them particularly advantageous for farmers and users alike.[56]

Pesticide microemulsions typically have particles below 30 nm and form spontaneously, without requiring high-pressure homogenizers or high-shear instruments. These microemulsions are resistant to degradation, having a high stability over time when stored.[26]

Zheng et al.[57] investigated the solubilization of organochlorine pesticides such as dichlorodiphenyltrichloroethane (DDT) or γ-hexachlorocyclohexane (γ-HCH) in oil-in-water microemulsions. These microemulsions consisted of nonionic surfactants such as Tween 80 or Triton X-100, plant oils like linseed oil or soybean oil, and a cosurfactant, particularly 1-pentanol. The study emphasized the importance of the cosurfactant-to-surfactant ratio in creating microemulsions. The results demonstrate that microemulsions improve DDT solubilization by 66.9-95.6% (Tween 80) and 51.9-80.9% (Triton X-100), respectively, in comparison to their respective surfactant solutions. Likewise, γ-HCH solubilization increased by 40.6-57.5% (Tween 80) and 43.0-65.8% (Triton X-100) in microemulsions when compared to surfactant solutions. Further investigation has shown that the solubilizing capacity of microemulsions is affected by the oil and cosurfactant content, with increased capacity evident when more oil or cosurfactant is present. These results highlight the role that microemulsion systems play in enhancing the solubilization of pesticides.

Leng et al.[58] developed a sustainable microemulsion against *Rhizoctonia solani*, utilizing carbendazim as the active antifungal ingredient, along with a mixture of surfactants as emulsifiers, and chloroform and acetic acid as solvents. Notably, this system increased the solubility of carbendazim to 30 g/L. This microemulsion showed noteworthy results by surpassing the applicable temperature range for pure carbendazim use, exhibiting a cloud point exceeding 90°C, and maintaining stability down to −18°C. Following pesticide microemulsion standards, it proved to have superior efficacy against *R. solani* when compared to a 0.2 g/L aqueous solution of pure carbendazim. The study demonstrated significant improvement in inhibiting *R.*

solani, which includes mycelium growth, sclerotia formation, and germination when compared to the carbendazim aqueous solution.

The nature of the organic phase also poses a challenge when preparing pesticide formulations in microemulsions. An alternative approach to increase safety and decrease toxicity is to utilize biodiesel as the organic phase. To investigate this possibility, Chin et al.[59] examined the potential of a microemulsion formulation loaded with lambda-cyhalothrin, utilizing biodiesel as a substitute solvent to transport and deliver the oil-soluble active ingredient. The study investigated the transition of microemulsions from water-in-oil to oil-in-water after diluting with the water phase. Furthermore, evaluations were conducted, comparing the performance and stability of the biodiesel-based formulation with commercial counterparts. Greenhouse tests indicate that biodiesel has favorable characteristics in forming microemulsions for cabbage plants without causing residue or phytotoxicity issues. The microemulsion formulation utilizing biodiesel demonstrated transparency and a broader range of temperature stability, achieving superior performance and stability compared to the commercial formulation. Importantly, the active ingredient's functionality was not compromised, making it safer than commercial products for crops and plants.

13.6 NANOEMULSIONS AS PESTICIDES

Nanoemulsions offer significant advantages in pesticide delivery regarding cost-efficiency and safety. First, O-W nanoemulsions, being water-based systems, require significantly less organic solvent during preparation compared to conventional emulsifiable concentrates.[44] In addition, the formulation of nanoemulsions requires a lower concentration of surfactant (typically between 3% and 10%) compared to microemulsions which require higher levels of surfactant (around 20% or more).[48] The small size of nanoemulsion droplets ensures uniform deposition on plant leaves or insect surfaces and enhances wetting, spreading, and permeation due to the low surface tension of the system.[60] This aspect significantly increases the bioavailability of poorly water-soluble pesticides when solubilized in small oil droplets, potentially enhancing pesticide efficacy.[61]

Wang et al.[61] prepared water-poly(oxyethylene) nonionic surfactant/methyl decanoate nanoemulsions loaded with the water-insoluble pesticide β-cypermethrin. They found that the incorporation of the active molecules does not significantly alter the phase behavior of the original nanoemulsion, at least for concentrations below 12% w/w. On the other hand, the application of the formulation by spraying does not affect its stability. On the contrary, when the same test was carried out with commercially available microemulsions containing β-cypermethrin, rapid destabilization of the sprayed droplets was observed, characterized by the precipitation of the pesticide. Therefore, the formulations implemented have the potential to reduce the required application rates compared to those of commercially available products. This makes the new formulation promising from both an environmental and economic perspective. More recently, Du et al.[62] prepared O/W nanoemulsions stabilized by mixtures of alkyl polyglycoside (APG) and polyoxyethylene 3-lauryl ether ($C_{12}E_3$) for loading β-cypermethrin. The effects of APG and $C_{12}E_3$ mixing ratios, oil weight fraction, and total surfactant concentration on the droplet size and distribution within the nanoemulsion were systematically investigated. Long-term stability assessments were performed for the nanoemulsions prepared with different surfactant mixing ratios, revealing Ostwald ripening as the primary driver of droplet size increase over time. Dynamic light scattering (DLS) and transmission electron microscopy (TEM) measurements confirmed the near-uniform droplet size distribution of the nanoemulsions (PDI <0.2), with the incorporation of β-cypermethrin showing no significant effect on the size and stability of the nanoemulsions. To assess the practical applicability, the dilution stability and spreading properties of the pesticide-loaded nanoemulsion were evaluated by DLS, contact angle, and dynamic surface tension measurements. Upon dilution, the nanoemulsion exhibited homogeneity, although there was a destabilization of the droplet size observed by DLS. However, contact angle and dynamic surface tension assessments showed excellent spreading properties of the optimized nanoemulsion.

Following the optimization of nanoemulsion formulations for pest control applications with a low environmental impact, Yan et al.[63] prepared oil-in-water nanoemulsions in biodiesel by dissolving the pesticide bifenthrin in a mixture of a nonionic surfactant and an anionic surfactant in a water system via the low-energy phase inversion composition method. The bifenthrin nanoemulsion demonstrated low interfacial tension and contact angle, promising easy absorption on plant leaf surfaces, and exhibited good shelf life, suggesting its potential as a pesticide formulation. Furthermore, the above research investigated the insecticidal effectiveness of bifenthrin nanoemulsions on cabbage maggots at various dilution levels. Mortality rates consistently rose during the exposure period, reaching over 90% after 20 h for all samples. Remarkably, at 100-fold and 500-fold dilutions, complete mortality was observed after 20 h of exposure. This finding demonstrates the strong insecticidal effectiveness of bifenthrin nanoemulsions, indicating their potential suitability for practical pest control applications.

Very recently, Wei et al.[64] optimized the emulsification process to produce a fenpropathrin nanoemulsion. The resulting nanoemulsion exhibited remarkable stability and extraordinary insecticidal efficacy against *Helicoverpa armigera*. Following established FAO guidelines, an ideal composition was developed, comprising 10% w/w fenpropathrin and 9% w/w ethoxylated castor oil surfactant, blended with a hydrocarbon solvent (8% w/w) and deionized water (making up the remaining 100% w/w). In addition, the nanoemulsions exhibited excellent wetting, adhesion, and spreading properties, as evidenced by their high adhesion and minimal contact angle on corn leaves. In comparison with a commercial emulsifiable concentrate, the fenpropathrin nanoemulsions displayed a lower lethal concentration to affect 50% of the individuals against the third instar larva of *H. armigera*, resulting in complete darkening, shriveling, and incapacity shortly after treatment. In addition, the fenpropathrin nanoemulsions led to a decreased cytotoxicity (apoptosis rate of 8.52% after 24 h) compared to the emulsifiable. The nanoemulsions also demonstrated a lower impact on cell viability (85.32% after 24 h and 68.95% after 48 h) compared to the emulsifiable concentrate (with an apoptosis rate of 63.18%). These results make a valuable contribution toward advancing the preparation and use of nanoemulsions for lipophilic pesticides. It enhances insecticidal potency while also reducing cytotoxicity and improving environmental compatibility. Equivalent results were reported by Feng et al.[65] for nanoemulsions loaded with abamectin.

13.7 NEW TRENDS IN EMULSION-BASED PESTICIDES: BIOPESTICIDES

Biopesticides are naturally occurring compounds derived from biological sources that are used to control various pests found in forests, gardens, farmland, and other areas. These compounds, classified into several types, have been developed from a variety of biological sources.[66] Biopesticides offer many advantages over their chemical counterparts and are expected to capture a significant share of the market in the foreseeable future. Biopesticides typically have lower intrinsic toxicity than conventional pesticides. They tend to affect only the target pest and strongly associated organisms, unlike broad-spectrum conventional pesticides, which can affect a wide range of organisms from birds to mammals. In addition, biopesticides often have high efficacy at low rates and degrade rapidly, reducing exposure and significantly reducing the environmental concerns associated with conventional pesticides. Integrating biopesticides into Integrated Pest Management (IPM) initiatives can significantly reduce reliance on conventional pesticides while maintaining robust crop yields.[67]

According to the Environmental Protection Agency (EPA, USA),[68] biopesticides can be classified into three major groups: (i) biochemical pesticides (plant extracts, pheromones, Kairomones); (ii) microbial pesticides (bacterium, fungus, viruses, and protozoan), and (iii) plant-incorporated-protectants (PIPs).[69] However, the natural organisms or substances derived from natural materials (such as animals, plants, bacteria, or certain minerals), including their genes or metabolites, need to be formulated to obtain a stable and effective product for use in an IPM. The first two groups will be discussed in the following section, as they contain natural organisms or substances that can be formulated as emulsions.

13.7.1 Botanicals Pesticides: Essential Oils, Their Components, and Botanicals Extracts

Oil-in-water (O/W) emulsions, particularly microemulsions, and nanoemulsions incorporating botanical extracts, usually essential oils (EOs) or their individual components, as active ingredients, have emerged as a promising strategy to enhance the dispersion, and bioavailability of these bioactive compounds in aqueous environments.[16] While essential oil emulsion-based delivery systems represent an emerging field in pest control, their application under real-field conditions remains under-researched. This poses a challenge for the government, particularly regulatory agencies, and the industry. Despite the widespread interest in the properties of essential oils, particularly in the control of pathogen proliferation in the food industry, many studies evaluating the efficacy of such formulations have focused primarily on model systems. Therefore, understanding their biological activity in real, complex matrices remains an ongoing challenge.[70] The growing interest in essential oil-based systems is exemplified by the work of Balasubramani et al.,[71,72] who investigated the antibacterial, antioxidant, and larvicidal activities of nanoemulsions containing essential oils of *Ocimum basilicum* and *Vitex negundo*, demonstrating highly effective performance in all three applications.[73]

13.7.1.1 Microemulsions

Microemulsions containing essential oils represent a promising frontier in pest control. The strategic fusion of microemulsion technology with the unique properties of essential oils or essential oil components (EOCs) offers a multifaceted approach to addressing pest-related challenges. A key advantage is the ability of microemulsions to effectively disperse Eos in an aqueous environment. This property is particularly valuable in pest control applications where precise dispersion in water is essential. Agricultural environments, public health interventions, and various pest management scenarios benefit from the efficient and uniform distribution achieved by microemulsion formulations. In addition, the extended duration of action offered by microemulsions is a critical aspect of pest control strategies. This sustained release capability ensures prolonged and consistent efficacy of active ingredients, providing lasting protection against targeted pests. In agricultural and public health initiatives where sustained and reliable pest control is paramount, microemulsions contribute significantly to the efficacy of formulated products. On the other hand, microemulsions act as a formidable protective shield, preventing the degradation of encapsulated compounds, particularly essential oils. This protective shield preserves the integrity and potency of the active ingredients over time, ensuring that the formulated products maintain their efficacy throughout the application period. By improving the solubility of various active ingredients, microemulsions play a key role in optimizing the dispersion and delivery of essential oils. This improved solubility ensures a homogeneous distribution of active ingredients, maximizing their impact on pests while minimizing environmental impact. The simplicity of the microemulsion preparation process adds to their appeal in the context of pest control. This straightforward method not only makes the formulation process easily scalable but also aligns with the practical requirements of large-scale production, a crucial aspect for widespread application in agriculture and public health interventions.[74]

A crucial consideration in the formulation of microemulsion-based systems is the definition of the pseudo-ternary phase diagram (water-oil-surfactant + cosurfactant) specific to the system under investigation. Chaisri et al.[75] carried out a study on microemulsions containing citronella oil in water stabilized with Tween 20 and propylene glycol in different weight ratios (3:1, 2:1, 1:1, 1:2, and 1:3). Their results showed that the optimal surfactant-cosurfactant mixture occurs for a 3:1 weight ratio. This combination maximized the compositional range conducive to microemulsion formation. The resulting microemulsions exhibited a transparent yellow color, with droplets averaging between 20 nm and 40 nm in size, good long-term stability, and low viscosity. In applications against *Rhipicephalus microplus*, the acaricidal properties of these microemulsions were remarkable. The concentration of essential oil required to achieve 100% larval mortality was four times lower than that required when free essential oil was used for the same purpose. In addition, adult mortality observed with the microemulsions was significantly higher than with free essential oil (almost two times higher) and even exceeded that associated with the synthetic insecticide cypermethrin. This increased efficacy can be attributed to the emergence of resistance to cypermethrin in *R. microplus*, positioning microemulsions as a promising tool to address pest management challenges related to resistance issues.

Xu et al.[76] carried out a study of the acaricidal activity of microemulsions, focusing specifically on formulations containing neem oil stabilized with a surfactant mixture of Tween 80 and sodium dodecyl benzene sulfonate (in a weight ratio of 4:1), with hexyl alcohol as cosurfactant. The preparation involved gentle homogenization of the components (stirring at 800 rpm for 15 min at 40°C), resulting in highly stable dispersions of spherical droplets. These microemulsions showed sustained stability even after storage at temperatures of 4°C and 54°C for 15 days. Notably, the emulsions also withstood multiple freeze-thaw cycles without any sign of destabilization. When *Sarcoptes scabie* (var. Cuniculi) larvae were exposed to these microemulsions containing neem oil, pronounced acaricidal activity was observed, quantified by the time required to achieve larval mortality. Navayan et al.[77] used a surfactant blend (Tween 80 and Span 20 in a 1:1 weight ratio) together with a cosurfactant (propylene glycol) to stabilize microemulsions containing *Eucalyptus globulus* essential oil. The resulting microemulsions had droplets with an average size of 16-66 nm and a viscosity in the range of 230-330 cps. Such emulsions present a strong repellent effect against *Culex pipiens* and *Ochlerotatus caspius* mosquitoes, almost three times greater than that observed with the free essential oil. This enhanced repellence can be attributed to the encapsulation of the essential oil within the emulsion droplets, which reduces the evaporation of the molecules and delays the release of the oil. As a result, the repellent material remains in the environment for a longer period, resulting in prolonged exposure of mosquitoes to the repellent molecules. While the protection associated with microemulsions is remarkably high, it falls short of that of conventional synthetic repellents. However, the use of microemulsions as a replacement for synthetic repellents minimizes skin irritation during application, making them a safer alternative to traditional essential oil-based repellents.

Laothaweerungsawat et al.[78] extensively examined various factors influencing microemulsion formation, encompassing surfactant and cosurfactant types, as well as their ratios. They constructed pseudo-ternary phase diagrams with diverse proportions of water, *Origanum vulgare* EO, different cosurfactants (butylene glycol, propylene glycol, and glycerine), and various polysorbate surfactants (Tween 60, Tween 80, and Tween 85). Their findings highlighted significant differences in the compositional range leading to microemulsion formation, largely contingent upon the specific chemical nature of the surfactant. Hydrophilic-lipophilic balance (HLB) and structural attributes emerged as critical parameters in governing the extent of the microemulsion region. Moreover, their research delved into the role of cosurfactants, revealing that increased hydrophobicity corresponded to larger microemulsion regions. This phenomenon was attributed to the improved integration of hydrophobic cosurfactants within the surfactant film's hydrophobic region, reducing rigidity and supporting microemulsion stability. The resulting microemulsions exhibited optical homogeneity, transparency, non-birefringence, and a yellowish hue. Additionally, these microemulsions displayed sustained stability after six cycles of heating-cooling, preserving their physicochemical properties (appearance, particle size, polydispersity, zeta potential, electrical conductivity, refractive index, pH, and viscosity). Biological testing revealed that encapsulation of *O. vulgare* EO improved carvacrol release profiles.

Hamed et al.[79] utilized the oil titration method to formulate alcohol-free microemulsions of clove oil and its primary phenolic component, eugenol, without any cosurfactant participation. The authors gradually dripped the oil into vials that contained a steady measure of an aqueous solution of surfactant (Tween 20.5% w/w) to establish the maximum oil quantity that the dispersion could accommodate for microemulsion creation. The maximum feasible oil content was 1.1% w/w for clove oil and 0.9% w/w for eugenol. As a result, monodisperse microemulsions were created, with droplet sizes averaging just below 10 nm. These microemulsions have a high water content (~94% w/w) and show potential as alternatives for antimicrobial purposes. Specifically, the formulations exhibit strong antibacterial activity against both Gram-positive and Gram-negative bacteria. However, the resistance of the latter was greater, with increased concentrations of phenolic compounds significantly enhancing susceptibility in both bacterial types. Purwasena et al.[80] studied Tween 20-stabilized clove oil microemulsions, discovering their partial capability in preventing *Pseudomonas aeruginosa* biofilm proliferation. Similar efficacy was noted in microemulsions containing *Salvia officinalis* and *Cinnamomum cassia* EOs, effectively hindering *Staphylococcus aureus* proliferation.[81] These microemulsions were examined for their antifungal activity against *Penicillium Digitatum* in orange fruit. It was found that including essential oils within microemulsions significantly increased inhibitory effects on spore germination and germ tube elongation compared to using pure EO.[82]

Cespi et al.[83] developed microemulsions of *Smyrnium olusatrum* EO by using a mixture of Tween 80 and ethanol (surfactant and cosurfactant, respectively) under continuous stirring conditions via the oil titration method. The detection of crystallization in the oil phase of the primary constituent, isofuranodiene, was observed, which was mitigated by the inclusion of ethyl oleate. This addition prevented the essential oils from crystallizing and enabled the creation of microemulsions with droplets averaging a hydrodynamic diameter of 20-40 nm. This feature ensured excellent long-term stability. Similarly, Pavela et al.[84] produced microemulsions containing only isofuranodiene, which were predominantly monodisperse, with droplets measuring around 20–30 nm in hydrodynamic diameter. The microemulsions displayed considerable effectiveness against *Culex quinquefasciatus* larvae, outperforming both the complete *S. olusatrum* EO and the free isofuranodiene. Furthermore, these preparations present limited ecotoxicity toward nontarget organisms. Pavela et al.[85] also reported isofuranodiene crystallization in the formulation of microemulsions with *Pimpinella anisum*, *Trachyspermum ammi*, and *Crithmum maritimum* EOs. In bioactivity tests against *C. quinquefasciatus* larvae, the microemulsions demonstrated acute and chronic toxicity, resulting in almost 90% larval mortality at relatively low doses (1.81-6.48 mL/L). It is worth noting that there was no significant dependence on the chemical composition of the oil phase for the bioactivity observed. Moreover, these microemulsions inhibited adult emergence and exhibited minimal toxicity to nontarget species, indicating their promise as a reasonably safe pest control alternative.

13.7.1.2 Nanoemulsions

The ratio between the weight fraction of the surfactant and oil phase affects the stability of nanoemulsions that contain essential oils or their individual components. Wu et al.[86] conducted research on nanoemulsions containing thyme oil, stabilized using a soluble soybean polysaccharide. The results indicated that increasing the surfactant concentration in relation to the oil phase decreased droplet size and prevented nanoemulsion creaming. Likewise, Anjali et al.[87] observed similar results for neem oil nanoemulsions stabilized by Tween 20. Their research showed that as the surfactant concentration increased, resulting in smaller droplets, there was an improvement in larvicidal activity against *C. quinquefasciatus*. Feng et al.[88] noted comparable tendencies in nanoemulsions that comprise limonene. These nanoemulsions maintain stability during extended storage at high and low temperatures and demonstrate significant effectiveness against several fungi. Additionally, Sogan et al.[89] scrutinized castor oil nanoemulsions steady by blends of Tween 80 and Span 20. They concluded that stable nanoemulsions are only possible when the surfactant concentration exceeds the concentration of the oil phase. The study found that including castor oil in the oil phase of O/W nanoemulsions significantly enhanced their bioactivity against *Anopheles culicifacies* mosquitoes, the vector for malaria. Sugumar et al.[90] also underline the importance of the weight fractions of surfactant and oil phases in stabilizing nanoemulsions of orange oil in water. The authors note that, although a minimal surfactant concentration is necessary for stabilization, an excessive amount may accelerate nanoemulsion destabilization.

Mossa et al.[91] conducted additional research on the effects of the weight ratio between the surfactant and oil phases on nanoemulsion stabilization. Their study focused on preparing garlic oil nanoemulsions using ultrasonic emulsification and revealed significant correlations between droplet size and nanodispersion stability with respect to the ratio between surfactant and oil phase concentrations. The research depicted that stable nanoemulsions were generated with surfactant concentrations slightly exceeding those of the oil phase. This study also assessed the acaricidal effectiveness of the nanoemulsions against *Aceria oleae* and *Tegolophus hassani*, which are the olive bud and olive rust mites, respectively. Results showed that the essential oil's acaricidal activity was improved when integrated as the oil phase in the nanoemulsion, due to its wider contact area facilitated by the distribution of the essential oil as nanodroplets.

The way in which the specific surfactant chemistry influences the stabilization of nanoemulsions containing essential oil components has been the subject of investigation by several researchers. Ghosh et al.[92] investigated this by analyzing the effect of different surfactants from the polysorbate family (Tween 20 and Tween 80) on the stabilization of eugenol in water nanoemulsions. Their results showed that the use of

Tween 80 resulted in the formation of smaller droplets with less variability in size compared to the use of Tween 20. This variability in droplet size was attributed to differences in the packing of the nonionic surfactants at the interface between the continuous phase and the droplets.

The way in which the specific surfactant chemistry influences the stabilization of nanoemulsions containing essential oil components has been the subject of investigation by several researchers. Ghosh et al.[92] investigated this by analyzing the effect of different surfactants from the polysorbate family (Tween 20 and Tween 80) on the stabilization of eugenol in water nanoemulsions. Their results showed that the use of Tween 80 resulted in the formation of smaller droplets with less variability in size compared to the use of Tween 20. This variability in droplet size was attributed to differences in the packing of the nonionic surfactants at the interface between the continuous phase and the droplets. Chang et al.[93] further investigated the influence of polysorbate family surfactants on the stability of nanoemulsions, specifically focusing on the effect of the hydrophilic-lipophilic balance (HLB) on the stabilization of carvacrol nanoemulsions in water. Their study showed that higher hydrophobicity in the surfactant contributed to improving the stability of the nanoemulsions, resulting in smaller droplet sizes. However, surfactants with similar hydrophobicity did not consistently produce nanoemulsions with identical droplet sizes. For example, Tween 60 and Tween 80, both with HLB values of 14.9 and 15, respectively, produced droplets with average diameters of around 84 nm and 55 nm, respectively. This difference was attributed to differences in the molecular structure of the surfactants. Despite having the same polar head, variations in their hydrophobic tails resulted in different interfacial packing, thereby affecting the degree of stabilization. Mazarei and Rafati[94] highlighted the importance of surfactant interfacial packing in stabilizing nanoemulsions of *Satureja khuzestanica* EO. In their study, different combinations of water-soluble (Tween 80 and Tween 20) and oil-soluble (Span 80 and Span 20) surfactants were investigated. They observed a synergistic effect when a rigid oil-soluble surfactant was combined with a flexible water-soluble surfactant, resulting in improved nanoemulsion stability. Interestingly, they found that the droplet size obtained during emulsification did not consistently correlate with the long-term stability of the nanoemulsions. Similar conclusions were drawn by Botas et al.,[95] who investigated the stabilization of nanoemulsions containing *Baccharis reticularis* and D-limonene using mixtures of Tween 80 and Span 80 in different weight ratios. Their results highlighted the key role of the HLB of the surfactant blend in regulating the physicochemical properties and stability of the nanoemulsions. In addition, both nanoemulsions containing *Baccharis reticularis* EO and D-limonene exhibited significant larvicidal activity against *Aedes aegypti* larvae.

Sabbour and Abd El-Aziz[96] investigated the impact of nanoemulsions containing purslane oil on moths of *Ephestia cautella* and *Sitophilus granaries*. They noted a strong sterilizing effect of these nanoemulsions, even after storing them for over four months. This confirmed that utilizing nanoemulsions is a beneficial approach to maintaining the integrity and biological activity of essential oils and their components. The nanoemulsions had an impact on both the viability of deposited eggs and the emergence of adults. In addition, nanoemulsions containing purslane oil displayed three times more effectiveness than the essential oil alone. This increase in potency is attributed to the heightened toxicity of the dispersed essential oils as nanodroplets, leading to a greater contact area. Nenaah et al.[97,98] showed the comparable impacts of diverse *Asteraceae* and *Achillea* essential oils on *Callosobruchus maculatus* (cowpea beetle) and *Tribolium castaneum*. They demonstrated that lower surface tension in nanoemulsions can be related to greater biological activity.

Sabbour and Abd El-Aziz[95,96] also investigated other essential oils (castor and mustard oils) and found them to have lower biological activity against insects. This suggests a possible relationship between the chemical composition of the essential oils present in the nanoemulsions and their observed biological effects, as also highlighted by the research of Moretti et al.[99] They carried out experiments with highly diluted nanoemulsions (with a water content of around 98% w/w) containing different essential oils and stabilized with Tween 80. They evaluated the effect of these formulations on *Lymantria dispar* L. (Lepidoptera: Lymantridae, gypsy moth) larvae (second and third instar) in laboratory bioassays. Their results showed a strong correlation between larval mortality and the specific chemical nature of the essential oils used in the oil phase. In particular, nanoemulsions containing *Thymus herba-barona* EO were the most effective larvicidal agents, resulting in 80% larval mortality within 72 h of treatment with the formulation. This increased toxicity was associated with particularly high levels of phenolic compounds, especially carvacrol.[100]

Lucia et al.[101] investigated the effect of nanoemulsions containing different components of essential oils on *Pediculus humanus capitis* and observed a strong correlation between the toxicity of the nanoemulsions and the specific chemical composition of the oil phase. Surprisingly, and contrary to the suggestions of other researchers, they found that the size and polydispersity of the droplets played a minor role in their bioactivity, with over 50% mortality observed for different essential oil components used in the oil phase. This observation was further supported by recent research investigating the larvicidal activity against *A. aegypti* using nanoemulsions containing blends of thymol and carvacrol as the oil phase.[102] The study showed that the larvicidal efficacy of the formulations increased significantly with higher levels of thymol in the oil phase, regardless of the size distribution of the emulsions. In addition, the combined biological activity of the two essential oils was found to be additive, whereas the replacement of carvacrol by eugenol resulted in an antagonistic biological interaction.[103] Oliveira et al.[104] expanded on the importance of the chemical composition of the oil phase in the biological activity of nanoemulsions. Their study focused on nanoemulsions containing *Lippia sidoides* EO (with thymol as the main component, accounting for 68% w/w) and thymol alone as the oil phase. Interestingly, they found that while the physicochemical properties of the nanoemulsions were minimally affected by the chemical composition of the oil phase, the biological activity against *Sitophilus zeamais* varied significantly. Nanoemulsions containing *L. sidoides* EO showed a twofold increase in insect mortality compared to those containing only thymol. The dose required to induce 50% mortality was 7.10 μg/mg and 17.08 μg/mg for nanoemulsions containing *L. sidoides* EO and thymol, respectively.

Nanoemulsions containing essential oils have also been used as insect repellents. Nuchuchua et al.[105] formulated nanoemulsions using a mixture of three different essential oils (citronella, hairy basil, and vetiver oils) as the oil phase and evaluated their repellent activity against *A. aegypti* mosquitoes. They found that smaller droplets were more repellent. This effect is attributed to the formation of more robust films that minimize the evaporation of the essential oils, thereby prolonging the repellent activity of the nanoemulsions. In addition, the reduction in droplet size contributes to the stability of the formulations, confirming the findings of Sakulku et al.[106] They prepared nanoemulsions with high citronella content (around 20% w/w) by high-pressure homogenization. They found that increasing the surfactant concentration resulted in smaller droplet sizes and reduced polydispersity, allowing for a more controlled release of citronella. Furthermore, the introduction of glycerol into the aqueous phase proved to be an effective strategy to modulate release of kinetics, duration of protection against mosquitoes, droplet size, polydispersity index, and stability of the nanoemulsions. Consequently, higher concentrations of glycerol and surfactant corresponded to longer protection times against *A. aegypti*.

13.7.2 Microbial Pesticides: Bacterium and Fungus

Microbial pesticides consist basically of bacteria, fungi, viruses, protozoa, and algae and/or their bioactive compounds, which are used in the control of different pests.[107] The major categories of microbial agents used as microbial pesticides include entomopathogenic bacteria (*Bacillus thuringiensis* var. *Kurstaki*, *B. thuringensis* var. *israelensis*, *B. sphaericus*) and entomopathogenic fungi, such as *Beauveria bassiana*, *Paecilomyces lilacinus*, *Hirsutella thompsonii*, *Metarhizium brunneum*, *Paecilomyces fumosoroseus Verticillium lecanii*, or *Lagenidium giganteum*).[108]

The activity, viability, and storage stability are the most important parameters of a microbial pesticide. Unlike conventional pesticides, the active agents are particulate and are characterized as suspensions in the formulation or spray tank, instead as solutions. (e.g., infective juvenile nematode, fungal conidia, baculovirus occlusion body, bacterial spore, protein crystals).[109] However, there are other types of biological formulations, such as wettable powders, granules, water-dispersible granules, emulsions, concentrate suspensions, microencapsulated suspension, tablets, floating tables, and effervescent tablets.[110]

13.7.2.1 Pickering Emulsions and W/O/W Multiple Emulsions

Pickering emulsions have been used to disperse *Bacillus thuringiensis serovar aizawai.* They prepared water-in-mineral oil emulsions stabilized by hydrophobic silica by using a high shear homogenization method. As a result, the *Bt* cells, spores, and crystals were encapsulated in the water droplets of the inverse Pickering emulsions, showing higher bioactivity than the nonencapsulated *Bt* against *Spodoptera littoralis* first instar larvae.[111] On the other hand, Leemon et al.[112] developed an oil-in-water emulsion of *Metarhizium anisopliae* spores, which was tested by spraying against the parasitic stages of *Rhipicephalus* (*Boophilus*) *microplus*. These formulations evidence an increase in the velocity for killing the target pest together with a significant reduction in egg production in the three days post treatment.

On the other hand, the storage stability of *Burkholderia gladioli* was improved by their encapsulation into a W/O/W emulsion containing rapeseed oil as the oil phase, polyglycerin polyriconolate (PGPR) and castor oil polyoxyethylene (COG 25) as the primary and secondary emulsifier and microcrystalline cellulose as an emulsion stabilizer. The storage stability test revealed that the viability of formulated cells in emulsion was higher than that of unformulated cells. Moreover, in activity test, formulated cells in emulsion were more effective in inhibiting the growth of pathogen than unformulated cells in culture broth.[113]

13.8 CONCLUDING REMARKS

The use of emulsions as pesticides is a breakthrough in pest management programs. This chapter provides an in-depth analysis of the various roles of emulsions play in transforming practices for controlling pests. Within this sphere, emulsions have become a crucial tool in the formulation of pesticides, thereby revolutionizing strategies for pest control. Their flexibility and versatility are among their key benefits. Emulsions facilitate the integration of various active ingredients, resulting in improved dispersion and effectiveness in pest control while minimizing negative impacts on the environment and nontarget organisms. Additionally, emulsions provide substantial environmental benefits by reducing the use of toxic solvents and potential harm to ecosystems caused by conventional pesticides. Therefore, emulsions represent a significant breakthrough in sustainable practices. Improving emulsion formulations to be more ecologically friendly while retaining efficacy facilitates the coexistence of pest control measures and ecological systems. This domain offers vast potential for innovation and research, with abundant opportunities for further exploration and enhancement. Prioritizing optimizing formulations could bolster their performance, stability, and safety in the future.

ACKNOWLEDGMENTS

The work by E.G. was supported by grants PID2023-147156NB-I00 funded by MCIN/AEI/10.13039/501100011033 (Spain) and PR12/24-31566 (Ayudas para la Financiación de Proyectos de Investigación UCM 2023), and the European Innovative Training Network-Marie Sklodowska-Curie Action NanoPaInt (grant agreement 955612), funded by the E.U. A.L. and A.F. was funded by Agencia Nacional de Promoción Científica y Técnica (ANPCyT) under grant PICT-2021-GRF-TI-00704.

REFERENCES

1. Khot, L.R.; Sankaran, S.; Maja, J.M.; Ehsani, R.; Schuster, E.W. Applications of nanomaterials in agricultural production and crop protection: A review. *Crop Prot.* **2012**, *35*, 64–70.

2. Kah, M.; Kookana, R.S.; Gogos, A.; Bucheli, T.D. A critical evaluation of nanopesticides and nanofertilizers against their conventional analogues. *Nature Nanotech.* **2018**, *13*, 677–684.
3. Nuruzzaman, Md.; Rahman, M.M.; Liu, Y.; Naidu, R. Nanoencapsulation, nano-guard for pesticides: A new window for safe application. *J. Agric. Food Chem.* **2016**, *64*, 1447–1483.
4. Athanassiou, C.G.; Kavallieratos, N.G.; Benelli, G.; Losic, D.; Usha Rani, P.; Desneux, N. Nanoparticles for pest control: Current status and future perspectives. *J. Pest Sci.* **2018**, *91*, 1–15.
5. Scott, N.R.; Chen, H.; Cui, H. Nanotechnology applications and implications of agrochemicals toward sustainable agriculture and food systems. *J. Agric. Food Chem.* **2018**, *66*, 6451–6456.
6. Nair, R.; Varghese, S.H.; Nair, B.G.; Maekawa, T.; Yoshida, Y.; Kumar, D.S. Nanoparticulate material delivery to plants. *Plant Sci.* **2010**, *179*, 154–163.
7. Xiang, Y.; Wang, M.; Sun, X.; Cai, D.; Wu, Z. Controlling pesticide loss through nanonetworks. *ACS Sustainable Chem. Eng.* **2014**, *2*, 918–924.
8. Sarlak, N.; Taherifar, A.; Salehi, F. Synthesis of nanopesticides by encapsulating pesticide nanoparticles using functionalized carbon nanotubes and application of new nanocomposite for plant disease treatment. *J. Agric. Food Chem.* **2014**, *62*, 4833–4838.
9. Liu, Z.; Qie, R.; Li, W.; Hong, N.; Li, Y.; Li, C.; Wang, R.; Shi, Y.; Guo, X.; Jia, X. Preparation of avermectin microcapsules with anti-photodegradation and slow-release by the assembly of lignin derivatives. *New J. Chem.* **2017**, *41*, 3190–3195.
10. Li, Y.; Zhou, M.; Pang, Y.; Qiu, X. Lignin-based microsphere: Preparation and performance on encapsulating the pesticide avermectin. *ACS Sustainable Chem. Eng.* **2017**, *5*, 3321–3328.
11. Jia, X.; Sheng, W.; Li, W.; Tong, Y.; Liu, Z.; Zhou, F. Adhesive polydopamine coated avermectin microcapsules for prolonging foliar pesticide retention. *ACS Appl. Mat. Interfaces* **2014**, *6*, 19552–19558.
12. Sheng, W.; Li, W.; Li, B.; Li, C.; Xu, Y.; Guo, X.; Zhou, F.; Jia, X. Mussel-inspired photografting on colloidal spheres: A generalized self-template route to stimuli-responsive hollow spheres for controlled pesticide release. *Macromol. Rapid Commun.* **2015**, *36*, 1640–1645.
13. Lucia, A.; Toloza, A.C.; Guzmán, E.; Ortega, F.; Rubio, R.G. Novel polymeric micelles for insect pest control: Encapsulation of essential oil monoterpenes inside a triblock copolymer shell for head lice control. *PeerJ.* **2017**, *5*, e3171.
14. Ding, K.; Shi, L.; Zhang, L.; Zeng, T.; Yin, Y.; Yi, Y. Synthesis of photoresponsive polymeric propesticide micelles based on PEG for the controlled release of a herbicide. *Polym. Chem.* **2016**, *7*, 899–904.
15. Chen, K.; Fu, Z.; Wang, M.; Lv, Y.; Wang, C.; Shen, Y.; Wang, Y.; Cui, H.; Guo, X. Preparation and characterization of size-controlled nanoparticles for high-loading λ-cyhalothrin delivery through flash nanoprecipitation. *J. Agric. Food Chem.* **2018**, *66*, 8246–8252.
16. Lucia, A.; Guzmán, E. Emulsions containing essential oils, their components or volatile semiochemicals as promising tools for insect pest and pathogen management. *Adv. Colloid Interface Scie.* **2021**, *287*, 102330.
17. Golden, G.; Quinn, E.; Shaaya, E.; Kostyukovsky, M.; Poverenov, E. Coarse and nano emulsions for effective delivery of the natural pest control agent pulegone for stored grain protection. *Pest Manag. Sci.* **2018**, *74*, 820–827.
18. Zhao, P.; Cao, L.; Ma, D.; Zhou, Z.; Huang, Q.; Pan, C. Translocation, distribution and degradation of prochloraz-loaded mesoporous silica nanoparticles in cucumber plants. *Nanoscale* **2018**, *10*, 1798–1806.

19. Xu, C.; Cao, L.; Zhao, P.; Zhou, Z.; Cao, C.; Li, F.; Huang, Q. Emulsion-based synchronous pesticide encapsulation and surface modification of mesoporous silica nanoparticles with carboxymethyl chitosan for controlled azoxystrobin release. *Chem. Eng. J.* **2018**, *348*, 244–254.
20. Li, Z.-Z.; Xu, S.-A.; Wen, L.-X.; Liu, F.; Liu, A.-Q.; Wang, Q.; Sun, H.-Y.; Yu, W.; Chen, J.-F. Controlled release of avermectin from porous hollow silica nanoparticles: Influence of shell thickness on loading efficiency, uv-shielding property and release. *J. Control. Release.* **2006**, *111*, 81–88.
21. Samar, S.I. Essential oil nanoformulations as a novel method for insect pest control in horticulture. In *Horticultural Crops*, Baimey, H.K., Hamamouch, N., Kolombia, Y.A., Eds.; IntechOpen: London, UK, 2020. https://doi.org/10.5772/intechopen.80747
22. Barradas, T.N.; de Holanda e Silva, K.G. Nanoemulsions as optimized vehicles for essential oils. In *Sustainable Agriculture Reviews: Pharmaceutical Technology for Natural Products Delivery*, Saneja, A., Panda, A.K., Lichtfouse, E., Eds.; Springer: Cham, Switzerland, 2020; Volume 44, pp. 115–167.
23. Jo, Y.-J.; Chun, J.-Y.; Kwon, Y.-J.; Min, S.-G.; Hong, G.-P.; Choi, M.-J. Physical and antimicrobial properties of trans-cinnamaldehyde nanoemulsions in water melon juice. *LWT Food Sci. Technol.* **2015**, *60*, 444–451.
24. Tenjarla, S. Microemulsions: An overview and pharmaceutical applications. *Crit. Rev. Ther. Drug Carr. Syst.* **1999**, *16*, 62.
25. Pavoni, L.; Perinelli, D.R.; Bonacucina, G.; Cespi, M.; Palmieri, G.F. An overview of micro- and nanoemulsions as vehicles for essential oils: Formulation, preparation and stability. *Nanomaterials* **2020**, *10*, 135.
26. Sultan Rana, A.; Nazeer, M.; Abd El-Gawad, H.H.; Inam, M.; Ibrahim, M.M.; El-Bahy, Z.M.; Faizan Nazar, M. Microemulsions as potential pesticidal carriers: A review. *J. Mol. Liquids* **2023**, *390*, 122969.
27. Liu, Y.; Wei, F.; Wang, Y.; Zhu, G. Studies on the formation of bifenthrin oil-in-water nano-emulsions prepared with mixed surfactants. *Colloids Surf. A* **2011**, *389*, 90–96.
28. Chin, C.-P.; Lan, C.-W.; Wu, H.-S. Application of biodiesel as carrier for insecticide emulsifiable concentrate formulation. *J. Taiwan Inst. Chem. Eng.* **2012**, *43*, 578–584.
29. da Costa, J.T.; Forim, M.R.; Costa, E.S.; De Souza, J.R.; Mondego, J.M.; Boiça Junior, A.L. Effects of different formulations of neem oil-based products on control zabrotes subfasciatus (Boheman, 1833) (Coleoptera: Bruchidae) on beans. *J. Stored Prod. Res.* **2014**, *56*, 49–53.
30. Singh, A.; Dhiman, N.; Kar, A.K.; Singh, D.; Purohit, M.P.; Ghosh, D.; Patnaik, S. Advances in controlled release pesticide formulations: Prospects to safer integrated pest management and sustainable agriculture. *J. Hazard. Mat.* **2020**, *385*, 121525.
31. Kumar, S.; Nehra, M.; Dilbaghi, N.; Marrazza, G.; Hassan, A.A.; Kim, K.-H. Nano-based smart pesticide formulations: Emerging opportunities for agriculture. *J. Control. Release* **2019**, *294*, 131–153.
32. Salvia-Trujillo, L.; Artiga-Artigas, M.; Molet-Rodríguez, A.; Turmo-Ibarz, A.; Martín-Belloso, O. Emulsion-based nanostructures for the delivery of active ingredients in foods. *Front. Sustain. Food Syst.* **2018**, *2*, 00079.
33. Teixé-Roig, J.; Oms-Oliu, G.; Odriozola-Serrano, I.; Martín-Belloso, O. Emulsion-based delivery systems to enhance the functionality of bioactive compounds: Towards the use of ingredients from natural, sustainable sources. *Foods* **2023**, *12*, 1502.
34. Bensaci, O.A.; Rouabah, K.; Aliat, T.; Lombarkia, N.; Plushikov, V.G.; Kucher, D.E.; Dokukin, P.A.; Temirbekova, S.K.; Rebouh, N.Y. Biological pests management for sustainable agriculture: Understanding the influence of cladosporium-bioformulated endophytic fungi application to control myzus persicae (Sulzer, 1776) in potato (Solanum Tuberosum L.). *Plants* **2022**, *11*, 2055.

35. Kah, M.; Kookana, R. Emerging investigator series: Nanotechnology to develop novel agrochemicals: Critical issues to consider in the global agricultural context. *Env. Sci.: Nano* **2020**, *7*, 1867–1873.
36. Deka, B.; Babu, A.; Baruah, C.; Barthakur, M. Nanopesticides: A systematic review of their prospects with special reference to tea pest management. *Front. Nutr.* **2021**, *8*, 686131.
37. Pradhan, S.; Mailapalli, D.R. Nanopesticides for pest control. In *Sustainable Agriculture Reviews*, Lichtfouse, E., Ed.; Springer: Cham, Switzerland, 2020; Volume 40, pp. 43–74.
38. McClements, D.J. Encapsulation, protection, and release of hydrophilic active components: Potential and limitations of colloidal delivery systems. *Adv. Colloid Interface Sci.* **2015**, *219*, 27–53.
39. Knowles, D.A. Formulation of agrochemicals. In *Chemistry and Technology of Agrochemical Formulations*, Knowles, D.A., Ed.; Springer Netherlands: Dordrecht, The Netherlands, 1998; pp. 41–79.
40. Santos, J.; Trujillo, L.A.; Calero, N.; Alfaro, M.C.; Muñoz, J. Physical characterization of a commercial suspoemulsion as a reference for the development of suspoemulsions. *Chem. Eng. Technol.* **2013**, *36*, 1883–1890.
41. Rizvi, S.A.H.; Xie, F.; Ling, S.; Zeng, X. Development and evaluation of emulsifiable concentrate formulation containing sophora Alopecuroides L. extract for the novel management of Asian citrus psyllid. *Environ. Sci. Pollut. Res.* **2019**, *26*, 21871–21881.
42. Hazra, D.K.; Karmakar, R.; Poi, R.; Bhattacharya, S.; Mondal, S. Recent advances in pesticide formulations for eco-friendly and sustainable vegetable pest management: A review. *Arch. Agric. Environ. Sci.* **2017**, *2*, 232–237.
43. Hans de Ruiter; Henk, J. Holterman; Corné Kempenaar; Hans, G.J. Mol; Jan, J. de Vlieger; Jan, C. van de Zande. *Influence of Adjuvants and Formulations on the Emission of Pesticides to the Atmosphere*, Wageningen: The Netherlands, 2003.
44. Knowles, A. Recent developments of safer formulations of agrochemicals. *Environmentalist* **2008**, *28*, 35–44.
45. Hazra, D.K.; Purkait, A. Role of pesticide formulations for sustainable crop protection and environment management: A review. *J. Pharmacognosy Phytother.* **2019**, *8*, 686–693.
46. Ford, R.E. The use of emulsions in the application of pesticides. In *Theory and Practice of Emulsion Technology*, Smith, A.L., Ed.; Academic Press: Cambridge, MA, USA, 1976; pp. 227–237.
47. Zimdahl, R.L. Herbicide formulation. In *Fundamentals of Weed Science*, Zimdahl, R.L., Ed.; Academic Press, Inc.: Cambridge, MA, USA, 2018; pp. 501–509.
48. Chin, C.-P.; Lan, C.-W.; Wu, H.-S. Application of biodiesel as carrier for insecticide emulsifiable concentrate formulation. *J. Taiwan Inst. Chem. Eng.* **2012**, *43*, 578–584.
49. Zhang, X.; Jing, T.; Zhang, D.; Luo, J.; Li, B.; Liu, F. Assessment of ethylene glycol diacetate as an alternative carrier for use in agrochemical emulsifiable concentrate formulation. *Ecotoxicol. Environ. Saf.* **2018**, *163*, 349–355.
50. Tran, N.H.; Hieu, T.V.T.; Dang-Bao, T.; Anh, T.T.K. Assessment of methyl ester as a green carrier solvent in pesticide emulsifiable concentrate formulation. *IOP Conf. Ser.: Earth Environ. Sci.* **2021**, *947*, 012031.
51. Arthur, F. Residual efficacy of a deltamethrin emulsifiable concentrate formulation against rhyzopertha dominica (F.) and sitotroga cerealella (Oliver) after partial treatment of brown rice. *Insects* **2019**, *10*, 95.
52. Sherin, H.M.S.; Mai, M.A.G.; Farag, E.M. Acaricidal efficiency of solar 50% new emulsifiable concentrate formulation against the Two-Spotted Spider Mite (TSSM) tetranychus urticaekoch (Acari: Tetranychidae) under laboratory and greenhouse conditions. *J. Appl. Nat. Sci.* **2022**, *14*, 1456–1464.

53. Randy Cush. Back to basics: A review of pesticide formulation types. *Golf Course Manag.* **2006**, *74*, 143–145.
54. Webster, G.R.; Bisset, N.B.; Cahill, D.M.; Jones, P.; Killick, A.; Hawley, A.; Boyd, B.J. Kinetic resolution of the interactions between agrochemical products and adjuvant systems upon mixing. *J. Agric. Food Chem.* **2016**, *64*, 6139–6147.
55. Tadros, Th. F. Disperse systems in pesticidal formulations. *Adv. Colloid Interface Sci.* **1990**, *32*, 205–234.
56. Shao, H.; Xi, N.; Zhang, Y. Microemulsion formulation of a new biopesticide to control the diamondback moth (Lepidoptera: Plutellidae). *Sci, Rep.* **2018**, *8*, 10565.
57. Zheng, G.; Zhao, Z.; Wong, J.W.C. Role of non-ionic surfactants and plant oils on the solubilization of organochlorine pesticides by oil-in-water microemulsions. *Env. Technol.* **2011**, *32*, 269–279.
58. Leng, P.; Zhang, Z.; Li, Q.; Zhao, M.; Pan, G. Microemulsion formulation of carbendazim and its in vitro antifungal activities evaluation. *PLoS One* **2014**, *9*, e109580.
59. Chin, C.-P.; Lan, C.-W.; Wu H.-S. Study on the performance of lambda cyhalothrin microemulsion with biodiesel as an alternative solvent. *Ind. Eng. Chem. Res.* **2012**, *51*, 4710–4718.
60. McClements, D.J. Nanoemulsions versus microemulsions: Terminology, differences, and similarities. *Soft Matter* **2012**, *8*, 1719–1729.
61. Wang, L.; Li, X.; Zhang, G.; Dong, J.; Eastoe, J. Oil-in-water nanoemulsions for pesticide formulations. *J. Colloid Interface Sci.* **2007**, *314*, 230–235.
62. Du, Z.; Wang, C.; Tai, X.; Wang, G.; Liu, X. Optimization and characterization of biocompatible oil-in-water nanoemulsion for pesticide delivery. *ACS Sustainable Chem. Eng.* **2016**, *4*, 983–991.
63. Yan, H.; Bao, C.; Chen, X.; Yu, C.; Kong, D.; Shi, J.; Lin, Q. Preparation of biodiesel oil-in-water nanoemulsions by mixed surfactants for bifenthrin formulation. *RSC Adv* **2019**, *9*, 11649–11658.
64. Wei, N.; Hou, C.; Liu, Z.; Liang, Q.; Lv, Z.; Meng, X.; Feng, J. Preparation of fenpropathrin nanoemulsions for eco-friendly management of helicoverpa armigera: Improved insecticidal activity and biocompatibility. *Colloids Surf. A* **2023**, *656*, 130442.
65. Feng, J.; Chen, W.; Liu, Q.; Chen, Z.; Yang, J.; Yang, W. Development of abamectin-loaded nanoemulsion and its insecticidal activity and cytotoxicity. *Pest Manag. Sci.* **2020**, *76*, 4192–4201.
66. Kumar, J.; Ramlal, A.; Mallick, D.; Mishra, V. An overview of some biopesticides and their importance in plant protection for commercial acceptance. *Plants* **2021**, *10*, 1185.
67. Hernandez-Tenorio, F.; Miranda, A.M.; Rodríguez, C.A.; Giraldo-Estrada, C.; Sáez, A.A. Potential strategies in the biopesticide formulations: A bibliometric analysis. *Agronomy* **2022**, *12*, 2665.
68. What are Biopesticides? *https://www.epa.gov/ingredients-used-pesticide-products/what-are-biopesticides* (accessed 22nd December 2023).
69. Hezakiel, H.E.; Thampi, M.; Rebello, S.; Sheikhmoideen, J.M. Biopesticides: A green approach towards agricultural pests. *Appl. Biochem. Biotechnol.* **2023**. https://doi.org/10.1007/s12010-023-04765-7
70. Amaral, D.M.F.; Bhargava, K. Essential oil nanoemulsions and food applications. *Adv. Food Technol. Nutr. Sci.* **2015**, *1*, 84–87.
71. Balasubramani, S.; Rajendhiran, T.; Moola, A.K.; Diana, R.K.B. Development of nanoemulsion from vitex negundo L. Essential oil and their efficacy of antioxidant, antimicrobial and larvicidal activities (Aedes Aegypti L.). *Environ. Sci. Pollut. Res.* **2017**, *24*, 15125–15133.

72. Sundararajan, B.; Moola, A.K.; Vivek, K.; Kumari, B.D.R. Formulation of nanoemulsion from leaves essential oil of ocimum basilicum L. and its antibacterial, antioxidant and larvicidal activities (Culex Quinquefasciatus). *Microb. Pathog.* **2018**, *125*, 475–485.
73. López, M.D.; Cantó-Tejero, M.; Pascual-Villalobos, M.J. New insights into biopesticides: Solid and liquid formulations of essential oils and derivatives. *Front. Agron.* **2021**, *3*, 763530.
74. Pavela, R.; Benelli, G.; Pavoni, L.; Bonacucina, G.; Cespi, M.; Cianfaglione, K.; Bajalan, I.; Morshedloo, M.R.; Lupidi, G.; Romano, D.; Canale, A.; Maggi, F. Microemulsions for delivery of apiaceae essential oils—towards highly effective and eco-friendly mosquito larvicides? *Ind. Crop Prod.* **2019**, *129*, 631–640.
75. Chaisri, W.; Chaiyana, W.; Pikulkaew, S.; Okonogi, S.; Suriyasathaporn, W. Enhancement of acaricide activity of citronella oil after microemulsion preparation. *Jpn. J. Vet. Res.* **2019**, *67*, 15–23.
76. Xu, J.; Fan, Q.-J.; Yin, Z.-Q.; Li, X.-T.; Du, Y.-H.; Jia, R.-Y.; Wang, K.-Y.; Lv, C.; Ye, G.; Geng, Y.; Su, G.; Zhao, L.; Hu, T.-X.; Shi, F.; Zhang, L.; Wu, C.-L.; Tao, C.; Zhang, Y.-X.; Shi, D.-X. The preparation of neem oil microemulsion (Azadirachta Indica) and the comparison of acaricidal time between neem oil microemulsion and other formulations in Vitro. *Vet. Parasitol.* **2010**, *169*, 399–403.
77. Navayan, A.; Moghimipour, E.; Khodayar, M.J.; Vazirianzadeh, B.; Siahpoosh, A.; Valizadeh, M.; Mansourzadeh, Z. Evaluation of the mosquito repellent activity of nano-sized microemulsion of eucalyptus globulus essential oil against culicinae. *Jundishapur J. Nat. Pharm. Prod.* **2017**, *12* (4), e55626.
78. Laothaweerungsawat, N.; Neimkhum, W.; Anuchapreeda, S.; Sirithunyalug, J.; Chaiyana, W. Transdermal delivery enhancement of carvacrol from origanum vulgare L. essential oil by microemulsion. *Int. J. Pharm.* **2020**, *579*, 119052.
79. Hamed, S.F.; Sadek, Z.; Edris, A. Antioxidant and antimicrobial activities of clove bud essential oil and eugenol nanoparticles in alcohol-free microemulsion. *J. Oleo Sci.* **2012**, *61*, 641–648.
80. Purwasena, I.A.; Astuti, D.I.; Taufik, I.; Putri, F.Z. The potential of clove essential oil microemulsion as an alternative biocide against pseudomonas aeruginosa biofilm. *J. Pure Appl. Microbiol.* **2020**, *14*, 261–269.
81. Campana, R.; Casettari, L.; Fagioli, L.; Cespi, M.; Bonacucina, G.; Baffone, W. Activity of essential oil-based microemulsions against staphylococcus aureus biofilms developed on stainless steel surface in different culture media and growth conditions. *Int. J. Food Microbiol.* **2017**, *241*, 132–140.
82. He, S.; Ren, X.; Lu, Y.; Zhang, Y.; Wang, Y.; Sun, L. Microemulsification of clove essential oil improves its in vitro and in vivo control of penicillium digitatum. *Food Control* **2016**, *65*, 106–111.
83. Cespi, M.; Quassinti, L.; Perinelli, D.R.; Bramucci, M.; Iannarelli, R.; Papa, F.; Ricciutelli, M.; Bonacucina, G.; Palmieri, G.F.; Maggi, F. Microemulsions enhance the shelf-life and processability of smyrnium olusatrum L. essential oil. *Flavour Fragr. J.* **2017**, *32*, 159–164.
84. Pavela, R.; Pavoni, L.; Bonacucina, G.; Cespi, M.; Kavallieratos, N.G.; Cappellacci, L.; Petrelli, R.; Maggi, F.; Benelli, G. Rationale for developing novel mosquito larvicides based on isofuranodiene microemulsions. *J. Pest Sci.* **2019**, *92*, 909–921.
85. Pavela, R.; Benelli, G.; Pavoni, L.; Bonacucina, G.; Cespi, M.; Cianfaglione, K.; Bajalan, I.; Morshedloo, M.R.; Lupidi, G.; Romano, D.; Canale, A.; Maggi, F. Microemulsions for delivery of apiaceae essential oils—towards highly effective and eco-friendly mosquito larvicides? *Ind. Crop Prod.* **2019**, *129*, 631–640.

86. Wu, J.-E.; Lin, J.; Zhong, Q. Physical and antimicrobial characteristics of thyme oil emulsified with soluble soybean polysaccharide. *Food Hydrocoll* **2014**, *39*, 144–150.
87. Anjali, C.; Sharma, Y.; Mukherjee, A.; Chandrasekaran, N. Neem oil (Azadirachta Indica) nanoemulsion—a potent larvicidal agent against culex quinquefasciatus. *Pest. Manag. Sci.* **2012**, *68*, 158–163.
88. Feng, J.; Wang, R.; Chen, Z.; Zhang, S.; Yuan, S.; Cao, H.; Jafari, S.M.; Yang, W. Formulation optimization of d-limonene-loaded nanoemulsions as a natural and efficient biopesticide. *Colloids Surf. A* **2020**, *596*, 124746.
89. Sogan, N.; Kapoor, N.; Kala, S.; Patanjali, P.K.; Nagpal, B.N.; Vikram, K.; Valecha, N. Larvicidal Activity of castor oil nanoemulsion against malaria vector anopheles culicifacies. *Int. J. Mosquito Res.* **2018**, *5*, 1–6.
90. Sugumar, S.; Singh, S.; Mukherjee, A.; Chandrasekaran, N. Nanoemulsion of orange oil with nonionic surfactant produced emulsion using ultrasonication technique: Evaluating against food spoilage yeast. *Appl. Nanosci.* **2016**, *6*, 113–120.
91. Mossa, A.-T.H.; Afia, S.I.; Mohafrash, S.M.M.; Abou-Awad, B.A. Formulation and characterization of garlic (Allium Sativum L.) Essential oil nanoemulsion and its acaricidal activity on eriophyid olive mites (Acari: Eriophyidae). *Environ. Sci. Pollut. Res.* **2018**, *25*, 10526–10537.
92. Ghosh, V.; Mukherjee, A.; Chandrasekaran, N. Eugenol-loaded antimicrobial nanoemulsion preserves fruit juice against, microbial spoilage. *Colloids Surf. B* **2014**, *114*, 392–397.
93. Chang, Y.; McLandsborough, L.; McClements, D.J. Physicochemical properties and antimicrobial efficacy of carvacrol nanoemulsions formed by spontaneous emulsification. *J. Agric. Food Chem.* **2013**, *61*, 8906–8913.
94. Mazarei, Z.; Rafati, H. Nanoemulsification of satureja khuzestanica essential oil and pure carvacrol; Comparison of physicochemical properties and antimicrobial activity against food pathogens. *LWT* **2019**, *100*, 328–334.
95. Botas, G.; Cruz, R.; de Almeida, F.; Duarte, J.; Araújo, R.; Souto, R.; Ferreira, R.; Carvalho, J.; Santos, M.; Rocha, L.; Pereira, V.; Fernandes, C. Baccharis Reticularia DC. and limonene nanoemulsions: Promising larvicidal agents for aedes aegypti (Diptera: Culicidae) control. *Molecules* **2017**, *22*, 1990.
96. Sabbour, M.M.; El-Aziz, S.E.A. Efficacy of three essential oils and their nano-particles against *sitophilus granarius* under laboratory and store conditions. *J. Entomol. Res.* **2016**, *40*, 229.
97. Nenaah, G.E. Chemical composition, toxicity and growth inhibitory activities of essential oils of three achillea species and their nano-emulsions against tribolium castaneum (Herbst). *Ind. Crop. Prod.* **2014**, *53*, 252–260.
98. Nenaah, G.E.; Ibrahim, S.I.A.; Al-Assiuty, B.A. Chemical composition, insecticidal activity and persistence of three asteraceae essential oils and their nanoemulsions against callosobruchus maculatus (F.). *J. Stored Prod. Res.* **2015**, *61*, 9–16.
99. Moretti, M.D.L.; Sanna-Passino, G.; Demontis, S.; Bazzoni, E. Essential oil formulations useful as a new tool for insect pest control. *AAPS PharmSciTech* **2002**, *3*, 64–74.
100. Karpouhtsis, I.; Pardali, E.; Feggou, E.; Kokkini, S.; Scouras, Z.G.; Mavragani-Tsipidou, P. Insecticidal and genotoxic activities of oregano essential oils. *J. Agric. Food Chem.* **1998**, *46*, 1111–1115.
101. Lucia, A.; Toloza, A.C.; Guzmán, E.; Ortega, F.; Rubio, R.G. Novel polymeric micelles for insect pest control: Encapsulation of essential oil monoterpenes inside a triblock copolymer shell for head lice control. *PeerJ.* **2017**, *5*, e3171.
102. Lucia, A.; Girard, C.; Fanucce, M.; Coviella, C.; Rubio, R.G.; Ortega, F.; Guzmán, E. Development of an environmentally friendly larvicidal formulation based on essential oil compound blend to control *Aedes Aegypti* larvae: Correlations between physicochemical properties and insecticidal activity. *ACS Sustainable Chem. Eng.* **2020**, *8*, 10995–11006.

103. Lucia; A.; Toloza, A.C.; Fanucce, M.; Fernández-Peña, L.; Ortega, F.; Rubio, R.G.; Coviella, C.; Guzmán, E. Nanoemulsions based on thymol-eugenol mixtures: Characterization, stability and larvicidal activity against Aedes Aegypti. *Bull. Insectol.* **2020**, *73*, 153–160.
104. Oliveira, A.P.; Santana, A.S.; Santana, E.D.R.; Lima, A.P.S.; Faro, R.R.N.; Nunes, R.S.; Lima, A.D.; Blank, A.F.; Araújo, A.P.A.; Cristaldo, P.F.; Bacci, L. Nanoformulation prototype of the essential oil of lippia sidoides and thymol to population management of sitophilus zeamais (Coleoptera: Curculionidae). *Ind. Crop Prod.* **2017**, *107*, 198–205.
105. Nuchuchua, O.; Sakulku, U.; Uawongyart, N.; Puttipipatkhachorn, S.; Soottitantawat, A.; Ruktanonchai, U. In vitro characterization and mosquito (Aedes Aegypti) repellent activity of essential-oils-loaded nanoemulsions. *AAPS PharmSciTech* **2009**, *10*, 1234.
106. Sakulku, U.; Nuchuchua, O.; Uawongyart, N.; Puttipipatkhachorn, S.; Soottitantawat, A.; Ruktanonchai, U. Characterization and mosquito repellent activity of citronella oil nanoemulsion. *Int. J. Pharm.* **2009**, *372*, 105–111.
107. Arthurs, S.; Dara, S.K. Microbial biopesticides for invertebrate pests and their markets in the United States. *J. Invertebr. Pathol.* **2019**, *165*, 13–21.
108. Ayilara, M.S.; Adeleke, B.S.; Akinola, S.A.; Fayose, C.A.; Adeyemi, U.T.; Gbadegesin, L.A.; Omole, R.K.; Johnson, R.M.; Uthman, Q.O.; Babalola, O.O. Biopesticides as a promising alternative to synthetic pesticides: A case for microbial pesticides, phytopesticides, and nanobiopesticides. *Front. Microbiol.* **2023**, *14*, 1040901.
109. Behle, R.; Birthisel, T. Formulations of entomopathogens as bioinsecticides. In *Mass Production of Beneficial Organisms*, Morales-Ramos, J.A., Rojas, M.G., Shapiro-Ilan, D.I., Eds.; Academic Press: London, UK, 2014; pp. 483–517.
110. Kala, S.; Sogan, N.; Agarwal, A.; Naik, S.N.; Patanjali, P.K.; Kumar, J. Biopesticides: Formulations and delivery techniques. In *Natural Remedies for Pest, Disease and Weed Control*, Egbuna, C., Sawicka, B., Eds.; Academic Press: London, UK, 2020; pp. 209–220.
111. Yaakov, N.; Kottakota, C.; Mani, K.A.; Naftali, S.M.; Zelinger, E.; Davidovitz, M.; Ment, D.; Mechrez, G. Encapsulation of bacillus thuringiensis in an inverse pickering emulsion for pest control applications. *Colloids Surf. B* **2022**, *213*, 112427.
112. Leemon, D.M.; Turner, L.B.; Jonsson, N.N. Pen studies on the control of cattle tick (Rhipicephalus (Boophilus) Microplus) with metarhizium anisopliae (Sorokin). *Vet. Parasitol.* **2008**, *156*, 248–260.
113. Hwa Jin Kim; Young Hee Cho; Eun Kyung Bae; Shin, T.S.; Choi, S.W.; Choi, K.H. Development of w/o/w multiple emulsion formulation containing burkholderia gladioli. *J. Microbiol. Biotechnol.* **2005**, *15*, 29–34.

14 Emulsions in Pharmaceutical Applications

Pablo G. Argudo

14.1 INTRODUCTION

In the field of pharmaceutical research and development, emulsions have emerged as powerful tools for delivering drugs and therapeutics. These multiphase systems usually consist of two liquid phases, where one is dispersed in the form of droplets within the other. Emulsions have proven their reliability in overcoming challenges in drug formulation and delivery. They are not just stable systems borrowed from other fields, like food or personal care. Rather, they are precisely complexes designed to work effectively under biological conditions with minimal side effects, while meeting specific therapeutic requirements. Thus, their importance is more than evident, as a wide range of medicines available on the market owe their effectiveness to their adjustable properties, including topical creams or intravenous solutions.

Over time, the role of emulsions in pharmaceuticals has evolved from being simple excipients to intricate delivery systems that offer unique advantages such as improved bioavailability, enhanced solubility, controlled release of drugs, stability, and targeted delivery. However, it is crucial to acknowledge the role played by materials science and engineering in their development. Advancements in nanotechnology and surface chemistry have significantly contributed to optimizing emulsion formulations. The ability to finely adjust droplet size, interfacial tension, and rheological properties directly impacts their optimization process and ultimate efficacy. Therefore, a multidisciplinary approach to the field is a must in order to fully develop its full potential.

This chapter aims to provide an overview of pharmaceutical emulsions and shed light on their crucial role in our society. For this purpose, it will be explored how they can enhance and broaden the effectiveness of drugs through their delivery properties. First, fundamental concepts of emulsions, which have been extensively covered in previous sections of this book, will be discussed. Next, the role of emulsions in a plethora of therapeutic applications will be introduced. By presenting examples from several research groups, the strengths of emulsions in the pharmaceutical field will be illustrated. Additionally, the regulatory and quality control aspects involved in their development, as well as the formulation challenges that these complex systems must overcome, before they can be made available to end users, will be discussed.

 DOI: 10.1201/9781032636184-17

Finally, research trends where emulsions shine and highlight their significance in shaping the future generation of pharmaceuticals will be outlined.

14.2 PHARMACEUTICAL EMULSION TYPES

According to the IUPAC, an emulsion is 'A fluid colloidal system in which liquid droplets and/or liquid crystals – which often exceed the usual limits for colloids in size – are dispersed in a liquid'.[1] To be stable, the tension between the two liquids is reduced, thus avoiding any coalescence process. To understand the relevance of emulsions in the pharmaceutical field, it is important to have an overview of the existing types, as this will be linked to their final applicability. The emulsification system chosen will determine, among other things, the solubility and release rate of the drug, the absorption of the droplet and its targeting properties, its stability, or its patient compliance. The most common pharmaceutical emulsions are oil-in-water (O/W), water-in-oil (W/O), and multiple ones (such as water-in-oil-in-water, W/O/W, or oil-in-water-in-oil, O/W/O), as represented in Figure 14.1. However, other systems can also be found, such as oil-in-oil (O/O) emulsions, where a polar and a nonpolar oil are commonly used.[2] For a more in-depth characterization of the behavior of these types of emulsions, we recommend reading previous chapters of this book. To stabilize the oil–water interface, several compounds can be used. In terms of surfactants, most of the formulations are based on nonionic ones, as they are more resistant to the effects of charge and pH changes in the medium.[3] Cationic surfactants, in some cases as in oral pharmaceutical formulations, are avoided due to their generally high toxicity.[4] Nanoparticles, as in Pickering emulsions, can also be used as stabilizers while promoting membrane permeability and targeted delivery.[5] They are typically based on inorganic particles such as silica or magnesium hydroxide, or organic particles such as poly(*N*-isopropylacrylamide) (pNIPAM), polystyrene (PS), or natural derivatives (i.e., polysaccharides).[6] Finally, the aqueous and/or oily media used are specifically tailored to the site of application and the drug being transported. As is usually the case when selecting oils, they not only make the system viable and stable, but also enhance its properties.

14.2.1 Oil-in-Water and Water-in-Oil Emulsions

Oil-in-water emulsions are characterized by an oily and an aqueous medium that behave as the dispersed and continuous phase, respectively. As the human body is

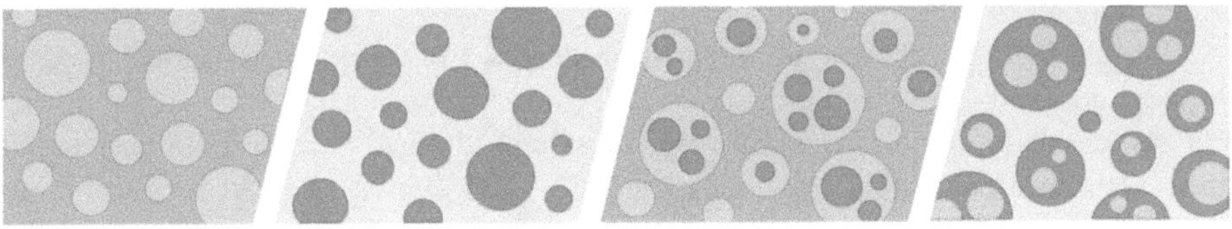

FIGURE 14.1 Classification of the most extended emulsions. *From left to right:* Oil-in-water (O/W), water-in-oil (W/O), water-in-oil-in-water (W/O/W), and oil-in-water-in-oil (O/W/O). The aqueous and oily phases are depicted in dark and light grey, respectively.

mainly composed of water, which is highly prevalent in regions such as the eyes, lungs, or brain,[7,8] and most of the proposed drugs have a hydrophobic character, O/W emulsions are the ideal systems to be used. Their predominantly aqueous composition allows them to diffuse easily into similar water-like environments, as the continuous phase can be adapted to the biological media of application. Also, compared to other systems, their contact makes them feel less greasy, ideal for topical applications. The presence of their oily phase can be used for the loading and delivery of hydrophobic or poorly water–soluble drugs. The lipophilic phase can act as a protective reservoir for the drug and even enhance its properties. Able to be stabilized for long periods of time by surfactants and the hydrophilic nature of the continuous phase, most pharmaceutical emulsions follow this formulation due to its ease of taste adjustment, administration, biocompatibility, versatility, and controlled release.

Water-in-oil emulsions are formulated with water droplets dispersed in a continuous oil phase. As previously stated, they are not as widely represented compared to their O/W counterparts. This could be attributed to the use of mostly hydrophobic drugs in the pharmaceutical industry, which can make the use, in some cases, of O/W nearly mandatory. Still, several hydrophilic drugs are now rising in popularity and can suffer from low stability or oxidation processes.[9] These challenges make W/O emulsions ideal for their application, as the drug can be protected until its delivery. Similar to O/W emulsions, this system can also be described by its biocompatibility, versatility, and controlled release.

14.2.2 Multiple Emulsions: W/O/W and O/W/O

Multiple emulsions consist of droplets of one dispersed liquid dispersed in another, also known as emulsions of emulsions. Compared to O/W or W/O emulsions, multiple emulsions require the use of two sets of emulsifiers to stabilize the final system. The most widely used are water-in-oil-in-water (W/O/W) emulsions. However, in some specific applications, O/W/O emulsions can also be employed. In W/O/W emulsions, hydrophilic drugs can be carried in the core aqueous phase while being protected by an oleic layer which acts as a liquid membrane in a final continuous aqueous phase. Conversely, a mirror O/W/O emulsion allows lipophilic drugs to be carried. One of their innate properties is the slow, sustained, and controlled release from the inner phase to the continuous phase, via the intermediate one. But also the reverse process, as they can entrap and remove any toxin or compound present in the continuous phase on the inner one with a similar nature. For a complete understanding of multiple emulsions, the lecture of previous chapters as well as other works are recommended.[10]

Multiple W/O/W emulsions are in high demand in the pharmaceutical field as they are generally safer and can show better immune responses than simple ones.[11] W/O/W emulsion vaccines containing Newcastle disease virus (NDV) hemagglutinin–neuraminidase (HN) glycoprotein have been shown to provide ≥80% protection against challenge exposure to virulent NDV in chickens four weeks after vaccination,[12] with no adverse local side effects, as is typical in W/O counterparts.[13] This type of emulsion can also be used to deliver plasmid DNA encoding the gp100 peptide antigen for melanoma immunotherapy to murine dendritic cells, resulting in suppression of tumor growth and induction of cell-mediated anti-tumor immunity.[14] Moreover, they have recently been used as an adjuvant against an influenza A virus in a porcine model.[15] For skin treatment, water and isopropyl palmitate-based emulsions have been highlighted for their high storage stability, greater skin permeability in vitro, and improved bioavailability in rabbits when compared to controlled DS creams, such as Dakangfen®.[16] Using β-cyclodextrin and Tween 80, Pickering double emulsions are also proposed, in this case for the permeation of vitamin C.[17]

In this line, W/O/W multiple emulsions have been used for drug delivery applications where the release kinetics are mostly controlled by swelling or shearing.[18,19] The hydrophilic drug can be delivered by the incorporation of concentration gradients leading to osmotic water flow, or swelling, and final breakup of the droplet, as depicted in Figure 14.2, left. This process can be studied by monitoring the viscosity over time,[20] as there are no concentration changes in an iso-osmotic dilution, whereas in a hypo-osmotic system they exist, usually leading to a rapid increase in viscosity followed by a slow decrease until equilibrium.[21] For instance, a W/O/W emulsion based on water as the aqueous phase and paraffin as the oily one was reported to carry sodium lactate, spironolactone, and chlorhexidine digluconate in the inner, intermediate, and outer phases, respectively. The release of sodium lactate could be controlled by osmotic gradient regulation, leading to the final disruption of the oil film and swelling of the inner droplets.[21] In the case of shearing or by the application of shear forces, resulting in deformation and bursting of the droplet, viscosification or thickening of the aqueous continuous phase is reported to be key to the final drug delivery efficacy. Unthickened emulsions may remain stable, showing no changes in the granulometric distribution as well as similar low delivery rates at different shear conditions. However, a change in the viscosity of the external water layer can improve their applicability, resulting in smaller particle sizes as a function of shear rate, thus leading to higher drug delivery.[22] Similar release control was achieved in this case by shearing a W/O/W emulsion based on water and isohexadecane. At

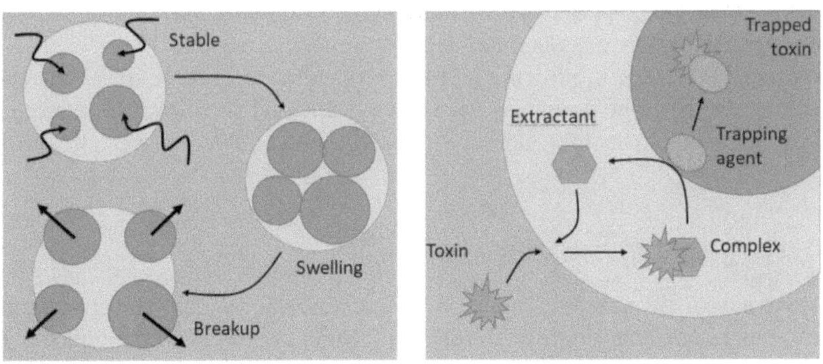

FIGURE 14.2 Mechanism of W/O/W multiple emulsions. *Left:* Drug delivery release by swelling and break up. *Right:* Target detoxification. The aqueous and oleic phases are depicted in dark and light grey, respectively.

human body temperature, a moderate shear rate of around 1,000 s^{-1}, which is the average reached during topical application, results in a high release rate of the ideal drug, which is otherwise stable for more than three months.[23]

Finally, reactive detoxifying emulsions (RDE) have also emerged as a fruitful way of removing toxins or undesirable particles from a medium. Briefly, a lipophilic molecule or an extractant present in the oily phase is able to recognize and bind the target toxin in the external aqueous medium. It is then transported to the internal aqueous compartment in the form of a complex, released, and captured by a trapping agent present there, as shown in Figure 14.2 right. In the pharmaceutical field, highly stable W/O/W RDEs have been shown in vitro to be a viable system against Nivaquine®, a drug widely accused of being used in suicide attempts, with a yield of 85% in 5 min. Bis(2-ethylhexyl)phosphoric acid, the extractant, selectively binds the chloroquine sulfate drug at the outer O/W interface by a cationic exchange mechanism. Later, the toxin is transported to the internal W/O interface and delivered to a mineral acid, HCl, which acts as the trapping agent.[24] The same extractant can be used for Paraquat®, another low-dose exposure drug. Using NaOH as a trapping agent, in vitro experiments under simulated gastric conditions (pH 2) achieved an extraction rate of 45% in the first 5–10 min after application, compared to 10% for the control emulsions. In vivo studies in rats denoted no adverse effects of the proposed multiple emulsion and a 100% increase in median survival time.[25]

O/W/O multiple emulsions also deserve special mention, as they demonstrate a high potential for prolonged release and prevent drug oxidation. They can be used as carriers, where they show higher pentazocine release rates than simple O/W emulsions in in vitro and in vivo tests in mice, where the tissue levels of the drug were prolonged.[26] They have also been proven to be the best at stabilizing vitamin A,[27] protecting and modulating the release after topical application of triterpene derivatives in vitro,[28] or Flumethrin®.[29] However, they are mostly used as templates for the development of capsules or particles.[30,31] O/W/O emulsions tend to be less stable, have less controlled release, and are more difficult to optimize than their counterparts. Moreover, W/O/W emulsions are generally preferred in the pharmaceutical field, as most places of application are aqueous and they can deliver both hydrophilic and hydrophobic drugs as well. Finally, as for simple emulsions, multiple emulsions can be tuned by surface modification. To improve the targeting of the droplets, they can be coated by specific ligands, such as an *o*-palmitoyl derivative of mannan–dextran for enhancement in lung uptake,[32] or Concanavalin-A for specific antitumor effects.[33]

14.3 THERAPEUTIC APPLICATIONS

Emulsions have evolved from simple physicochemical entities to versatile therapeutic tools capable of meeting a wide range of clinical needs. Their properties make them invaluable assets in the pharmaceutical toolkit. This section provides an overview of the most advanced areas of research in which emulsions shine, highlighting their traditional roles and innovative implementations.

14.3.1 Topical Emulsions

Topical administration involves spreading the emulsion on top of an external surface, such as the skin or mucous membrane. For ease of use, gels, lotions, creams, foams, or aerosols, among others, have been developed for their treatment in recent years. However, emulsions are one of the best systems to use in this field. They allow for the normal evaporation of aqueous media, making them comfortable at the time of application, are less greasy, and have a high affinity with the skin, since the stratum corneum is mainly based on lipids, proteins, and water. Furthermore, for proper efficacy, they can be adjusted to reach the treated target, from adherence to penetration into the epidermis, while maintaining the functionality of the biological membrane.

14.3.1.1 Drug Release, Partition, and Transport

After emulsion application, the transported drug goes through a series of steps: it is released from the dispersed phase, delivered to the skin, and diffuses into the underlying tissues at an appropriate rate and amount. The drug release involves its transfer through the O/W interface. In an ideal system and considering homogeneous spherical droplets, the amount of drug released to the skin in a short time is as follows:[34,35]

$$\frac{M_t}{M_\infty} = \frac{6D^{1/2}t^{1/2}}{\pi^{1/2}r} \tag{14.1}$$

and in longer times,

$$\frac{M_t}{M_\infty} = 1 - \frac{6}{\pi^2}\exp\left(-\frac{\pi^2 Dt}{r^2}\right), \tag{14.2}$$

where t is the time, M_t is the amount of drug released at time t, M_∞ is the total amount of drug, D is the diffusion coefficient of the solute, and r is the radius of the sphere. However, if a phase boundary is present, closer to a real system, a slow interfacial transfer from the dispersed to the continuous phase appear. Emulsions, characterized by droplets with a radius lower than 1 mm, are controlled by this slow interfacial transfer. Having k, which determines the heterogeneous rate constant between the two phases, M_0 the initial drug concentration, and A the surface area of the spheres, the amount of drug released at short times can be described as follows:

$$M_t = AM_0kt \tag{14.3}$$

and in longer times,

$$M_t = \frac{1}{3}AM_0r\left[1 - \exp\left(-\frac{3kt}{r^2}\right)\right], \tag{14.4}$$

where controlling these kinetics is crucial for an appropriate drug delivery. Subsequently, the drug diffuses into the skin. Transdermal drug delivery is one of the most desirable routes due to the avoidance of hepatic metabolism and

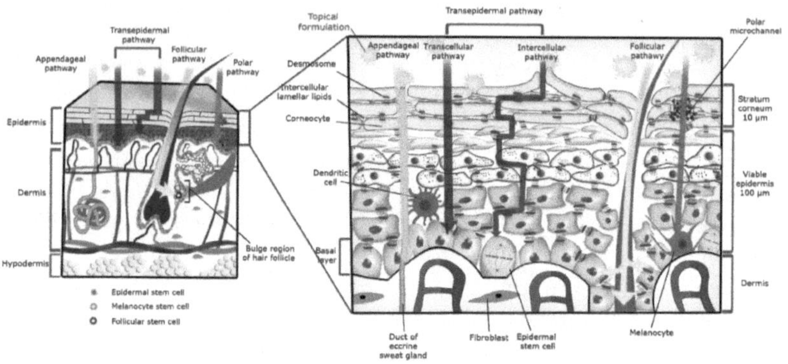

FIGURE 14.3 Schematic of the structure of the skin and its drug penetration pathways. Reproduced from Ahmadi-Ashtiani et al.,[38] with permission under Creative Common Attribution License 4.0 [CC-BY 4.0].

gastrointestinal side effects of drug administration.[36] After reaching the epidermis, the drug must pass through the stratum corneum, an approximately 10 μm thick hydrophobic barrier composed mainly of ceramides and corneocytes.[37] Its structure can be represented as a 'brick wall' made of cornified dead cells filled with keratin with the spaces filled with multiple stacks of lipid lamellae. To cross it, the drug mainly goes through a transcellular, intercellular, appendageal, or follicular route, as shown in Figure 14.3.[38] Small hydrophobic drugs prefer a transcellular route, while hydrophilic drugs go through an intercellular one, crossing the corneocytes.[39] The contribution of the appendageal route is nearly negligible, as the hair coverage of the skin is below 1% of the area. The partition coefficient of the drug between the stratum corneum and the dispersion, K_{SC}, is one of the proposed parameters to calculate the efficacy of the drug, estimated by the Hildebrand solubility, while others suggest the use of the relative polarity index.[40] However, only estimations can be obtained. Commonly used methods based on Fick's law estimate the diffusion experimentally, by applying variables such as diffusion distance, skin thickness, flux, or lag time.[41,42] Finally, if permeable, the drug can reach the epidermis, dermis, and hypodermis, made of fatty tissues. However, a proper design of the emulsion is required as the epidermis is based on two layers: a secondary living hydrophilic layer and a hydrophobic one.[43] Also, effects such as skin irritation must be avoided. It should be remarked that alternative experimental skin models behave differently than human ones. For example, mouse skin is four times thinner and has fewer hair follicles, leading to higher percutaneous absorption. Guinea pig is the closest one, as its thickness of stratum corneum and epidermis is in line with human skin.[44]

14.3.1.2 Formulation Challenges and Examples

A proper topical formulation is characterized by great stability and efficacy while preserving the biological conditions of the treated surface. In this line, the surfactants employed must enhance skin permeation, where their composition and concentration will be directly related to the emulsion behavior.[45] Generally, nonionic surfactants (such as sorbitan, lecithin, polyoxyethylene sorbitan, or polyoxyl lauryl ether derivatives) are used due to their low toxicity and increased penetration through skin solubilization and keratin binding.[46] Anionic surfactants like higher fatty alcohols (sodium cetyl/stearyl sulfate) can be used because of their stronger interaction with keratin and lipids, the generation of additional bonding sites in the membrane, and increased skin hydration. Also, cationic surfactants, such as long-chain quaternary ammonium compounds, can enhance skin permeability by binding to anionic components present in the stratum corneum.[36] Finally, mixtures of these surfactants and other cosurfactants can also be found.[47] For example, middle-chain alcohols reduce the interfacial tension and thus the phase separation, while long-chain ones promote a closed water domain inside a continuum of a hydrocarbon layer.[48] Cetyl, stearyl, or cetostearyl alcohol can be added to produce a cream texture, glycerol and paraffin may also enhance the gel structure, and a ratio of alcohol to surfactant 9:1 can be used to prevent aging. In parallel, Pickering emulsions offer higher surface areas and tailored surface functionality compared to classical coating methods, and improved dermal drug permeation, as particle-enhancers with this property can be used. For instance, emulsions based on silica particles lead to higher permeation efficiencies compared to average classical coatings, and chitosan-based Pickering emulsions improve wound healing.[6]

Focused on the aqueous phases, permeability is one of the key factors to consider in terms of efficacy. Water is the primary skin penetration enhancer, as it leads to wetting and untightening of the lipid barrier.[49] Related to the oily phase, long-chain triglycerides (LCTs) or medium-chain triglycerides (MCTs) can be used. Oleic acid is the most common oil phase employed, enhancing absorbance of water of the stratum corneum. However, there are not enough reports indicating the need to use LCTs to gain lymphatic access, so extensive research is required. However, numerous publications have confirmed that unsaturated fatty acids have superior dermal lipid-disrupting properties compared to saturated fatty acids. To name a few, natural olive and rice bran oils are known skin penetration enhancers with no side effects when used under the right conditions. Penetration modifiers or emollients can also be employed. While glycerol may induce retarding effects, diethylene glycol monoethyl ether, isopropyl myristate, terpenes, short-chain alcohols, azone®, or propylene glycol are used as enhancers by altering the lipid barrier organization.[36,47]

Skin hydration can be achieved by its occlusion or by providing water to it by using W/O or O/W emulsions, respectively. For instance, alpha-lipoic acid, an organosulfur with strong antioxidant properties, was used in the preparation of W/O emulsions. Based on an ionic liquid strategy, a recent line of research for enhancing the delivery of active pharmaceutical ingredients,[50] it could be properly dissolved and stabilized in the inner water phase.[51] Acyclovir, used to treat herpes infections of the skin, also showed an enhanced delivery and stability in ionic liquid-in-oil microemulsions. Using isopropyl myristate as a continuous oil phase, choline oleate and Span-20 as surfactants, and several choline derivatives as the nonaqueous polar phases in the core, histological tests confirmed the nontoxicity and in vitro skin irritation properties of the system.[52] However, most drugs are poorly soluble in water, making them ideal to be transported in an oleic medium such as in O/W emulsions.

For example, imidazolium ionic liquid can be added to the aqueous phase as a transdermal enhancer. Using an oil-based phase containing lidocaine–ibuprofen, artemisinin can be loaded and delivered through the skin in nanoemulsions based on commercial surfactants.[53] O/W emulsions using alkyl polyglucosides as carriers and spironolactone as a drug were also developed for potential skin hydration, which is important in the treatment of acne.[54] Based on a 4:1 ratio of polysorbate to sorbitan emulsions using safflower oil as the internal phase, artemether, an agent applied to treat acute malaria, can be delivered. In the form of nanoemulsions, the drug can be distributed to the stratum corneum–epidermis of human skin.[55] Stable emulsions containing also a natural oil, sunflower oil, glycerin, and vitamin B3 have been developed to treat inflammatory skin conditions, prevent acanthosis, and for potential wound healing. Sunflower oil repairs the skin barrier, replenishes cellular lipids, creates occlusive seals, improves subcutaneous hydration, reduces inflammatory response, and decreases microbial growth ratio. Glycerin reduces the transepidermal water loss and enhances the antimicrobial properties, and vitamin B3 controls the postinflammatory epidermal hyperpigmentation.[56] Oil-in-water emulsions containing liquid crystals, in this case of mixed cetyl alcohol and Polysorbate 60, are also under research. These crystalline structures, especially the lamellar ones, are similar to those present in the stratum corneum, increasing their affinity, and can be used to enhance the stability and storage of drugs, such as vitamin C, against its degradation.[57] Focused on Pickering emulsions, recent works enhance the use of calcium carbonate particles as they are pH and shear thinning human compatible. These biocompatible inorganic particles can lead to a strong network structure and thus a viscous component, enabling an easy application system. Moreover, they can be modified by chitosan or cyclodextrins.[58] Chitosan–collagen peptide nanoparticles are also evaluated as green surfactant-free vehicles for topical delivery. Cannabidiol, with anti-inflammatory and anti-itching effects on the skin, was dissolved in olive oil and exhibited elastic gel-like properties, long-term stable storage, and high retention in the stratum corneum.[59] Poly(lactic-*co*-glycolic acid) nanoparticles can also stabilize not one, but two active pharmaceutical ingredients. An immunosuppressive agent, either cyclosporine A or tacrolimus, along with an anti-inflammatory drug, calcitriol, can be encapsulated and delivered without altering the viability of human keratinocytes, opening up the treatment of skin diseases such as psoriasis.[60] Alternative methods for transdermal delivery can also be found. Microneedle-assisted transdermal delivery can be used in topical photodynamic therapy, a clinical treatment for cutaneous squamous cell carcinoma. These commercially available microneedle rollers can be used to pierce the stratum corneum and create transient microchannels, enhancing the final efficiency of the system. The photosensitizer 5-ALA, using a perfluorotripropylamine-based oxygenated emulsion, can easily pass through the stratum corneum and epidermal layers, replenishing oxygen to the hypoxic tumor.[61]

14.3.2 Oral Emulsions

The mouth route has been used for centuries to treat patients. Oral emulsions are particularly valuable for drugs that are poorly soluble or have irregular absorption characteristics. In addition, the emulsions can be tailored to have a desired flavor, making them easier to take. On the other hand, this pathway can reach undesired delivery targets, as after entering the gastrointestinal tract and subsequently the bloodstream, the emulsion can interact with several undesired organs, tissues, or receptors. Also, extreme changes in media conditions, such as pH or enzymes, make the stability and controlled release of emulsions a challenge.

14.3.2.1 Self-Emulsifying Drug Delivery System

Self-emulsifying drug delivery systems (SEDDS), which cover self-microemulsifying and self-nanoemulsifying drug delivery systems (SMEDDS/SNEDDS), have gained recent interest in peptide and protein delivery, as these compounds are characterized by limited absorption and solubility. Economical and easy to prepare and scale-up, these emulsions are formed spontaneously by moderate stirring and dilution in a water phase, conditions that characterize the gastrointestinal tract. The free energy required to form them is low or even negative, being crucial for their final stability and size distribution. Compared to conventional emulsions, they avoid instability issues such as creaming, coalescence, or breaking. Also, they are stable under slight temperature changes and can be applied in solid dosage forms. The incorporation of novel multifunctional auxiliary agents can enhance their properties. Microcrystalline cellulose can be used to complement their gastric residence, polymeric precipitator inhibitors to improve their intestinal solubility, or chitosan for enhanced intestinal drug permeability.[62] Moreover, these systems can also be tuned to deliver hydrophilic macromolecular drugs, as shown in Figure 14.4.[63]

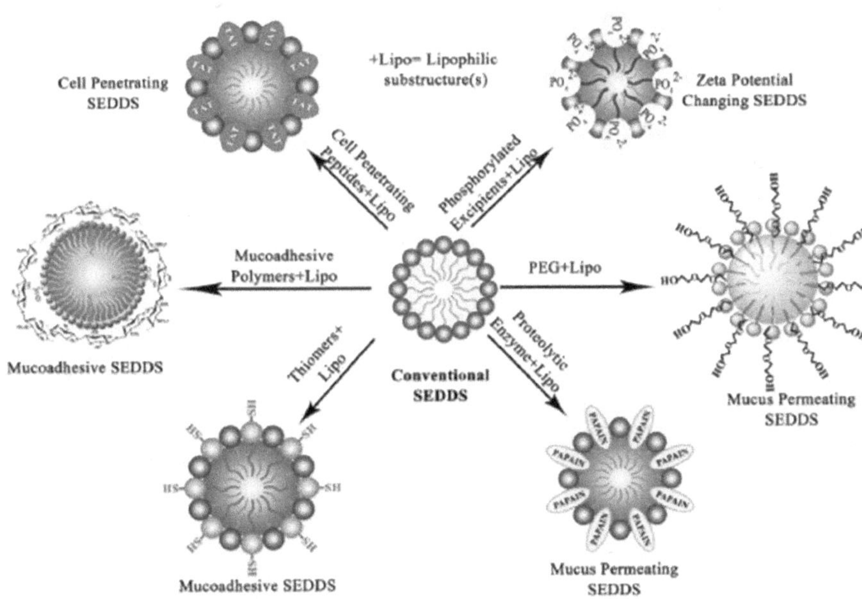

FIGURE 14.4 Novel SEDDS for improved binding and permeation properties. Reproduced from Mahmood and Bernkop-Schnürch,[63] Copyright (2019), with permission from Elsevier.

14.3.2.2 Formulation Challenges and Examples

A wider range of emulsifying agents can be used compared to other applications. For oral treatment, SEDDS are preferably based on nonionic compounds because of their low toxicity, even though they can lead to irreversible changes in the permeability of the gastrointestinal wall. Surfactants, such as ethoxylated polyglycolyzed glycerides and polyoxyethylene 20 oleate with a hydrophilic–lipophilic balance (HLB) value above 12, are preferred, as they can achieve better emulsification. As in other therapeutic applications, polysorbate, propylene glycol monocaprylate, sorbitan monooleate, glycerol monooleate, or polyoxyl/castor oil derivatives are some of the most widespread surfactants.[64] In parallel, Pickering emulsions can be formulated by using natural sources, such as chitosan, cellulose, or gelatin. Alternatively, synthetic products like polylactic acid (PLA), polyglycolic acid (PLGA), and their derivative polymers such as polylactic-*co*-glycolic acid (PLGA), glycerol monostearate, or sorbitan monooleate, can be found in the literature. Also, drug nanocrystals can act as stabilizers.[5] Starch or silybin-based nanocrystals are some of the formulations being investigated. As oily phase, natural edible oils remain as the desired ingredient. MCTs and their derivatives are preferred because of their better solubilizing ability, self-emulsification capacity, and increased drug transport through the portal vein. They stimulate the release of endogenous biliary lipids from the gallbladder, promoting the solubilization and absorption of the emulsion droplet in the intestinal tract by the formation of micelles. LCTs are used due to their enhanced lymphatic transport, while still possessing drawbacks as they are harder to emulsify.[65] Thus, a mixture of them is highly desirable. Recently, modified vegetable oils are also applied, as they show enhanced drug solubility and biocompatible degradation products, being able to replace classical MCTs.[62]

Tritolide, a diterpenoid triepoxide with pancreatic anticancer activity but also organ toxicity, can be target-delivered along diammonium glycyrrhizinate, a cytoprotective agent, and lipidic surfactants. Pharmacokinetic studies in rats revealed a higher tritolide bioavailability and longer residence time in the bloodstream when placed in an emulsion system than in solution, with a relatively low liver damage.[66] Coenzyme Q10, an antiaging supplement, has an enhanced delivery compared to its powder form when stabilized by oxa acid derivatives, with similar cell viability to commercial surfactants such as polyoxyethylene or sorbitan monooleate.[67] Cinnamaldehyde, an unstable drug with several pharmacological applications such as anti-inflammatory, antipyretic, analgesic, or antitumor, reported higher in vitro release kinetics and in vivo pharmacokinetics when placed in an emulsion compared to its free form. Protected by Tween 80, Span 80, MCTs, and lecithin, the O/W submicrometer emulsion improved the T_{max}, C_{max}, and $AUC_{0-\infty}$ pharmacokinetic parameters in rats.[68] Similar can be said for carotenoids, which enhance the immune system against various infections. Their solubility and stability can be increased when charged in emulsion-based delivery systems, where several commercial emulsifiers and vegetable oils have denoted their viability.[69] Also, Pickering emulsions are stabilized by cellulose nanocrystals.[70] A plethora of nanoparticle-based emulsified systems have recently been developed in this line of research, ranging from liquid to gel or dry emulsions. Non-emulsifying lipid-based formulations loaded with fenofibrate can lead to stable O/W emulsions. Using stearic acid and silica nanoparticles, it forms a gel when cellulose biopolymers are added.[71] The same drug can be delivered in a dry form stabilized by silica and lipids as well. In vitro and in vivo studies showed improved bioavailability results due to the inhibition of serum triglyceride and cholesterol concentrations compared to plain fenofibrate.[72] In a more traditional approach, $CaCO_3$ nanoparticles can be used alongside MCTs for the delivery of lipophilic drugs and simultaneous oral supplementation of calcium because of their biocompatibility and digestibility.[73] Ginkgolide B nanocrystals have been reported to self-stabilize Pickering nanoemulsions. Utilized to treat cardiovascular and cerebrovascular ailments, in vitro and in vivo results denoted extended absorption and slower elimination profiles, compared to ginkgolide B in powder form.[74] Finally, focused on the field of SEDDS, copaiba oil-resin was demonstrated to have a more efficient and safer treatment when administered orally in contrast to intravenously. This anti-leishmanial compound showed in vitro antiproliferative activity and low cytotoxicity against macrophages.[75] Also, compared to spray-dried dispersions, SEDDS can lead to higher exposure levels, indicating their oral pharmacokinetic efficacy.[76] Several peptide and protein derivatives can be stabilized, such as salmon calcitonin, to treat chronic diseases on a daily form basis while avoiding proteolysis. When stabilized in complexes by hydrophobic ion pairing method, the emulsion efficiency is directly related to their pK_a and lipophilicity.[77] Curcumin can be released by SMEDDS to treat depression. In vivo results revealed an improved brain permeability and further increase in biological activity of the drug when applied in emulsion form, compared to nonencapsulated curcumin and control groups.[78] Irinotecan/rapamycin-loaded oral lecithin-based self-emulsifying nanoemulsions are employed for the treatment of pancreatic cancer. Using a mixture of lecithin, Tween 80, cremophor EL, and propylene glycol as the surfactant system, and capryol-90 as the oil phase, both drugs exhibit a strong synergistic effect in tumor inhibition and cell viability than in powder.[79]

14.3.3 Parenteral Emulsions

These emulsions, which bypass the digestive system, allow for a more direct administration of the drug, usually intramuscularly, intravenously, or subcutaneously. They can provide an immediate onset of action, low toxicity, drug solubility, low degradation, and sustained release. Emulsion droplets, covered by lipids, are able to move through the human body while remaining stable. The incorporated hydrophobic or hydrophilic drugs within the dispersed phase can be sequestered from direct contact with any body, fluid or tissues until reaching the desired target. However, they are not free of drawbacks, as in Pickering emulsions, where the particle size can be a disadvantage when avoiding phagocytic processes.[80]

14.3.3.1 Nutritional Fats

The body requires regular intake of carbohydrates, fats, amino acids, salts, and vitamins. A direct incorporation of them into a large central vein can be life-saving in some circumstances. Among them, fats provide the highest amount of energy, are present in the cell membrane, and are needed for cellular signaling, precursor modulation, and biosynthesis processes. Thus, O/W emulsions are emerging as a reliable tool to maintain the required amounts of fat for proper body behavior and to avoid essential fatty acid deficiency. For instance, mammals are unable to synthesize long-chain fats such as omega-3 or omega-6. These compounds can be delivered in the form of dispersed droplets by emulsions.[81] After administration, several processes occur, such as protein coating, phospholipid exchange of the droplet surface, or uptake by hepatic and Kupffer cells. On average, 50% are taken up by the muscle tissue, 25% by the spleen, 14% by the myocardium, and 13% by the subcutaneous tissue. However, the size, composition, and charge of the droplets determine their final fate.[34]

14.3.3.2 Emulsion-Adjuvanted Vaccines

The intramuscular route is the most efficient route for emulsion application as most of the systems are designed to pass quickly between muscle fibers without interacting with them. After injection, emulsions induce an increase in monocytes, eosinophils, neutrophils, and dendritic cells at the site of application and later in the draining lymph node, leading to the accumulation of antigen-loaded monocytes in the subcapsular region. Followed by a cascade of responses, the human immune system responds with an increase in vaccine-specific T cells, a broad repertoire of memory B cells, and high-functioning antibody titers. Thus, although each adjuvant could be characterized by a different immune induction profile, emulsion adjuvants can be described by their rapid diffusion of the adjuvant from the muscle to the draining lymph node, without secondary effects such as 'depot effect'. They also promote a fast and powerful mobilization of immune cells in the draining lymph node, and a final enhanced immune response. Figure 14.5 shows this mechanism.[82]

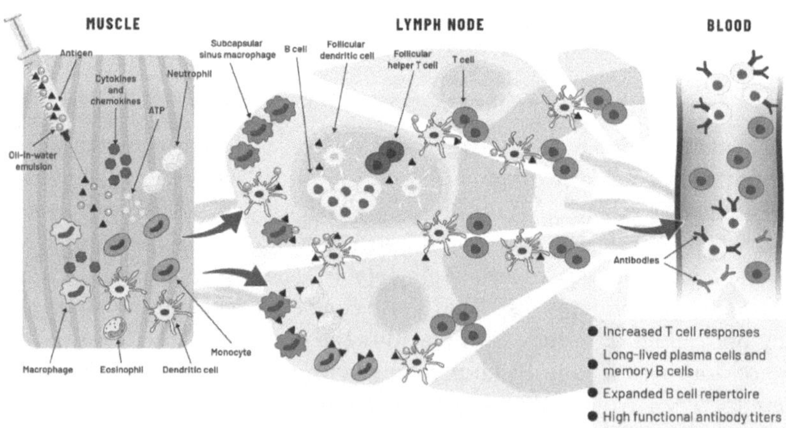

FIGURE 14.5 Action mechanism of emulsion-adjuvanted vaccines after intramuscular injection. Reproduced from O'Hagan et al.,[82] with permission under Creative Common Attribution 4.0 International License [4.0].

14.3.3.3 Formulation Challenges and Examples

In parenteral emulsions, nonionic emulsifiers are used, most commonly lecithin, because of its relative safety and stability depending on its source and degree of purification. While the phosphatidylethanolamine (PE) and phosphatidylcholine (PC) components present remain uncharged at physiological pH, phosphatidylserine and phosphatidylglycerol are ionized, resulting in negative zeta-potential droplet surface charges, as in the blood. Synthetic nonionic agents such as pluronics (poloxamers or polyoxyethylene/polyoxypropylene derivatives) can also be found, albeit in small amounts due to 'overloading syndrome' secondary effects.[83] Moreover, the molar mass will determine their removal from the bloodstream. Surfactants with higher molar mass will be more challenging to remove when compared to their counterparts. Focused on both liquid phases, oil is commonly used as the dispersed phase. With scale-up, purity, and production costs in mind, initial studies tend to focus on LCTs or MCTs.[84] In some commercial emulsions, combinations of them have been reported, although MCTs are more soluble in water than LCTs. Apart from natural oils such as soybean, safflower, sesame, castor, or corn oil, other alternatives like squalene, triolein, or purified fish oil continue to be investigated in the field, even though further research is still required.[82,83] The continuous aqueous phase is also adjusted in this case, by buffers, antioxidants, and preservatives. Glycerol is the most widely used agent as it reduces the droplet size and improves the creaming of the emulsion.[85] In this line, α-tocopherol is commonly used as an antioxidant and immune-potentiating component, or methyl and butyl derivatives of *p*-hydroxybenzoic acid are employed as preservatives.[86,87] In contrast, sodium chloride and glucose are avoided due to problems with emulsion stability,[83] lecithin should be used carefully,[88] and W/O emulsions have not achieved remarkable results compared to their counterparts, mostly because they are less tolerated.

In parenteral nutrition, most of the intravenous lipid emulsions used are based on phospholipids, phosphatidylcholines, lysophosphatidylcholines, and triacylglycerols.[89] Voriconazole, a poorly water-soluble antifungal drug can be delivered using all-in-one parenteral nutrition admixtures, while being stable in a wide range of storage temperatures and concentrations for up to a week.[90] With a similar approach, aminoglycosides, a broad-spectrum bactericidal antibiotic, are suggested to also be added to commercial parenteral nutrition formulations.[91] Apart from nutrition, intralipid emulsions (ILEs) can also be used to treat drug toxicities, such as β-blockers, in cases of overdose. This compound is one of the most common causes of poison-induced cardiogenic shock and can be eliminated by the formation of β-blocker-ILE complexes. Lower hydrophobic β-blockers have an enhanced adsorption and binding constant with ILEs.[92] Still, new formulations are needed in the field of total parenteral nutrition, as they could lead to parenteral nutrition-associated liver disease in patients, characterized by secondary effects on the hepatic expression of genes in antioxidative stress and in drug metabolism.[93] For instance, using α-tocopherol (vitamin E),[94] or a fish oil lipid-based emulsion instead of a soybean oil intravenous lipid emulsion to treat children.[95] Emulsion adjuvants have also risen in popularity in recent years after facing pandemic challenges, with squalene being the most used component. AS03 or MF59 are the most historically known vaccines based on squalene and α-tocopherol, or squalene and surfactants only, respectively.[82] More complex emulsions can also be used, such as perfluorocarbon-based ones (FC). These hydrophobic perfluorochemicals are capable of dissolving many gases, including oxygen and carbon dioxide, and bring properties such as reduced allogeneic blood transfusions or improved tissue oxygenation.[96] An O/W emulsion based on squalene, LCTs, MCTs, and FC-40, a mixture of perfluorocarbon compounds, denoted higher antibody titers and more abundant long-lived plasma cells when squalene was present in the system.[97] *Lycium barbarum* polysaccharides-loaded particulate alum, with similar safety behavior to squalene, can be used via Pickering emulsion to adsorb antigens at the interface of the droplets. In vivo experiments denote an enhanced recruitment of antigen-presenting cells at the injection sites, activation of the dendritic cells in the lymph nodes, promotion of expression and activation of T lymphocytes, and induction of a strong immune memory T cell response.[98] Moreover, novel cationic microemulsions are also proposed to prevent influenza virus infections, as an alternative to the classical squalene-based O/W emulsions. CTAB, along with Polyoxyl 35 castor oil, d-alpha tocopheryl polyethylene glycol 1000 succinate, and macrogol 15 hydroxystearate, is capable of stabilizing wheat germ oil droplets. Animals immunized with this drug denoted an enhanced humoral and cellular immune response against influenza virus, showing its potential.[99] Finally, various biomimetic nanoemulsions are being proposed for treating tumors. For instance, incorporating a target-specific protein with a photosensitizer, which can lead to subcutaneous breast tumor growth inhibition, while the perfluorocarbon present in the nanoemulsion can provide oxygen as the source of photodynamic therapy.[100]

14.3.4 Other Administration Routes

While over three out of four emulsified systems are delivered orally, topically, or parenterally, other routes are also viable, such as ocular, intranasal, pulmonary, or intrathecal/brain delivery, which rely on small droplet sizes.[3] O/W emulsions are the preferred formulation to deliver hydrophobic ophthalmic drugs. They provide the best contact time of the drug with the eye by decreasing its drainage away by tears from the precorneal area. Current studies are focused on the treatment or targeting of different parts of the eye, such as cornea, conjunctiva, sclera, or choroid, as shown in Figure 14.6.[101,102] Cationic emulsions are predominantly prepared due to the negative nature of the ocular surface. They have a higher retention time and reduced water evaporation, promoting the integrity of the lacrimal film. Surfactants such as polysorbates, cremophors, poloxamers, or tyloxapol can be found in most systems, while stearylamine, oleylamine, polyethylenimine, or polylysine derivatives are avoided due to toxicity issues. MCTs, mineral, or vegetable oils are used as oleic phase, due to their biocompatibility, adequate viscosity, and low concentration requirement.[103] Cyclosporine A (CsA) is one of the most common hydrophobic drugs used to treat ocular inflammatory diseases such as dry eye diseases, uveitis, or keratoconjunctivitis. Several vegetable-based O/W emulsions have been designed to enhance its dispersion in aqueous media. Its application after allogeneic hematopoietic stem cell transplantation not only fights graft-versus-host disease sequelae but also prevents

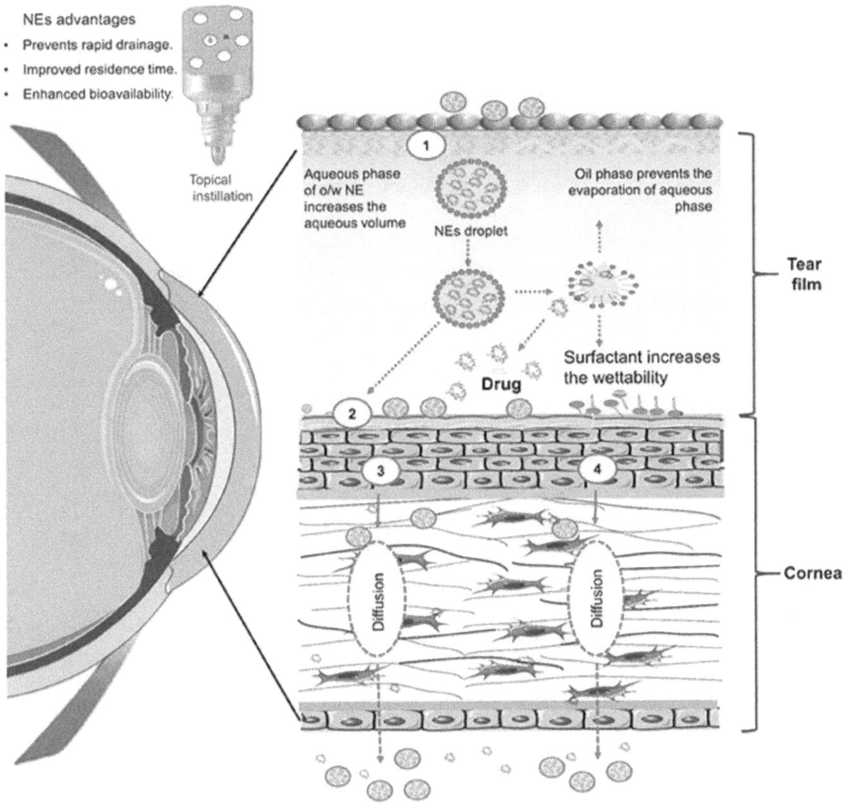

FIGURE 14.6 Drug interaction and penetration pathways in the eye. From the external tear film (1), and the mucin layer (2), to paracellular (3) and transcellular (4) diffusion from the epithelium to the endothelium. Reproduced from Singh et al.,[102] Copyright (2020), with permission from Elsevier.

it.[104] For instance, in Restasis®, CsA is dissolved in castor oil with Polysorbate 80 as the emulsifying agent, while Cyclokat® or Ikervis® are based on a cationic nano-emulsion.[105] Difluprednate, a corticosteroid used for the treatment of postoperative ocular inflammation and pain, can also be found dispersed in the same dispersing agents. However, the design of in vitro studies to accurately predict the behavior of emulsions after application, during tear turnover, and blinking, are still ongoing and required.[106]

The nasal route is also explored. However, as it is the first barrier of defense, several enzymes prevent the drug from being delivered. Droplets must reach the concha, while not being removed by the ciliated epithelial lining and survive alkaline pHs and lower temperatures than average human body conditions. Despite this challenge, this route allows for high absorption rates of hydrophobic drugs and facilitates high-efficacy administration. Figure 14.7 depicts the pathway made by the drug.[107] Moreover, due to the penetration of the cribriform plate above the concha, it allows

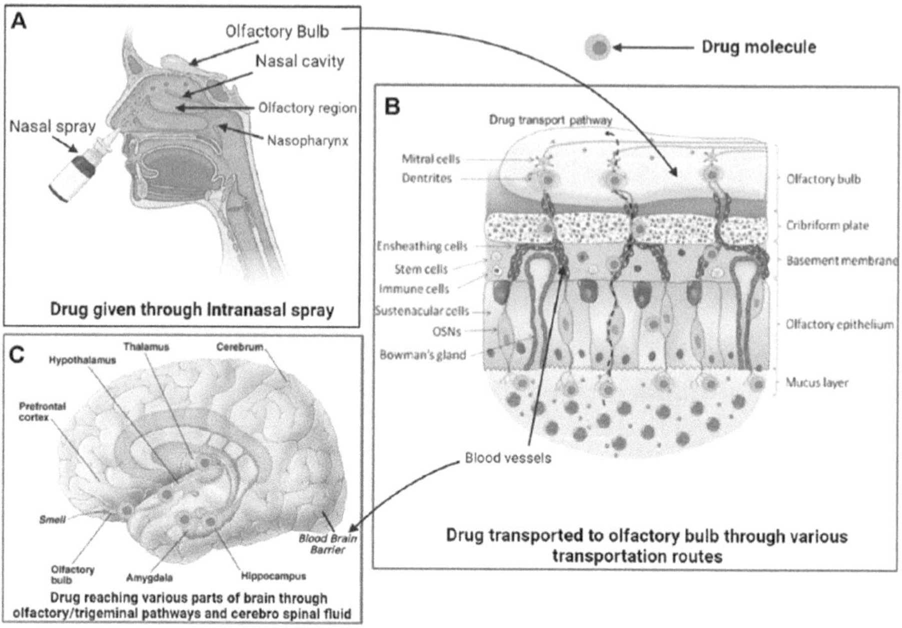

FIGURE 14.7 Nose-to-brain delivery pathway of neurotherapeutic drugs. Reproduced with permission from Rajput et al.,[107] with permission under Creative Common 4.0 Attribution International License [CC-BY 4.0].

the drug to be distributed to the nervous system or the brain after crossing the blood–brain barrier.[108] This barrier protects the central nervous system, which is involved in the proper functional behavior of neurons, and is the target of most present therapies.[109] Poly(D,L-lactide-*co*-glycolide) and chitosan are the two most commonly used polymers for drug encapsulation and controlled release, while cyclodextrins are used as absorption enhancers for nasal emulsions. Sorbitan monolaurate, polyoxyethylene, or lecithin are chosen in nasal-to-brain nanoemulsions. Cosurfactants such as ethanol or glycol derivatives are employed to improve system stabilization, along with natural oils as the dispersed phase. Simple O/W nanoemulsions based on eucalyptus oil, chitosan, and polysorbate 80 have been found to be viable in vitro and in vivo for the delivery of sulforaphane or tea polyphenols,[110] as chitosan-based nanoemulsions have risen in popularity in nose-to-brain drug delivery due to their enhanced blood–brain barrier permeability.[109] Insulin and pramlintide, peptide drugs capable of controlling blood glucose concentrations and providing an alternative therapeutic approach for Alzheimer's disease treatment, can also be delivered. These polyelectrolyte complexes were prepared using dextran sulfate and chitosan and then applied to stabilize Pickering emulsions along glycerol, MCTs, sorbitan monooleate, and polysorbate 80. Negatively charged dextran/pramlintide particles improve mucodiffusion properties, leading to a more constant and controlled permeation through the nasal epithelium. Positively charged chitosan/insulin particles promote a faster mucosal permeation of the peptide compared to its free form.[111] Phenytoin, an anti-convulsant drug used to treat neuropathic pain, can also be distributed in emulsions based on Miglyol® 812 (a MCT), polysorbate 80, and diethylene glycol monoethyl ether. The viscosity of the system was reported to be directly related to its release rate. A 10% anhydrous phase proportion leads to an immediate release profile, while 40% leads to a prolonged one.[112]

14.3.5 Other Considerations

While emulsions offer a wealth of benefits for drug delivery and therapeutic applications in the pharmaceutical field, their formulation is not without its challenges. Several variables must be considered when developing the ideal emulsion, apart from choosing the surfactants and phases to be used. This section focuses on various key points encountered in formulating pharmaceutical emulsions, highlighting general parameters, and how they are approached in some of the most relevant fields where they are successfully applied.

14.3.5.1 Droplet Size

Emulsion size is crucial for the physical stability, safety, applicability, and efficacy of the final system. Thus, it must be monitored during the emulsion manufacturing process.[113] For instance, parenteral emulsion systems require controlling and homogenizing the emulsion to maintain recommended droplet sizes below 500 nm. Globules in the 4–6 μm range can lead to emboli and changes in blood pressure. Also, their removal from the bloodstream is determined by the reticuloendothelial system and the uptake of microphages. Within the 1–20 nm range, they show greater average circulation half-times, while droplets of 30–100 nm avoid leakage into capillaries.[114] Larger particles are easily found, eliminated, and can result in vascular occlusion and pulmonary embolism, among other issues. Moreover, a blend of various sizes, always within the safe application range, can be used to include the desired release kinetics chemotherapy treatment designs.[115] In respiratory drug delivery, inhaled particles ranging from 1 μm to 5 μm are deposited with ease, exhibiting greater pulmonary penetration in the alveolar and bronchial areas, while smaller particles tend to be exhaled. However, the size will depend on the deposition mechanism (inertial impaction, sedimentation, and Brownian diffusion) and the targeted region. For example, bronchodilators are targeted by upper-range droplets,[116] while smaller ones are suitable for the distal lung.[117] In average nasal sprays, droplets are distributed in a range of 5–23 μm, although higher particles have also been reported.[118] Similarly, ophthalmic emulsions are recommended below 25 μm, to ensure better tolerance and reduce the risk of eye irritation. Delivery of topical and transdermal drugs depends on the penetration pathway. Focused solely on size, larger particles remain on the surface of the skin, whereas those below 3 μm can easily penetrate follicles and the stratum corneum, and even smaller ones result in percutaneous absorption via the follicular route. Finally, higher particle sizes can be employed in oral forms, where the final target and stability are key at the time of design.

14.3.5.2 pH, Temperature, and Osmolality Considerations

The pH and osmolality of an emulsion will influence its stability, tolerability, and therapeutic efficacy, depending on the final application. High temperatures may be necessary to form a stable emulsion, but cannot be applied to the patient. For example, human blood has a pH of 7.4, plasma is approximately 290 mOsm/L, and the temperature is 37°C. The use of either low or high pH can result in cellular damage, depending on the type of tissue exposed and the duration of exposure, while the use of hypo- or hypertonic solutions can lead to hemolysis or cell shrinking, respectively. Although not tested on humans, animal models suggest that there is a direct relationship between the use of undesirable pH and osmolarity levels of an infusate and the development of phlebitis.[119] In parenteral emulsions, the optimal pH is adjusted by adding small amounts of salt. Buffering agents are not usually used because of the potential for buffer catalysis of lipid hydrolysis, thereby altering the pH of the system. In ophthalmic solutions, the pH should be around 7.4 for maximum comfort and to avoid irritation, although unbuffered or slightly buffered emulsions can tolerate a wider range of pH. The osmolality should be around 300 mOsm/L, even though this can change due to evaporation processes with the eye open. Therefore, values below 100 or above 640 mOsm/L must be avoided.[103] For oral liquid medications, acidic to neutral pH, within the normal range of the mouth, and osmolality values between 100 mOsm/kg and 400 mOsm/kg are recommended, while extreme values are avoided to prevent gastrointestinal upset and diarrhea. For instance, hypertonic solutions of >400 mOsm/kg have been reported to cause gastrointestinal damage in newborns.[120] Finally, the skin thrives in a mildly acidic environment and 280 mOsm/L, and so are the recommended topical emulsions to be employed.[121]

14.3.5.3 Sterility and Preservative Challenges

The selection of appropriate sterilization and preservation methods is necessary for the correct use of emulsions to protect the patient from potential side effects such as infections, as the lipid components present in the formulated system can promote the growth of microorganisms. After homogenization, the emulsion can be filtered to remove any particles larger than the droplet size or receive a moist heat sterilization step. In addition, the emulsion can be stirred during these processes to maintain its integrity. However, there may be some limitations due to unwanted particles having the same size range as the emulsion droplets or instability due to temperature changes. The latter can lead to hydrolysis of lipids and lecithins, resulting in the formation of free fatty acids, and changing the pH of the system.[122] Alternative protocols, such as adding preservatives or sterilizing each component independently and aseptically prior to assembling the emulsion, can also be followed. However, they require a costly process and are difficult to scale up. Commercially, the final emulsion is typically packaged in USP Type I or II glass bottles, which may be siliconized to provide a hydrophobic surface in contact with the emulsion.[123] Although minimal traces of non-PVC plastics may be present without secondary effects, plastic containers are generally undesirable as they are permeable to oxygen and may contain oil-soluble plasticizers which can be extracted by the emulsion. Finally, the emulsion is placed in a nitrogen-based environment, where oxygen is removed from the headspace to prevent oxidation.

14.4 REGULATORY AND QUALITY CONTROL ASPECTS

Regulatory agencies worldwide closely monitor the use of pharmaceutical emulsions. These organizations establish measures to primarily ensure patient health. For that reason, their efficacy, quality, and performance are controlled in various aspects such as composition, stability, administration, and potential side effects. The Food and Drug Administration (FDA) in the United States,[124] the European Medicines Agency (EMA) in Europe,[125] the Central Drugs Standard Control Organization (CDSCO) in India,[126] the National Medical Products Administration (NMPA) in China,[127] and the African Medicines Agency (AMA) in Africa[128] are some of the most recognized agencies in charge of approving pharmaceutical products based on emulsions. However, while these agencies have similar requirements, each region may have specific considerations regarding drug viability. For instance, the NMPA and CDSCO place emphasis on their traditional medicines, leading to different approaches in medical treatment. On the other hand, the AMA focuses more on diseases that are more prevalent in its region, compared to others, such as human immunodeficiency virus (HIV) or tuberculosis (TB).[129,130] This section offers a summary and overview of the guidelines established by these agencies for the final marketing of pharmaceutical emulsions, as depicted in Figure 14.8. Our aim is to act as a connection between pharmaceutical scientists and manufacturers for emulsion development. However, for an updated and detailed view of the requirements, it is recommended to consult these organizations separately.

In terms of emulsions, it is generally preferable for them to have small sizes. They are highly recommended to have a size distribution ranging from 10 nm to 1000 nm, although larger values can be achieved by some systems. In that case, any deviations from this range should be strictly justified based on the application and formulation used. Particles below 100 nm can be efficiently taken up by cells and penetrate tissues while being less likely to be recognized and removed by macrophages.[131] It is also crucial to conduct stability studies for products over an average period of six months or longer, under appropriate storage conditions. During this time frame, their properties should also be thoroughly tested. For example, ophthalmic emulsions should ideally have a pH value around 7.4, similar to tears.[132] The osmolarity should be around 300 mOsm/kg, as the composition of lacrimal fluid is nearly

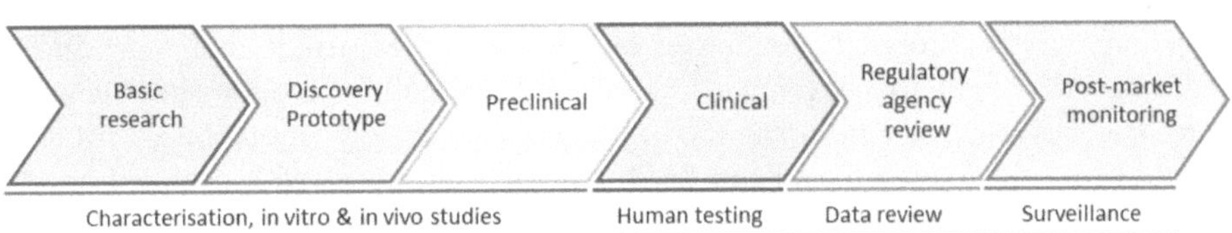

FIGURE 14.8 Average stages of drug development and approval.

isotonic, a 0.9% sodium chloride solution.[133] Additionally, it is important for the zeta potential values to deviate from neutrality.[134] While formulations outside these specified ranges may still receive a final approval, they require extensive stability testing. Finally, sterility and endotoxin tests are also performed during the full evaluation of the investigational product.

Once the system characterization is complete, preclinical studies are performed to assess cellular responses using in vitro tests. Subsequently, in vivo studies at various doses are carried out in animals with a distribution across genders and types, including rodents and nonrodents, to determine the viability of the compound. Depending on the administration route and drug formulation, specific toxicological studies may also be necessary, such as determining the maximum tolerated dose (MTD) or no observed adverse effect level (NOAEL), which include acute, subchronic, or reproductive toxicity evaluations, among others.[135,136] Once satisfactory preclinical results are obtained and there is an understanding of the kinetic and dynamic behavior of drugs, it progresses toward clinical trials.[137,138]

To briefly summarize, this process is typically divided into four phases. The first step, which focuses on ensuring the safety of the drug, involves determining a range of doses that are considered safe and to identify any side effects experienced by humans. This helps to understand the MTD without reaching toxicity levels. The second step is to refine the understanding of how the drug behaves in humans. Its efficacy and any side effect appearance in a larger number of participants are closely monitored. This phase also helps to determine whether the drug effectively treats the disease or condition it was developed for, demonstrating its therapeutic benefits. In phase 3, the outcomes from phase 2 are confirmed in a wider group of participants. This allows identifying effects and compare its effectiveness with existing standard treatments, if any. Finally, after submitting a New Drug Application (NDA) to the respective agency and obtaining the approval, it moves into phase 4 or post-marketing surveillance. There, the long-term behavior and effects of the drug on the population are monitored.

From a scientific perspective, the rigorous analysis of these organizations highlights why pharmaceutical researchers need to develop accurate formulations. As advancements are made in drug delivery systems, adhering to these guidelines not only safeguards health but also instills confidence in any new system proposed. Thus, it is key for scientists, manufacturers, and regulators to maintain continuous collaboration.

14.5 TRENDS AND INNOVATIONS

As already demonstrated in previous sections of the chapter, the development of emulsions has not stopped growing and adapting to the current challenges in the pharmaceutical field. Moreover, it is also expected to continue evolving until it establishes itself as one of the pillars in the future of drug delivery. In this line, as of now, several lines of research have been developed and optimized with new and unique approaches to any biological inconvenience at the time of treatment of a patient. Here, we highlight some of the most prominent focuses of research where emulsions are being developed and applied, providing a broad view of their existing and incoming potential.

14.5.1 Novel Drug Delivery Systems Using Nanoemulsions

One of the most notable advances has been the proliferation of nanoemulsion platforms, which are at the center of drug delivery applications and are likely to remain this way in the near future.[139–141] Characterized by droplet sizes in the nanometer range, typically around 200 nm, they are suitable candidates for enhancing drug solubility, especially for those drugs that are hydrophobic in nature.[142] Due to their small droplet size, nanoemulsions offer an increased surface area, which enhances their penetration and absorption. In addition, their intrinsic properties allow them to be more effective, easier to administer, and more patient-friendly than their traditional larger counterparts. Such advanced drug delivery systems can provide improved bioavailability, minimized side effects, superior thermodynamic stability over microemulsions, and more precise targeting of pathological sites. Still, it should be noted that other emulsification systems have their own unique characteristics.

Of all diseases, cancer remains as one of the leading causes of mortality.[143] However, advances in nanotechnology have had a rapid impact on its treatment. From simple water/Tween 80/castor oil systems[144] to complex ones based on miRNA treatments,[145] nanoemulsions have proven to be key due to their innate properties. For instance, the pulmonary delivery of an osimertinib-loaded perfluoro-15-crown-5-ether nanoemulsion showed excellent efficacy in lung cell cancer. By rapidly distributing the nanoemulsion in lung tissue and tumors without short- and long-term toxic effects, the nanocarrier is able to properly release the drug under the guidance of low-intensity focused ultrasound (LIFU), as shown in Figure 14.9.[146] In the case of breast cancer, the overexpression of folate receptors is exploited by targeting them, as achieved in vitro and in vivo by folate/PEG-DSPE-based nanoemulsions.[147] On the other hand, the incorporation of paclitaxel, perillyl alcohol, or tributyrin is another routine method to increase the cytotoxicity of nanoemulsions against cancer cells, selectively.[148] Intraductal drug delivery, which mainly comes from the ductal lining, can specifically target neoplastic and pre-neoplastic lesions, thus reducing systemic side effects. A positively charged O/W nanoemulsion below 50 nm size has been shown to encapsulate and transport C6 ceramide at a concentration

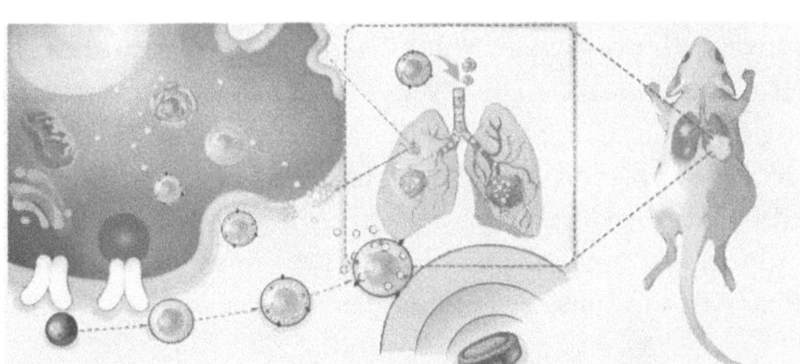

FIGURE 14.9 Schematic of the pulmonary delivery of an osimertinib-loaded perfluoro-15-crown-5-ether nanoemulsion guided by magnetic resonance imaging and improved delivery by low-intensity focused ultrasound. Reproduced with permission from Yang et al.,[143] Copyright 2022, American Chemistry Society.

enough to decrease breast cancer cell viability to 50% (EC_{50}), supporting the benefits of nanoemulsions in enabling localization to breast tissue. Moreover, other drugs, such as tributyrin, can also be added to the oleic phase to enhance its efficacy.[149] A similar approach is used to treat central nervous system (CNS) diseases, where the blood–brain barrier needs to be crossed. Several nanoemulsions have been proven to be effective when used for direct nose-to-brain delivery, and we encourage the reader to consult the cited works, as this is a broad area of research.[150–153] In the realm of ocular therapeutics, the challenges associated with drug delivery are enormous due to the eye's complex anatomy and protective barriers.[154,155] For example, new nanoemulsions based on Tween 80, Soluphor®, ethyl oleate, and water have shown promising results in vitro and ex vivo, where this O/W system enhances the therapeutic effect of moxifloxacin and can therefore be used as a safe and effective delivery vehicle for ophthalmic therapies.[156] Moreover, cationic nanoemulsions are key to the treatment of dry eye disease (DED), which affects millions of people globally. Non-steroidal anti-inflammatory drugs (NSAIDs), such as ibuprofen, can be loaded into chitosan-coated O/W systems to enhance their effectiveness due to their mucoadhesive character, as confirmed by in vitro studies, and excellent biocompatibility, based on model and ex vivo porcine corneas.[157] Finally, among other research areas, the transdermal properties of nanoemulsions are being explored for applications such as pain management or disease treatment, offering controlled release and enhanced permeation through the skin barrier.[158] Superficial tinea infection is reported to be treated by SNEDDs, based on clove oil loaded with naftifine. The synergetic anti-inflammatory and antifungal properties of the oil, along with the drug, showed high activity, permeability, diffusion, and reasonable stability with no or mild signs of skin sensitivity in rats.[159] Curcumin, a hydrophobic polyphenol known for its therapeutic properties, has historically faced challenges due to its poor solubility. However, recent research has highlighted the enhanced solubility and bioavailability of curcumin when incorporated into nanoemulsions. By using eucalyptol or pinene as chemical penetration enhancers, a nanoemulsion formulated with polysorbate 80 and soybean lecithin demonstrated superior in vivo performance in delivering curcumin into the skin. This efficacy is largely attributed to the interfacial modification induced by the monoterpene compounds.[160] Tacrolimus, a potent immunosuppressive with poor water solubility used to treat psoriasis, can be found in propylene carbonate/paraffin. However, Protopic® can be described as having a greasy texture and being difficult to apply and spread on the skin. Similar emulsions based on lecithin, polysorbate 80, Precirol® ATO 5, and Capryol 90 have also shown enhanced delivery in vitro.[161] Alternatively, nanoemulsions based on kalonji oil, polyethene glycol 400, and Cremophor RH 40 have been reported to denote a high in vivo efficacy of the same compound.[162]

14.5.2 Biopolymer-Based Emulsions for Drug Delivery

At present, pharmaceutical emulsions are mostly stabilized by synthetic surfactants. However, their substitution for biocompatible, bioactive, sustainable, and mostly less toxic emulsifiers is necessary. Biopolymers, mainly derived from animals and plants, are already widely used in food products to stabilize emulsions because of their amphiphilic behavior, as they reduce the interfacial tension at the oil–water interface. Thus, their incorporation would be a major breakthrough.[163] Still, these emulsifiers are not as implemented in the pharmaceutical field as it would be desired. Even though these emulsifiers bring an astonishing amount of benefits, issues such as their systematic and reproducible production, final average emulsion sizes, or possible allergenicity, bacteriological, and storage concerns keep them still under trial, compared to the actual well-established ones.

Several polysaccharides have gained popularity in recent years, even though most of them do not show surface activity. Xanthan gum is an anionic polysaccharide commonly used in food but also in the pharmaceutical industry due to its emulsifying, thickening, and gelling properties. It is composed of pentasaccharide repeating units consisting of mannose, glucose, and glucuronic acid, and can be present as a single, double, or triple helix. Its high low-shear viscosity and strong shear-thinning character denote its better properties as an emulsifier than other alternatives such as β-lactoglobulin or gum arabic for stabilizing almond oil–based O/W emulsions. While the latter is only adsorbed at the interface, xanthan gum also thickens the continuous phase.[164,165] Due to that gelling property, it can also be used in the form of nanoparticles to stabilize Pickering emulsions. In conditions of up to 40% oil content, the droplet size can be controlled and decreased in the presence of the polysaccharide.[166] Carrageenan and its derivatives, extracted from seaweed, can be used to develop gels and W/O emulsions due to their skin permeation ability,[167] as well as hyaluronan,[168] characterized by a high molar mass and good water solubility. Hypromellose or hydroxypropylmethylcellulose is characterized by hydrophobic (methoxyl) and hydrophilic (hydroxypropyl) groups and is used for controlled drug release. Low usage ratios of 0.45 to 1% (w/w) have been reported in eye drops and artificial tear solutions.[169] In the form of nontoxic inexpensive particles, they can stabilize emulsions for months at low concentrations. The hydrophobicity and wetting properties of the in situ formed particles strongly influenced the dispersion stability.[170] Nevertheless, hypromellose is primarily employed as a tablet binder or matrix in oral pharmaceutical products, and further research is required. Chitosan, a linear polysaccharide consisting of D-glucosamine and *N*-acetyl-D-glucosamine, has denoted its effectiveness as an emulsion stabilizer for nasal vaccines at physiological pH in the presence of oleic acid. At pH values above 6.5, the amine groups of chitosan are partially deprotonated and can interact with hydrophobic groups.[171] Pectin, mainly composed of esterified D-galacturonic acid residues in an α-(1-4) chain, is found in the cell walls of plants and used in treatments such as diarrhea, constipation, or obesity, and also as an emulsifier.[169] High methoxy pectins (HMP) are reported to result in smaller droplet sizes compared to low methoxy pectins (LMP). By high-pressure homogenization, highly stable O/W emulsions with a low creaming index can be achieved using 0.5–1.5% pectins and 20% (w/w) rice bran oil.[172] HMPs have been proposed for the delivery of Itraconazole in the form of SEDDS to treat various pathogens such as *Candida albicans* or *Cryptococcus neoformans*, major causes of mortality in HIV patients.[173] In addition, pectin-based nanoparticles can be functionalized, for example, by dopamine, to form 30-day stable, highly pH-responsive, and biocompatible Pickering emulsions for the delivery of indomethacin, a water-insoluble drug commonly used as an anti-inflammatory.[174] Mixtures of polysaccharides can also be found, such as the use of gum arabic and chitosan nanoparticles to form O/W emulsions for curcumin delivery. The designed Pickering emulsion is sensitive to pH and ionic strength but stable to temperature changes. With a nanoparticle concentration as low as 0.75%, a curcumin-controlled embedding rate of 94% was achieved.[175] Protein-based emulsions can also be found, as they are able to adsorb at the liquid–liquid interface and reduce the interfacial tension due to their hydrophilic and hydrophobic residues, and thus regions, after their assembly and reorganization. Lecithins are mixtures of phospholipids derived from vegetable and animal origins characterized by having a hydrophobic region that helps to stabilize the water–oil mixture. These proteins have mostly been applied in O/W emulsions, such as soybean,[176] sunflower,[177] or corn[178] lecithin. Microemulsions based on soybean lecithin have shown improved in vitro and in vivo transdermal permeation of ketoprofen, a potent nonsteroidal anti-inflammatory drug, compared to conventional formulations using penetration enhancers such as oleic acid.[179] The same transdermal behavior has also been exploited as a cosurfactant for the delivery of 5-fluorouracil, an anticancer drug,[180] or fluconazole to avoid its gastrointestinal side effects.[181] However, in recent years, protein nanocages have been attracting interest for stabilizing Pickering emulsions, as they can act as co-carriers, targeting entities, and stabilizers.[182] Representative protein nanocages are illustrated in Figure 14.10. For instance, the supramolecular E2 protein nanocage, a dodecahedral cage-structured protein with an average diameter of 27 nm, was used to stabilize a Pickering

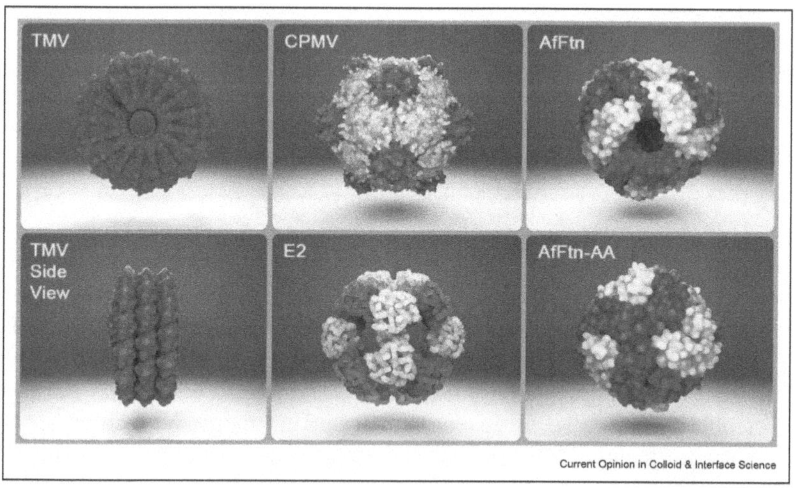

FIGURE 14.10 Schematic of protein nanocages used in Pickering emulsions. *From top left to bottom right:* Tobacco mosaic virus, cowpea mosaic virus, *Archaeoglobus fulgidus* ferritin, tobacco mosaic virus side view, *Geobacillus stearothermophilus* E2, and AfFtn closed-pore mutant. Reproduced from Lim and Salentinig,[182] with permission under Creative Common 4.0 Attribution International License [CC-BY 4.0].

emulsion characterized by 400 nm negatively charged O/W droplets of rosemary oil. In addition, it was engineered to load the antitumor drug Doxorubicin® in the internal cavity and disassemble in response to pH changes, releasing both its inner drug and that present in the oleic phase, as it also breaks the emulsified system.[183]

In parallel, more complex systems have been developed using mixtures of proteins and polysaccharides. The resulting conjugate has improved surface, solubility, and emulsifying properties, among others. Xanthan gum, locust bean gum, and lecithin are used to develop W/O/W emulsions. The presence of polysaccharides improves the long-term stability against coalescence, while lecithin is only required in low amounts of 1% (w/w), allowing the final delivery of polyphenols.[184] Wheat gluten nanoparticle–xanthan gum complexes can stabilize Pickering emulsions over a wide range of pH values (4–8) and salt levels (0–1,000 mM NaCl). Moreover, in vitro experiments denoted their enhanced β-carotene stability and bioaccessibility for nutraceutical delivery, compared to wheat gluten alone.[185] Using the layer-by-layer technique, O/W emulsions based on whey protein and either chitosan or carrageenan indicate high in vitro skin absorption properties and salicylic acid delivery. The charge of the emulsion droplets plays a key role in dermal and transdermal delivery.[186] Complexation occurs directly at the droplet surface through attractive electrostatic interactions. This layering process can be repeated using oppositely charged biopolymers. A triple shell based on β-lactoglobulin/chitosan/pectin can be formed to stabilize oil droplets. Due to the presence of the polysaccharides and their charge, the system demonstrates good stability against aggregation from pH 7 to 4. In vitro, digestive models also demonstrated how these coatings control the rate of digestibility, which is crucial for their application.[187] Moreover, the addition of the protein layer before the polysaccharide layer seems to be crucial for achieving the final stability. The combination of whey protein and gum arabic leads to better stability than separately. The protein moieties are more surface-active and their preferential adsorption is significant for the interfacial layer built on the oil droplets.[188]

14.5.3 Stimuli-Responsive Emulsions

The long-term stability of emulsions is one of the key factors that makes them ideal. However, in some practical applications, it is necessary to break or invert the emulsion under certain conditions to release the drug. As an alternative to traditional methods such as shearing, as described in the section on multiple emulsions, an intelligent response to external stimuli is desired, i.e., stimuli-responsive emulsions. For example, the stability of the emulsions can be controlled by changing the behavior of the stabilizers. Currently, several variables such as temperature, pH, radiation, or magnetic field can be used, separately or together, to tune up emulsion properties and behavior.[189,190]

Temperature-responsive emulsions are one of the most developed systems, as temperature changes occur in living organisms. Branched copolymer surfactants based on diethylene glycol methyl ether methacrylate (DEGMA), polyethene glycol methacrylate (PEGMA), ethylene glycol dimethacrylate (EGDMA), and dodecanethiol (DDT) have been used to develop O/W emulsions with a temperature-responsive gelation at 32°C, making them suitable for in situ gelation after human body contact.[191] A similar approach has been used to deliver copaiba oil resin, an anti-inflammatory and wound-healing aid. At 35°C, a system composed of 18–20%, 0.25%, and 15% (w/w) of F127, Carbopol®, and copaiba oil resin, respectively, showed a great thermal stimulus-responsiveness, resulting in a wide viscosity range before and after skin administration.[192] In terms of pH, Pickering emulsions stabilized by an iron oxide/chitosan nanocomposite were applied. The superior stability of the emulsion was observed at lower pH values (1.5–4) compared to pH 7.4. The difference could be related to the particle–particle electrostatic repulsion induced by the deprotonation of the amino group of chitosan and the hydroxyl group of iron oxide at higher pH values.[193] Light-responsive emulsions also offer a wide range of applications. It is a nondisruptive factor whose area, duration, and intensity can be easily controlled during application. W/O Pickering emulsions have been stabilized with tuned TiO_2 nanoparticles. When treated with UV, this particle's hydrophilicity changes due to the adsorption of hydroxyl groups on its surface, disabling its activity.[194] Similar behavior has been described for biocompatible photothermally responsive O/W emulsions based on polydopamine bowl-shaped mesoporous nanoparticles for the delivery of dopamine,[195] or W/O/W emulsion capsules for the encapsulation and release of doxorubicin.[196] *Alcaligenes faecalis* ATCC 8750 bacteria can also be delivered using this property through W/O Pickering emulsions, following an inversion protocol. Photochromic sipropyran undergoes a hydrophilic–lipophilic conformational change under NIR/visible light, allowing the stabilization of O/W emulsions.[197] Induction magnetic response has also been a focus of emulsion development. The system can be destabilized by an external field and later restored by mechanical mixing.[198] However, this area requires further development.

Nevertheless, the future of these emulsions lies in dual- or multi-stimulus systems, emulsion stabilizers can respond to at least two of these variables. Therefore, sophisticated nanoparticle and copolymer emulsifiers are being developed and applied. Based on poly(*N*-isopropylacrylamide) (PNIPAM) microgel particles, an octanol-in-water emulsion could be destabilized by lowering the pH at a fixed temperature, or by heating at a given pH. Both stimuli lead to the protonation of carboxyl groups in the particles, increasing the hydrophobicity and interparticle repulsion, reducing the droplet coverage, and promoting the coalescence.[199] Using silica/PNIPAM nanoparticles loaded with Nile Red, a final O/W Pickering emulsion was developed and charged with (6)-carboxyfluorescein diacetate (CFDA). An increase in temperature releases the Nile Red, while a decrease in pH do the same for the CFDA.[200] pH and magnetically bi-responsive emulsions have also been established as the way to go. Magnetic Fe_2O_3 hydroxyapatite nanoparticles lead to stable and reversible mineral oil–water emulsions at pH 10. A decrease in pH is associated with an increase in droplet size and final demulsification, while the emulsion stability is reduced and a droplet coalescence is induced by the application of a magnetic field, as shown in Figure 14.11.[201] Moreover, the pH can be used to control the droplet size using chitosan pH-responsive properties, while the magnetic component directs the motion.[202] The magnetic field can also be applied along a temperature response. O/W emulsions have been stabilized by Fe_3O_4/PNIPAM,[203] or Fe_3O_4/PNIPAM-*co*-HEMA-co-MAA

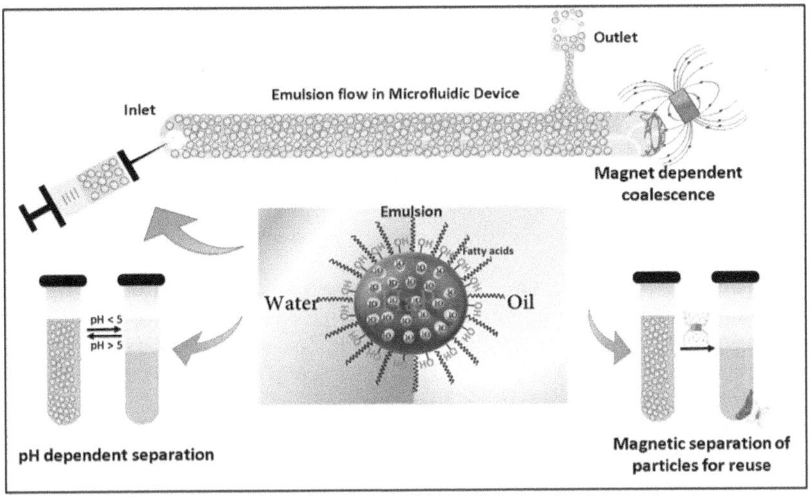

FIGURE 14.11 Schematic of pH and magnetic dual-stimulus-responsive Pickering emulsions based on hydroxyapatite nanoparticles. Reproduced with permission from Mendiratta et al.,[201] Copyright 2022, American Chemistry Society.

temperature-sensitive magnetic nanoparticles.[204] Heating the emulsions with a high-frequency magnetic field leads to phase separation while the oil droplets can be separated from the continuous phase by applying a static magnetic field. Finally, it is important to mention alternative dual-stimulus responsive systems, such as cross-linked star polymers based on 2-methoxyethyl acrylate and poly(ethylene glycol) acrylate, that can be found for temperature and salt response,[205] or P(MAA-*co*-MMA-*co*-VM) for pH and light response.[206]

14.6 CONCLUDING REMARKS

In the field of drug delivery systems and pharmaceutical formulations, emulsions have undoubtedly become indispensable. From simple oil-in-water and water-in-oil to more complex systems, emulsions are not just a basic focus of academic research. Their intrinsic physicochemical properties and versatility make them ideal for their use in current and future therapeutic applications. Ease of design, fine-tuning, biocompatibility, drug delivery, enhanced solubility, targeting efficacy, or controlled release kinetics and responsiveness are just a few of these properties. However, despite this profound potential, their very nature as a dispersion of two immiscible liquids presents several formulation difficulties. Stability, sterility, pH considerations, drug-excipient interactions, or even regulatory aspects are some of the conditions on the other side of the coin. In summary, the future of emulsions in pharmaceutical applications is filled with both challenges and opportunities. As research continues to push the limits of what is possible with emulsions, their importance in the pharmaceutical landscape is set to grow exponentially.

REFERENCES

1. McNaught, A.D.; Blackwell, A.W. *The IUPAC Compendium of Chemical Terminology*, International Union of Pure and Applied Chemistry (IUPAC): Research Triangle Park, NC, USA; 2019.
2. Zia, A.; Pentzer, E.; Thickett, S.; Kempe, K. Advances and opportunities of oil-in-oil emulsions. *ACS Appl. Mater. Interfaces* **2020**, *12*, 38845–38861.
3. Callender, S.P.; Mathews, J.A.; Kobernyk, K.; Wettig, S.D. Microemulsion utility in pharmaceuticals: implications for multi-drug delivery. *Int. J. Pharm.* **2017**, *526*, 425–442.
4. Cserháti, T.; Forgács, E.; Oros, G. Biological activity and environmental impact of anionic surfactants. *Environ. Int.* **2002**, *28*, 337–348.
5. de Carvalho-Guimarães, F.B.; Correa, K.L.; de Souza, T.P.; Rodríguez Amado, J.R.; Ribeiro-Costa, R.M.; Silva-Júnior, J.O.C. A review of pickering emulsions: Perspectives and applications. *Pharmaceuticals* **2022**, *15*, 1413.
6. Harman, C.L.G.; Patel, M.A.; Guldin, S.; Davies, G.L. Recent developments in pickering emulsions for biomedical applications. *Curr. Opin. Colloid Interface Sci.* **2019**, *39*, 173–189.
7. Popkin, B.M.; D'Anci, K.E.; Rosenberg, I.H. Water, hydration, and health. *Nutr. Rev.* **2010**, *68*, 439–458.
8. Mitchell, H.H.; Hamilton, T.S.; Steggerda, F.R.; Bean, H.W. The chemical composition of the adult human body and its bearing on the biochemistry of growth. *J. Biol. Chem.* **1945**, *158*, 625–637.
9. Yadav, K.S.; Soni, G.; Choudhary, D.; Khanduri, A.; Bhandari, A.; Joshi, G. Microemulsions for enhancing drug delivery of hydrophilic drugs: Exploring various routes of administration. *Med. Drug Discov.* **2023**, *20*, 100162.
10. Aserin, A. *Multiple Emulsions: Technology and Applications*, Aserin, A., Ed.; John Wiley & Sons, Inc.: Hoboken, NJ, USA, 2007.
11. Bozkir, A.; Hayta, G.; Saka, O.M. Comparison of biodegradable nanoparticles and multiple emulsions (water-in-oil-in-water) containing influenza virus antigen on the in vivo immune response in rats. *Pharmazie* **2004**, *59*, 723–725.
12. Fukanoki, S.I.; Iwakura, T.; Iwaki, S.; Matsumoto, K.; Takeda, R.; Ikeda, K.; Shi, Z.; Mori, H. Safety and efficacy of water-in-oil-in-water emulsion vaccines containing Newcastle disease virus haemagglutinin-neuraminidase glycoprotein. *Avian Pathol.* **2001**, *30*, 509–516.
13. Mori, H.; Nakazawa, H.; Shirai, N.; Shibata, N.; Sumida, M.; Matsubara, F. Foreign gene expression by a baculovirus vector with an expanded host range. *J. Gen. Virol.* **1992**, *73*, 1877–1880.
14. Kalariya, M.; Amiji, M.M. Systemically administered Gp100 encoding DNA vaccine for melanoma using water-in-oil-in-water multiple emulsion delivery systems. *Int. J. Pharm.* **2013**, *453*, 400–407.
15. Pliasas, V.C.; Menne, Z.; Aida, V.; Yin, J.H.; Naskou, M.C.; Neasham, P.J.; North, J.F.; Wilson, D.; Horzmann, K.A.; Jacob, J.; Skountzou, I.; Kyriakis, C.S. A novel neuraminidase virus-like particle vaccine offers protection against heterologous H3N2 influenza virus infection in the porcine model. *Front. Immunol.* **2022**, *13*, 1–19.
16. Han, L.; Yang, R.; Yuan, S.; Ding, S.; Wu, Z.; Wu, Z.; Qi, X. Soft multiple emulsions demonstrating reversible freeze-thawing capacity and enhanced skin permeability of diclofenac sodium. *Colloid Polym. Sci.* **2018**, *296*, 471–481.
17. Wang, Q.; Zhu, X.; Shi, F.; Liu, Y.; Hu, W.; Yang, Z. Influence of β-Cyclodextrin concentration on the physicochemical properties and skin permeation behavior of vitamin c-loaded pickering water-in-oil-in-water (w1/o/w2) double emulsions. *J. Drug Deliv. Sci. Technol.* **2022**, *72*, 103368.

18. Khadem, B.; Khellaf, M.; Sheibat-Othman, N. Investigating swelling-breakdown in double emulsions. *Colloids Surf. A* **2020**, *585*, 124181.
19. Luhede, L.; Wollborn, T.; Fritsching, U. Stability of multiple emulsions under shear stress. *Can. J. Chem. Eng.* **2020**, *98*, 186–193. h
20. Grossiord, J.L.; Seiller, M.; Puisieux, F. Apport Des Analyses Rheologiques Dans l'6tude d' Mulsions Multiples H / L / H. *Rheol. Acta* **1993**, *180*, 168–180.
21. Raynal, S.; Grossiord, J.L.; Seiller, M.; Clausse, D. A topical w/o/w multiple emulsion containing several active substances: Formulation, characterization and study of release. *J. Control. Release* **1993**, *26*, 129–140.
22. Muguet, V.; Seiller, M.; Barratt, G.; Ozer, O.; Marty, J.; Grossiord J.. Formulation of shear rate sensitive multiple emulsions. *J. Control. Release* **2001**, *70*, 37–49.
23. Olivieri, L.; Seiller, M.; Bromberg, L.; Besnard, M.; Duong, T.-N.-L.; Grossiord, J.-L. Optimization of a thermally reversible w/o/w multiple emulsion for shear-induced drug release. *J. Control. Release* **2003**, *88*, 401–412.
24. Hamoudeh, M.; Seiller, M.; Chauvierre, C.; Auchère, D.; Lacour, B.; Pareau, D.; Stambouli, M.; Grossiord, J.L. Formulation of stable detoxifying w/o/w reactive multiple emulsions: In vitro evaluation. *J. Drug Deliv. Sci. Technol.* **2006**, *16*, 223–228.
25. Frasca, S.; Couvreur, P.; Seiller, M.; Pareau, D.; Lacour, B.; Stambouli, M.; Grossiord, J.L. Paraquat detoxication with multiple emulsions. *Int. J. Pharm.* **2009**, *380*, 142–146.
26. Mishra, B.; Pandit, J.K. Prolonged release of pentazocine from multiple o/w/o emulsions. *Drug Dev. Ind. Pharm.* **1989**, *15*, 1217–1230.
27. Yoshida, K.; Sekine, T.; Matsuzaki, F.; Yanaki, T.; Yamaguchi, M. Stability of vitamin a in oil-in-water-in-oil-type multiple emulsions. *J. Am. Oil Chem. Soc.* **1999**, *76*, 1–6.
28. Laugel, C.; Baillet, A.; Ferrier, D.; Grossiord, J.L.; Marty, J.P. Incorporation of triterpenic derivatives within an o/w/o multiple emulsion: Structure and release studies. *Int. J. Cosmet. Sci.* **1998**, *20*, 183–191.
29. Benichou, A.; Aserin, A.; Garti, N. O/W/O double emulsions stabilized with wpi-polysaccharide conjugates. *Colloids Surf. A* **2007**, *297*, 211–220.
30. Ma, W.L.; Mou, C.L.; Chen, S.H.; Li, Y.D.; Deng, H.B. A mild method for encapsulation of citral in monodispersed alginate microcapsules. *Polymers.* **2022**, *14*, 1165.
31. Zhang, M.; Nowak, M.; Malo De Molina, P.; Abramovitch, M.; Santizo, K.; Mitragotri, S.; Helgeson, M.E. Synthesis of oil-laden poly(ethylene glycol) diacrylate hydrogel nanocapsules from double nanoemulsions. *Langmuir* **2017**, *33*, 6116–6126.
32. Khopade, A.J.; Mahadik, K.R.; Jain, N.K. Targeting of multiple emulsions to the lungs. *Pharmazie* **1996**, *51*, 558–562.
33. Khopade, A.J.; Jain, N.K. Surface-modified fine multiple emulsions for anticancer drug delivery. In *Multiple Emulsions,* Aserin, A., Ed.; Wiley: Hoboken, NJ, USA, 2007; pp. 235–256.
34. Tadros, T.F. 2. Pharmaceutical emulsions. *Pharm. Cosmet. Pers. Care Formul.* **2018**, *2*, 19–44.
35. Guy, R.H.; Hadgraft, J.; Kellaway, I.W.; Taylor, M. Calculations of drug release rates from particles. *Int. J. Pharm.* **1982**, *11*, 199–207.
36. Shaker, D.S.; Ishak, R.A.H.; Ghoneim, A.; Elhuoni, M.A. Nanoemulsion: A review on mechanisms for the transdermal delivery of hydrophobic and hydrophilic drugs. *Sci. Pharm.* **2019**, *87*, 17.
37. Vavrova, K.; Kovačik, A.; Opalka, L. Ceramides in the skin barrier. *Eur. Pharm. J.* **2017**, *64*, 28–35.
38. Ahmadi-Ashtiani, H.R.; Bishe, P.; Baldisserotto, A.; Buso, P.; Manfredini, S.; Vertuani, S. Stem cells as a target for the delivery of active molecules to skin by topical administration. *Int. J. Mol. Sci.* **2020**, *21*, 2251.

39. Maibach, H.I.; Dragicevic, N. *Percutaneous Penetration Enhancers Chemical Methods in Penetration Enhancement*, Springer: Berlin, Germany, 2015.
40. Lu, G.W.; Gao, P. Emulsions and microemulsions for topical and transdermal drug delivery. In *Handbook of Non-Invasive Drug Delivery System*, Kulkarni, V.S., Ed.; Elsevier: Amsterdam, The Netherlands, 2010; pp. 59–94.
41. Schulz, R.; Yamamoto, K.; Klossek, A.; Rancan, F.; Vogt, A.; Schütte, C.; Rühl, E.; Netz, R.R. Modeling of drug diffusion based on concentration profiles in healthy and damaged human skin. *Biophys. J.* **2019**, *117*, 998–1008.
42. Shaker, D.S.; Ghanem, A.H.; Li, S.K.; Warner, K.S.; Hashem, F.M.; Higuchi, W.I. Mechanistic studies of the effect of hydroxypropyl-β-cyclodextrin on in vitro transdermal permeation of corticosterone through hairless mouse skin. *Int. J. Pharm.* **2003**, *253*, 1–11.
43. Bolzinger, M.A.; Briançon, S.; Pelletier, J.; Chevalier, Y. Penetration of drugs through skin, a complex rate-controlling membrane. *Curr. Opin. Colloid Interface Sci.* **2012**, *17*, 156–165.
44. Rai, V.K.; Mishra, N.; Yadav, K.S.; Yadav, N.P. Nanoemulsion as pharmaceutical carrier for dermal and transdermal drug delivery: Formulation development, stability issues, basic considerations and applications. *J. Control. Release* **2018**, *270*, 203–225.
45. Mei, Z.; Chen, H.; Weng, T.; Yang, Y.; Yang, X. Solid lipid nanoparticle and microemulsion for topical delivery of triptolide. *Eur. J. Pharm. Biopharm.* **2003**, *56*, 189–196.
46. de Freitas Araújo Reis, M.Y.; de Araújo Rêgo, R.I.; Rocha, B.P.; Guedes, G.G.; de Medeiros Ramalho, Í.M.; de Medeiros Cavalcanti, A.L.; Guimarães, G.P.; de Lima Damasceno, B.P.G. A general approach on surfactants use and properties in drug delivery systems. *Curr. Pharm. Des.* **2021**, *27*, 4300–4314.
47. Otto, A.; Du Plessis, J.; Wiechers, J.W. Formulation effects of topical emulsions on transdermal and dermal delivery. *Int. J. Cosmet. Sci.* **2009**, *31*, 1–19.
48. Kegel, W.K.; Lekkerkerker, H.N.W. Phase behavior of an ionic microemulsion system as a function of the cosurfactant chain length. *Colloids Surf. A* **1993**, *76*, 241–248.
49. Albèr, C.; Buraczewska-Norin, I.; Kocherbitov, V.; Saleem, S.; Lodén, M.; Engblom, J. Effects of water activity and low molecular weight humectants on skin permeability and hydration dynamics – A double-blind, randomized and controlled study. *Int. J. Cosmet. Sci.* **2014**, *36*, 412–418.
50. Gomes, A.; Aguiar, L.; Ferraz, R.; Teixeira, C.; Gomes, P. The emerging role of ionic liquid-based approaches for enhanced skin permeation of bioactive molecules: A snapshot of the past couple of years. *Int. J. Mol. Sci.* **2021**, *22,* 1991.
51. Zhou, Z.; Liu, C.; Wan, X.; Fang, L. Development of a w/o emulsion using ionic liquid strategy for transdermal delivery of anti – aging component α – lipoic acid: Mechanism of different ionic liquids on skin retention and efficacy evaluation. *Eur. J. Pharm. Sci.* **2020**, *141*, 105042.
52. Islam, M.R.; Chowdhury, M.R.; Wakabayashi, R.; Kamiya, N.; Moniruzzaman, M.; Goto, M. Ionic liquid-in-oil microemulsions prepared with biocompatible choline carboxylic acids for improving the transdermal delivery of a sparingly soluble drug. *Pharmaceutics* **2020**, *12*, 392.
53. Zhang, Y.; Cao, Y.; Meng, X.; Li, C.; Wang, H.; Zhang, S. Enhancement of transdermal delivery of artemisinin using microemulsion vehicle based on ionic liquid and lidocaine ibuprofen. *Colloids Surf. B* **2020**, *189*, 110886.
54. Ilic, D.; Cvetkovic, M.; Tasic-Kostov, M. Emulsions with alkyl polyglucosides as carriers for off-label topical spironolactone–safety and stability evaluation. *Pharm. Dev. Technol.* **2021**, *26*, 373–379.

55. van Jaarsveld, E.; du Plessis, J.; du Preez, J.L.; Shahzad, Y.; Gerber, M. Formulation and characterisation of artemether-loaded nano-emulsion for topical applications. *J. Drug Deliv. Sci. Technol.* **2022**, *73*, 103449.
56. Freitas, L.M.; Antunes, F.T.T.; Obach, E.S.; Correa, A.P.; Wiiland, E.; de Mello Feliciano, L.; Reinicke, A.; Amado, G.J.V.; Grivicich, I.; Fialho, M.F.P.; Rebelo, I.N.; de Souza, A.H. Anti-inflammatory effects of a topical emulsion containing helianthus annuus oil, glycerin, and vitamin b3 in mice. *J. Pharm. Investig.* **2021**, *51*, 223–232.
57. Caritá, A.C.; Resende de Azevedo, J.; Vinícius Buri, M.; Bolzinger, M.A.; Chevalier, Y.; Riske, K.A.; Ricci Leonardi, G. Stabilization of vitamin c in emulsions of liquid crystalline structures. *Int. J. Pharm.* **2021**, *592*, 120092.
58. Marto, J.; Nunes, A.; Martins, A.M.; Carvalheira, J.; Prazeres, P.; Gonçalves, L.; Marques, A.; Lucas, A.; Ribeiro, H.M. Pickering emulsions stabilized by calcium carbonate particles: A new topical formulation. *Cosmetics* **2020**, *7*, 62.
59. Sharkawy, A.; Silva, A.M.; Rodrigues, F.; Barreiro, F.; Rodrigues, A. Pickering emulsions stabilized with chitosan/collagen peptides nanoparticles as green topical delivery vehicles for Cannabidiol (CBD). *Colloids Surf. A* **2021**, *631*, 127677.
60. Beladjine, M.; Albert, C.; Sintès, M.; Mekhloufi, G.; Gueutin, C.; Nicolas, V.; Canette, A.; Trichet, M.; Tsapis, N.; Michel, L.; Agnely, F.; Huang, N. Pickering emulsions stabilized with biodegradable nanoparticles for the co-encapsulation of two active pharmaceutical ingredients. *Int. J. Pharm.* **2023**, *637*, 122870.
61. Zhang, J.; Feng, Q.; Xiao, J.; Zhang, J.; Lin, J.; Wang, J. Microneedle-assisted transdermal delivery of perfluorotripropylamine-based oxygenated emulsion gel loaded with 5-aminolevulinic acid for enhanced photodynamic therapy of cutaneous squamous cell carcinoma. *Eur. J. Pharm. Sci.* **2023**, *188*, 106493.
62. Gursoy, R.N.; Benita, S. Self-Emulsifying Drug Delivery Systems (SEDDS) for improved oral delivery of lipophilic drugs. *Biomed. Pharmacother.* **2004**, *58*, 173–182.
63. Mahmood, A.; Bernkop-Schnürch, A. SEDDS: A game changing approach for the oral administration of hydrophilic macromolecular drugs. *Adv. Drug Deliv. Rev.* **2019**, *142*, 91–101.
64. Gibaud, S.; Attivi, D. Microemulsions for oral administration and their therapeutic applications. *Expert Opin. Drug Deliv.* **2012**, *9*, 937–951.
65. Buya, A.B.; Beloqui, A.; Memvanga, P.B.; Préat, V. Self-nano-emulsifying drug-delivery systems: From the development to the current applications and challenges in oral drug delivery. *Pharmaceutics* **2020**, *12*, 1194.
66. Mu, L.; Wu, P.; Zhang, Y.; Li, S.; Yang, R.; Wang, S. Development of a novel oral complex lipid emulsion containing triptolide for targeting pancreatic cancer. *Pharm. Dev. Technol.* **2022**, *27*, 881–891.
67. Sato, Y.; Yokoyama, S.; Yamaki, Y.; Nishimura, Y.; Miyashita, M.; Maruyama, S.; Takekuma, Y.; Sugawara, M. Enhancement of intestinal absorption of coenzyme Q10 using emulsions containing oleyl polyethylene acetic acids. *Eur. J. Pharm. Sci.* **2020**, *142*, 105144.
68. Yong, Z.; Xingqi, W.; Jie, H.; Rongfeng, H.; Xiaoqin, C. Formulation, production, in vitro release and in vivo pharmacokinetics of cinnamaldehyde sub-micron emulsions. *Pharm. Dev. Technol.* **2020**, *25*, 676–685.
69. Boonlao, N.; Ruktanonchai, U.R.; Anal, A.K. Enhancing bioaccessibility and bioavailability of carotenoids using emulsion-based delivery systems. *Colloids Surf. B* **2022**, *209*, 112211.
70. Qi, W.; Zhang, Z.; Wu, T. Encapsulation of β-carotene in oleogel-in-water pickering emulsion with improved stability and bioaccessibility. *Int. J. Biol. Macromol.* **2020**, *164*, 1432–1442.

71. Johannesson, J.; Pathare, M.M.; Johansson, M.; Bergström, C.A.S.; Teleki, A. Synergistic stabilization of emulsion gel by nanoparticles and surfactant enables 3d printing of lipid-rich solid oral dosage forms. *J. Colloid Interface Sci.* **2023**, *650*, 1253–1264.
72. Jadhav, N.V.; Pawar, M.A.; Vavia, P.R. Pickering dry emulsion system for improved oral delivery of fenofibrate. *AAPS PharmSciTech* **2022**, *23*, 16–18.
73. Guo, X.; Li, X.; Chan, L.; Huang, W.; Chen, T. Edible CaCO3 nanoparticles stabilized pickering emulsion as calcium-fortified formulation. *J. Nanobiotechnology* **2021**, *19*, 1–16.
74. Liu, Y.; Zhang, C.; Cheng, L.; Wang, H.; Lu, M.; Xu, H. Enhancing both oral bioavailability and anti-ischemic stroke efficacy of ginkgolide b by preparing nanocrystals self-stabilized pickering nano-emulsion. *Eur. J. Pharm. Sci.* **2024**, *192*, 106620.
75. de Oliveira, M.C.; Balbinot, R.B.; Villa Nova, M.; Gonçalves, R.S.; Bidóia, D.L.; Caetano, W.; Nakamura, C.V.; Bruschi, M.L. Development of environmentally responsive self-emulsifying system containing copaiba oil-resin for leishmaniasis oral treatment. *Pharmaceutics* **2023**, *15*, 2127.
76. Potharaju, S.; Mutyam, S.K.; Liu, M.; Green, C.; Frueh, L.; Nilsen, A.; Pou, S.; Winter, R.; Riscoe, M.K.; Shankar, G. Improving solubility and oral bioavailability of a novel antimalarial prodrug: Comparing spray-dried dispersions with self-emulsifying drug delivery systems. *Pharm. Dev. Technol.* **2020**, *25*, 625–639.
77. Jirwankar, P.; Shah, D.; Shao, J. Protection of protein drugs by self-emulsified nanoemulsion against proteolysis. *J. Pharm. Sci.* **2020**, *109*, 2615–2621.
78. Aswar, M.; Bhalekar, M.; Trimukhe, A.; Aswar, U. Self-Microemulsifying Drug Delivery System (SMEDDS) of curcumin attenuates depression in olfactory bulbectomized rats. *Heliyon* **2020**, *6*, e04482.
79. Liu, Y.H.; Chen, L.C.; Cheng, W.T.; Wei, P.S.; Hsieh, C.M.; Sheu, M.T.; Lin, S.Y.; Ho, H.O.; Lin, H.L. Synergistic combination of irinotecan and rapamycin orally delivered by nanoemulsion for enhancing therapeutic efficacy of pancreatic cancer. *Pharmaceutics* **2023**, *15*, 473.
80. Blanco, E.; Shen, H.; Ferrari, M. Principles of nanoparticle design for overcoming biological barriers to drug delivery. *Nat. Biotechnol.* **2015**, *33*, 941–951.
81. Fell, G.L.; Nandivada, P.; Gura, K.M.; Puder, M. Intravenous lipid emulsions in parenteral nutrition. *Adv. Nutr.* **2015**, *6*, 600–610.
82. O'Hagan, D.T.; van der Most, R.; Lodaya, R.N.; Coccia, M.; Lofano, G. "World in motion" – Emulsion adjuvants rising to meet the pandemic challenges. *npj Vaccines* **2021**, *6*, 158.
83. Floyd, A.G. Top ten considerations in the development of parenteral emulsions. *Pharm. Sci. Technol. Today* **1999**, *2*, 134–143.
84. Müller, R.H.; Heinemann, S. Fat emulsions for parenteral nutrition. IV. Lipofundin MCT/LCT regimens for Total Parenteral Nutrition (TPN) with high electrolyte load. *Int. J. Pharm.* **1994**, *107*, 121–132.
85. Chanana, G.D.; Sheth, B.B. Particle size reduction of emulsions by formulation design. I: Effect of polyhydroxy alcohols. *J. Parenter. Sci. Technol.* **1993**, *47*, 130–134.
86. Levy, M.Y.; Benita, S. Short- and long-term stability assessment of a new injectable diazepam submicron emulsion. *J. Parenter. Sci. Technol.* **1991**, *45*, 101–107.
87. Levy, M.Y.; Benita, S. Design and characterization of a submicronized o/w emulsion of diazepam for parenteral use. *Int. J. Pharm.* **1989**, *54*, 103–112.
88. Haidar, I.; Harding, I.H.; Bowater, I.C.; Eldridge, D.S.; Charman, W.N. The role of lecithin degradation on the pH dependent stability of halofantrine encapsulated fat nanoemulsions. *Int. J. Pharm.* **2017**, *528*, 524–535.

89. Banstola, B.; Gamage, P.L.; Jiang, W.; Mudalige, T. Analysis of phospholipids and triacylglycerols in intravenous lipid emulsions. *J. Pharm. Biomed. Anal.* **2023**, *222*, 115112.
90. Reber, E.; Neyer, P.; Schönenberger, K.A.; Saxer, C.; Bernasconi, L.; Stanga, Z.; Huber, A.; Hammerer-Lercher, A.; Mühlebach, S. Physicochemical stability and compatibility testing of voriconazole in all-in-one parenteral nutrition admixtures. *Pharmaceutics* **2021**, *13*, 1447.
91. Gostyńska, A.; Dettlaff, K.; Jelińska, A.; Stawny, M. Improving the safety of clinical management of covid-19 patients receiving aminoglycosides and parenteral nutrition: Y-site compatibility studies. *J. Pharm. Sci.* **2023**, *112*, 2597–2603.
92. Ravald, H.; Jaikishan, S.; Samuelsson, J.; Sukhova, A.; Šolínová, V.; Fornstedt, T.; Kašička, V.; Wiedmer, S.K. Capillary electrokinetic chromatography for studying interactions between β-blockers and intralipid emulsion. *J. Pharm. Biomed. Anal.* **2023**, *234*, 115554.
93. Ferrucci-Da Silva, C.; Zhan, L.; Shen, J.; Kong, B.; Campbell, M.J.; Memon, N.; Hegyi, T.; Lu, L.; Guo, G.L. Effects of total parenteral nutrition on drug metabolism gene expression in mice. *Acta Pharm. Sin. B* **2020**, *10*, 153–158.
94. Amran, M.H.H.; Zulfakar, M.H.; Danik, M.F.; Abdullah, M.S.P.; Shamsuddin, A.F. A new alternative for intravenous lipid emulsion 20% w/w from superolein oil and its effect on lipid and liver profiles in an animal model. *DARU, J. Pharm. Sci.* **2019**, *27*, 191–201.
95. Gura, K.M.; Premkumar, M.H.; Calkins, K.L.; Puder, M. Fish oil emulsion reduces liver injury and liver transplantation in children with intestinal failure-associated liver disease: A multicenter integrated study. *J. Pediatr.* **2021**, *230*, 46–54.
96. Spahn, D.R. Blood substitutes: Artificial oxygen carriers: Perfluorocarbon emulsions. *Crit. Care* **1999**, *3*, 93–97.
97. Fox, C.B.; Baldwin, S.L.; Duthie, M.S.; Reed, S.G.; Vedvick, T.S. Immunomodulatory and physical effects of oil composition in vaccine adjuvant emulsions. *Vaccine* **2011**, *29*, 9563–9572.
98. Fan, Z.; Zhang, Y.; Jiao, L.; Zhu, T.; Feng, Z.; Liu, Z.; Yang, Y.; Wang, D. Lycium barbarum polysaccharides-loaded particulate alum via pickering emulsion as an adjuvant to enhance immune responses. *Int. J. Pharm.* **2023**, *630*, 122418.
99. Lamaisakul, S.; Tantituvanont, A.; Lipipun, V.; Ritthidej, G. Development of novel cationic microemulsion as parenteral adjuvant for influenza vaccine. *Asian J. Pharm. Sci.* **2020**, *15*, 591–604.
100. Zhang, Y.; Liao, Y.; Tang, Q.; Lin, J.; Huang, P. Biomimetic nanoemulsion for synergistic photodynamic-immunotherapy against hypoxic breast tumor. *Angew. Chemie – Int. Ed.* **2021**, *60*, 10647–10653.
101. Chang, J.N. Recent advances in ophthalmic drug delivery. In *Handbook of Non-Invasive Drug Delivery System*, Kulkarni, V.S., Ed; Elsevier: Amsterdam, The Netherlands, 2010; pp. 165–192.
102. Singh, M.; Bharadwaj, S.; Lee, K.E.; Kang, S.G. Therapeutic nanoemulsions in ophthalmic drug administration: Concept in formulations and characterization techniques for ocular drug delivery. *J. Control. Release* **2020**, *328*, 895–916.
103. Dhahir, R.K.; Al-Nima, A.M.; Al-Bazzaz, F.Y. Nanoemulsions as ophthalmic drug delivery systems. *Turkish J. Pharm. Sci.* **2021**, *18*, 652–664.
104. Tong, L.; Lim, E.W.L.; Yeo, S.W.J.; Hou, A.; Linn, Y.C.; Ho, A.; Than, H.; Quek, J.K.S.; Hwang, W.Y.K.; Lim, F.L.W.I.; Lim, L. Conjunctival T cell profile in allogeneic hematopoietic stem cell transplant patients after instilling topical cyclosporine-A 0.1% cationic emulsion. *Ophthalmol. Ther.* **2023**, *12*, 1547–1567.

105. Meng, T.; Kulkarni, V.; Simmers, R.; Brar, V.; Xu, Q. Therapeutic implications of nanomedicine for ocular drug delivery. *Drug Discov. Today* **2019**, *24*, 1524–1538.
106. Jurišić Dukovski, B.; Ljubica, J.; Kocbek, P.; Safundžić Kučuk, M.; Krtalić, I.; Hafner, A.; Pepić, I.; Lovrić, J. Towards the development of a biorelevant in vitro method for the prediction of nanoemulsion stability on the ocular surface. *Int. J. Pharm.* **2023**, *633*, 122622.
107. Rajput, A.; Pingale, P.; Dhapte-Pawar, V. Nasal delivery of neurotherapeutics via nanocarriers: Facets, aspects, and prospects. *Front. Pharmacol.* **2022**, *13*, 979682.
108. McDonough, J.; Dixon, H.; Ladika, M. Nasal delivery of micro- and nano-encapsulated drugs. In *Handbook of Non-Invasive Drug Delivery System*, Kulkarni, V.S., Ed; Elsevier: Amsterdam, The Netherlands, 2010; pp. 193–208.
109. Bonferoni, M.C.; Rossi, S.; Sandri, G.; Ferrari, F.; Gavini, E.; Rassu, G.; Giunchedi, P. Nanoemulsions for "nose-to-brain" drug delivery. *Pharmaceutics* **2019**, *11*, 84.
110. Salawi, A.; Almoshari, Y.; Sultan, M.H.; Madkhali, O.A.; Bakkari, M.A.; Alshamrani, M.; Safhi, A.Y.; Sabei, F.Y.; Al Hagbani, T.; Ali, M.S.; Alam, M.S. Production, characterization, and in vitro and in vivo studies of nanoemulsions containing St. John's wort plant constituents and their potential for the treatment of depression. *Pharmaceuticals* **2023**, *16*, 490.
111. Zuglianello, C.; Martins Silva, N.G.; Lemos-Senna, E. Polysaccharide-peptide complexes stabilized around nanoemulsion droplets: A new approach for nasal delivering of pramlintide and insulin. *J. Drug Deliv. Sci. Technol.* **2023**, *85*, 104527.
112. Pires, P.C.; Peixoto, D.; Teixeira, I.; Rodrigues, M.; Alves, G.; Santos, A.O. Nanoemulsions and thermosensitive nanoemulgels of phenytoin and fosphenytoin for intranasal administration: Formulation development and in vitro characterization. *Eur. J. Pharm. Sci.* **2020**, *141*, 105099.
113. Shekunov, B.Y.; Chattopadhyay, P.; Tong, H.H.Y.; Chow, A.H.L. Particle size analysis in pharmaceutics: Principles, methods and applications. *Pharm. Res.* **2007**, *24*, 203–227.
114. Yukuyama, M.N.; Kato, E.T.M.; Lobenberg, R.; Bou-Chacra, N.A. Challenges and future prospects of nanoemulsion as a drug delivery system. *Curr. Pharm. Des.* **2016**, *23*, 495–508.
115. Feng, S.S. Nanoparticles of biodegradable polymers for new-concept chemotherapy. *Expert Rev. Med. Devices* **2004**, *1*, 115–125.
116. Usmani, O.S.; Biddiscombe, M.F.; Barnes, P.J. Regional lung deposition and bronchodilator response as a function of β2-agonist particle size. *Am. J. Respir. Crit. Care Med.* **2005**, *172*, 1497–1504.
117. Martin, R.J. Therapeutic significance of distal airway inflammation in asthma. *J. Allergy Clin. Immunol.* **2002**, *109*, S447–S460.
118. Garmise, R.J.; Mar, K.; Crowder, T.M.; Hwang, C.R.; Ferriter, M.; Huang, J.; Mikszta, J.A.; Sullivan, V.J.; Hickey, A.J. Formulation of a dry powder influenza vaccine for nasal delivery. *AAPS PharmSciTech* **2006**, *7*, 1–7.
119. Stranz, M.; Kastango, E.S. A review of pH and osmolarity. *Int. J. Pharm. Compd.* **2002**, *6*, 216–220.
120. Cutaia, K.; Chablani, L.; Zhao, F. Basics of compounding: Vehicles for compounded oral liquid medications: A review. *Int. J. Pharm. Compd.* **2018**, *22*, 480–489.
121. Pan, W.; Roccabianca, S.; Basson, M.D.; Bush, T.R. Influences of sodium and glycosaminoglycans on skin oedema and the potential for ulceration: A finite-element approach. *R. Soc. Open Sci.* **2019**, *6*, 182076.
122. Herman, C.J.; Groves, M.J. Hydrolysis kinetics of phospholipids in thermally stressed intravenous lipid emulsion formulations. *J. Pharm. Pharmacol.* **1992**, *44*, 539–542.

123. Freitas, C.; Müller, R.H. Effect of light and temperature on zeta potential and physical stability in Solid Lipid Nanoparticle (SLN®) dispersions. *Int. J. Pharm.* **1998**, *168*, 221–229.
124. FDA. Drug review process: Ensuring drugs are safe and effective. https://www.fda.gov/drugs/information-consumers-and-patients-drugs/fdas-drug-review-process-ensuring-drugs-are-safe-and-effective (accessed Sep 13, 2023).
125. EMA. European Medicines Agency: Authorisation of medicines. https://www.ema.europa.eu/en/about-us/what-we-do/authorisation-medicines (accessed Sep 13, 2023).
126. CDSCO. Central Drugs Standard Control Organization: New drugs. https://cdsco.gov.in/opencms/opencms/en/Drugs/New-Drugs/ (accessed Sep 13, 2023).
127. NMPA. National Medical Products Administration (English version). http://english.nmpa.gov.cn/ (accessed Sep 13, 2023).
128. AMA. African Medicines Agency. https://www.nepad.org/microsite/african-medicines-agency-ama (accessed Sep 13, 2023).
129. World Health Organization. *Global Tuberculosis Report*, 2022.
130. World Health Organization. HIV and AIDS factsheet. https://www.who.int/news-room/fact-sheets/detail/hiv-aids (accessed Sep 13, 2023).
131. Danaei, M.; Dehghankhold, M.; Ataei, S.; Hasanzadeh Davarani, F.; Javanmard, R.; Dokhani, A.; Khorasani, S.; Mozafari, M.R. Impact of particle size and polydispersity index on the clinical applications of lipidic nanocarrier systems. *Pharmaceutics* **2018**, *10*, 57.
132. Ansel, C.; Allen, L.V.; Popovich, N.G. Ansel's pharmaceutical dosage form and drug delivery systems. *B.I Publ.* **2005**, *70*, 193–243.
133. Kang, H.; Cha, K.H.; Cho, W.; Park, J.; Park, H.J.; Sun, B.K.; Hyun, S.M.; Hwang, S.J. Cyclosporine a micellar delivery system for dry eyes. *Int. J. Nanomedicine* **2016**, *11*, 2921–2933.
134. Liu, Y.; Wang, Y.; Yang, J.; Zhang, H.; Gan, L. Cationized hyaluronic acid coated spanlastics for cyclosporine a ocular delivery: prolonged ocular retention, enhanced corneal permeation and improved tear production. *Int. J. Pharm.* **2019**, *565* (100), 133–142.
135. Ye, J.; Li, L.; Yin, J.; Wang, H.; Li, R.; Yang, Y.; Guan, Y.; Xia, X.; Liu, Y. Tumor-targeting intravenous lipid emulsion of paclitaxel: Characteristics, stability, toxicity, and toxicokinetics. *J. Pharm. Anal.* **2022**, *12*, 901–912.
136. Fidler, J.M.; An, J.; Carter, B.Z.; Andreeff, M. Preclinical antileukemic activity, toxicology, toxicokinetics and formulation development of triptolide derivative MRx102. *Cancer Chemother. Pharmacol.* **2014**, *73*, 961–974.
137. Giannaccare, G.; Rossi, C.; Borselli, M.; Bonzano, C.; Carnovale Scalzo, G.; Nicolò, M.; Scorcia, V.; Traverso, C.E.; Vagge, A. Clinical outcomes of topical 0.1% ciclosporin cationic emulsion used on label in children with vernal keratoconjunctivitis. *Ophthalmol. Ther.* **2023**, *12*, 1787–1793.
138. Binder, C.; Giordano, V.; Thanhaeuser, M.; Kreissl, A.; Huber-Dangl, M.; Longford, N.; Haiden, N.; Berger, A.; Repa, A.; Klebermass-Schrehof, K. A mixed lipid emulsion containing fish oil and its effect on electrophysiological brain maturation in infants of extremely low birth weight: A secondary analysis of a randomized clinical trial. *J. Pediatr.* **2019**, *211*, 46–53.
139. Gupta, A.; Eral, H.B.; Hatton, T.A.; Doyle, P.S. Nanoemulsions: Formation, properties and applications. *Soft Matter* **2016**, *12*, 2826–2841.
140. McClements, D.J. Nanoemulsions versus microemulsions: Terminology, differences, and similarities. *Soft Matter* **2012**, *8*, 1719–1729.
141. Anton, N.; Vandamme, T.F. Nano-emulsions and micro-emulsions: Clarifications of the critical differences. *Pharm. Res.* **2011**, *28*, 978–985.

142. Date, A.A.; Nagarsenker, M.S. Design and evaluation of Self-Nanoemulsifying Drug Delivery Systems (SNEDDS) for cefpodoxime proxetil. *Int. J. Pharm.* **2007**, *329*, 166–172.
143. Siegel, R.L.; Miller, K.D.; Wagle, N.S.; Jemal, A. Cancer statistics, 2023. *CA. Cancer J. Clin.* **2023**, *73*, 17–48.
144. Javanshir, A.; Karimi, E.; Maragheh, A.D.; Tabrizi, M.H. The antioxidant and anti-cancer potential of *Ricinus communis* L. essential oil nanoemulsions. *J. Food Meas. Charact.* **2020**, *14*, 1356–1365.
145. Nagachinta, S.; Bouzo, B.L.; Vazquez-Rios, A.J.; Lopez, R.; de la Fuente, M. Sphingomyelin-Based Nanosystems (SNs) for the development of anticancer MiRNA therapeutics. *Pharmaceutics* **2020**, *12*, 189.
146. Yang, J.; Li, Y.; Sun, J.; Zou, H.; Sun, Y.; Luo, J.; Xie, Q.; Rong, A.; Wang, H.; Li, X.; Wang, K.; Yang, L.; Ma, T.; Wu, L.; Sun, X. An osimertinib-perfluorocarbon nanoemulsion with excellent targeted therapeutic efficacy in non-small cell lung cancer: Achieving intratracheal and intravenous administration. *ACS Nano* **2022**, *16*, 12590–12605.
147. Song, B.; Wu, S.; Li, W.; Chen, D.; Hu, H. Folate modified long circulating nano-emulsion as a promising approach for improving the efficiency of chemotherapy drugs in cancer treatment. *Pharm. Res.* **2020**, *37*, 242.
148. Salata, G.C.; Lopes, L.B. Phosphatidylcholine-based nanoemulsions for paclitaxel and a p-glycoprotein inhibitor delivery and breast cancer intraductal treatment. *Pharmaceuticals* **2022**, *15*, 1110.
149. Migotto, A.; Carvalho, V.F.M.; Salata, G.C.; da Silva, F.W.M.; Yan, C.Y.I.; Ishida, K.; Costa-Lotufo, L.V.; Steiner, A.A.; Lopes, L.B. Multifunctional nanoemulsions for intraductal delivery as a new platform for local treatment of breast cancer. *Drug Deliv.* **2018**, *25*, 654–667.
150. Bahadur, S.; Pardhi, D.M.; Rautio, J.; Rosenholm, J.M.; Pathak, K. Intranasal nanoemulsions for direct nose-to-brain delivery of actives for CNS disorders. *Pharmaceutics* **2020**, *12*, 1–27.
151. Misra, S.K.; Pathak, K. Nose-to-brain targeting via nanoemulsion: Significance and evidence. *Colloids and Interfaces* **2023**, *7*, 23.
152. Meirinho, S.; Rodrigues, M.; Santos, A.O.; Falcão, A.; Alves, G. Self-emulsifying drug delivery systems: An alternative approach to improve brain bioavailability of poorly water-soluble drugs through intranasal administration. *Pharmaceutics* **2022**, *14*, 1–28.
153. Karami, Z.; Saghatchi Zanjani, M.R.; Hamidi, M. Nanoemulsions in CNS drug delivery: Recent developments, impacts and challenges. *Drug Discov. Today* **2019**, *24*, 1104–1115.
154. Gawin-Mikołajewicz, A.; Nartowski, K.P.; Dyba, A.J.; Gołkowska, A.M.; Malec, K.; Karolewicz, B. Ophthalmic nanoemulsions: From composition to technological processes and quality control. *Mol. Pharm.* **2021**, *18*, 3719–3740.
155. Choradiya, B.R.; Patil, S.B. A comprehensive review on nanoemulsion as an ophthalmic drug delivery system. *J. Mol. Liq.* **2021**, *339*, 116751.
156. Shah, J.; Nair, A.B.; Jacob, S.; Patel, R.K.; Shah, H.; Shehata, T.M.; Morsy, M.A. Nanoemulsion based vehicle for effective ocular delivery of moxifloxacin using experimental design and pharmacokinetic study in rabbits. *Pharmaceutics* **2019**, *11*, 230.
157. Jurišić Dukovski, B.; Juretić, M.; Bračko, D.; Randjelović, D.; Savić, S.; Crespo Moral, M.; Diebold, Y.; Filipović-Grčić, J.; Pepić, I.; Lovrić, J. Functional ibuprofen-loaded cationic nanoemulsion: Development and optimization for dry eye disease treatment. *Int. J. Pharm.* **2020**, *576*, 118979.

158. Roy, A.; Nishchaya, K.; Rai, V.K. Nanoemulsion-based dosage forms for the transdermal drug delivery applications: A review of recent advances. *Expert Opin. Drug Deliv.* **2022**, *19*, 303–319.
159. Alghaith, A.F.; Alshehri, S.; Alhakamy, N.A.; Hosny, K.M. Development, optimization and characterization of nanoemulsion loaded with clove oil-naftifine antifungal for the management of tinea. *Drug Deliv.* **2021**, *28*, 343–356.
160. Nikolic, I.; Mitsou, E.; Pantelic, I.; Randjelovic, D.; Markovic, B.; Papadimitriou, V.; Xenakis, A.; Lunter, D.J.; Zugic, A.; Savic, S. Microstructure and biopharmaceutical performances of curcumin-loaded low-energy nanoemulsions containing eucalyptol and pinene: Terpenes' role overcome penetration enhancement effect? *Eur. J. Pharm. Sci.* **2020**, *142*, 105135.
161. Savić, V.; Ilić, T.; Nikolić, I.; Marković, B.; Čalija, B.; Cekić, N.; Savić, S. Tacrolimus-loaded lecithin-based nanostructured lipid carrier and nanoemulsion with propylene glycol monocaprylate as a liquid lipid: Formulation characterization and assessment of dermal delivery compared to referent ointment. *Int. J. Pharm.* **2019**, *569*, 118624.
162. Sahu, S.; Katiyar, S.S.; Kushwah, V.; Jain, S. Active natural oil-based nanoemulsion containing tacrolimus for synergistic antipsoriatic efficacy. *Nanomedicine* **2018**, *13*, 1985–1998.
163. Bouyer, E.; Mekhloufi, G.; Rosilio, V.; Grossiord, J.L.; Agnely, F. Proteins, polysaccharides, and their complexes used as stabilizers for emulsions: Alternatives to synthetic surfactants in the pharmaceutical field? *Int. J. Pharm.* **2012**, *436*, 359–378.
164. Bouyer, E.; Mekhloufi, G.; Huang, N.; Rosilio, V.; Agnely, F. β-lactoglobulin, gum arabic, and xanthan gum for emulsifying sweet almond oil: Formulation and stabilization mechanisms of pharmaceutical emulsions. *Colloids Surf. A* **2013**, *433*, 77–87.
165. Reufer, M.; Machado, A.H.E.; Niederquell, A.; Bohnenblust, K.; Müller, B.; Völker, A.C.; Kuentz, M. Introducing diffusing wave spectroscopy as a process analytical tool for pharmaceutical emulsion manufacturing. *J. Pharm. Sci.* **2014**, *103*, 3902–3913.
166. Li, Z.; Zheng, S.; Zhao, C.; Liu, M.; Zhang, Z.; Xu, W.; Luo, D.; Shah, B.R. Stability, microstructural and rheological properties of pickering emulsion stabilized by xanthan gum/lysozyme nanoparticles coupled with xanthan gum. *Int. J. Biol. Macromol.* **2020**, *165*, 2387–2394.
167. Toumi, S.; Yahoum, M.M.; Lefnaoui, S.; Hadjsadok, A. Synthesis, characterization and potential application of hydrophobically modified carrageenan derivatives as pharmaceutical excipients. *Carbohydr. Polym.* **2021**, *251*, 116997.
168. Juhaščik, M.; Kováčik, A.; Huerta-Ángeles, G. Recent advances of hyaluronan for skin delivery: From structure to fabrication strategies and applications. *Polymers (Basel).* **2022**, *14,* 4833.
169. Rowe, R.C.; Sheskey, P.J.; Quinn, M.E. *Handbook of Pharmaceutical Excipients*, Sixth edit.; Rowe, R.C., Sheskey, P.J., Quinn, M.E., Eds.; Pharmaceutical Press: London, UK, 2009.
170. Wege, H.A.; Kim, S.; Paunov, V.N.; Zhong, Q.; Velev, O.D. Long-term stabilization of foams and emulsions with in-situ formed microparticles from hydrophobic cellulose. *Langmuir* **2008**, *24*, 9245–9253.
171. Vemireddy, S.; Preethi, P.P.; Halmuthur M., S.K. Chitosan stabilized nasal emulsion delivery system for effective humoral and cellular response against recombinant tetravalent dengue antigen. *Carbohydr. Polym.* **2018**, *190*, 129–138.
172. Juttulapa, M.; Piriyaprasarth, S.; Takeuchi, H.; Sriamornsak, P. Effect of high-pressure homogenization on stability of emulsions containing zein and pectin. *Asian J. Pharm. Sci.* **2017**, *12*, 21–27.
173. Burapapadh, K.; Kumpugdee-Vollrath, M.; Chantasart, D.; Sriamornsak, P. Fabrication of pectin-based nanoemulsions loaded with itraconazole for pharmaceutical application. *Carbohydr. Polym.* **2010**, *82*, 384–393.

174. Zhang, J.; Li, H.; Wang, J.; Liu, S.; Li, J.; Li, D. Dopamine-functionalized pectin-based pickering emulsion as an oral drug delivery system. *Colloids Surf. A* **2023**, *673*, 131807.
175. Han, J.; Chen, F.; Gao, C.; Zhang, Y.; Tang, X. Environmental stability and curcumin release properties of pickering emulsion stabilized by chitosan/gum arabic nanoparticles. *Int. J. Biol. Macromol.* **2020**, *157*, 202–211.
176. Zarnowski, R.; Jaromin, A.; Certik, M.; Czabany, T.; Fontaine, J.; Jakubik, T.; Iqbal, M.C.M.; Grandmougin-Ferjani, A.; Kozubek, A.; Pietr, S.J. The oil of *Adenanthera pavonina* L. seeds and its emulsions. *Zeitschrift fur Naturforsch. – Sect. C J. Biosci.* **2004**, *59*, 321–326.
177. Cabezas, D.M.; Madoery, R.; Diehl, B.W.K.; Tomás, M.C. Emulsifying properties of different modified sunflower lecithins. *JAOCS, J. Am. Oil Chem. Soc.* **2012**, *89*, 355–361.
178. Liu, H.; Liu, T.; Fan, H.; Gou, M.; Li, G.; Ren, H.; Wang, D.; Cheng, Z. Corn lecithin for injection from deoiled corn germ: Extraction, composition, and emulsifying properties. *Eur. J. Lipid Sci. Technol.* **2018**, *120*, 1–11.
179. Paolino, D.; Ventura, C.A.; Nisticò, S.; Puglisi, G.; Fresta, M. Lecithin microemulsions for the topical administration of ketoprofen: Percutaneous adsorption through human skin and in vivo human skin tolerability. *Int. J. Pharm.* **2002**, *244*, 21–31.
180. Elmeshad, A.N.; Tadros, M.I. Transdermal delivery of an anti-cancer drug via w/o emulsions based on alkyl polyglycosides and lecithin: Design, characterization, and in vivo evaluation of the possible irritation potential in rats. *AAPS PharmSciTech* **2011**, *12*, 1–9.
181. Jadhav, K.; Kadam, V.; Pisal, S. Formulation and evaluation of lecithin organogel for topical delivery of fluconazole. *Curr. Drug Deliv.* **2009**, *6*, 174–183.
182. Lim, S.; Salentinig, S. Protein nanocage-stabilized pickering emulsions. *Curr. Opin. Colloid Interface Sci.* **2021**, *56*, 101485.
183. Sarker, M.; Lee, H.; Gonçalves, R.A.; Lam, Y.M.; Su, H.; Lim, S. Supramolecular protein assembly retains its structural integrity at liquid–liquid interface. *Adv. Mater. Interfaces* **2020**, *7*, 1–10.
184. Tian, H.; Xiang, D.; Li, C. Tea polyphenols encapsulated in w/o/w emulsions with xanthan gum–locust bean gum mixture: Evaluation of their stability and protection. *Int. J. Biol. Macromol.* **2021**, *175*, 40–48.
185. Fu, D.; Deng, S.; McClements, D.J.; Zhou, L.; Zou, L.; Yi, J.; Liu, C.; Liu, W. Encapsulation of β-carotene in wheat gluten nanoparticle-xanthan gum-stabilized pickering emulsions: Enhancement of carotenoid stability and bioaccessibility. *Food Hydrocoll.* **2019**, *89*, 80–89.
186. Combrinck, J.; Otto, A.; Du Plessis, J. Whey protein/polysaccharide-stabilized emulsions: Effect of polymer type and PH on release and topical delivery of salicylic acid. *AAPS PharmSciTech* **2014**, *15*, 588–600.
187. Li, Y.; Hu, M.; Xiao, H.; Du, Y.; Decker, E.A.; McClements, D.J. Controlling the functional performance of emulsion-based delivery systems using multi-component biopolymer coatings. *Eur. J. Pharm. Biopharm.* **2010**, *76*, 38–47.
188. Klein, M.; Aserin, A.; Svitov, I.; Garti, N. Enhanced stabilization of cloudy emulsions with gum arabic and whey protein isolate. *Colloids Surf. B* **2010**, *77*, 75–81.
189. Wu, Y.; Zeng, M.; Cheng, Q.; Huang, C. Recent progress toward physical stimuli-responsive emulsions. *Macromol. Rapid Commun.* **2022**, *2200193*, 1–15.
190. Tang, J.; Quinlan, P.J.; Tam, K.C. Stimuli-responsive pickering emulsions: Recent advances and potential applications. *Soft Matter* **2015**, *11*, 3512–3529.

191. Rajbanshi, A.; Mahmoudi, N.; Murnane, D.; Pavlova, E.; Slouf, M.; Dreiss, C.A.; Cook, M.T. Combining branched copolymers with additives generates stable thermoresponsive emulsions with in situ gelation upon exposure to body temperature. *Int. J. Pharm.* **2023**, *637*, 122892.
192. Campanholi, K. da S.S.; da Silva, J.B.; Batistela, V.R.; Gonçalves, R.S.; Said dos Santos, R.; Balbinot, R.B.; Lazarin-Bidóia, D.; Bruschi, M.L.; Nakamura, T.U.; Nakamura, C.V.; Caetano, W. Design and optimization of stimuli-responsive emulsion-filled gel for topical delivery of copaiba oil-resin. *J. Pharm. Sci.* **2022**, *111*, 287–292.
193. Sia, C.S.; Lim, H.P.; Lin, Y.N.; Beh, L.C.; Tey, B.T.; Goh, B.-H.; Low, L.E. PH-Controllable stability of iron Oxide@chitosan nanocomposite-stabilized magnetic pickering emulsions. *Eur. Polym. J.* **2023**, *186*, 111870.
194. Bai, R.X.; Xue, L.H.; Dou, R.K.; Meng, S.X.; Xie, C.Y.; Zhang, Q.; Guo, T.; Meng, T. Light-triggered release from pickering emulsions stabilized by TiO2 nanoparticles with tailored wettability. *Langmuir* **2016**, *32*, 9254–9264.
195. Acter, S.; Vidallon, M.L.P.; King, J.P.; Teo, B.M.; Tabor, R.F. Photothermally responsive pickering emulsions stabilised by polydopamine nanobowls. *J. Mater. Chem. B* **2021**, *9*, 8962–8970.
196. Hu, S.H.; Fang, R.H.; Chen, Y.W.; Liao, B.J.; Chen, I.W.; Chen, S.Y. Photoresponsive protein-graphene-protein hybrid capsules with dual targeted heat-triggered drug delivery approach for enhanced tumor therapy. *Adv. Funct. Mater.* **2014**, *24*, 4144–4155.
197. Chen, Z.; Zhou, L.; Bing, W.; Zhang, Z.; Li, Z.; Ren, J.; Qu X. Light controlled reversible inversion of nanophosphor-stabilized pickering emulsions for biphasic enantioselective biocatalysis. *J. Am. Chem. Soc.* **2014**, *136*, 7498–7504.
198. Melle, S.; Lask, M.; Fuller, G.G. Pickering emulsions with controllable stability. *Langmuir* **2005**, *21*, 2158–2162.
199. Ngai, T.; Behrens, S.H.; Auweter, H. Novel emulsions stabilized by PH and temperature sensitive microgels. *Chem. Commun.* **2005**, 41, 331–333.
200. Cong, Y.; Li, Q.; Chen, M.; Wu, L. Synthesis of dual-stimuli-responsive microcontainers with two payloads in different storage spaces for preprogrammable release. *Angew. Chemie – Int. Ed.* **2017**, *56*, 3552–3556.
201. Mendiratta, S.; Ali, A.A.A.; Hejazi, S.H.; Gates, I. Dual stimuli-responsive pickering emulsions from novel magnetic hydroxyapatite nanoparticles and their characterization using a microfluidic platform. *Langmuir* **2021**, *37*, 1353–1364.
202. Raju, R.R.; Liebig, F.; Klemke, B.; Koetz, J. PH-responsive magnetic pickering janus emulsions. *Colloid Polym. Sci.* **2018**, *296*, 1039–1046.
203. Chen, Y.; Bai, Y.; Chen, S.; Ju, J.; Li, Y.; Wang, T.; Wang, Q. Stimuli-responsive composite particles as solid-stabilizers for effective oil harvesting. *ACS Appl. Mater. Interfaces* **2014**, *6*, 13334–13338. https://doi.org/10.1021/am504124a.
204. Brugger, B.; Richtering, W. Magnetic, Thermosensitive microgels as stimuli-responsive emulsifiers allowing for remote control of separability and stability of oil in water-emulsions. *Adv. Mater.* **2007**, *19*, 2973–2978.
205. Chen, Q.; Xu, Y.; Cao, X.; Qin, L.; An, Z. Core Cross-Linked Star (CCS) polymers with temperature and salt dual responsiveness: Synthesis, formation of High Internal Phase Emulsions (HIPEs) and triggered demulsification. *Polym. Chem.* **2014**, *5*, 175–185.
206. Wang, F.; Yu, X.; Yang, Z.; Duan, H.; Zhang, Z.; Liu, H. Dual PH-and Light-responsive amphiphilic random copolymer nanomicelles as particulate emulsifiers to stabilize the oil/water interface. *J. Phys. Chem. C* **2018**, *122*, 18995–19003.

Index

I

L

M

O

P

R

S

U

W

FSC
www.fsc.org
MIX
Papier | Fördert gute Waldnutzung
FSC® C083411

Kolibri 360 GmbH
Ferdinand-Jühlke-Straße 7
99095 Erfurt, Deutschland
produktsicherheit@kolibri360.de